Powder Metallurgy: An Overview

Powder Metallurgy: An Overview

Edited by

I. Jenkins and J. V. Wood

The Institute of Metals
1991

Book Number 492

Published in 1991 by The Institute of Metals
1 Carlton House Terrace, London SW 1Y 5DB

and
The Institute of Metals
North American Publications Center
Old Post Road, Brookfield VT 05036
U S A

British Library Cataloguing in Publication Data

Powder metallurgy: an overview.
I. Jenkins, I. (Ivor)
II. Wood, J. V. (John)

671.37

ISBN 0-901462-81-0

American Library of Congress Cataloging in Publication Data

Applied for

Compiled from original typescripts and illustrations supplied by the authors.

Cover illustration: Porous sintered metal filter components (courtesy GKN Ltd, Powder Metallurgy Division).
Cover design: Brian Roll.

Printed and bound in Great Britain.

Contents

Preface

For some time there has been concern, more especially in Europe, from students through to practising engineers, about the lack of educational material in the field of Powder Metallurgy. The Working Group on Education of the European Powder Metallurgy Federation reported in its survey of the European educational scene that there are very few text books or educational aids which are up-to-date and easily available. The Group recommended that this omission should be rectified with some urgency.

The science and practice of Powder Metallurgy has developed over the past twenty years in such a comprehensive and multi-disciplinary manner that, unlike many other branches of metallurgy, it would be difficult for one person authoritatively, critically and completely to cover the scientific, technological and engineering aspects of such a diverse subject. In addressing this problem, the Editors have planned a series comprising three complementary volumes which will, respectively, cover the educational needs of the student as well as those wishing to up-date themselves in the individual technologies of the subject.

The first volume is directed, primarily, to meet the needs of undergraduates, and presents the basic scientific and technological concepts of powder metallurgy, supported by appropriate examples of industrial practise. The main emphasis on technological procedures in industry, however, is in a separate volume which provides a selection of case studies covering most of the processes involved in powder metallurgy, without being exhaustive. In particular, the examples chosen and provided by contributors from industry and elsewhere demonstrate the application of the basic principles described in the previous volume to practical problems and will be of value both to the undergraduate and to the materials and design engineers in industry. The third volume is a reference book providing authoritative and critical reviews by international experts of the important advances which are being made in the science and practice of this rapidly developing technology. The specific contributions cover the broad spectrum of the powder metallurgy process, ranging from powder production through consolidation, sintering, applications and design. It will be of value as a general text book to the research worker in academe and industry and to the practising powder metallurgist and design engineer as well as to those wishing to be up-dated on a specific sector of the subject.

Professor Malcolm Waldron died in December 1988. He had taken the lead in coordinating and almost completing the preparatory editorial work involved with the series and had brought to bear so effectively not only his own profound knowledge of the subject but also his characteristic dedication to an accepted task. The book series, it will be hoped, will prove to be a fitting memorial to him, to the high standards which he set himself and to the great esteem with which he was held internationally.

Special thanks are accorded to Mrs J. A. Abrahams and Mrs G. Robinson for their excellent secretarial contributions and support of the editorial work involved.

Ivor Jenkins
John Wood

Introduction

The traditional powder metallurgy process and industrial variations on it are well-documented and will not be described in detail here. However, the following interposed outline may be helpful to the reader who is unfamiliar with the technology.

In the traditional process, a metal, alloy or ceramic powder in the form of a mass of dry particles, normally less than 150 microns in diameter, is converted into an engineering component of pre-determined shape and possessing properties which allow it to be used in most cases without further processing. The basic steps in the production of sintered engineering components are those of powder production; the mechanical compaction of the powder into a handleable preform; and the heating of the preform to a temperature below the melting point of the major constituent for a sufficient time to permit the development of the required properties.

Metal powders can be produced by mechanical comminution but the main industrial processes have either a chemical or a physical basis. The thermal dissociation of a metal compound such as, for example, a carbonyl (nickel and iron) or a hydride (titanium) has been long established industrially but the most widely used chemical processes involve the reduction of metal oxides by either hydrogen or carbonaceous agents. Typical products include iron, copper, molybdenum and tungsten powders. Both elemental and alloy powders are produced in substantial quantities by the process designated "atomization". In this process a stream of molten metal is broken up into droplets which are rapidly quenched to powder by high velocity jets of either water or a gas. Powders covering a wide range of alloy steels, copper alloys and light alloys are manufactured in this way. Powders of high purity are produced electrolytically but the process is restricted mainly to products of high intrinsic value such as cobalt, copper and certain precious metals.

The behaviour of a powder during the subsequent consolidation processes is determined by both particle and bulk properties. A powder is characterized, therefore, not only by chemical composition but also by particle shape, size and surface chemistry and in bulk by compressibility, flowability and apparent density. Powders are blended to meet the required specifications and mixed with a lubricant - usually a metal stearate - which acts both as a binder and as a lubricant during the compaction. The compression of the powder in a closed die reduces the voids between the particles by about one third to one half of the original volume of the uncompressed powder.

The pressed compact is given a thermal treatment designated "sintering" under a protective atmosphere or in vacuo, during which the lubricant is evaporated and porosity is reduced by metal transport involving surface and volume diffusion. As sintering proceeds, the original inter-connected porosity is reduced, closed pores are formed, and the overall shrinkage is controlled to ensure that the final dimensions of the compact come within the required engineering tolerances.

Variations on the foregoing process are made to meet particular product requirements. Thus, metal filters and porous membranes for chemical separation processes are produced by the sintering of un-compacted powder. Virtually completely dense bodies are produced by the process of "infiltration" in which a lightly pressed and sintered compact is brought into contact with a molten metal of lower melting point which is absorbed to fill the interconnected porosity. Molten metal is also an essential feature of "liquid-phase" sintering. In this process, a minor component of the compacted powder melts at the sintering temperature. Rapid and virtually complete densification occurs by material transport through the liquid phase. A range of important industrial products are made in this way including cemented carbide, high density 'heavy alloy', high speed steel preforms and the copper-tin self-lubricating bearing.

The origins of powder metallurgy are lost in the mists of antiquity, but it is reasonable to assume that powder processing was used by the earliest civilizations in advance of their knowledge of the smelting of ores and of the melting of iron and copper alloys. It is believed that the development began in the Near East where man first learned the art of reducing iron oxide with carbonaceous material and of using the new metal for tools, weapons and ornaments. The technique spread to the eastern Mediterranean countries, into central and northern Europe and eastwards to India and Asia. In due time, it reached the people of Africa but whilst in Europe and Asia iron metallurgy advanced, with carburizing, hardening and tempering being used by Egyptians and Romans during the first millenium BC, in the more remote regions of Africa the technology remained static. In 1954, Gardi published an account of his exploration to the Mandara mountain country in the North Cameroons where he found a primitive iron culture using powder metallurgy techniques. The source of powder was magnetite obtained from alluvial deposits in the nearby river. This was reduced by charcoal in a vertical shaft furnace with much ceremony and incantations by the smith of the tribe and the resulting sponge iron was broken up to a powder to free it from excess charcoal and gangue. The iron powder was tamped into clay cups which were sealed with clay and heated in a reducing atmosphere generated by a charcoal fire. The resulting sintered iron was then forged in a charcoal fire with stone hammers to weapons, ornaments and implements. The product was remarkably free from inclusions, a feature which has been observed in photomicrographs of iron dowels and pins used for the keying of the marble blocks of the Parthenon in Athens (440 BC), evidence in support of the use of a powder processing technique at that time. The process was undoubtedly in widespread use at the inception of the Christian era and continued well into the first millenium AD, with the massive iron pillar of Delhi, dating from 300 AD, being an outstanding example. On the other side of the world, also, the technique was being used by the Incas in the decorative application of gold powder.

In the advancing civilisations in Europe and Asia, however, the process was being rapidly overtaken and eventually by-passed by advances in the smelting of ores and the refining of metals particularly those based on copper and iron. There was a revival in industrial powder metallurgy activity at the end of the 18th century centred around the need for malleable platinum and this led in due course to the classical work of Wollaston in Britain and of Soboleviskiy in Russia who produced chemically fine platinum powder which was pressed into a cake and hot forged. The process had a limited industrial life, however, being superceded in the mid-19th century by a fusion process. The dramatic developments in engineering of the late 18th and 19th centuries made considerable demands upon materials development and these were dominated entirely by cast and wrought products and the extension of the industrial materials base to include alloys and alloy steels in particular. A revival in powder metallurgy technology in the early part of the present century was spasmodic and was concerned entirely with products which were unique to the process. The development of the electric lamp led to the search for a filament material of high melting point, a low evaporation rate and adequate conductivity which could provide an efficient use of electrical energy and be available at reasonable cost. The basic property requirements were met by tungsten which, however, could not be fabricated into wire because of its brittleness. A solution to the problem was discovered in 1910 by Coolidge who demonstrated that when a pressed bar of tungsten powder is sintered at 3000°C by direct electric resistance heating and then worked sufficiently hot, as by swaging for example, its ductility is improved to permit continued working at progressively lower temperatures until a wire of the required filament diameter can be cold-drawn. The process is still the classical method used world wide for the manufacture of incandescent lamp filaments. The technique of direct resistance heating of pressed powders was extended to the production of other metals of high melting point such as molybdenum, niobium and tantalum, being superceded in due course by novel melting techniques such as the vacuum-arc and electron beam processes.

Further industrial innovation in powder technology took place in the 1920s with the development of the self-lubricating porous bronze bearing and cemented carbide. The former product is made from a mixture of 90% copper and 10% tin powders to which a little graphite may be added, and which is pressed into compacts close to the finished dimensions of the bearing. Sintering produces a 10% tin bronze alloy product, with little dimensional change, and which may contain up to 35% of fine, interconnected porosity. The final close dimensional tolerances of the bearing are achieved by pressing the sintered compact in a "sizing" die, after which it is impregnated with oil.

In the mid-1920s, cemented carbides were invented by Schröter of the German company, Krupp. The product is produced from a mixture of tungsten carbide and cobalt powders which are pressed into a "green" compact and heated to above the melting point of the cobalt when rapid and virtually complete densification occurs, giving a final structure of carbide particles in a matrix of 10%-15% of a cobalt/tungsten carbide eutectic. Used originally for wire drawing dies, cemented carbides soon found extensive application in metal machining and where high wear resistance is required as in rock drilling, metal hot working dies, etc. Since the original invention many new compositions have been developed involving, for example, additions of titanium and tantalum carbides. Diamond is also used in industry for fine wire-drawing dies and the abrasive properties of the material are exploited in diamond grinding and cutting wheels which were developed a few years after cemented carbide as powder metallurgy products in which diamond particles are dispersed in a metallic matrix.

At the outbreak of the Second World War, the range of industrial powder metallurgy products was still, in the main, those which were unique to the technology and which could not be made by other means. Interest was being shown, however, in the production of structural parts by the process, stimulated in part by the exigencies of the War and, in part, by those who recognised in the process for producing the porous bearing, a basis of a technology for making net-shape engineering components. The early products of the industry, mainly based on iron, were required to meet a modest engineering performance and the market attraction was the ability of the process to produce components to required dimensional specifications and at relatively low cost. Slowly and cautiously the industry began to grow and to extend its product base but it is only during the last 30 years that substantial expansion has occurred, dominated tonnage-wise by sintered iron and steel products, but also including components in the refractory metals, stainless steel, copper, nickel and aluminium alloys. Contributing significantly to the growth has been a progressive improvement in the manufacturing technology stimulated by a better understanding of the under-pinning science of the process; improved mechanical properties resulting from the development of more compressible powders which increase the densification of the sintered part; the extension of the materials base to include alloy steels which can be heat treated; and improvements in the control of processing parameters involving the automation of the pressing and sintering operations. Equally important, however, have been the increasingly effective dialogue between the powder metallurgy technologist and the design engineer and the results of their joint efforts in optimizing the use of the process to provide high quality, precision components with a degree of reproducibility, homogeneity and level of properties which often are superior to those achievable in the equivalent cast or wrought product.

Full densification of metal powders can be achieved by the use of conventional metal fabrication techniques. Thus, alloy strip is produced by the continuous forming and compaction either of lubricated powder or of a metallic paste which, after sintering, is densified by rolling. Powders, vacuum-sealed in metal containers, may be extruded hot and simultaneously sintered, after which the container cladding is stripped away; whilst compacted powder billets, after sintering, may be rolled, extruded or forged. A process which continues to attract substantial industrial investment is the hot forging of sintered preforms, which is capable of yielding a densified product close to final shape and with a high material yield. For technical and economic reasons the process has yet to reach its full potential. An important development and one of considerable industrial potential also is that of combining the compaction and sintering stages in one operation by the use of hot pressing techniques. The process is not new, having been used for many years for the production from powder of fine-grained beryllium. The product shows some degree of ductility as compared with the brittle coarse-grained cast beryllium. Further development of the process includes the hot isostatic pressing of metal powders using flexible moulds. The injection moulding of a very fine metal powder blended with a thermoplastic binder which is subsequently removed, continues to arouse industrial interest with its promise of greater flexibility in the production of shapes but it requires a significant reduction in processing costs before it becomes more widely accepted.

Aside from structural parts, a number of specialised products are produced by powder processing techniques. These include the "heavy alloys" of tungsten with up to 15% of either nickel and copper or nickel and iron which are liquid-phase sintered at about 1450°C to a fully dense product. Developed originally in 1937 to meet a therapeutic demand for radium containers to replace the more cumbersome lead version, the alloys have found important additional applications for mass-balancing in gyroscopes and aircraft control systems. Soft magnetic parts such as pole pieces and relay cores are produced, often as complex parts, in substantial quantities, whilst sintered Alnico permanent magnets have been marketed since 1940 and have now been extended to include the high energy rare-earth types. Precious metal light duty electrical contacts are produced by powder metallurgy as well as the infiltrated tungsten-copper and tungsten-silver heavy duty variety.

The powder metallurgy process lends itself to the manufacture of metal/non-metal components as in the case of the copper-graphite commutator brush which provides superior electrical conductivity and wear resistance to that of graphite and is widely used in electric motors. The flexibility of the process is further demonstrated by the "Cermet" range of products which comprise sintered mixtures of metallic and non-metallic inorganic compounds and which have been developed to meet specific engineering requirements. These include friction materials for vehicle brake linings and clutch plates which consist of a high friction material such as a silicate dispersed in a metal matrix of good thermal conductivity such as copper or iron; molybdenum-chromium oxide for hot extrusion dies; and a range of nuclear fuels based upon the oxide, carbide or nitride of uranium. All of the foregoing are examples of products which could only be produced by a powder metallurgy process.

Until relatively recently, scientific studies into the powder metallurgy process and its products had been restricted in the main to the refractory metals and nuclear materials. The situation has changed markedly during the past 15 years or so arising from the continuing search for improved materials of high integrity for advanced engineering applications such as aerospace. A particular inducement has been the ability of the process to produce metal products with closely controlled microstructural features and with homogeneous and reproducible properties. During this time there has been substantial growth world wide in scientific research into all aspects of the production, characterisation, handling and consolidation of powders of a wide range of engineering alloys. The consequent accumulation of scientific understanding of the process and the related developments in processes and products have been of considerable benefit to all sectors of the powder metallurgy industry.

An important development arising from the upsurge in scientific effort has been that of a powder processing route for the manufacture of superalloys for gas turbine applications in order to overcome the difficulties met in the forging and heat treating of the advanced highly alloyed nickel base materials. Related to these developments are those of dispersion-strengthened alloys in which a very finely dispersed second phase, usually an oxide, provides a metal matrix with improved microstructural stability and enhanced resistance to creep deformation in particular. Industrial products in this group include dispersion strengthened nickel and lead, both produced fully dense by the hot consolidation of the respective powder mixtures.

More recently, there have been exciting developments involving the production of very fine particles with refined microstructural features which are retained when the powder is consolidated. Such products possess properties which cannot be produced by ingot metallurgy, with a general enhancement of mechanical properties at room and elevated temperatures, often accompanied by improvements in corrosion resistance. One such process is that designated "mechanical alloying" in which metal powders are subjected to high energy milling. Non-metallic additives such as stable oxides or carbides become coated with the softer metal and are re-distributed as a very fine dispersion through the repeated fracture and re-welding of the composite powder particles. Developed originally for super alloys for high temperature service, the process has been extended to light alloys for aerospace applications and steels for creep resistance.

Novel microstructures are also developed by the consolidation of very fine powders produced by the gas "atomization" of molten metal. During quenching by the atomizing gas, the metal droplets can solidify so rapidly that they can exhibit non-equilibrium behaviour such as high solubility extensions of certain alloying elements, together with microcrystallinity or even an amorphous structure. Several aluminium alloys processed in this way have reached a commercial stage whilst investigations continue on magnesium, titanium and copper alloys.

An important extension of rapid solidification technology is that of spray-forming (the Osprey process) in which the atomized metal stream is directed onto a substrate where it solidifies and is allowed to build-up to the required form. The process is applicable to a wide range of engineering materials including both steels and light alloys.

The foregoing techniques are used in the production of metal matrix composites in which small diameter non-metallic fibres and/or particulates of stable non-metallics reinforce the metallic base. The resulting refined microstructure and reinforcement confer enhanced strength, stiffness and elevated temperature properties. Aside from conventional powder metallurgy techniques and mechanical alloying, the manufacturing processes include spray forming in which the non-metallic additive is injected into the molten metal as it is sprayed onto a substrate. Aluminium alloy matrix composites are already in production, whilst those based on magnesium, titanium and high temperature alloys are under development.

Among the advanced materials of the present era is a new class of ceramics produced from powders and based on oxides, nitrides, borides or carbides. Manufactured from powders of high purity and carefully processed to avoid both contamination and the introduction of structural defects, certain products are challenging the traditional markets of wear-resisting materials whilst others are already well-established in the electrical field in micro-electronics, dielectrics and solid state electrolytes to mention but a few of the many uses. Engineering applications include heat exchangers, pump parts in corrosive environments and various components and linings in petrol and diesel engines. Further important engineering developments can be expected as advanced ceramics continue to challenge the traditional roles of metal components in engineering design, whilst the new high temperature ceramic superconductors offer new opportunities for industrial innovation.

The science and practice of powder metallurgy has advanced considerably since the introduction of cemented carbide and the porous bearing and today the industry is well established as a supplier of products of the highest quality and reliability in a wide range of materials and components to meet market requirements of considerable diversity. In the chapters which follow contributors, recognized internationally as experts in their chosen subject, provide state of the art reviews on the many facets of the science and practice of powder processing. The reviews provide a first class reference to the various sectors of the subject and should be of value as much to the processor of powders as to the design engineer who wishes to be kept fully briefed on advances in materials and manufacturing processes.

Further Reading

1. W.D. Jones, Fundamental Principles of Powder Metallurgy, Edward Arnold (Publishers) Ltd., London, 1960.
2. R.M. German, Powder Metallurgy Science, Metal Powder Industries Federation, Princeton, New Jersey, 1984.
3. H.H. Hausner and M. Kumar Mal, Handbook of Powder Metallurgy, Chemical Publishing Co.Inc., New York, 2nd ed., 1982.
4. Powder Metallurgy: State of the Art, (eds.) W.J. Huppmann, W.A. Kaysser and G. Petzow, Verlag Schmid Gmbh, Freiburg, vol.2, 1986.
5. F.V. Lenel, Powder Metallurgy: Principles and Applications, Metal Powder Industries Federation, Princeton, New Jersey, 1980.
6. Powder Metallurgy: The Process and Its Products, British Powder Metallurgy Federation, Wolverhampton.
7. Kenneth J.A. Brookes, Cemented Carbides for Engineers and Tool Users, International Carbide Data, East Barnet,

Section 1

Powder Production

1

Atomization of Metal Powders

J. J. DUNKLEY
Davy McKee, Sheffield, UK

Atomization can be defined as the production of metal powders by the breaking of molten metal into droplets and their subsequent freezing into solid particles. It thus has a fundamental advantage over such techniques as reduction or electrolysis in that it can produce alloy powders of any composition that can be melted. This, together with economic factors which have become favourable to atomization as development has taken place, has lead to a steady increase in the quantities of powders produced by atomization, which now amount to around 1,000,000t/yr. [For other reviews of the subject see Refs. 1, 3, 26, 30 and 48.]

While the theme of this series is largely the use of metal powders in the production of high performance components, it must not be forgotten that there is a very wide range of applications for metal powders, some of which may seem fairly low in their technical demands, but many of which are at the leading edge of various "hi-tech" industries. These other applications make many different demands on powder production technology. Thus thermal spraying of surface coatings requires powders of a wide range of compositions with strictly defined chemistry, particle shape and particle size. Likewise electronic solder creams make severe demands on purity, shape and size. Rocket fuels require aluminium powder of spherical shape and narrow size range around 10-40 μm.

There are numerous other applications for metal powders, but it will be understood that powder metallurgy is far from being the only demanding market where new and improved atomization techniques are needed.

1. Earlier Status

By the late 1960s both gas and water atomization were established techniques, which we understand to mean that they were in use by a number of plants to produce many different powders for different markets. The level of technology, as evidenced by the yields, consistency, quality and costs of the processes used, was very variable, and in many cases poor. Plants were built on an empirical basis and some extremely low efficiencies were tolerated as unavoidable some plants giving yields of "good" product sub 250 μm of as little as 50-60% and of more demanding products with narrow size range, as low as 10-20%. Obviously the impact of such yields on costs was severe, and quality, whether defined as oxygen content, particle size distribution, apparent density (shape), cleanliness or porosity was low and variable.

The aim of recent developments has thus been twofold:

(i) To increase quality
(ii) To reduce costs

It is generally true to say that advances in understanding of the process of atomization tend to allow both these objectives to be achieved. However, despite a lot of work by academic investigators, it cannot be said that our theoretical knowledge of gas or water atomization approaches that of more readily analysed processes such as centrifugal atomization. The exceedingly complex mathematics of turbulent flow and subsonic/supersonic momentum and heat transfer rule out any physical modelling that can give precise predictions. Empirical advances and intuitive or qualitative understanding are all we can safely claim as yet.

2. The Basic Processes

Both techniques reviewed here are "two fluid" atomization, i.e. the atomization of liquid metal by a fluid. Both liquids and gases can be used, and there are two basic designs of atomization jets (Fig. 1). Open or "freefall" jets are normally used for liquid atomization with oil or water. Closed jet designs are often used with gas atomization as the rate of decay of the energy in jets of gas is so rapid that freefall designs tend to give rather low efficiencies, especially when finer powders, less than 100 microns, are required.

A complete plant (Fig. 2) typically consists of a source of liquid metal (ladle, furnace, etc.) which pours into a tundish to provide a consistent metal supply through a ceramic nozzle to the atomizing jet or jets. There is then a vessel to collect the atomizing fluid and powder following which the powder is separated (e.g. dewatered or filtered from the gas).

When it is desired to minimize oxidation during atomization, it is normal to seal the tundish to the atomizing vessel and to purge the vessel with inert gas. For extremely demanding applications or for very reactive alloys it is necessary to carry out the melting and transfer operations under vacuum or inert gas cover.

Such plants can be operated continuously, but outputs are then very high and only lower melting point metals such as Pb, Sn, Zn, Al and Cu are practical to melt continuously. Generally alloys are batch melted to assure compositional control and plant flexibility.

The rate of atomization practically achievable is rather different for gas and water atomization. For gas atomization rates from 1-70kg/min are used, while for water atomization outputs can be varied more widely from 1-500kg/min. Thus annual rates (based on 5000 hours operation) range from 300-21,000t/yr for gas atomizers and from 300-150,000t/yr for water atomizers. It is only in the largest gas atomizing plants, such as Nyby in Sweden, that these rates are inadequate and double atomizing streams and jets are used to boost output.

3. Water Atomization

A search of the recent literature reveals that water (and other liquid) atomization is the subject of far fewer publications than gas atomization, despite accounting for about half of world atomizing capacity. A ratio of around four gas atomizing papers to one water atomizing paper was found. This is certainly not due to the process of water atomization being better understood than gas atomization. Rather it reflects the major funding over recent years of research programmes aimed at improving aluminium and other reactive alloys by rapid solidification processing (RSP). Such alloys are too reactive for water atomization.

Almost all reports on water atomizing, and all production plants, use a freefall nozzle configuration with a fall height typically from 100-300mm. Two basic types of water jet are used - annular and flat jet (Fig. 3). Literature reports cover both but it is believed that flat jet designs are generally used more in plants producing special alloys of a wide range of sizes. Annular jets are rugged and simple to use, but very critical to design and expensive to make. Thus they tend to be used in large scale continuously operated plants where atomizing conditions are seldom changed.

4. Particle Size

The particle size distribution of a powder is one of its principal qualities. Unfortunately it has only recently been accepted that the rational way to describe an atomized powder size distribution is to use the median particle size and standard deviation of the log-normal distribution which is generally a very good fit to the experimental data (prior to screening). Certain authors have also quoted such data after "scalping" off oversize particles, which clearly distorts the data. In this chapter all particle sizes will be medians (i.e. 50% finer than) and σ will represent the standard deviation of the distribution

$$\sigma = \frac{d84.1\%}{d50\%} = \frac{d50\%}{d15.9\%} \tag{1}$$

The majority of experimental data which are fully reported conforms to a log-normal distribution (Fig. 4); major deviations generally indicate instability caused by material escaping from the jets. Sampling errors, fines losses and other experimental problems may also be suspected.

Liquid metal properties influence the particle size [e.g. for Cu-Sn see Ref. 6] but as surface tension and viscosity data are not widely available, equations involving such parameters [15] are of little practical use. It is certain that reducing surface tension generally reduces particle size [12].

The practical parameter used to control particle size is water pressure. In Ref. 4 it is shown that a general relationship of the form:

$$Dm = KP^{-n} \tag{2}$$

where Dm is median particle size, P pressure and K and n are constants for a given system can hold good over pressures from 100kPa - 20MPa with values of n typically 0.6-0.8. [8,11,16,17] Other authors report different relationships for annular jets.

The role of water flow/metal flow ratio is obscure. Practical systems tend to operate with a mass ratio from 5-10:1. It is possible to reduce this to as low as 1:1 and good results can be achieved at 2.5:1. However published data is fragmented and contradictory, possibly because annular and flat jet systems have very different responses and variation of water/metal ratio by changing metal flow at constant water flow or water flow at constant metal flow [5] can yield different results. In extreme cases it is possible for metal streams to "punch" through very thin water jets, even at high pressures (e.g. 10MPa) and only be partially atomized to a biomodal distribution. With some jet designs, high metal flow rates result in metal escaping from the jets or even being thrown back upwards.

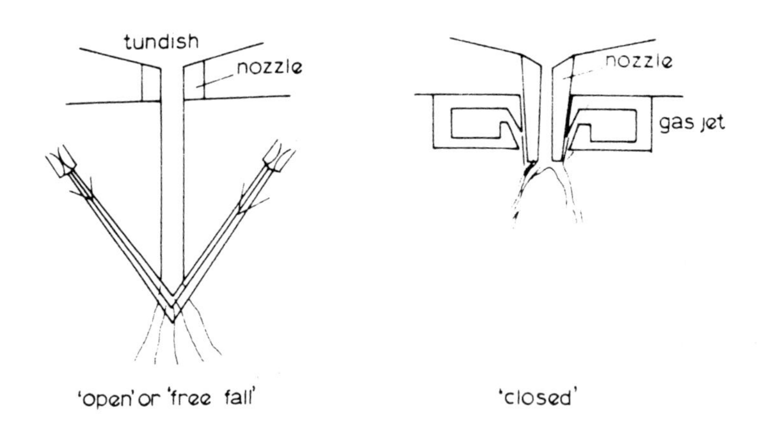

Figure 1 'Open' and 'closed' designs in two-fluid atomising systems.

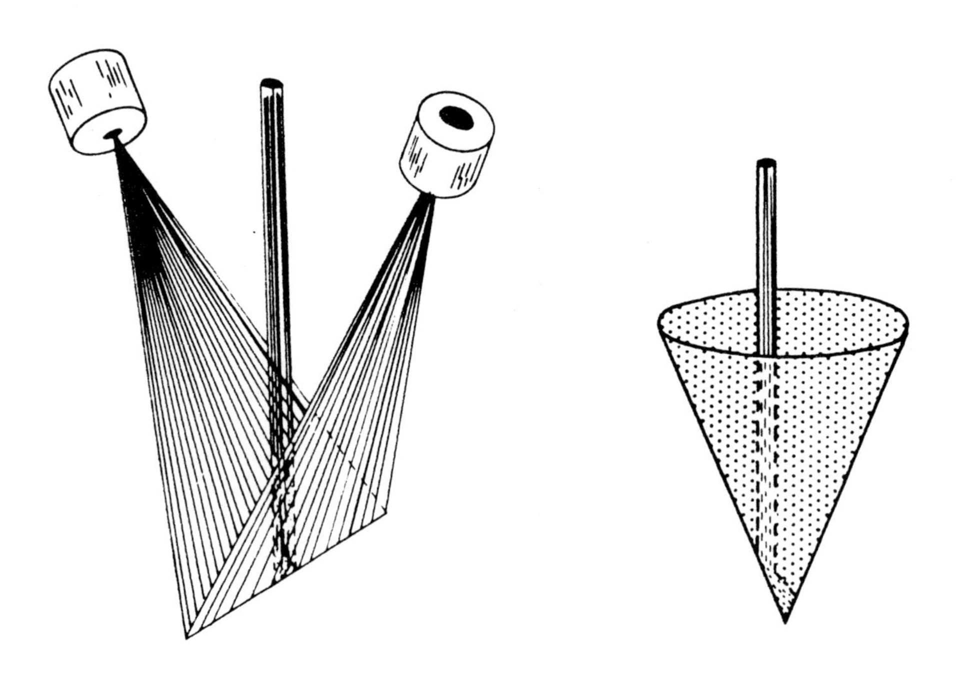

Figure 3 Annular and flat jets for water atomisation.

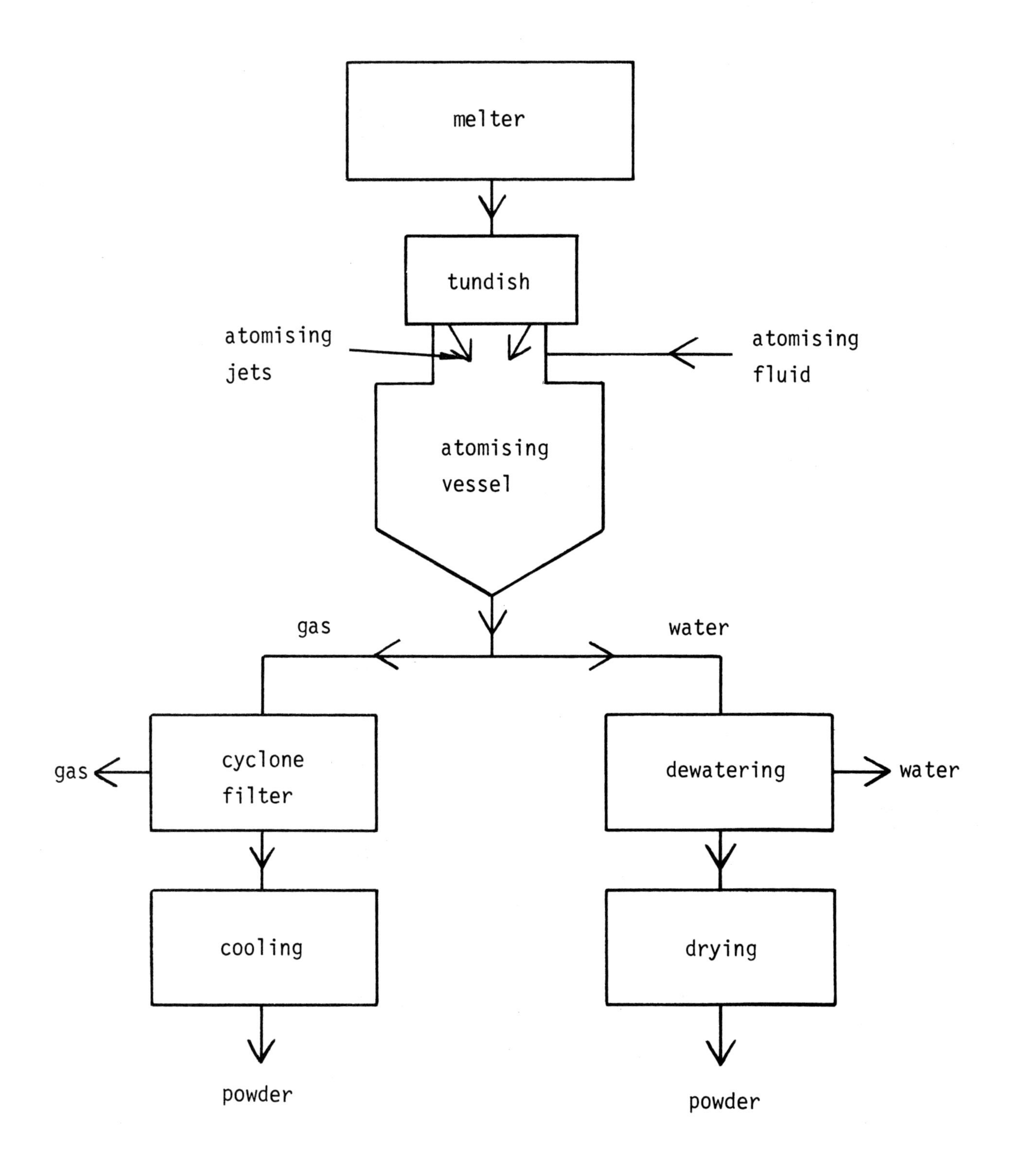

Figure 2 Schematic flow sheet for atomisation systems.

In general, experience shows that very low water/metal ratios rapidly give coarser powders [5] but that increasing the water/metal ratio above 6:1 gives little reduction in size.

Metal superheat influences particle size, partly through reducing surface tension, and at very low melting points (below 500°C) by avoiding premature freezing of particles [4,5,6,14]. At higher temperatures (e.g. melting points over 500°C) an increase of 200°C in superheat can reduce median particle size by 10-20% [4,6].

Atomizing jet design also influences particle size. A high inter-jet angle is desirable [14,16,17] but too high an angle gives rise to problems of metal rejection (Fig. 5). Likewise a short water jet length is desirable, but this is limited in practice.

5. Standard Deviation

This parameter is not universally reported in the literature but it can be stated that there is no clear evidence of any systematic change with variation in median particle size in the range below 500 microns (Fig. 4). Very coarse (e.g. 2000 micron) powders tend to lower standard deviations as coarser droplets become unstable.

In general, standard deviations are sensitive to atomizing geometry. However, if "good" atomization occurs, values tend to a minimum in the range 1.8-2.3. Certain alloys give lower values of 1.8-2.0 (e.g. Ni-B-Si, Fe-Si). These seem to be characterized by a fluxy oxide and spherical particle shape. Alloys containing high levels of Cr, Al, etc. which form refractory oxides and give irregular particles, tend to have a higher value (2.0-2.3).

In the case of iron, standard deviation is highly sensitive to deoxidation and it may be that secondary break-up due to carbon monoxide evolution is responsible .

6. Particle Shape

Particle shape can be measured in many ways, especially now that sophisticated quantitative microscopy is available. However, the only index of shape widely used is apparent density so we shall concentrate on this parameter. Klar and Fesko [2] point out the major importance of collision events where semi-liquid particles stick together prior to freezing. They also show that a low surface tension in the melt tends to favour low apparent density [see also Tamura Ref. 12].

Dunkley [4] shows that the melting point of the metal, related to the nucleate/film boiling transition of water (approximately 300-600°C) has a major influence with low melting point metals such as Pb, Sn, giving very low densities [see also Ref. 6]. Manipulation of atomizing parameters to reduce the intensity of cooling, or increasing the metal superheat likewise to increase freezing time, raises apparent density. Nichiporenko's classic analysis [13] gives a quantitative basis and estimated spherodization times for 100 µm particles are of the order of 0.1-10 microseconds while freezing times are of the order of 100-1000 microseconds for high melting alloys.

The fact that spherodization times are commonly 2-3 orders of magnitude shorter than freezing times makes it hard to explain the development of shape based solely on freezing and spherodization taking place simultaneously. However, they do not start at the same time; atomization is not instantaneous and it takes time for the particle to be formed, to be accelerated, and to travel out of the highly turbulent atomization zone. This zone is at least 20mm long so even at 100m/sec it takes at least 200 microseconds to pass through it, during which time spherodization is prevented by the turbulence of repeated impacts by water (and perhaps metal) droplets [see Klar, refs. 1, 2].

Thus if freezing is completed in the turbulent zone, spherodization cannot even get started. An extreme example is reported by T. Sato [9] where Fe-B-Si alloy is produced with a density of 0.62g/ml and using an "action tube" to increase turbulence. High water jet angle also reduces density [11].

The effects of alloy chemistry are also significant, as pointed out by Nichiporenko [13]. The addition of elements forming refractory oxides can prevent spherodization by producing an oxide film with sufficient strength to oppose surface tension forces. Thus Mg additions to Cu are widely used and reduce its density (under constant atomizing conditions) from 3.5 to 2.0g/ml. Likewise the addition of Cr to Ni reduces powder densities from 4.0 to 2.6g/ml. [For other data see Brewin *et al.*, Ref. 8.]

7. Purity

The purity of a powder would be 100% if every particle had the same composition as the ideal specification. Impurities in metal powders are of several types and their importance varies in different applications:

- Dissolved impurities such as S and P in iron.
- Surface impurities, such as oxide films.
- Ceramic or other foreign particles.
- Contamination by other metal powders.

In certain cases particles with incorrect size or shape can be significant "impurities".

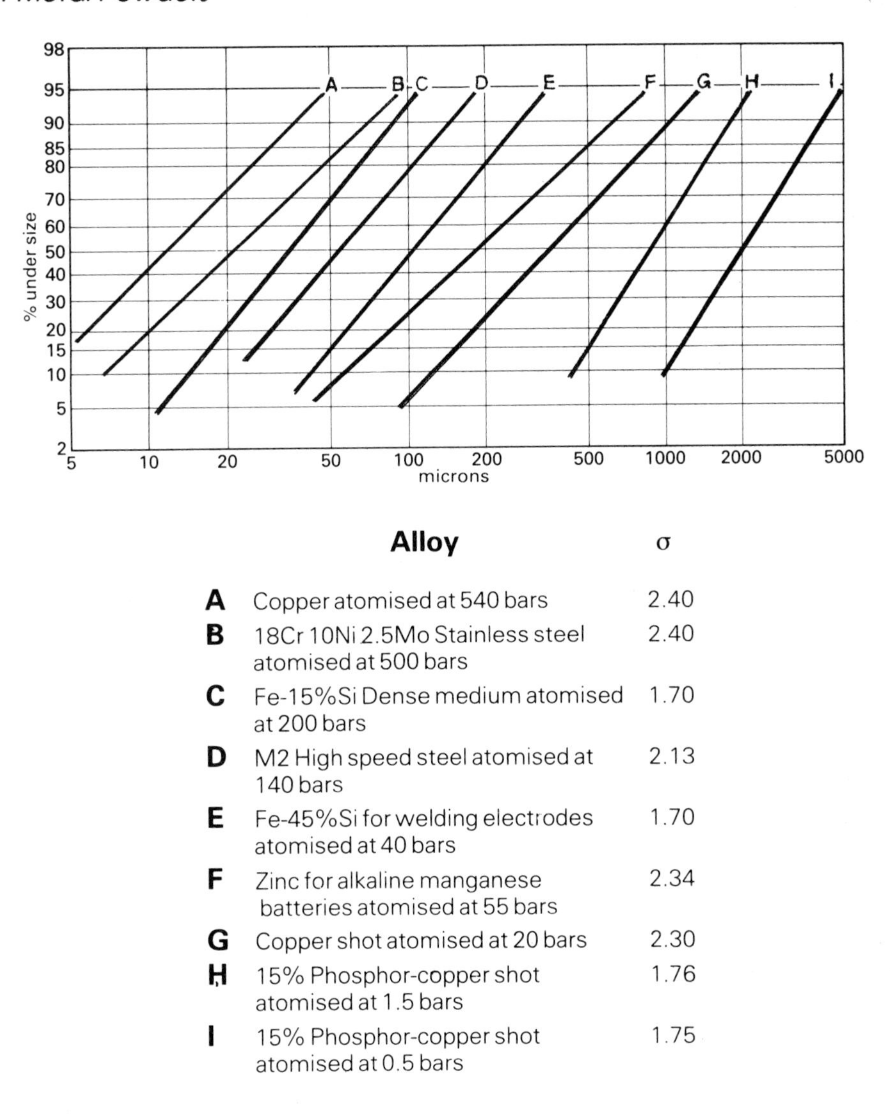

	Alloy	σ
A	Copper atomised at 540 bars	2.40
B	18Cr 10Ni 2.5Mo Stainless steel atomised at 500 bars	2.40
C	Fe-15%Si Dense medium atomised at 200 bars	1.70
D	M2 High speed steel atomised at 140 bars	2.13
E	Fe-45%Si for welding electrodes atomised at 40 bars	1.70
F	Zinc for alkaline manganese batteries atomised at 55 bars	2.34
G	Copper shot atomised at 20 bars	2.30
H	15% Phosphor-copper shot atomised at 1.5 bars	1.76
I	15% Phosphor-copper shot atomised at 0.5 bars	1.75

Figure 4 Typical water atomised particle size distributions.

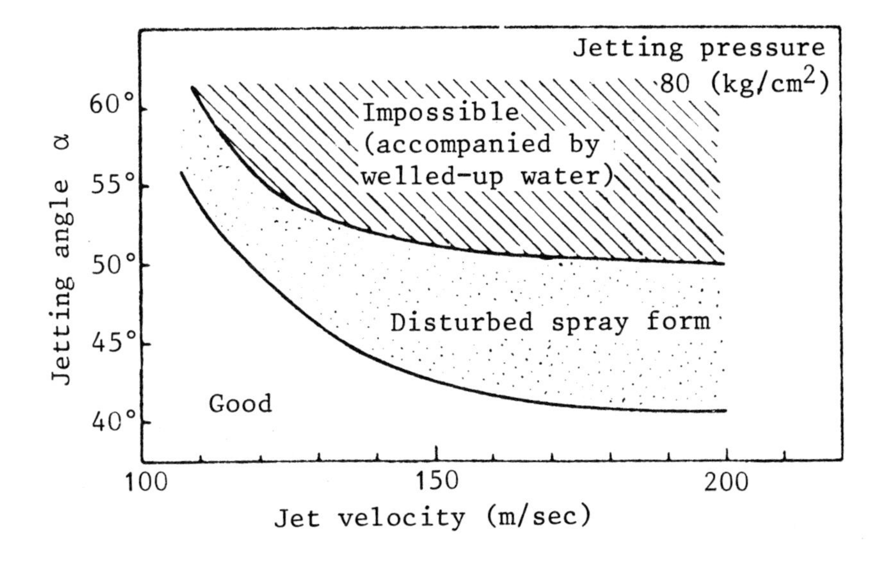

Figure 5 Effect of jet angle and water velocity on stability of water atomisation (Ref. 17) .

Atomization is incapable of removing impurities dissolved in the molten metal, so it is necessary to resort to pyrometallurgical means such as oxygen blowing, basic slags, ladle metallurgy, or vacuum metallurgy to purify the liquid metal.

Oxide films are frequently significant in water atomized powders, but levels of oxide are very variable. Once the effect of oxygen in the atomizing atmosphere is eliminated [17], the oxide content of powders strongly depends on the composition of the melt [7, 18]. Typical values are shown in Table 1.

General rules are hard to formulate but two factors are important:

1. The ΔG of the reaction

$$M + H_2O \rightarrow MO + H_2$$

This is normally determined by reference to the "Ellingham diagram" of free energy of formation of oxides against temperature.

2. The properties of the oxide film itself.

Thus metals with very low affinities for oxygen, such as gold, silver, etc. can be water atomized with oxide levels of as little as 40-300ppm. Metals such as Cu, Ni, Sn are typically higher, from 200-800ppm. However, pure iron rises to about 3000ppm and the addition of the more reactive silicon reduces this rapidly to 1000ppm or even 500ppm. The less reactive manganese raises it to 30,000ppm. The reason for this may be that silica is a stable, refractory and impermeable oxide, while iron and manganese are polyvalent and have low melting point oxides which allow rapid diffusion of oxygen [7].

Inclusion levels in water atomized powders are inherently no higher than in other atomized products (unless hard water is used for atomization). Cross contamination depends very strongly on equipment design and operating practise. Suitable designs of atomizer can be readily cleaned by water jetting and the system literally flushed out when changing grades. However, large continuous systems can harbour massive amounts of trapped powder so that iron powder plants, for instance, are sometimes very difficult to clean out.

8. Equipment Design

Much improved equipment is now available for water atomization compared with some earlier designs. It has been found possible to build sealed purged units which allow the oxygen content of powders to be held down to a consistent minimum level. A range of specially small units has been developed for precious metals and R & D work where low losses and easy cleaning for flexibility are very important (Fig. 6). Continuous units which allow a slurry of powder to be pumped to automatic dewatering equipment are also available and can operate for many hours with minimum attention (Fig. 7). [The options are discussed in Ref. 52.]

The atomizer itself is only a part of the system and advances in the design of high pressure pumps, where size and cost have been reduced substantially, allow smaller, cheaper plants. Induction melting is now very widely used, not just for Fe, Ni and Co alloys, but for Cu, Ag, Au, Pt, Zn and Al. Computer control of the plant is possible, but generally only desirable on complex continuous plants where cyclones, centrifuges, filters and dryers, together with conveyors, weighing systems etc can be fully integrated under central control.

The aim of these improvements in design is to improve quality, especially consistency, and to reduce the cost of energy, refractories, capital (equipment and buildings) and labour. Critical to cost reduction is a consistently high yield of useful product. This will be discussed later.

9. Ultra-fine Powders by Water Atomization

One significant development in the last decade has been the use of extremely high water pressures (500-1000 bar or 50-100MPa) to atomize powders with median particle sizes reported to be in the range 5-20μm. These are ideal for use in metal injection moulding. It has been found that it is no simple matter to take designs of atomizing jets which operate well at conventional pressures of 10-20MPa and simply raise pressures. Extensive changes are necessary if equation (2) is to be adhered to.

The leading research institute in the field is the Japanese NRIM. Japanese companies now making such powders include Kobe Steel using 100MPa on steel to 6μm [19], Pacific Metals (100MPa - stainless steels to 8μm), Mitsubishi Metals (10μm), Nippon atomizer (100MPa, 10μm steels). It is reported that Hoganaes of USA have also obtained a licence.

It is understood that the product is still very expensive compared with conventional powders. The cost of the pumps does not account for this as they only amount to 5-10% of plant costs. The problem seems to be that the current designs are limited to a metal stream diameter in the range of 2-3mm. This means atomizing speeds of 2-5kg/min and great problems with metal supply systems and refractories when dealing with steels. Thus there is still great scope for development here, although published data are scanty.

The addition of long chain polyethylene oxide to reduce particle size is mentioned by Japanese authors and was patented by BSC in 1976 [20]. There are little published data on its effects and the economics of its use may be doubtful.

10. Oil Atomization

Oil atomisation shares most of the characteristics of water atomization, and can readily be carried out in adapted water atomizers [10].

The advantage over water atomization is that very low oxygen contents can be achieved; however carburization can occur instead. Larson [21] quotes powder oxygen contents below 100ppm but worked mainly on coarse (500 μm) powder using low pressure (500KPa) oil. Workers at Sumitomo [11,22,23] used higher pressure (14MPa) and an annular jet to produce "moulding grade" steel powders with median particle sizes around 70 μm.

Production of low alloy Cr-Mn steel powders by this process ceased in 1988. The economics were unfavourable because:

- The atomization process itself is more complex.
- It is necessary, after atomization, to carry out a wet hydrogen anneal to decarburize the powder, followed by a dry hydrogen reduction to reduce the oxide level. This raises annealing costs and the advantage of low oxide content of the atomized powder is largely lost.

However, the work shows that the process is practicable and Larson [21] continues to produce powder this way for use without heat treatment as an inoculant in casting. Mordike [10] speculates on other potential applications in rapid solidification of high-carbon alloys or alloys inert to carbon .

11. Gas Atomization

Publications abound on inert gas atomization, although its far more widely used sub-class, air atomization, is seldom mentioned. [Useful reviews are presented by Klar in refs. 1 and 30.] Once again, the methodology of research in the field is not yet universally agreed and many reports neglect vital parameters or quote results that are ill defined. Theoretical modelling of the process appears to have advanced only modestly since Lubanska's paper [27]. We are very far indeed from being able to do more than interpolate empirical data, and sometimes to extrapolate.

The principal concerns of published research have been to produce rapidly solidified, high purity powder. This in turn leads to a need to make finer powders, with 10 microns a common target. Once again, in practice both freefall and closed jet systems are used, but for fine powder closed jets are preferred and thus most publications concern them.

12. Ultrasonic Gas Atomization

This technique was at one time claimed to produce very fine powders. The principle is to use "whistling" atomizing jets with ultrasonic frequency and supersonic velocity, although there has been confusion between ultra and super in some cases. Despite very extensive work in at least 20 laboratories, no production plant is known using the principle, unless at Tula in the USSR where Abramov carried out trials [47].

The evidence of any unique properties due to ultrasonic vibration, or even of the vibration itself, is scanty [44-47], so USGA will be reviewed with other similar nozzle types below.

13. Conceptual Models of the Break-up Process

Various simplified models of the stages through which the continuous metal stream breaks up have been put forward. [A number of these are reviewed by Rao in Ref. 26.] Refs. 25, 26, 29, 31, 36 present observations of freefall jets of tin etc. alloys atomized to very coarse powders typically 100-500 μm. The wave-ligament model (Fig. 8 from Ref. 43) can be squared with such observations but photographs of frozen "ligaments" are not conclusive and the mathematics of the model do not permit precise predictions, partly because a velocity term is always involved and there is never a single calculable velocity in a practical system, leading to the need for numerous empirically derived constants.

Other workers looking at closed atomizers, producing much finer powders, have been unable to see such detail and have postulated a single step atomization process or some similar mechanism [32,33,43]. Attempts to observe the phenomenon of break-up photographically on closed jet systems have not proved fruitful [43]. The fundamental analysis of the stability of a liquid particle being accelerated in a gas stream is quoted in refs. 43 and 48, but the difficulty of modelling the gas flow and predicting peak relative metal/gas velocities has made it of doubtful application.

Figure 6 Small 5-10kg capacity (D5/1) water atomiser.

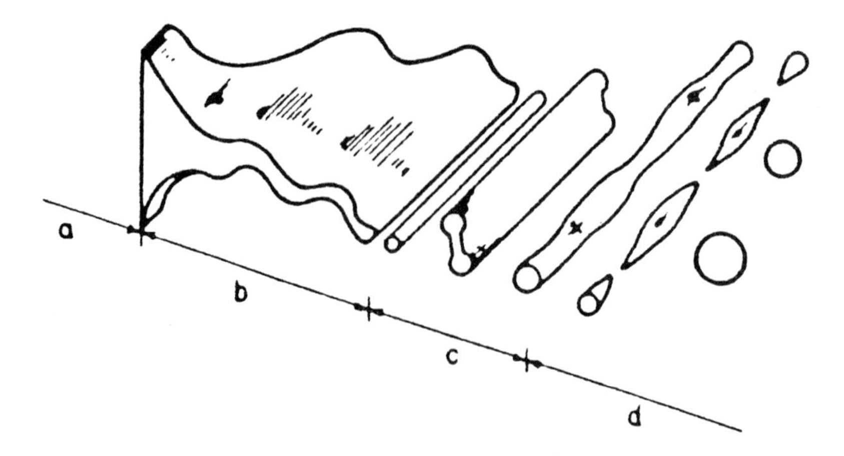

Figure 8

Figure 7 D25 continuous water atomiser processing silver scrap in Sweden.

Some of the best published photographs are by Unal (Fig. 9, Ref. 38), but even an 0. 5 microsecond flash reveals little of the mechanism of particle formation beyond typical "filming" nozzle behaviour [see also Refs. 30,39].

14. Particle Size

The work of Lubanska [27] who attempted a correlation of data for various nozzles and metals is often quoted. Her equation is:

$$\frac{Dm}{d} = K\left[\frac{\nu m}{\nu g\, W} \times \left(1 + \frac{M}{A}\right)\right]^{1/2} \qquad (3)$$

where:
Dm - median particle size
d - metal stream diameter
K - a constant "for particular conditions"
νm, νg - kinematic viscosities of metal and gas
M - mass flow rate of metal
A - mass flow rate of gas

$W = \frac{\rho V^2 d}{\gamma}$ the Weber number

where:
V - relative velocity of gas/metal impact
ρ - density of metal
γ - surface tension of metal

By plotting the parameter on graphs having 5 log decades, scatter appears modest while being in fact an order of magnitude. The expression deserts Ockham's razor which states that entities (parameters?) are not to be multiplied beyond necessity. The urge to analyse the problem in dimensionless groups leads to extra parameters, for which no experimental evidence was available, being added to an already complex equation. Rao [26] found great difficulty in fitting data to this equation.

More experienced, but no less qualified authors such as Klar [30] state that "all practically useful correlations are of an empirical nature".

A lot of studies attempt to explain particle size by analysing the gas flow pattern of closed, supersonic nozzles e.g. for shock waves [28, 32, 34, 41, 42, 43]. Unfortunately, even if a complete analysis allowed accurate knowledge of the entire flow-field when cold, the perturbation induced by heat transfer doubling or tripling the gas volume would be horrendous. Even then, we do not know how to predict particle size if we know the flow field. Thus more basic, but well designed experiments using simple numerical input and output parameters seem more fruitful.

Unal [38,39] shows the importance of gas/metal ratio. In Fig 10, data from refs. 24, 30, 33, 38, 44 are plotted as a function of gas/metal ratio. They all follow a very similar pattern, while plots against metal flow rate showed huge variations. In Fig. 11 (from Unal [39],) it is shown that the relationship of particle size to gas/metal ratio is modified by operating pressure for a given nozzle but that there exists an optimum pressure. Thus increasing pressure does *not* always give finer powder, either absolutely, or at a given gas/metal ratio.

It is possible to attempt a summary of the principal parameters governing particle size.

1. Gas/metal Flow Ratio

This is sometimes quoted as a mass ratio (kg/kg) or as a cu.m/kg ratio. It is probably the most important single parameter for a given nozzle type (see Fig. 10). Ratios (for nitrogen) and dense metal (7-9Mg/cu.m) of less than 0.5cu.m/kg give coarse powders over 100m (as well as exhaust temperatures of over 300°C). Thus higher flow ratios are common, with 1cu.m/kg allowing particle sizes of around 50µm. For very fine powder e.g. less than 20-30µm, flow rates of 2cu.m/kg or more may be advantageous, but extreme levels of 5-10 cu.m/kg show rapidly diminishing returns [see also refs. 24, 30,33,44].

2. Gas Velocity

While a high gas velocity promotes finer particles in every theoretical model, measurement of velocity is not often practical [but see Ref. 28]. It must be understood that gas pressures above 1 bar (100KPa) allow sonic velocities and raising pressure much above 1.0MPa allows only marginal increases in velocity [33]. The fact that gas velocity decays very rapidly in a free jet influences nozzle designers to reduce the gas nozzle/metal impact distance.

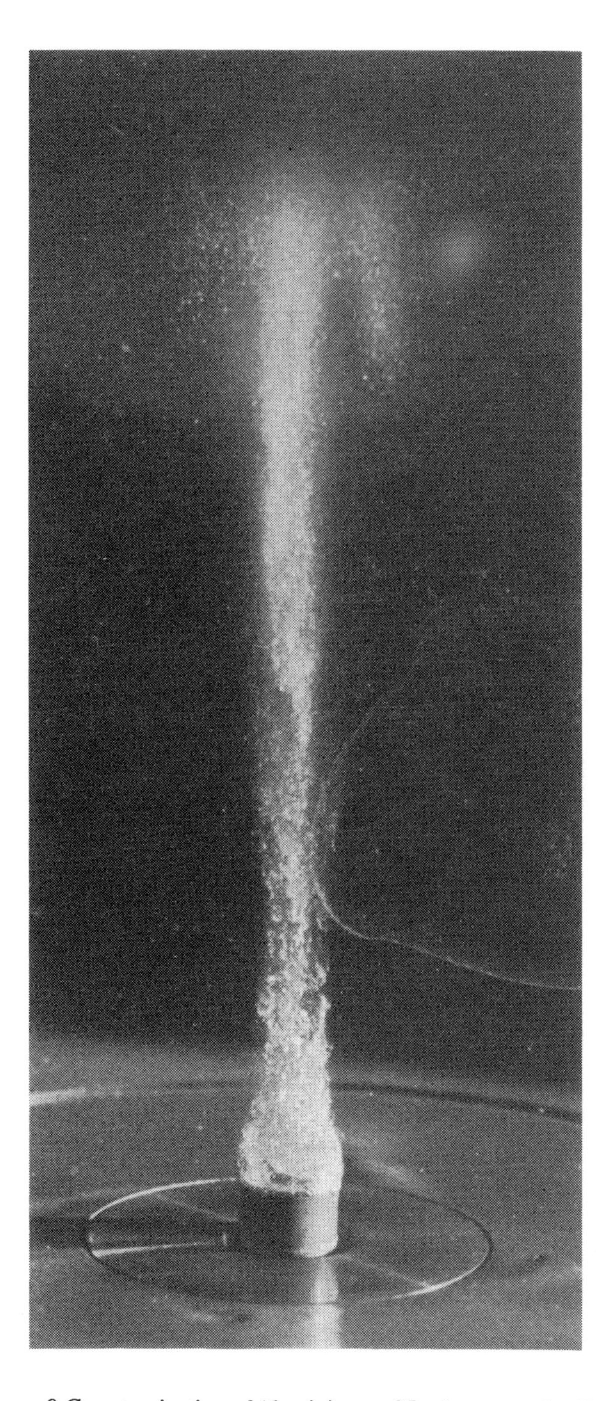

Figure 9 Gas atomisation of Aluminium to 25 micron powder (Ref. 38).

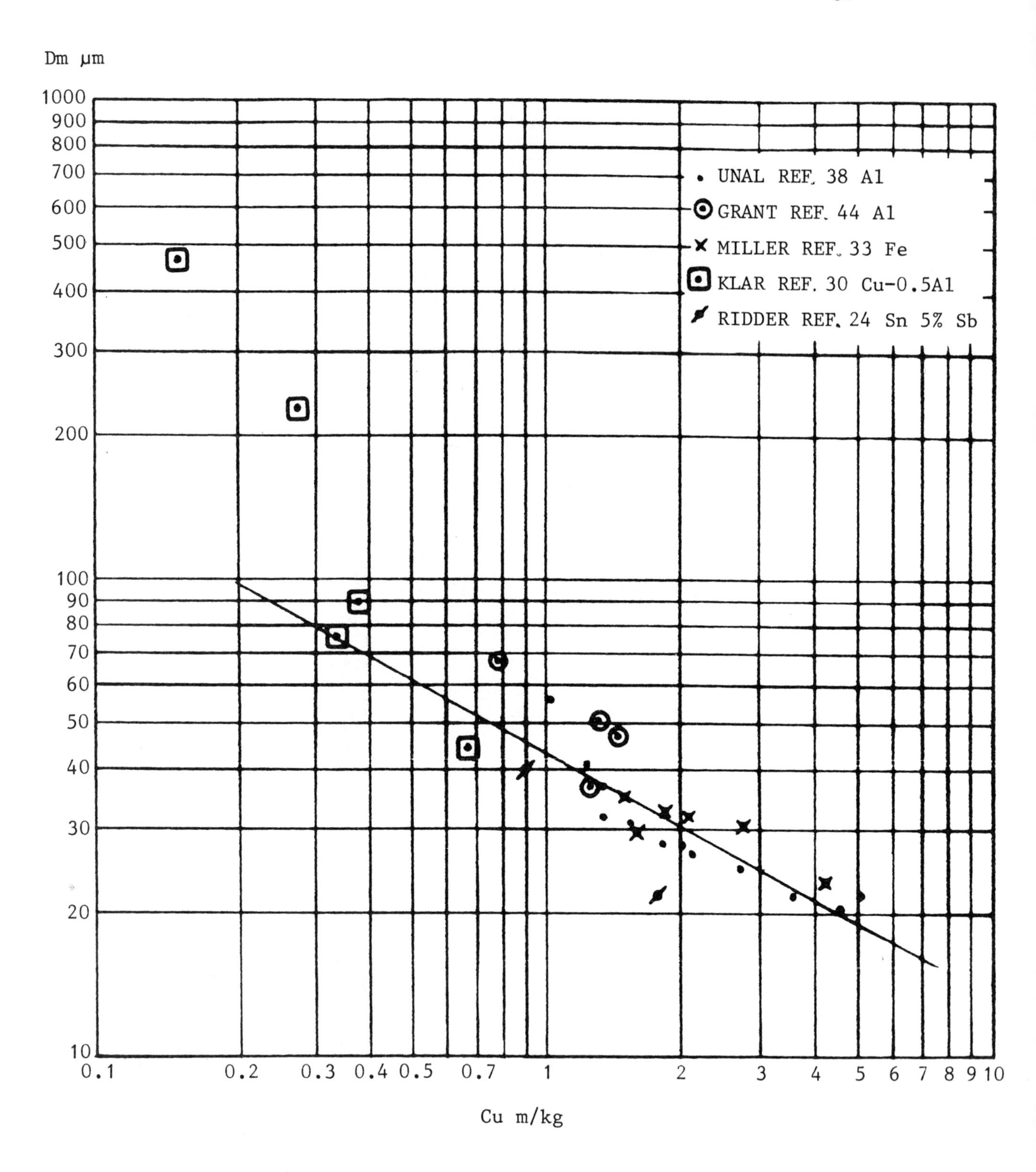

Figure 10 The effect of gas/metal ratio on median particle size.

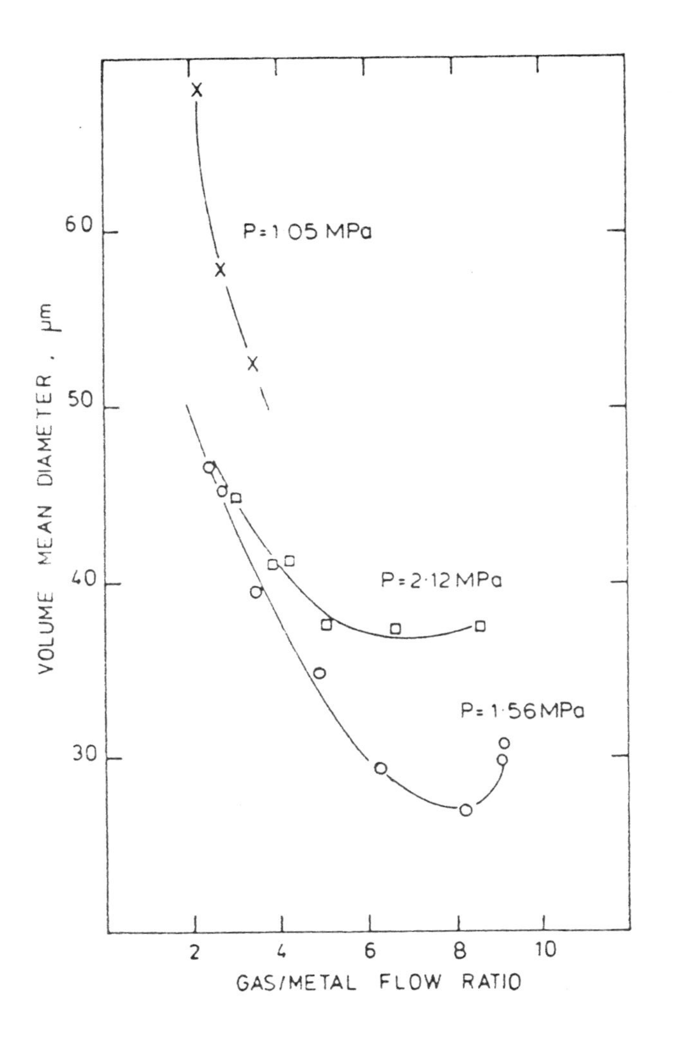

Figure 11 The effect of operating pressure on the gas/metal ratio: particle size relationship (Ref. 39).

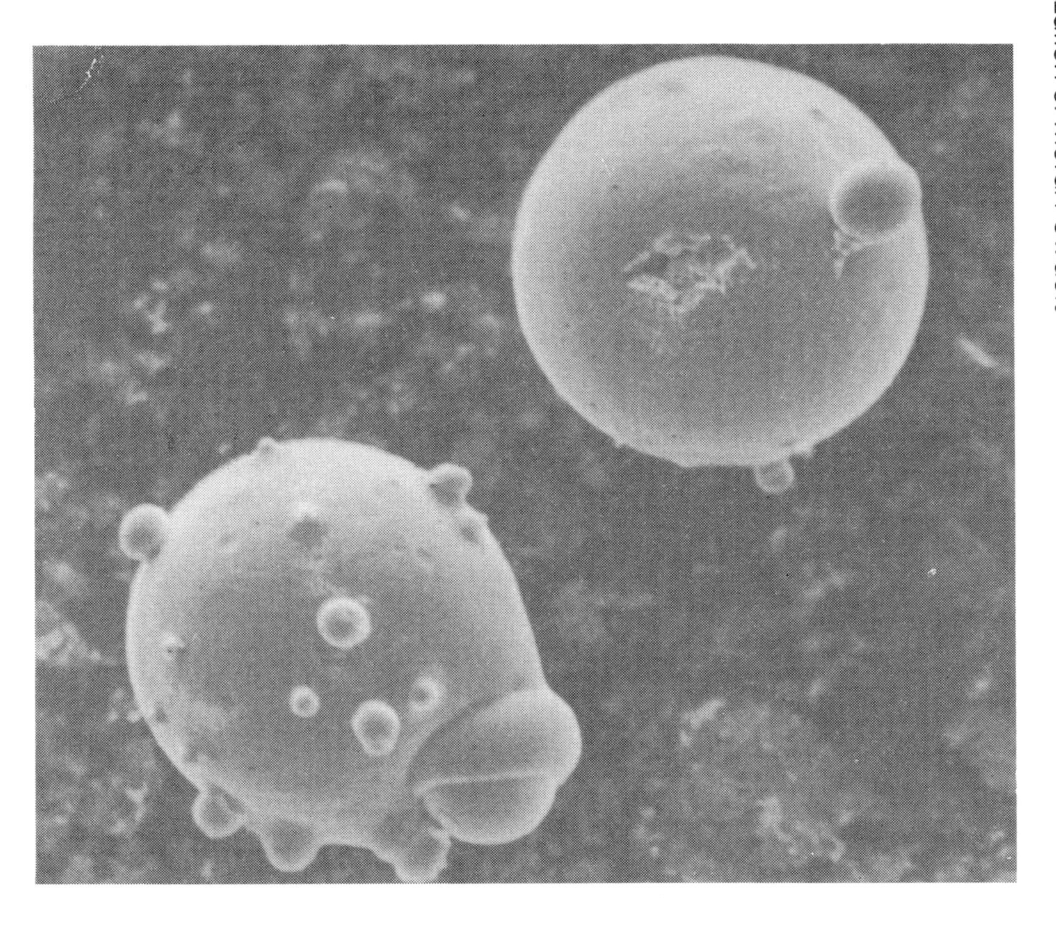

Figure 12 Satellites on gas atomised cobalt alloy.

3. Gas Pressure

The use of pressures much above 1.0MPa, as mentioned above, gives little advantage in velocity. However, it does allow the use of a compact nozzle design by reducing high pressure gas velocities in the feed system. It may perhaps also allow the density of the gas at impact to be higher. In practice, despite much work on the use of pressures up to 20MPa [32, 40, 42, 53], firm evidence of any benefit is scanty. In Ref. 46 it is reported that an air atomizer working at 200psi (1.3MPa) produced much finer powder than a helium USGA unit at 1200psi (8.2MPa) .

Reports showing a series of tests on a given nozzle producing finer powders as pressure rises often reflect more the importance of the increased gas/metal ratio than of pressure [31].

Practitioners opinion is that pressures much above 2-3MPa are of dubious value.

4. Gas Properties

As for pressure, the gas properties are difficult to investigate. Argon, nitrogen and air are very similar, only helium which is very costly, is very different. Not only are its density and viscosity radically changed, but sonic velocity is around 1000m/sec. Thus if He is used on a nozzle also tested on nitrogen, flow rises by three times. Thus the gas/metal ratio (cu.m/kg) changes greatly and reports of very fine powders may be specific to certain nozzles. Experience shows that Lubanska's equation (3) which predicts He to produce (through changes in velocity and viscosity) ten times finer powder, is not correct. As Ref. 46 shows, air can be used to make very fine powders.

The real influence of gas properties thus remains unclear.

5. Metal Properties

As for water atomization, surface tension and other metal properties are important. Thus tin and lead are routinely air atomized to around 10m, while steels are difficult to reduce below 30m.

6. Scale of Operation

While Lubanska includes stream diameter in her equation, and free-fall systems certainly give coarser products as scale increases, filming nozzles [such as Ref. 38] may not be affected so much by scale. Certainly there are few reports of data at rates over 10kg/min [27,30,31,45], but most commercial operations need to run at 5-50kg/min. Multiple streams and atomizing nozzles are used for some metals (e.g. Al, Zn).

7. Nozzle Design

Much of the detail remains proprietary but all designs aim to minimize gas use while achieving the correct particle size. This involves getting the gas exit close to the metal and raising gas velocity. Practical problems such as thermal shock, wear, fouling, etc. are very important and reproducible behaviour is vital for commercial success.

15. Particle Shape

In the absence of oxygen giving rise to oxide films, gas atomized powders tend to be spherical. The incidence of "satellites" (see Fig. 12) is the main variable shape factor. These are caused by collisions between particles in flight and become worse as finer powders are made due to the finest particles being drawn up into the atomizing plume [37]. Reports on this topic are few as numerical analysis of shape is difficult.

16. Hollow Particles

Hollow particles are found in both gas and water atomized powders, but are of greatest concern in the gas atomization of alloys for HIPing where argon filled porosity is a major problem. In Ref. 49 it is shown that argon dissolved in the melt raises porosity in nickel alloy. The definitive publication is by L'Estrade in Ref. 50. Overheating and hydrogen pick-up are definitely a common problem. Some spectacular examples of hollow particles are shown in Ref. 1.

17. Design of Equipment

As gas atomising is used for alloys ranging from tin to steels, gas atomizer designs vary very greatly. For low melting alloys (Pb, Sn, Zn, Al) continuous operation is common and vertical upward [37,38] horizontal or vertical downward designs are all in use, with product removed or collected in various ways. For higher melting alloys (Cu, Co, Ni, Fe) vertical downward atomizers are the rule, ranging from tiny units for very fine precious metals (Fig. 13) to the huge 10,000t/yr Anval Nyby unit over 10m tall. Coarse products may be removed by gravity, but recent trends are to use pneumatic conveying to a cyclone to save headroom [51]. Vessel size depends on particle freezing time and thus on:

- Particle size (maximum).
- Alloy freezing range.
- Cooling conditions.

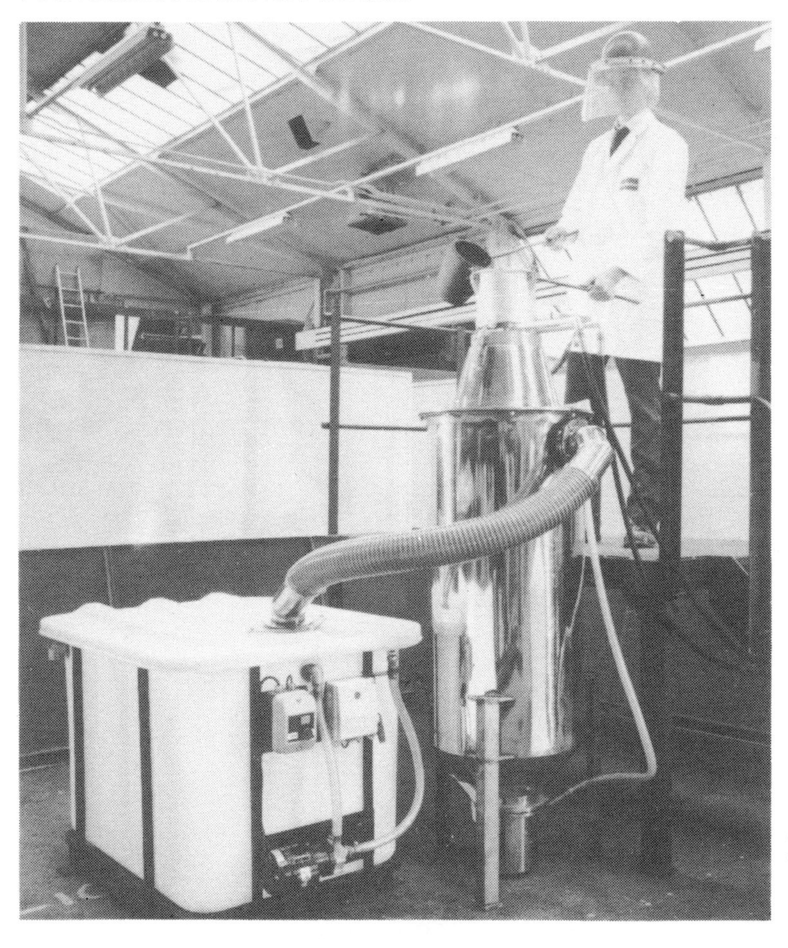

Figure 13 Small 5-10kg capacity (F5) gas atomiser with water quenching.

Thus some hard facing alloys such as Ni-Cr-Cu-Mo-B-Si with very wide freezing ranges are notoriously prone to splatting on the walls. The use of helium speeds freezing but is too costly.

Generally if mean particle sizes less than 100μm are required on alloys with freezing ranges of 100°C or less, a vessel 3m tall is adequate (see Fig. 14). For very wide freezing range alloys, water quenching is commonly used.

One area of interest is automatic control of the process as it is sensitive to minor variations in metal flow rate. Ref. 24 discusses one approach using laser sizing. This could be of great value in continuous production systems, e.g. for aluminium.

18. Economics of Gas and Water Atomization

There is little published data on costs [52-4]. In Ref. 52 the impact of operating scale is discussed in detail. Costs are very sensitive to output. Automation reduces operating costs but reduces flexibility. In batch melting plants, batch size governs costs to a large extent.

The foregoing remarks apply to gross production costs. The net cost of a product is very sensitive to the yield of product achieved.

Thus L'Estrade [53,54] shows that costs can vary by a factor of 50 depending on specifications regarding purity and yield. Data are given on the relative costs of water, nitrogen and argon atomizing and the selected melting condition (air, inert gas, vacuum) and handling requirements. Vacuum melting is four times as costly as air melting and is thus only used when it is essential. Dry argon atomizing is three times the cost of dry nitrogen atomizing while water atomizing is 0.6 times the cost of dry nitrogen atomizing.

The yields for different particle size fractions are discussed in Ref. 57. Figure 15 shows a "Master Yield Curve" where the yield available in a range of given size ratio (Max÷Min) is given for a range of standard deviations. Generally a low standard deviation is highly desirable, except for optimum packing density for HIPing where a bimodal distribution is preferable. The ability to control median particle size and get a consistent product is a vital one and SPC techniques are being increasingly applied to gas and water atomization.

The economics of gas atomization when gas recycling is feasible, are quite attractive and close to water atomization. Gas recycling is quite cheap unless purification of the gas is needed. Thus for less critical purity applications cooling, filtration and compressing are acceptable and greatly reduce gas costs which, at $1m^3/kg$, are around $200/t. Clearly a simple recycling system, costing $1-200,000, is attractive for outputs of over 500t/yr.

One cost often ignored is of obtaining low cross-contamination levels. [This is discussed in Ref. 35 as well as 53, 54.]

Figure 14 30kg capacity DG10 dry inert gas atomiser.

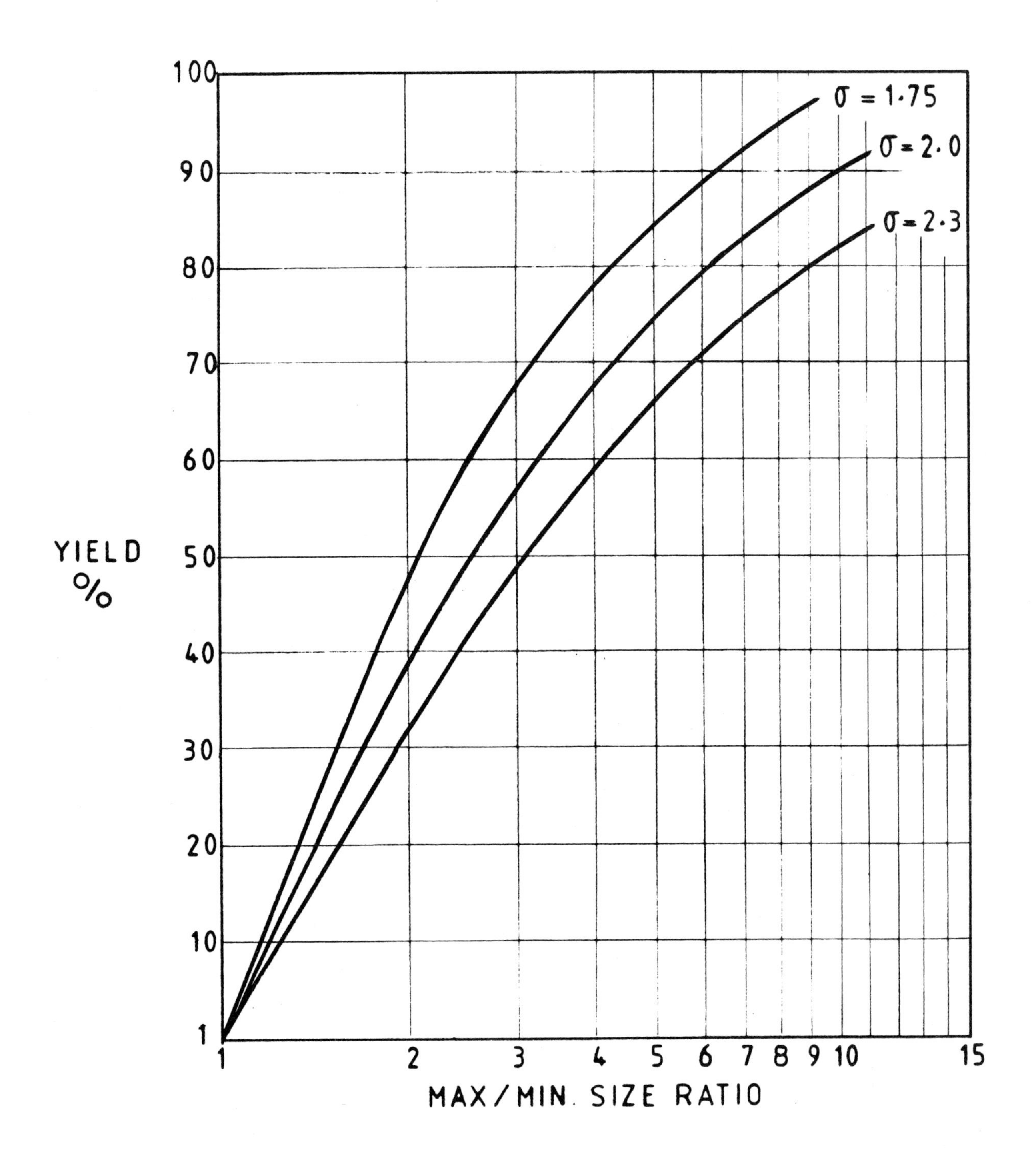

Figure 15 "Master Yield Curve" showing yield between two sizes as a function of size ratio and standard deviation (σ).

References

1. E. Klar and J.W. Fesko, Gas and Water Atomisation, Metals Handbook 9th Edition, AMS, 7, pp.25-39, 1984.
2. E. Klar and J.W. Fesko, On the Particle Shape of Atomised Metal Powders, Progress in PM 37, pp.47-66, 1981.
3. E. Klar, Commercial Water Atomisation of Metals, Metal Powder Report, pp.7-18, January 1985.
4. J.J. Dunkley, The Production of Metal Powders by Water Atomisation, PM International, 10, 1, pp 38-41, 1978.
5. J.J. Dunkley and J.D. Palmer, Factors Affecting the Particle Size of Atomised Metal Powder, Powder Metallurgy, 29, 4, pp.287-290, 1986.
6. J.J. Dunkley, The Influence of Composition on Powder Properties in the Copper-Tin System, Progress in PM, 39, pp.101-108, 1983.
7. J.J. Dunkley, The Factors Determining the Oxygen Content of Water Atomised 304 L Stainless Steel Powder, Progress in PM, 37, pp.39-45, 1981.
8. P.R. Brewin *et al.*, Production of High Alloy Powders by Water Atomisation, Powder Metallurgy, 29, 4, pp.281-285, 1986.
9. T. Sato, Rapid Quenching Water Atomisation Process for Amorphous Alloy Powders, Metal Powder Report, pp.428-430, June 1987.
10. K.U. Kainer and B.L. Mordike, Oil Atomisation, Metal Powder Report, pp.28-31, January 1989.
11. Yishinaga *et al.*, New Atomisation Process and Properties of Powder, SMI Technical Report, 40, No 1.
12. K. Tamura and S. Wanikawa, On the Atomising Variables in the Production of Metal Powder by Liquid Atomisation, J Jap Soc Powder Met, 15, pp.302-307, Oct. 1968.
13. O.S. Nichiporenko, Shaping of Powder Particles during the atomisation of a Melt with Water, Porosh Met 9, 165, pp.5-10, September 1976.
14. A.A. Kostyrya, Yu I Naida, O.S. Nichiporenko and V.A. Tsyban, Effect of Process Parameters on some properties of powder produced by the atomisation of melts with water, Porosh Met 4, 172, pp.21-25, April 1977.
15. O.S. Nichiporenko, A.B. Medvedovskii and Yu I Naida, Disintegration of Melts by Water, Porosh Met 8, 164, pp.6-9, August 1976.
16. R.J. Grandzol, Water Atomisation of 4620 Steel and Other Metals, PhD Thesis, Drexel University 1973.
17. T. Kato *et al.*, Manufacture of Steel Stainless Steel Powder, Denki Seiko, 46, 1, pp.4-10, 1975.
18. K. Tamura, T. Takeda, On the Manufacture of 13 Cr Stainless Steel Powder by Liquid Atomisation, J Jap Soc PM, pp.25-32, February 1965.
19. Honma *et al.*, Characteristics and Applications of High Pressure Water Atomised Fine Powder, Kobe Steel R&D, 37, 3.
20. D. Brennan, I. Stewart, BP 1574431.
21. U.R. Larson, US Patent 4, 124, 377.
22. M. Ichidate, T. Kubo, US Patent, 4, 385, 929.
23. T. Kubo *et al.* US Patent, 4, 448, 746.
24. S.D. Ridder and F.S. Biancaniello, Process Control During High Pressure Atomisation, Materials Science and Engineering, 98, pp.47-51, 1988.
25. J.B. See, G.H. Johnston, Interactions between Nitrogen Jets and Liquid Lead and Tin Streams, Powder Technology, 21, pp.119-133, 1978.
26. K.P. Rao, S.P. Mehrotra, Effect of Process Variables on Atomisation of Metals and Alloys, Modern Dev in PM, 12, pp.113-130, 1980.
27. H. Lubanska, Correlation of Spray Ring Data for Gas Atomisation of Liquid Metals, BISRA Open Report, P/2/69 also J Metals, 22, 2, p.45, 1970.
28. Y.I. Naida, O.S. Nichiporenko, M.A. Fenko, Selection of Atomisation Parameters for the Preparation of Powders of the Required Particle Size with a High Yield, Porosh Met, 6, 150, pp.14-19, June 1975.
29. C.E. Seaton, H. Henein, M. Glatz, Atomisation of Molten Metals Using the Coanda Effect, Powder Met, 30, pp.37-47, 1987.
30. E. Klar, W.M. Shafer, High Pressure Gas Atomisation of Metals in Powder Metallurgy for High Performance Applications, pp.57-68, 1972.
31. G. Matei *et al.*, Melt Disintegration During Air-atomisation of Ni-Cr-B-Si Alloys, Proc Int Conference PM, Dusseldorf, pp.33-36, 1986.
32. I.E. Anderson, R.S. Figliola, Observations of Gas Atomisation Process Dynamics, Mod Dev PM, 20, pp.205-223.
33. S.A. Miller, Close-coupled Gas Atomisation of Metal Alloys, Proc Int Conf PM, Dusseldorf, pp.29-32, 1986.
34. M.S. Stoichev, Aerodynamic Study of Gas Flow from a Molten Metal Atomising Nozzle, Mod Dev PM, 20, pp.713-721.
35. H.W. Meinhardt, P. Kunert, The Influence of Process Variables and Equipment Characteristics on the Purity of Atomised Metal Powders, Mod Dev PM, 20, pp.225-239.
36. P. Rao, Shape and Other Properties of Gas Atomised Metal Powders, PhD Thesis, Drexel University, 1973.

37. A. Unal, D.C. Robertson, Pilot Plant Gas Atomiser for Rapidly Solidified Metal Powders, Int Jnl Rapid Sol, 2, pp.219-229, 1986.
38. A. Unal, Influence of Nozzle Geometry in Gas Atomisation of Rapidly Solidified Aluminium Alloys, Mat Sci & Tech, 4, pp.909-915, October 1988.
39. A.Unal, Effect of Processing Variables on Particle Size in Gas Atomisation of Rapidly Solidified Aluminium Powders, Mat Sci & Tech, 3, pp.1029-1039, December 1987.
40. J. Liu *et al.*, Mass and Heat Transfer During Ultrasonic Gas Atomisation, PM International, 20, 2, pp.17-22, 1988.
41. J Liu *et al.*, Fluid Flow etc in Ultrasonic Gas Atomisation, Swedish Inst Metals Res Report, IM2178, 1987.
42. G. Rai, E. Lavernia, N.J. Grant, Powder Size and Distribution in Ultrasonic Gas Atomisation, J Met, pp.22-26, August 1985.
43. E.Y. Ting, N.J. Grant, Metal Powder Production by Gas Atomisation, Progress in PM, 41, pp.67-85, 1985.
44. D.H. Ro, H. Sunwoo, A Comparative Study of Conventional and Ultrasonic Gas Atomisation of Aluminium Alloys, Progress in PM, 39, pp.109-124, 1983.
45. M. Hohmann, S. Jönsson, Pros and Cons of Ultrasonic Gas Atomisation, ATB Metallurgie XXVII, 2-3, pp 85-87, 1987.
46. P.R. Bridenhaugh *et al.*, Particulate Metallurgy in Rapid Solidification, Light Metal Age, pp.18-26, October 1985.
47. O.V. Abramov *et al.*, Development and Study of Pneumoacoustic Nozzle for Atomising Molten Metals, Acoustic Journal (Moscow), 27, 6, pp.801-807, 1981.
48. A. Lawley, An Overview of Powder Atomisation Processes and Fundamentals, Int Jnl PM Tech, 13, 3, pp.169-188, 1977.
49. Y.F. Ternovoi *et al.*, Pore Formation in Atomised Powders, Porosh Met 1, 265, pp.10-44, January 1985.
50. L. L'Estrade *et al.*, Internal Porosity of Gas Atomised Powders, Mod Dev PM, 20, pp.187-204, 1988.
51. R.I. Howells *et al.*, Production of Gas Atomised Metal Powders and their Major Industrial Uses, Powder Met, 31, 4, pp.259-286, 1988.
52. J.J. Dunkley, The Impact of System Design on Metal Powder Production Costs, Int J PM & PT, 19, 2, pp.107-119, 1983.
53. L. L'Estrade, H. Hallen, Production of Water and Gas Atomised Powders, Metallurgia, 54, 11 November 1987.
54. L. L'Estrade, R. Koos, Properties of Gas Atomised Metal Powders, Proc Dusseldorf Conference, PM 1986, pp.22-28.

2

Spray Casting: A Review of Technological and Scientific Aspects

P. MATHUR AND D. APELIAN*

Research Professor, Department of Materials Engineering, College of Engineering, Drexel University, Philadelphia, USA

**Provost of Worcester Polytechnic Institute, Worcester, Connecticut, USA*

Abstract

Spray casting (e.g. the Osprey™ process) is emerging as an attractive technology to produce net or near-net-shaped components of a variety of materials. The process involves sequential atomization and droplet consolidation at deposition rates in excess of 0.25kg/s. In this way, it is possible to directly fabricate disks, billets, tubes and strips/sheets by suitable manoeuvring of the substrate under the spray of droplets. This paper is a review of technological and scientific aspects governing the shape, microstructure and yield of the preforms produced via spray casting. It addresses phenomena during droplet atomization, transfer of droplets in the spray, droplet consolidation at the substrate, solidification of the consolidated material, and shape/geometry of the preform produced. The knowledge base evolving from this analysis, in conjunction with appropriate sensor and control technology, provides a means to optimize and control the Osprey process.

1. Introduction

In response to increasing global competition, major changes in manufacturing philosophy are taking place within the materials processing industries. The traditional strategy of utilizing high plant capacity and mass production is now being replaced or complemented by flexible manufacturing methods. This new climate has resulted in a strong interest in net- or near-net shape manufacturing (NNSM) processes which are materials and energy efficient. Experts predict that NNSM practices will increase yields, improve efficiency, enhance product quality and generate higher profits [1-3]. The primary incentive is cost reduction, achieved by circumventing intermediate steps in the production process.

Spray casting is emerging as an attractive technology to produce net or near-net-shaped components of a variety of alloys. In principle, spray casting consists of sequential atomization and droplet consolidation. Currently there are two approaches to spray casting: the Osprey™ Process [4-6] and Liquid Dynamic Compaction (LDC) [7-9]. The primary difference between the two processes is in the mode of atomization; LDC utilizes high velocity pulsed gas jets from an ultrasonic gas atomizer (USGA), while the Osprey process uses a patented gas atomizer of conventional design. Low pressure plasma spraying/deposition (LPPD) [10] is another technique of droplet consolidation, wherein the starting material is in the form of solid powder particles. Upon injection into the hot plasma, the particles melt and impact a substrate to form a deposit. The rate of metal deposition in LPPD (0.2-0.5 kg/min. in R.F. plasma and 0.5-1.0 kg/min. in D.C. plasma) is significantly lower than in spray casting processes (10-150 kg/min).

1.1 The Osprey™ Process

Since the pioneering work of Singer in the early 1970s on spray casting [11-13], the first (and only) viable process for bulk fabrication of spray cast preforms was developed by Osprey Metals Ltd. [4-6,14,15]. Initial efforts were targeted towards the production of preforms suitable for subsequent hot forging into finished shapes. Today, the Osprey process is being investigated to fabricate disks, billets, tubes and strips/sheets on a commercial scale.

The Osprey spray casting process is shown schematically in (Fig. 1). Specific designs of the Osprey unit may vary depending on the size and geometry of preform being fabricated. However, the essential components of any Osprey unit include:

- a melting and dispensing unit,
- a gas atomizer,

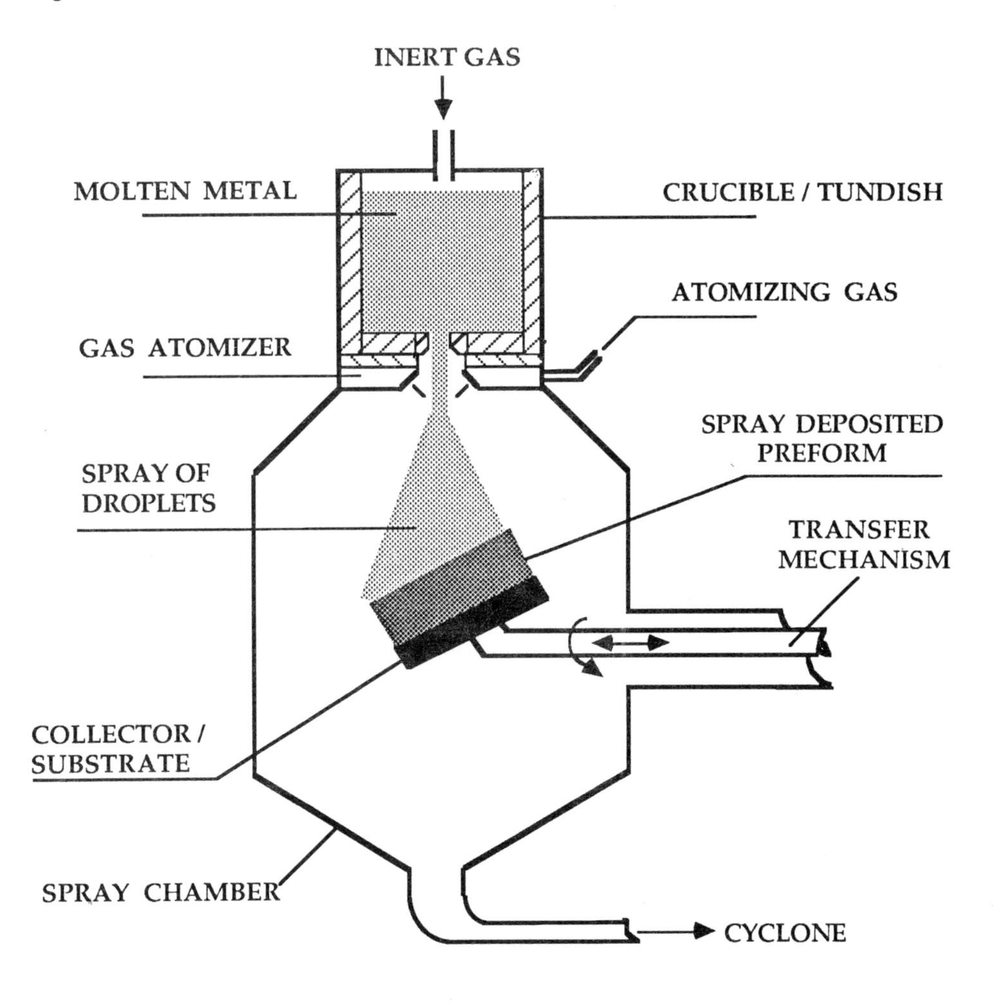
INERT GAS
MOLTEN METAL
CRUCIBLE / TUNDISH
ATOMIZING GAS
GAS ATOMIZER
SPRAY DEPOSITED PREFORM
SPRAY OF DROPLETS
TRANSFER MECHANISM
COLLECTOR / SUBSTRATE
SPRAY CHAMBER
CYCLONE

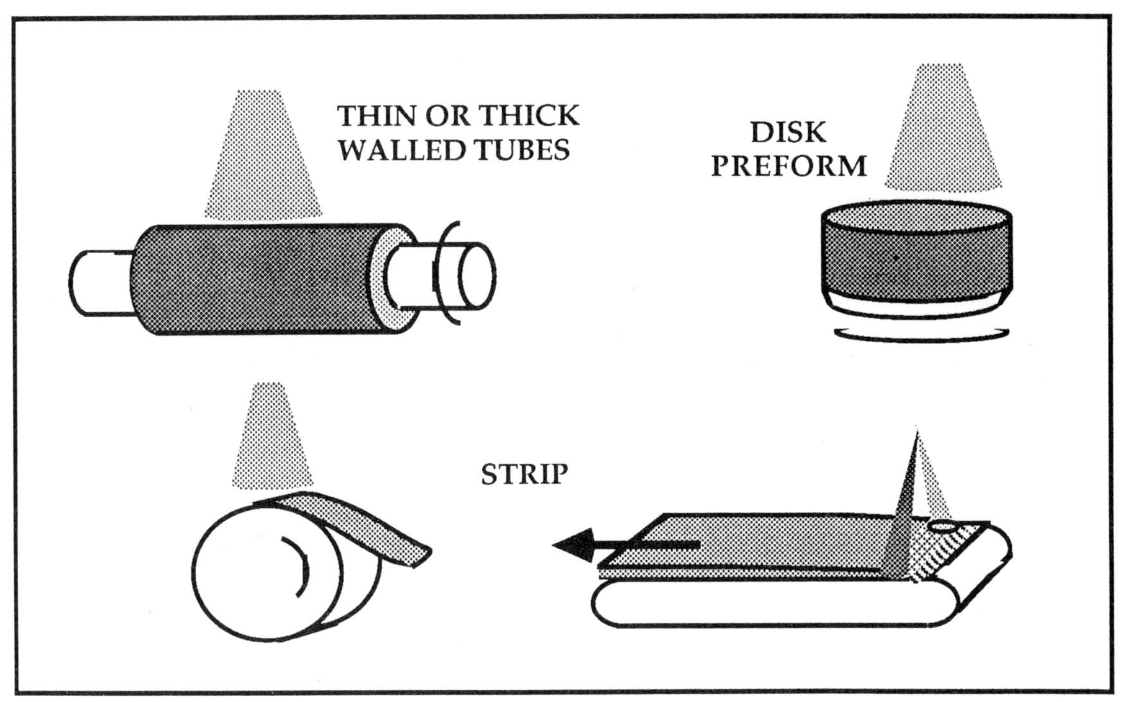
THIN OR THICK WALLED TUBES
DISK PREFORM
STRIP

Figure 1

- a spray chamber,
- a substrate mechanism,
- control panels for the atomization and substrate motion, and
- a gas distribution system [4,5,15].

Typically, the alloy charge is melted in a crucible located on top of the spray chamber, Fig. 1. During melting, the chamber is purged with inert gas and an over pressure of gas is also fed into the sealed crucible to prevent oxidation of the melt. When molten, the alloy exits through a refractory nozzle in the bottom of the crucible at a superheat in the range 50-150°C. In the atomizing zone below the crucible, the stream of molten metal is comminuted into a spray of droplets by the atomizing gas; either nitrogen or argon is used for atomization at a pressure of 0.6 - 1.0 MPa. In the metal spray, the droplets are cooled by the atomizing gas and accelerated towards the substrate (collector) which is positioned at a suitable distance (≥400mm) below the atomization zone. The droplets impinge and consolidate on the substrate to form a near-net shaped preform.

A variety of preform shapes can be produced by appropriate manoeuvring of the substrate beneath the spray [4,5,15], (Fig. 1). For example, Sandvik Steel in Sweden is producing stainless steel tubing of 100-440mm dia. and up to 8m in length by spraying onto a rotating, preheated mandrel [15,16]. Similarly, Sumitomo Heavy Industries in Japan is utilizing the Osprey process to manufacture large diameter rolls. Billets/disks of 100-250mm dia. are produced by spraying onto a rotating disk collector which is inclined to the spray axis and translated back and forth under the spray. Disk preforms are being evaluated by Osprey Metals Ltd. [4-6,14,15], by Alcan, Pechiney and Alusuisse for aluminum alloys [15,16], and by Howmet and General Electric Corporation for Ni-base superalloys [17,18]. Spray deposition onto a roller and/or an endless belt allows strip to be produced in a semi-continuous fashion [4,19,20]. A majority of the current production and development of strip and sheet products is being conducted by Mannesmann-Demag Hüettentechnik who have spray deposited strip in the thickness range 10-20mm, up to 1m in width and several metres in length [4,15,21]. Numerous other investigations have assessed the viability of spray casting as an alternative to currently employed material processing routes; however only a limited number of results have been published [22-26].

The major advantage of spray casting is that a fine grained, near-net shaped product can be fabricated in a single operation directly from the melt at deposition rates in the range 0.25-2.5 kg/s. Metallurgically, the product is characterized by a uniform distribution of fine, equiaxed grains (20μm-200μm), no macroscopic segregation of alloying elements, uniform distribution of second phases, low oxide content and the absence of particle boundaries [4-6,14-18,22-26]. The preforms can be thermo-mechanically treated and their mechanical properties are isotropic and at least comparable to products of conventional processes. Spray casting can also be used to fabricate composite materials by injecting particulates into the spray of molten droplets [4,6,14-16], or to fabricate dispersion-strengthened alloys by selectively reacting the droplets during flight [27].

2. Technological Aspects

Despite the apparent attractiveness of spray casting as a NNSM process, the materials processing industry has been slow to implement it on a commercial scale. Economic viability dictates that the process will compete in the manufacture of high technology materials where the components must not only have the final shape needed, but they must also meet stringent property requirements. Reproducibility and reliability of the final component mandates adequate knowledge of the effect of each process parameter on the shape, microstructure and yield of the preform produced. To date, these variables have been optimized on the basis of trial and error from which empirical relationships have been derived; this approach is labour and/or cost intensive and must be repeated for different materials being sprayed.

At least eight *independent process parameters* (IPPs) must be optimized under the current mode of operation of the Osprey process to achieve the desired preform shape with accompanying metallurgical integrity. These IPPs, and their regimes of influence in the Osprey process, are shown in Fig. 2. The relationship between any IPP and resultant preform quality can be simplified by identifying two *critical dependent parameters* (CDPs) which are shown schematically in Figs. 2 and 3:

CDP_1: The state of the spray just prior to consolidation. This is characterized in terms of % liquid in the spray (%L), the fraction and size distribution of solidified vs. liquid droplets, and the distribution of droplet mass in the spray.

CDP_2: The physical and thermal state of the surface on to which the droplets impact; this includes the fraction of solid and the surface roughness of the deposit.

It is envisaged that IPP_1 through IPP_6 in Fig. 2 combine and entirely determine CDP_1. Similarly, the parameter CDP_2 is entirely determined by CDP_1, IPP_7 and IPP_8. Knowledge of the optimal values of CDPs 1 and 2 will simplify the control scheme, and their values will depend upon the specific application. For example, the primary criteria for strip production are uniformity in thickness across the width of the strip (<2% variation in thickness) and minimization of surface-

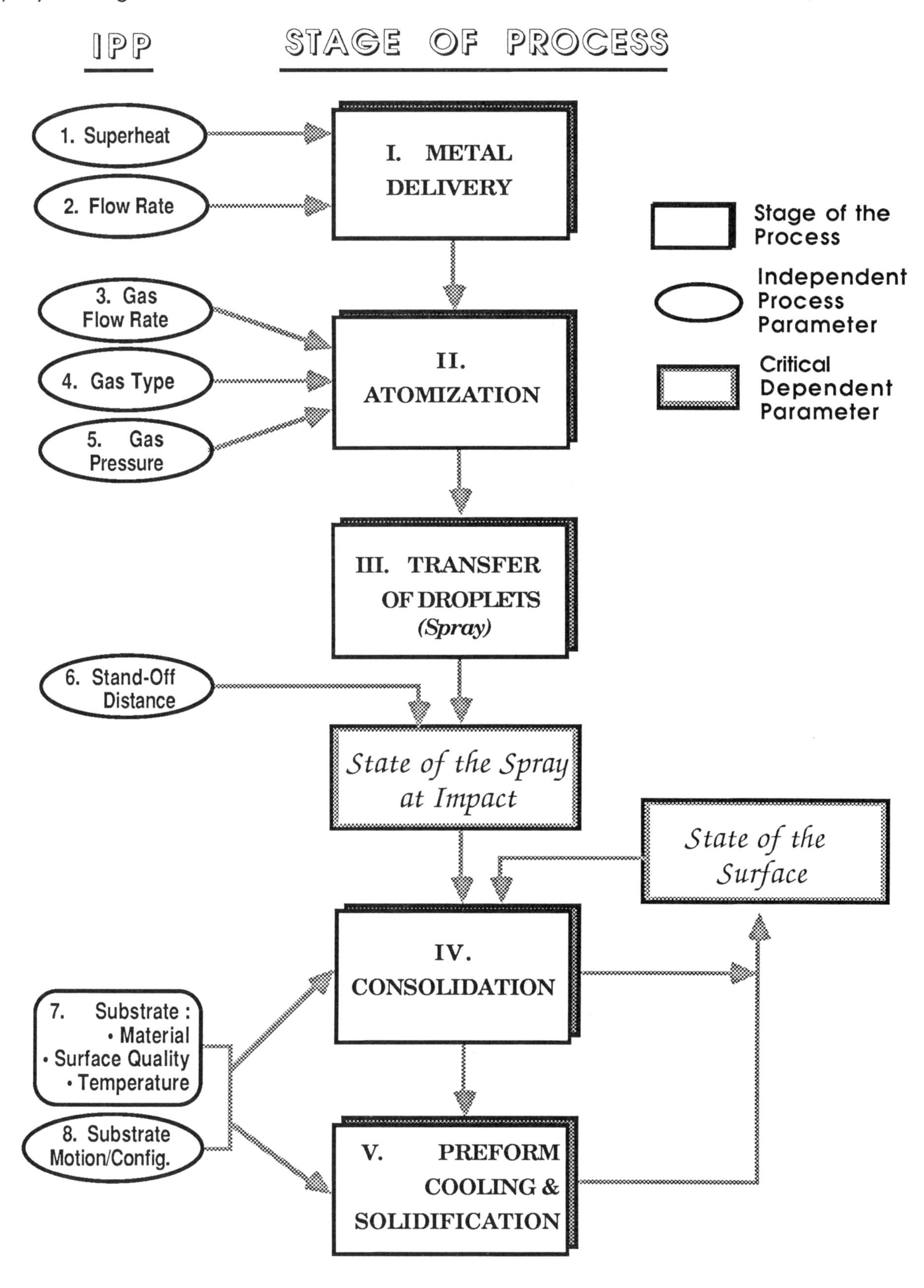
IPP
STAGE OF PROCESS
1. Superheat
2. Flow Rate
3. Gas Flow Rate
4. Gas Type
5. Gas Pressure
6. Stand-Off Distance
7. Substrate :
• Material
• Surface Quality
• Temperature
8. Substrate Motion/Config.
I. METAL DELIVERY
II. ATOMIZATION
III. TRANSFER OF DROPLETS (Spray)
State of the Spray at Impact
State of the Surface
IV. CONSOLIDATION
V. PREFORM COOLING & SOLIDIFICATION
Stage of the Process
Independent Process Parameter
Critical Dependent Parameter

Figure 2

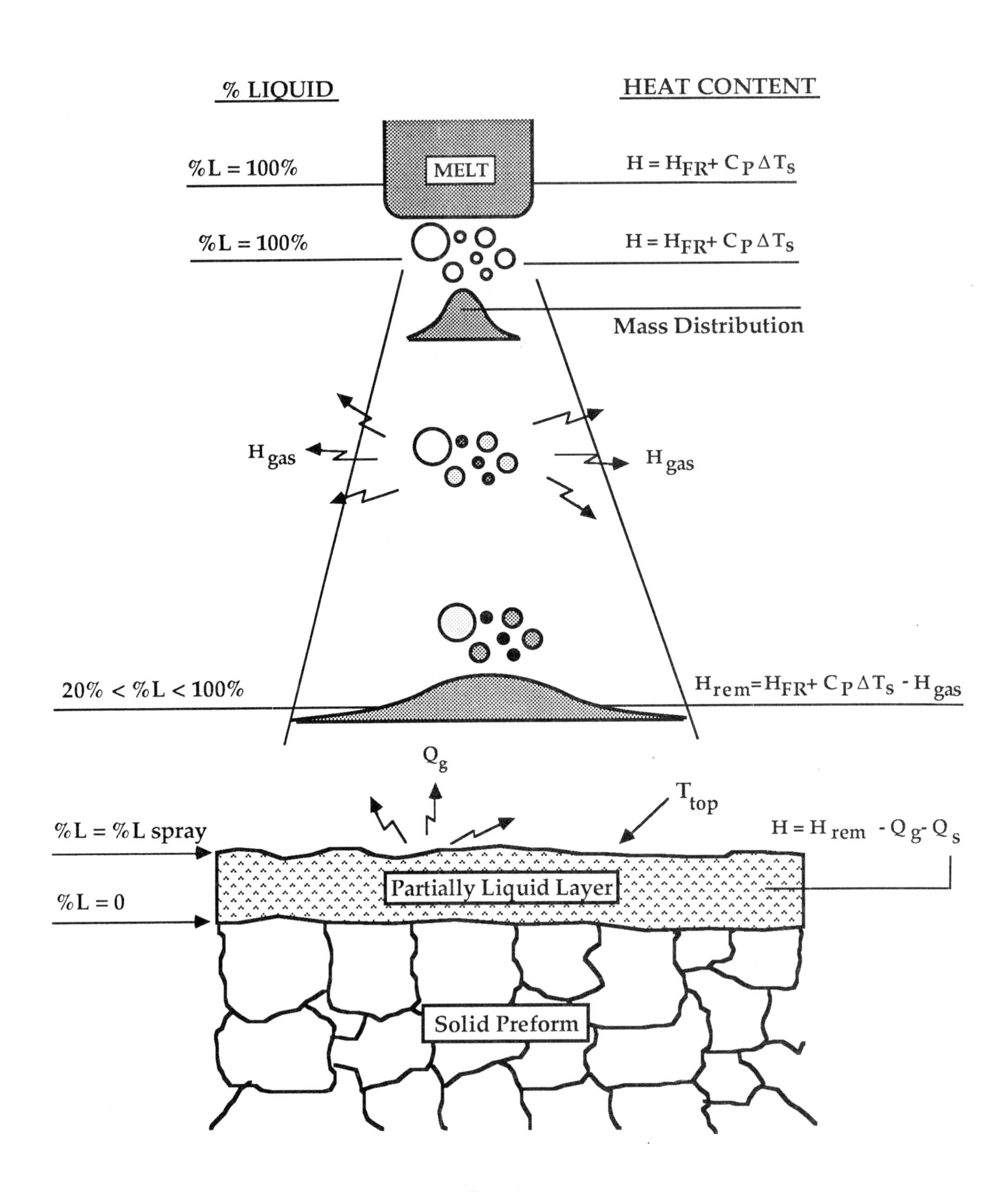
% LIQUID
HEAT CONTENT
%L = 100%
MELT
H = HFR+ CP ΔTs
%L = 100%
H = HFR+ CP ΔTs
Mass Distribution
H gas
H gas
20% < %L < 100%
Hrem=HFR+ CP ΔTs - Hgas
Qg
Ttop
%L = %L spray
H = H rem - Q g- Q s
%L = 0
Partially Liquid Layer
Solid Preform

Figure 3

connected porosity. Hence the shape of the spray and parameters of substrate motion/temperature are most important. Manufacture of bulk preforms such as disk or billet necessitates close control over the top surface temperature and thickness of the partially liquid layer in order to maintain conditions of *incremental solidification* [5,28]. Thus the amount of liquid in the spray at impact must be carefully controlled (Fig. 3): if the spray contains a high fraction of solid upon impact (%L→0), a majority of the droplets will be solidified and no coherent deposit will be formed; the process will then resemble powder production. On the other hand, spray casting will be analogous to conventional casting if the spray contains a high fraction of liquid at impact (%L→100). The "ideal" fraction of liquid is a value inbetween the two extremes cited above and must contain sufficient liquid to flow and fill the interstices between presolidified droplets. In addition, this ideal fraction of liquid must ensure that the impinging droplets "stick" to the surface of the growing deposit and maximize deposit yield.

Preform *yield* is dictated by the product of two efficiencies: (i) target efficiency, Π_t, which represents the fraction of droplets arriving at the substrate surface, and (ii) sticking efficiency, Π_s, which represents the fraction of droplets which adhere to the surface and contribute to preform growth. These factors are addressed in §3.3.1. The *shape* or geometry of the preform is determined by the combined effects of (i) the spatial distribution of droplet mass in the spray, (ii) the sticking efficiency of the droplets on arrival at the substrate, and (iii) motion of the substrate and/or spray. This is discussed in §3.3.2. Finally, the *microstructure* in spray cast preforms is determined by (i) the condition of the spray at impact, (ii) the spatial distribution of solid particles after impact, and (iii) the time required for complete solidification of the preform. Preform yield, shape and microstructure must be controlled during the spray casting operation, however, the difficulty of process control is magnified by:

- the complex interdependencies between independent process parameters (e.g. melt superheat, atomizing gas pressure,...) and the dependent variables (e.g. droplet size distribution, droplet temperature profiles, preform density and microstructure,...), and
- the need for sophisticated, non-invasive sensors due to the microscopic size of the droplets and the time scale of events which occur during the process.

3. Scientific Aspects

In this section, the intermediate stages of the process are analyzed and a scheme is developed to relate the IPPs to the CDPs and subsequently to preform integrity in terms of yield, shape and microstructure. A "standard" set of independent process parameters is selected for the analysis based upon current operating conditions of the Osprey process on a pilot scale; this set is listed in Table I and it is utilized as a benchmark for all comparisons.

3.1 Atomization

The size distribution of atomized droplets is an important dependent parameter in spray casting since it governs the remainder of the deposition process. Typically, the mass-median droplet diameter is in the range 40-120µm depending upon the material and process parameters used, and the log-normal standard deviation is in the range 1.75-2.25.

Prior attempts to predict the size distribution of atomized droplets have met with limited success due to the complexity of the disintegration process [29-33]. Furthermore, the mean droplet sizes predicted by a fundamental analysis such as Bradley's [32] are not in satisfactory agreement with experimentally measured values. Therefore, it is convenient to determine the size distribution experimentally by collecting solidified droplets from the spray (by removing the substrate) and subjecting them to a size analysis [16,34,35]. Data on droplet sizes can then be·correlated to the material and processing parameters during atomization by Lubanska's empirical relationship [33]:

$$d_m = K_L D \left[(1+1/GMR) (\nu_m/\nu_g) (1/We) \right]^{m_L} \approx 13 \cdot \sigma^3 \tag{1a}$$

$$We = \rho_m V^2 D / \gamma_{LV} \tag{1b}$$

where:
d_m - the mass-median droplet diameter,
σ - the standard deviation of the log-normal droplet size distribution,
K_L and m_L - constants specific to atomizer design,
D - the diameter of the metal stream,
GMR - the ratio of gas:metal mass flow rates,
ν - the kinematic viscosity,
We - the Weber number,
V - the gas velocity,
γ_{LV} - the surface tension of liquid metal.

The atomizer-specific parameters K_L and m in equation (1a) were found to be ~100 and 0.5, respectively, for the Osprey atomizer. This correlation facilitates computation of d_m and σ under different processing conditions when experimental data is unavailable. d_m can be decreased by increasing the melt superheat or gas:metal ratio, while substitution of argon in place of nitrogen as the atomizing gas is found to produce no significant effect on the droplet size distribution. Knowing d_m and σ, the probability of finding a given droplet size, $P(d_i)$, for a log-normal distribution is given by [36]:

$$P(d_i) = 1/(\sqrt{2\pi}.\sigma).\exp\{(-1/2\sigma^2)\cdot\log(d_i/d_m)\} \quad (2)$$

3.2 Transfer of Droplets (Metal Spray)

The metal spray comprises hot metal droplets surrounded by a high velocity gas jet. The gas accelerates and simultaneously cools/solidifies the droplets during their flight towards the substrate, (Fig. 3). The shape of the spray can be approximated by a cone whose apex is at the point of atomization and its base is a circular area over which the droplets are deposited.

3.2.1 Droplet Velocity

Upon atomization, each droplet is accelerated towards the substrate by the surrounding high velocity atomizing gas. Therefore, the velocity of each droplet increases with flight distance until a point in flight when the droplet velocity equals the velocity of the decaying gas jet. Beyond this flight distance, the droplet will travel faster than the gas and consequently it will be decelerated.

The velocity profile of droplets in flight has been modelled by several investigators [9,13,37,38,39-41]. Most of these models are based on the expression by Ranger and Nicholls [42] derived from the momentum equation for the acceleration of water droplets in air:

$$F = m\frac{dV_d}{dt} = \frac{1}{2} C_D\rho_g V_r^2 \cdot A_c + mg \quad (3)$$

where:
F - the force on the droplet,
m - its mass,
V_d - droplet velocity,
C_D - the drag coefficient,
ρ - the density of the gas,
$V_r = |V_g - V_d|$ - the relative velocity between the gas and the droplet,
A - the droplet's cross-sectional area,
g - the acceleration due to gravity.

Data for the gas flow field are available in Refs. 40,43-45,46 and the expression for the drag coefficient is adopted from Ref.47.

Droplet velocities just prior to consolidation at a flight distance of 400mm are displayed in Fig. 4 for droplets of Al, Al-4.5%Cu, Cu, Ni and Fe; these were calculated from the model described in Refs.37,38,35. Depending primarily upon droplet mass, the velocities at consolidation are in the range 10-100m/s. The curves are bell-shaped because droplets smaller than a critical diameter (^) have reached their peak velocity and are being decelerated by the gas, while droplets with dia. > ^ are accelerating in flight. The value of ^ increases with increasing flight distance and is a function of droplet mass (hence the curve is shifted to the right for Al)

Limited experimental measurements to date [46,48,49] have shown that the predicted velocity profiles are correct at least to an order of magnitude. However, with the current state-of-the-art, it is not possible to precisely measure the velocity of a single droplet since both velocity *and* diameter of the droplet must be measured simultaneously along the flight trajectory.

3.2.2 Droplet Temperature/Extent of Solidification

The variation of a droplet's temperature with flight distance can be divided into four stages, as shown schematically in Fig. 5a. Upon atomization at flight distance X=0, the liquid droplet is at a temperature T_i (= $T_L + \Delta T_s$), where T_L is the liquidus temperature and ΔT_s is the melt superheat. In Stage 1, the droplet cools primarily by losing heat to the surrounding gas via forced convection. Therefore the cooling rate is governed by the convective heat transfer coefficient, h_g, which in turn is a strong function of the difference in velocity between the droplet and the gas [50]:

$$h_g = K_g(2+0.6Re^{0.5}Pr^{0.33})(C_{g(avg)}/C_g)^{0.26}/d \quad (4a)$$

$$Re = d\,V_r/\nu \quad (4b)$$

where:
K_g - thermal conductivity of the gas,
Re - Reynold's number,
Pr - Prandlt number,
C_g - specific heat of the gas,
d - droplet diameter,
V_r - relative velocity,
ν - kinematic viscosity of the gas.

Knowing V_r from the droplet's velocity profile, the value of h is computed from Eq. 4 and it is employed in a simple heat balance to predict the variation of droplet temperature with flight distance [5,9,37,38,39]:

$$dT_d/dt = GMR \cdot h_g A_s (T_d - T_g) / C_P \tag{4c}$$

where:
T - temperature,
t - time,
GMR - ratio of gas:metal mass flow rate,
A_s - droplet's cross-sectional area,
C_P - specific heat of the droplet and
subscripts d and g - droplet and gas, respectively.

The liquid droplet continues to cool in Stage I until the nucleation temperature T_n (= $T_L - \Delta T_n$) is reached, where ΔT_n is the degree of undercooling. Droplet undercooling and solidification must be included in these models in order to provide a realistic description of the droplet temperature profile in flight [9,35,37,39]. The dependence of the degree of undercooling on droplet size is given in [35,37]; it is found that the degree of undercooling decreases exponentially with increasing droplet volume.

Following nucleation at T_n, the droplet begins to solidify rapidly (Stage II - solidification during recalescence). Models for droplet solidification assume that nucleation occurs on the droplet's surface and a hemispherical solid-liquid interface advances across its volume as a function of time [51-53], (Fig. 5a). The temperature *increases* during solidification (i.e. the droplet "recalesces") since the release of latent heat is faster than the rate of heat extraction by the surrounding gas. Recalescence continues until an arrest temperature is reached close to the liquidus temperature. Subsequent solidification (Stage III - "normal" solidification) occurs with attendant drop in temperature and is governed by the rate of heat extraction by the gas. The cooling rate in Stage III may be derived from equation (4c) by introducing an additional term to account for the latent heat of fusion. Solidification is terminated at the solidus temperature (or at eutectic temperature for non-equilibrium solidification according to the Scheil equation). In the final stage of cooling (Stage IV), the droplet cools in the solid state by forced convection and equation (4c) is applicable.

The predicted variation of temperature with flight distance is shown in Fig. 5b for three droplets of Al-4.5%Cu alloy of diameters: d_m=59μm, d_m/σ=37μm and $d_m \cdot \sigma$=95μm. Similarly, the degree of solidification of each droplet (i.e. % liquid in droplet) is plotted in Fig. 6a. It is observed that the distance required for complete solidification increases with increasing droplet diameter. Similar profiles of temperature and solidification are computed for a number of droplets with diameters in the range $d_m/3\sigma$ to $d_m \cdot 3\sigma$. The extent of solidification of each droplet at X=400mm (i.e. just prior to consolidation) was selected from these results and is plotted in Fig. 6b. The juxtaposition of the curves indicates that *droplets of high melting materials* (e.g. Fe) *will cool faster than those of low melting materials* (e.g. Al-4.5%Cu) due to a greater difference in temperature between the metal and atomizing gas. From the figure, *the condition of droplets arriving at deposition (400mm below the gas nozzles)* can be summarized as follows:

(i) droplets of diameter less than a critical value d^* are completely solidified upon impact ($\%L_d = 0$). Typically d^* is predicted to be in the size range 30μm-125μm for the alloys described, and increases with increasing melting point of the alloy. This corresponds to a significantly large fraction of presolidified droplets due to the bi-modal population distribution [37].

(ii) droplets of diameter greater than d^* impact the deposition surface in a "mushy" condition with varying fraction of liquid. These droplets comprise a solidified dendritic skeleton, as observed from glass slides [35,49,51].

(iii) only droplets greater than about 300μm for Al and 900μm for the other metals arrive at the deposition surface in a completely liquid state.

*H_{FR} = heat in the freezing range of the alloy = latent heat (H_f) + specific heat ($C_P \Delta T_f$).

3.2.3 State of the Spray (CDP_1)

The models described above retain the identity of each droplet and describe the spray on a microscopic scale, i.e. at a high level of resolution. For purposes of process control, it is advantageous to combine characteristics of individual droplets and, at the expense of resolution, derive average values which represent the state of the spray at a macroscopic level. For example, the fraction of liquid in individual droplets may be averaged to determine the ratio of solid:liquid in the spray ($\%L_{spray}$ in Fig. 3). The value of $\%L_{spray}$ at deposition is a critical parameter in spray casting (CDP_1). Similar averaging may be carried out with droplet enthalpy and mass, as discussed below.

(a) % Liquid in Spray and Spray Enthalpy

The fraction of liquid in the spray at any flight distance is the weighted average of the fraction of liquid in the individual droplets:

$$\%L_{spray}(X) = \sum_{i=1}^{n} [1-f_S(d_i,X)]\cdot P(d_i)\cdot 100 \tag{5a}$$

where $\%L_{spray}$ is the percentage of liquid in the spray at flight distance X, f_s is the fraction solid in a droplet of diameter d_i, and $P(d_i)$ is the weight fraction of droplets in the size range d_i to d_{i+1} (Eq.1). Summation is carried out from $d_i = d_m/3\sigma$ to $d_m \cdot 3\sigma$ in order to cover 99% of the range of droplet diameters produced during atomization. Similarly, spray enthalpy H_{spray} is computed by averaging the enthalpy of individual droplets:

$$H_{spray} = \Sigma H_d(d_i,X)\cdot P(d_i) \tag{5b}$$

where H_d is the enthalpy of a given droplet of diameter d_i at flight distance X.

The variation of $\%L_{spray}$ with X is plotted in Fig. 7a for five alloys under a fixed set of parameters in order to compare the effect of material properties. The shallow slope of the curves in the initial stages of flight (0<X<25mm) is due to the release of superheat and undercooling of the droplets prior to solidification. Thereafter the solidification rate reaches its highest level after a distance of ~50mm and decreases monotonically with flight distance. Among the *material* parameters, *the melting point, T_m, and the value of $H_{FR}{}^*$ have the most significant effect on the thermal condition of the spray*. This is evidenced by copper which solidifies most rapidly since it has the smallest value of H_{FR}. On the other hand, aluminum and Al-4.5%Cu possess high values of H_{FR} *and* a low T_m; hence their solidification is significantly slower than the other materials. The influence of *process parameters* (IPPs), such as the type of atomizing gas and the gas:metal ratio, is illustrated in Fig. 7b for the spray deposition of Al-4.5%Cu alloy. It is predicted that even a four-fold increase in the flow rate of argon is not as effective as nitrogen in cooling the spray due to its lower thermal diffusivity/conductivity.

The substrate must be positioned at a suitable distance below the gas nozzles (depending upon the material and process parameters) so that the spray contains the "desired" percentage of liquid at impact. During deposit build up, the effective spray height decreases with time and results in a corresponding increase in the amount of liquid carried into the deposit, (Fig. 7). The change in $\%L_{spray}$ with deposit build up becomes significant at operative distances less than about 300mm for the aluminum alloys under these conditions. This effect is compounded by the increase in deposition rate with decreasing flight distance. Therefore, in order to maintain uniformity in deposition conditions, it is necessary to continuously increase the distance between the emplaced substrate and the point of atomization in order to maintain a constant spray height at the deposition surface.

(b) Droplet Mass Distribution within the Spray

The flux of droplets in the spray is its maximum at the axis of the spray and decreases towards its periphery. This variation, i.e. the *deposition profile*, is governed by IPP_1 to IPP_6 and is independent of substrate configuration and motion. Since it is difficult to measure the droplet flux directly, the flux may be estimated indirectly by measuring its effect, i.e. the rate of deposit growth, $\dot{Z}$ (mm/s). $\dot{Z}$ is defined as the thickness of a deposit that would build up per unit time on a flat, stationary substrate positioned normal to the spray axis. For example, the traces from a video recording of deposit growth are shown in Fig. 8 at different time intervals from the start of deposition [35,37,38]. The dependence of $\dot{Z}$ on the radial distance from the spray axis, r, is derived from this figure and is found to follow a normal "Gaussian" curve:

$$\dot{Z}(r) = dZ/dt = \dot{Z}_0 \exp(-\beta r^2) \tag{6}$$

where:

Z - thickness of the deposit measured parallel to the spray axis (z direction),

$\dot{Z}_0$ - maximum growth rate (at the spray axis, r=0),

β - radial distribution coefficient that governs the spread of the spray.

$\dot{Z}_0$ and β were measured to be ~5.25mm/s and 0.0005 mm^{-2}, respectively, at a distance of 400mm below the gas nozzles.

During the production of billets/disk preforms via the Osprey process, it is beneficial to have a narrow, focussed spray

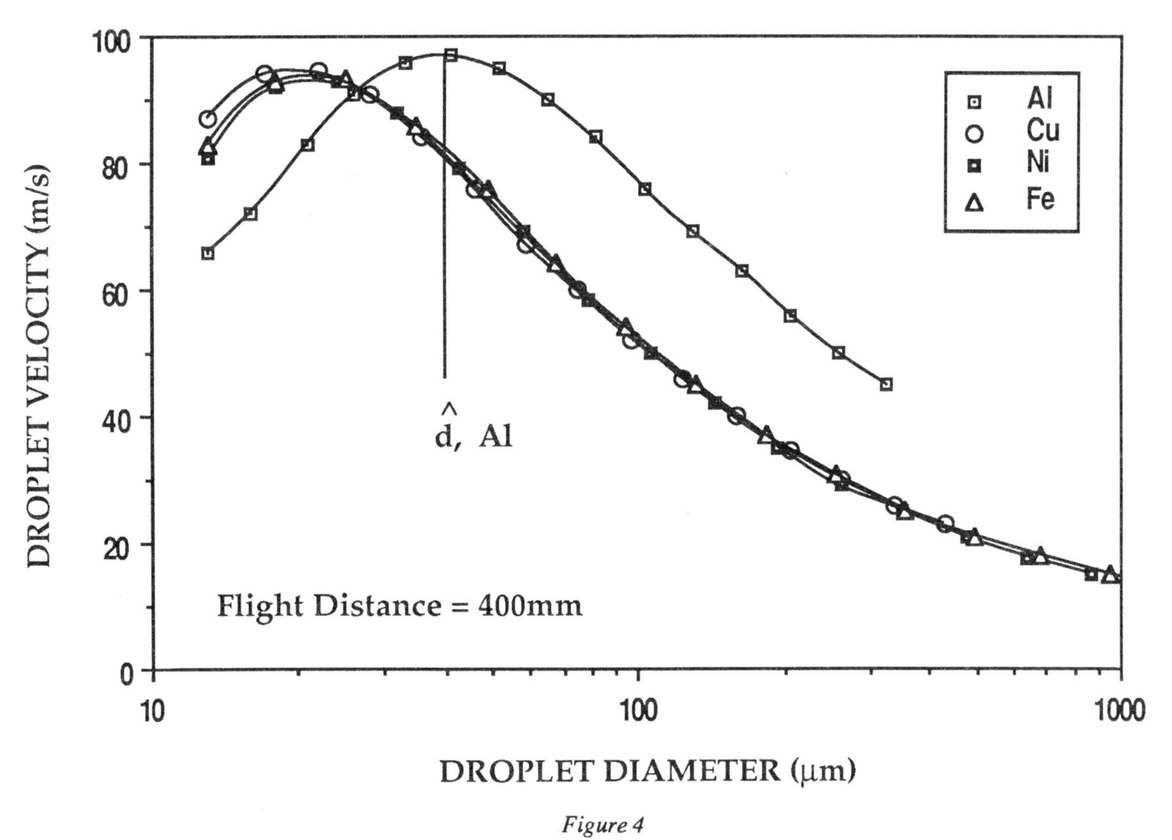

Figure 4

Table I : Process Parameters used in the Computations

Parameter	Std. Value	Varied
Melt Superheat	100°C	√
Gas Pressure	8 bar	x
Gas Type	Nitrogen	√
Gas Temperature	25°C	x
Metal Stream Dia.	5 mm	x
Metal Flow Rate	32 cm^3/s	√
Gas:Metal Ratio	2.60	√
Chamber Temperature	50°C	x
Spray Height	400 mm	√
Substrate Condition	25°C, water cooled	√
Substrate Motion	-	√

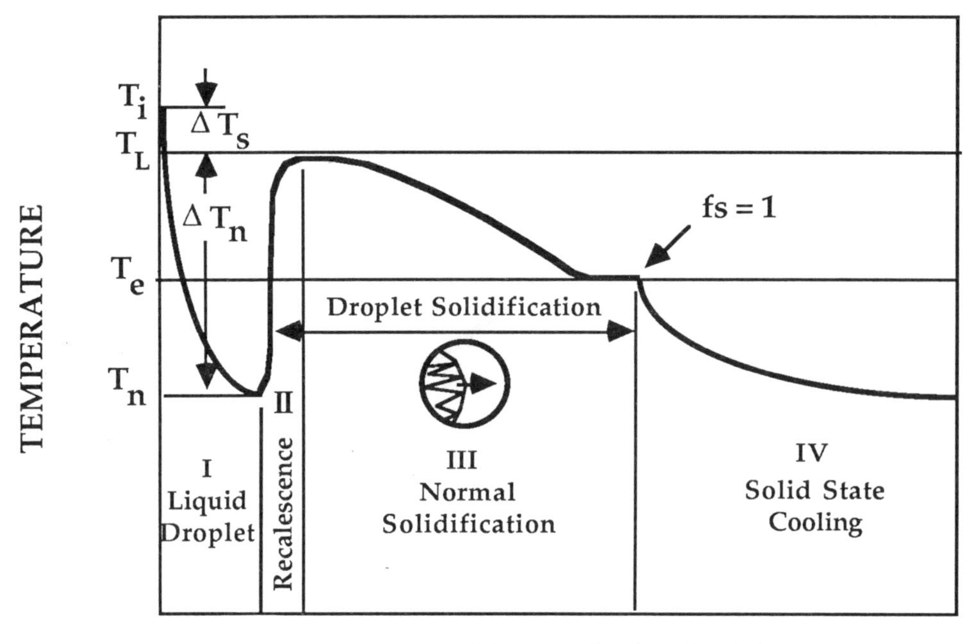
TEMPERATURE
T_i
ΔT_s
T_L
ΔT_n
T_e
T_n
fs = 1
Droplet Solidification
II
I
Liquid
Droplet
Recalescence
III
Normal
Solidification
IV
Solid State
Cooling
FLIGHT DISTANCE (X)

(a)

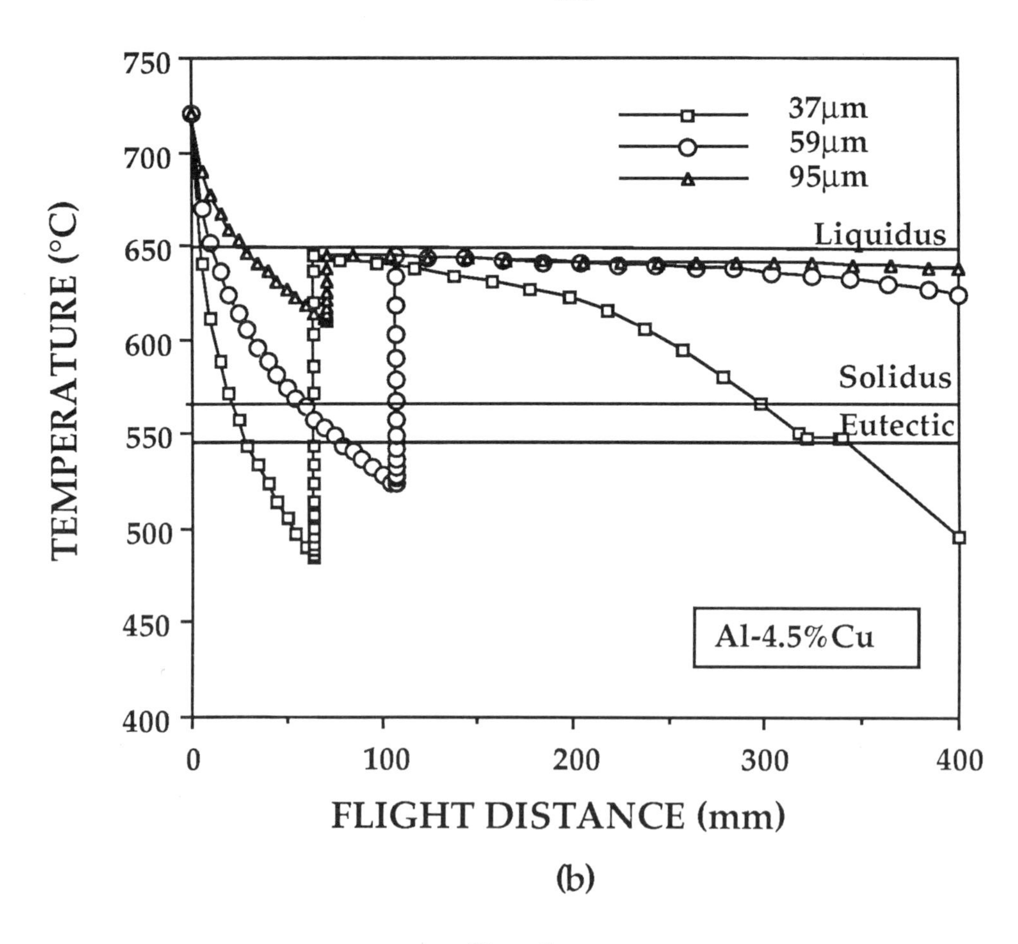
TEMPERATURE (°C)
750
700
650
600
550
500
450
400
37µm
59µm
95µm
Liquidus
Solidus
Eutectic
Al-4.5%Cu
0
100
200
300
400
FLIGHT DISTANCE (mm)

(b)

Figure 5

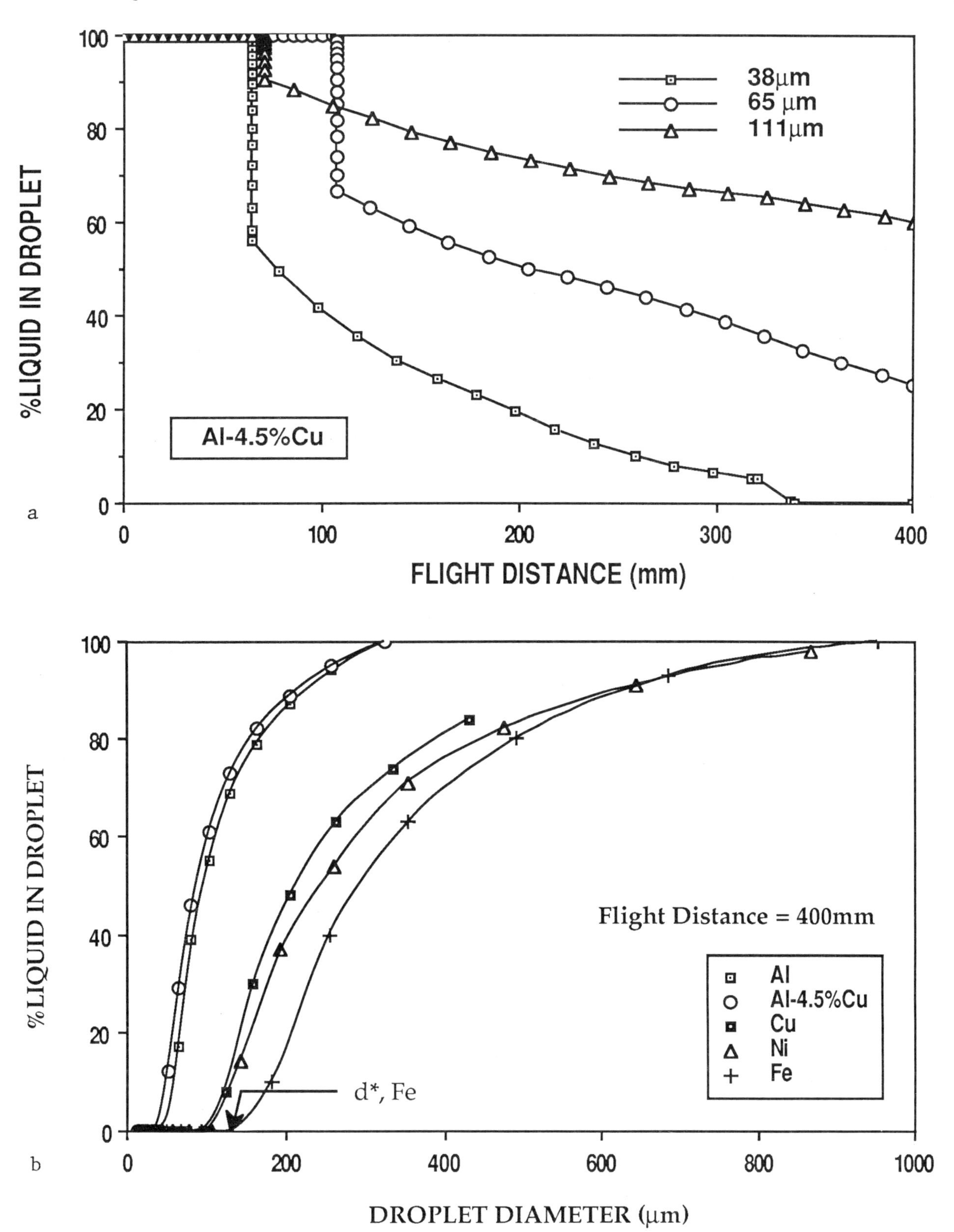
%LIQUID IN DROPLET
FLIGHT DISTANCE (mm)
38μm
65 μm
111μm
Al-4.5%Cu
a
%LIQUID IN DROPLET
DROPLET DIAMETER (μm)
Flight Distance = 400mm
Al
Al-4.5%Cu
Cu
Ni
Fe
d*, Fe
b

Figure 6

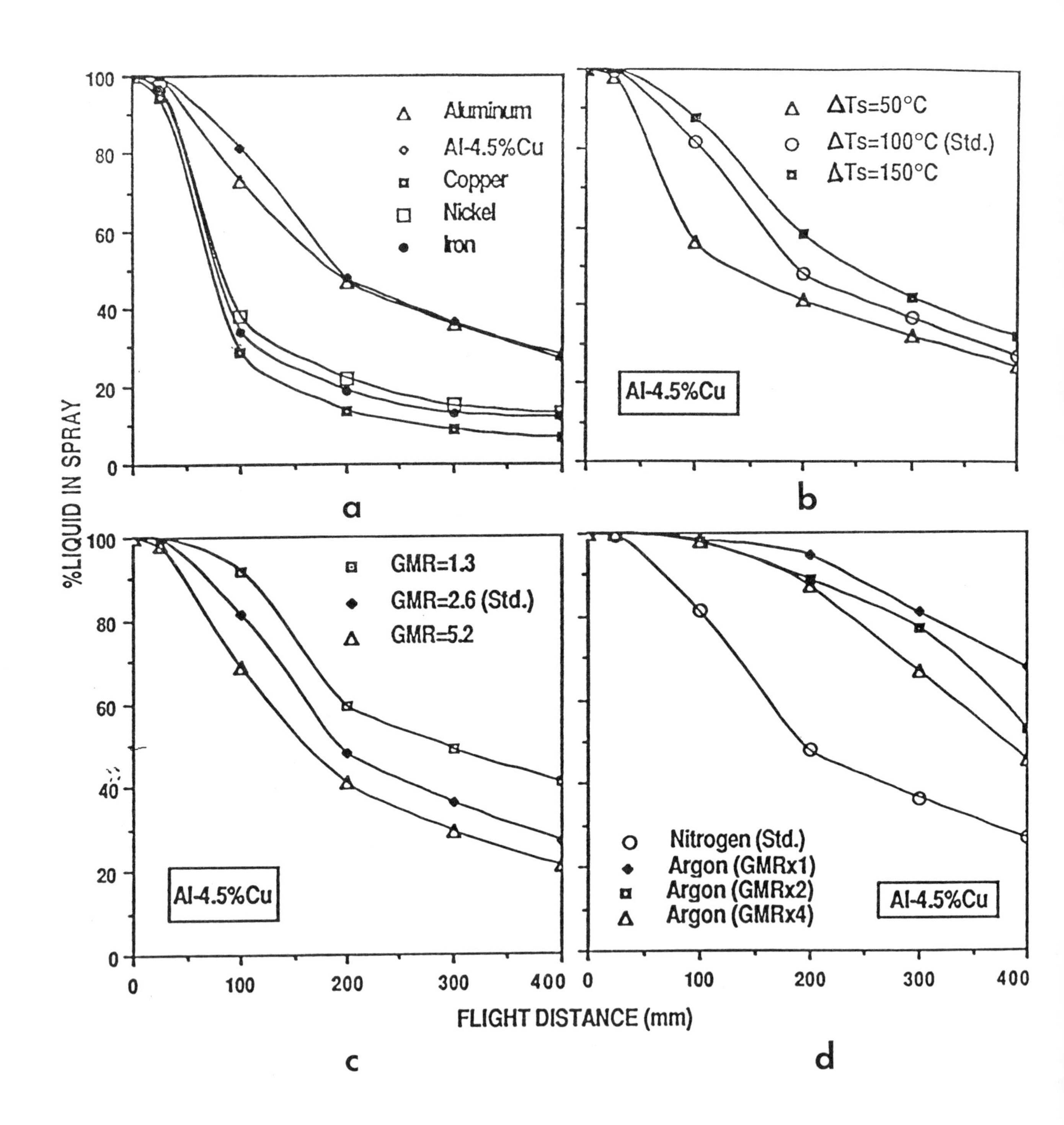

%LIQUID IN SPRAY
FLIGHT DISTANCE (mm)
Aluminum
Al-4.5%Cu
Copper
Nickel
Iron
a
ΔTs=50°C
ΔTs=100°C (Std.)
ΔTs=150°C
Al-4.5%Cu
b
GMR=1.3
GMR=2.6 (Std.)
GMR=5.2
Al-4.5%Cu
c
Nitrogen (Std.)
Argon (GMRx1)
Argon (GMRx2)
Argon (GMRx4)
Al-4.5%Cu
d

Figure 7

in order to increase the yield on a small collector area. Therefore, the coefficient 'β' in equation (6) must be kept at a minimum value. On the other hand, it is desirable to transform the deposition profile into an ellipse during strip casting in order to increase the width and uniformity in thickness of the strip produced. This can be achieved by employing a "*linear*" atomizer wherein the metal delivery nozzle and the surrounding gas jets have a rectangular cross section [54]. The resulting deposition profile across an elliptical spray can be described by the following (general) form of equation (6): $\dot{Z} = \dot{Z}_{max} [\beta_1 x^2 + \beta_2 y^2]$. It is also possible to tailor the spray to any desired shape (and deposition profile) by the utilization of *tertiary gas jets* [55] and/or magneto-hydrodynamic forces (MHD).

3.3 Formation of the Preform

3.3.1 Droplet Consolidation

When a substrate (i.e. target) is introduced into the spray of droplets, the droplets impinge and consolidate on the substrate to form a deposit. Only a portion of the droplets reach the target; this fraction is defined as the *target efficiency* of the spray, Π_t. Furthermore, only a fraction of the droplets arriving at the substrate will "stick" and contribute to deposit growth; the remaining droplets will "bounce-off". The fraction which adheres is termed the *sticking efficiency*, Π_s. Deposit yield is determined by the product of these two efficiencies:

$$\text{Yield \%} = (\Pi_t \cdot \Pi_s) \cdot 100 \tag{7}$$

Π_t depends upon the size of the substrate, the shape of the spray, the stand-off distance, the distribution of droplets within the spray and the substrate motion. Π_s is a metallurgical parameter which is governed by the state of the spray at impact (CDP_1), the state of the top surface (CDP_2), and by the substrate configuration/motion. Estimation of CDP_1 was described in §3.2.3 and the determination of the latter two parameters will be addressed in the following sections. A flow chart depicting the procedure adopted is provided in Fig. 9.

3.3.2 Preform Shape

The shape of the preform is governed by the mass distribution in the spray at impact, the sticking efficiency and the parameters of substrate motion (e.g. rotation speed, translation speed, limits of translation, etc.). Optimum values of the variables of substrate motion can be determined *a priori* by a mathematical model in which the state of the spray at consolidation (i.e. Eq. 6) is coupled with a model for substrate motion [22,35,38]. For a given substrate (i.e. a disk collector, a tubular mandrel, or a roller), this model selects a location P(x,y) on its surface and computes the growth of thickness during deposition. The calculations are then repeated for different locations which are defined by a grid on the substrate surface.

The magnitude of thickness increments during each time step at any location P(x,y) is governed by the radial distance $R_{(x,y)}$ between P(x,y) and the axis of the spray (Eq. 6). If the substrate is stationary, the value of $R_{(x,y)}$ is constant during the entire deposition cycle. If the substrate is non-stationary, $R_{(x,y)}$ changes with time depending upon the motion imparted, i.e. on the locus of P with respect to the spray axis. Summation of the thickness increments over the deposition cycle yields the final thickness at each location on the substrate. Shape or geometry of the preform is defined by a bounding surface which connects all points (x,y,Z) where Z is the calculated thickness at the grid point (x,y) on the substrate surface.

A three-dimensional representation of the final deposit shape after 40s of spraying onto a horizontal stationary substrate is shown in Fig. 10a. Similarly, the build up of a deposit on a stationary substrate inclined at an angle of 35° to the horizontal is shown in Fig. 10b; inclination of the substrate facilitates the build up of deposits with edges nearly perpendicular to the substrate surface.

The predicted geometry of an axisymmetric disk preform, produced by spraying for 40s on a 120mm dia. collector, is represented in Fig. 10c. The collector was rotated at ω = 200 rpm and simultaneously translated back and forth under the spray such that the spray axis reciprocated between the limits x_1=-60mm and x_2=-35mm (i.e. stroke length x_s=25mm). The speed of reciprocation was set at V_x= 20 mm/s, such that the spray traversed approximately one stroke length for every four rotations of the collector. From the volume of metal deposited, the target efficiency was calculated as ~72%.

This model to predict preform shape is also being utilized to predict the geometry of tubes, billets and/or strip [20,22,56] produced via spray casting. In order to produce wide strip (~1m in width), a "scanner atomizer" is used to oscillate the spray at a frequency of 3-7 Hz, (Fig. 1).

3.3.3 Preform Microstructure

Grain size and porosity are the primary indices of microstructure. Prior research [37,35,22] suggests that grain size in spray cast preforms is determined by (i) the size distribution of solidified particles arriving from the spray, (ii) the spatial distribution of these particles after impact, and (iii) the time required for complete solidification, t_f. At a given spray height, the size distribution of solid particles is obtained from the size distribution of droplets in the spray and the diameter of the largest solidified droplet (d^*) at that flight distance. Predicted values of d^* at X=400mm can be obtained from Fig. 6b. The spatial distribution of solid particles after impingement is strongly dependent upon their spatial distribution in the spray *and* upon secondary effects such as droplet fragmentation, bouncing-off, etc. which are difficult to quantify.

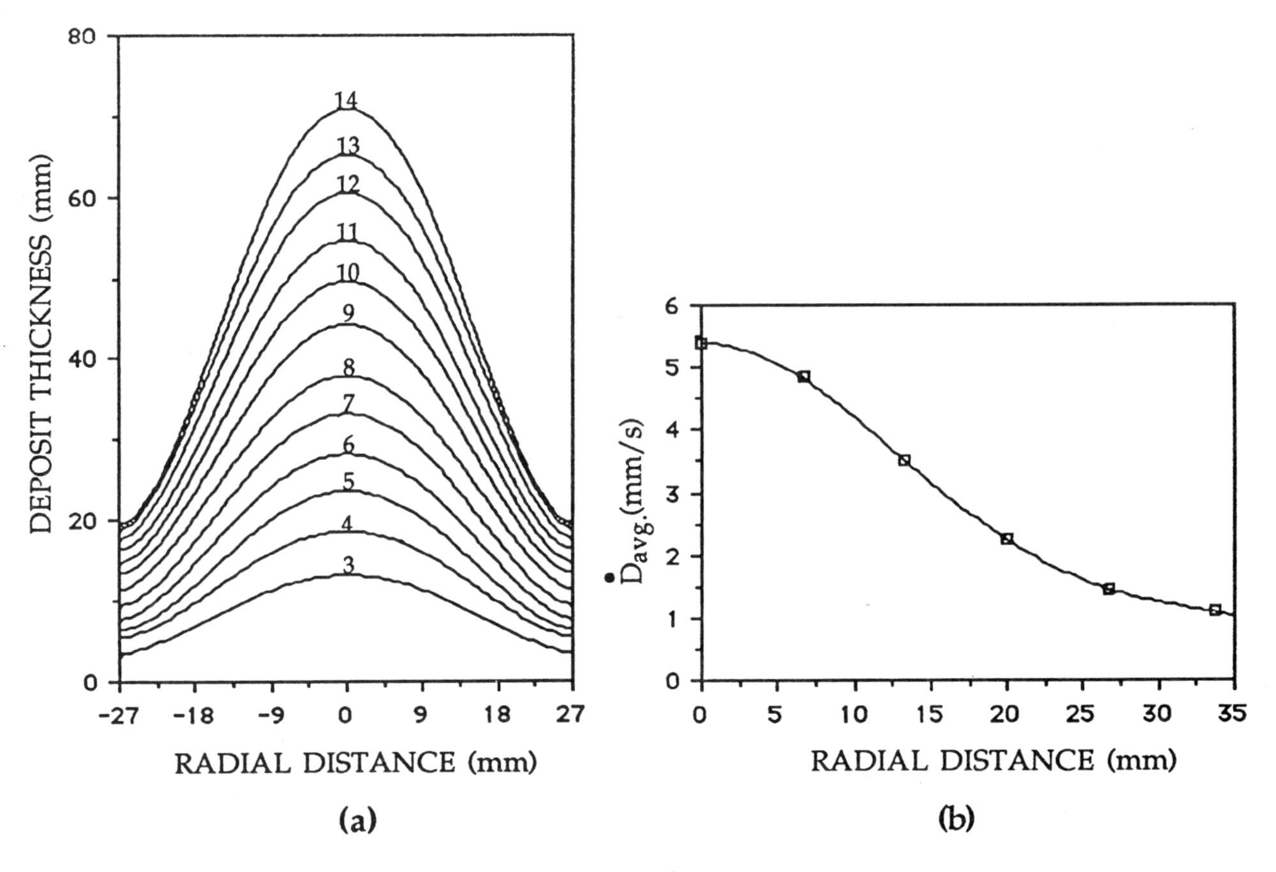

Figure 8

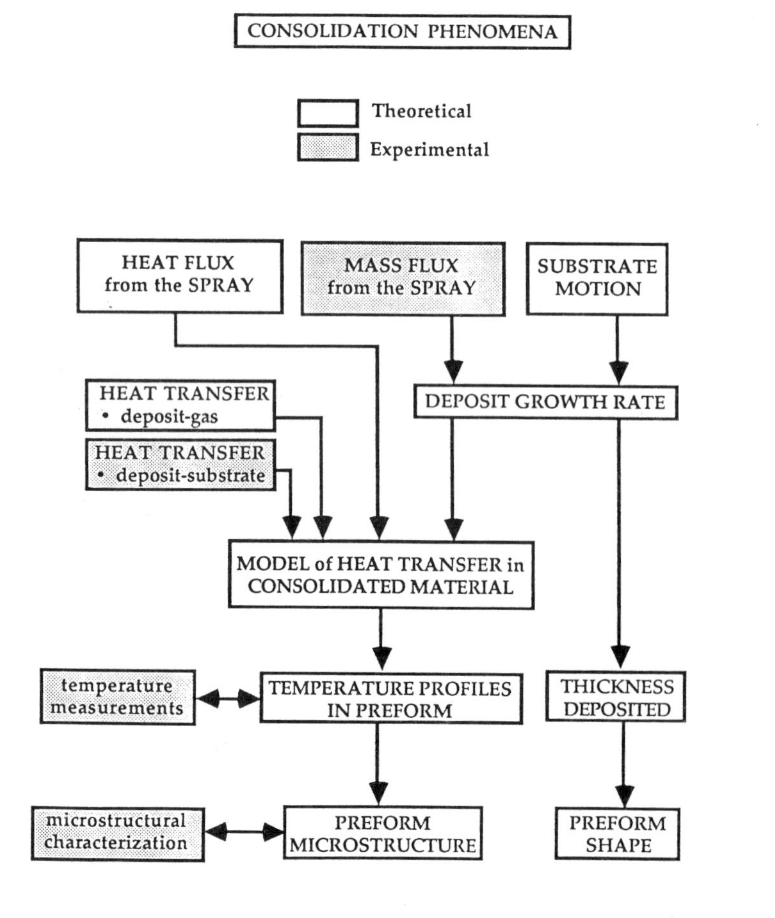

Figure 9

3.3.4 9(a)Preform Solidification and Cooling (CDP_2)

In order to effect complete solidification, a total amount of heat, H_S ($H_S = C_P \cdot \Delta T_s + H_{FR}$) must be removed from the metal, where C_P is the specific heat, ΔT_s is the melt superheat at atomization and H_{FR} is the heat contained in the freezing range of the alloy. Let H_{gas} represent the amount of heat removed by the gas during flight, as shown schematically in Fig. 3. Thus, the amount of heat which must be removed *after* deposition in order to complete the solidification is given by H_{rem}, (Fig. 3):

$$H_{rem}(X) = (C_P \cdot \Delta T_s + H_{FR}) - H_{gas}(X) \quad (8)$$

where X is spray height.

(b)Continuum Models

Two approaches have been adopted to calculate the solidification time, t_f, required to remove the heat H_{rem} from the preform. In the *continuum* approach [9,37,38,57], t_f is computed by a macroscopic energy balance between the heat influx from the spray H_{rem}, the mass influx Z, heat extracted by the substrate Q_s, and heat extracted by the atomizing gas Q_g. This is shown schematically in Fig.11a. Since solidification time varies with location within the preform, the thickness of the preform at any substrate location P(x,y) is divided into small volume elements arranged along the growth direction. A moving boundary transient heat conduction equation is solved numerically to obtain the enthalpy, temperature and fraction solid within the volume elements as a function of time [35,37,38]:

$$\rho\,(dH/dt) = (1/Z^2)\cdot(d/d\eta\,(K\,dT/d\eta)) + (\rho\eta/Z)\cdot(dH/d\eta)\cdot\dot{Z} \quad (9)$$

boundary conditions:

$$\text{bottom surface: } (K/Z)\,(dT/d\eta) = Q_s = h_s\,(T_b - Ts) \quad (9a)$$

$$\text{top surface: } \rho\,(H_{spray} - H)\,(dZ/dt) = (K/Z)\,(dT/d\eta) + Q_g \quad (9b)$$

where:

ρ - density,

H - enthalpy,

Z - thickness of the deposit at any instant,

η - z/Z,

z - distance into the deposit measured from the substrate upwards,

$\dot{Z}$ - deposit growth rate and t is elapsed time from the start of deposition.

Solidification is completed when the amount of heat removed from each volume element is equal to H_{rem}.

Heat flux from the deposit to the substrate has been measured with a heat flux sensor embedded in the substrate [35]; measured values of the heat transfer coefficient are a maximum of ~10^5 W/m^2/K at the start of deposition and decrease rapidly to ~500 W/m^2/K within a few seconds. This decrease is attributed to the formation of an "air gap" between the deposit and the substrate due to contractional stresses upon solidification. Similarly, the heat transfer coefficient for the gas cooling at the top surface of the deposit is estimated to be ~200 W/m^2/K.

Sample predictions of the continuum model are shown in Fig. 12 for the spray deposition of the Al-4.5%Cu disk whose shape is shown in Fig. 10c. Deposition was carried out onto a water-cooled, circular substrate which was simultaneously rotated and translated under the spray to achieve a cylindrically shaped preform, as shown in Figs. 1 and 10c. The "deposition line" in Fig. 12a is the predicted growth of thickness at the *centre* of the circular substrate. This growth of thickness with time depends upon the locus of the disk under the spray, and it was derived from the model to predict preform shape. It is observed that the disk grows to a height of ~7cms during 40s of deposition and remains at a constant height thereafter.

The isothermal lines in Fig. 12a provide the variation of temperature and volume fraction of solid (f_s) in the disk along its height as a function of time. During deposition (i.e. the first 40s), it is observed that the top surface of the disk is hotter than the bottom due to the influx of heat and mass from the spray. With continued deposition, the fraction of solid at the top surface decreases to a minimum value of ~83%; this is marginally greater than the 80% solid in the spray, due to gas cooling at the top surface. Once deposition is completed, the top surface begins to cool/solidify rapidly due to gas cooling at the top surface, and the last liquid to freeze is at a height of ~4.5 cms above the substrate surface. The interval between the deposition line and the f_s=1 contour provides the local solidification time, t_f. It is observed that the disk preform solidifies over a period of 70-100s and experiences a cooling rate <20°C/s. However, it must be emphasized that the solidification process is not entirely "slow" but occurs in two stages: a majority of the solidification occurs in flight under conditions of rapid solidification and only a small fraction of the liquid (10-40%) which is carried into the deposit

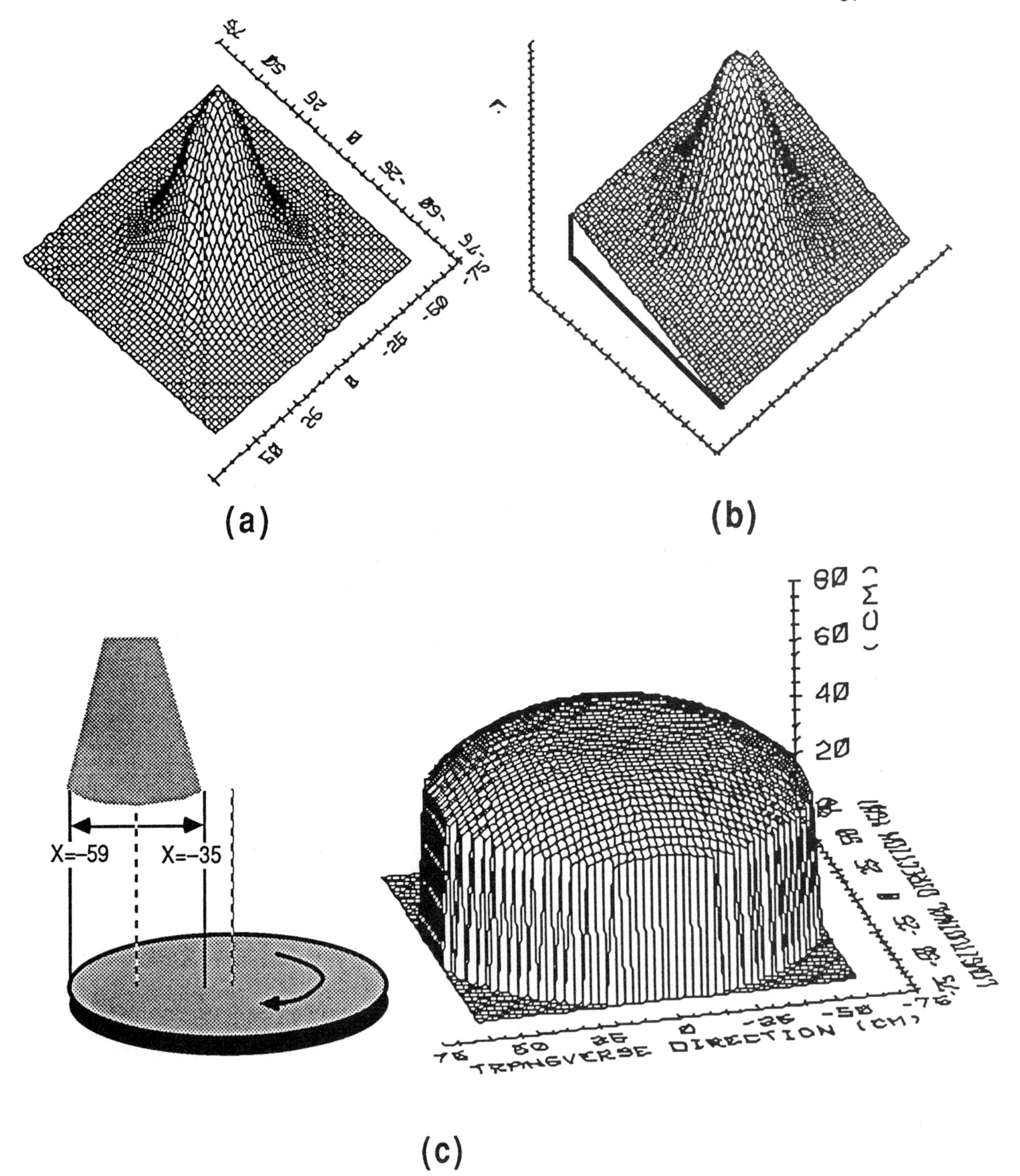

Figure 10

undergoes relatively slow cooling. Therefore it is desirable to maximize the heat removal during flight to the extent that the amount of liquid carried into the deposit is just sufficient to consolidate the presolidified droplets.

The partially liquid layer ($f_s<1.0$) forms on the surface of the preform during deposition due to a slight imbalance between the deposition rate and the rate of heat extraction, Fig. 12a. This is considered beneficial to the microstructure of spray cast preforms since the resultant increase in local solidification time allows time for fluid flow into interstices between presolidified droplets. However, the thickness of the layer must be controlled in order to maintain conditions of "incremental solidification" [28] and hence achieve uniformity in structure throughout the thickness of the deposit. The fraction of solid in the layer is estimated to be >60%; liquid fractions greater than ~0.4 may not be feasible since the apparent fluid viscosity declines sharply beyond this value [58] and a mould will be required to contain the flow. In addition, a second constraint is placed on the minimum viscosity of the fluid by the high velocity of the gas and centrifugal force imposed by rotation of the substrate.

The predicted variation of temperature with time at the top, middle and bottom of the disk are shown in Fig. 12b. Similar results have been obtained experimentally by measuring the temperature in the preforms with thermocouples at selected heights above the substrate surface [35,46]. It is observed that the temperature of the top surface of the disk preform oscillates in a cyclic manner corresponding to each pass under the spray. The frequency of these oscillations is governed by the reciprocation/rotation speed and the limits of reciprocation (i.e. stroke length), while the temperature range over which the oscillations occur depends upon the deposit-gas heat transfer coefficient and the thermal properties of the alloy. Oscillations in temperature are undesirable, particularly if they occur over a temperature range over which the surface temperature decreases below the solidus. Under these conditions, the surface will completely solidify in the interval between two successive passes under the spray; this results in poor bonding between the "layers" thus produced.

Grain size and segregate spacing in the preform are larger than those in atomized powders but smaller than the values based on empirical correlations of dendrite arm spacing and cooling rates predicted by the continuum model [37,57,59]:

$$d_s = d_0 \exp (t_f)^m = d_1 \exp (T)^n \tag{10}$$

where coefficients d_0 and d_1, and the exponents m and n, are dependent on the material. Although the predicted grain size is found to follow the same trend as the experimental data [9,37,57], it is higher in magnitude by a factor >2. Over-estimation of the grain size and/or segregate spacing from such equations may be due to (i) the two-stage solidification process in spray casting, (ii) retarded coarsening of the dendrite arms at a high volume fraction of solid [60], and (iii) nucleation by presolidified droplets from the spray. Therefore knowledge of the temperature/cooling rate in the preform is not sufficient to predict the grain size.

(c)Discrete Event Models

The second, non-continuum approach to model preform solidification has been developed recently [56,57]. Unlike the continuum model in which deposit growth is assumed to occur continuously, the non-continuum/discrete-event models are more realistic in the sense that deposit growth is assumed to occur in discrete steps by the addition of splats, (Fig. 11b). Individual droplets impact on the pre-existing surface and spread to form a splat in microseconds [61-63]. The splat then undergoes cooling and solidification via conduction through the bottom surface and via convection at the top surface. This continues until the next droplet arrives after a time interval δt between successive impacts. This sequence of events continues for every new splat arriving at the top surface, Fig. 10b.

While conceptually more accurate than the continuum approach, the drawbacks of the discrete-event model are that it is computationally intensive and there is uncertainty in the values of input parameters required to facilitate the computations (e.g. splat thickness, interval between splats,etc.). Values of splat thickness δx are estimated writing $d_{splat}= \xi d_m$, where ξ typically has a value between 3 and 6, assuming a random distribution of spherical droplets of average diameter, d_m [56,61]. A mean droplet diameter, d_m of 100µm yields splat thicknesses in the range 7.5-15µm and a splat thickness of 10µm is used. The calculation for δt is made incorporating the measured log-normal droplet size distribution. For the observed droplet size distributions, estimates of δt lie in the range 0.8-4ms and an order of magnitude value of 1 ms of δt is utilized.

The discrete-event model predicts cooling rates of the order of 10^3°C/s during the initial stages of deposition, which are higher than the predictions of the continuum models. However, both continuum and discrete-event models predict the formation of a partially liquid layer after a short time interval (~0.5s) and their predictions appear to converge after this period. Implications are that the predictions of the continuum approach are reasonable and valid after the time corresponding to the formation of the partially liquid layer.

(d) Degree of Porosity in the Preform

The predicted values of $\%L_{spray}$ upon deposition are at the low end of the range, i.e 10%-40%. This suggests that only a small fraction of liquid is required to consolidate the presolidified droplets during deposition, and that the mechanism of droplet consolidation in spray deposition resembles liquid phase sintering [63,64]. Therefore, the level of porosity in the preforms is determined primarily by a balance between the solid:liquid ratio being deposited and by the efficiency

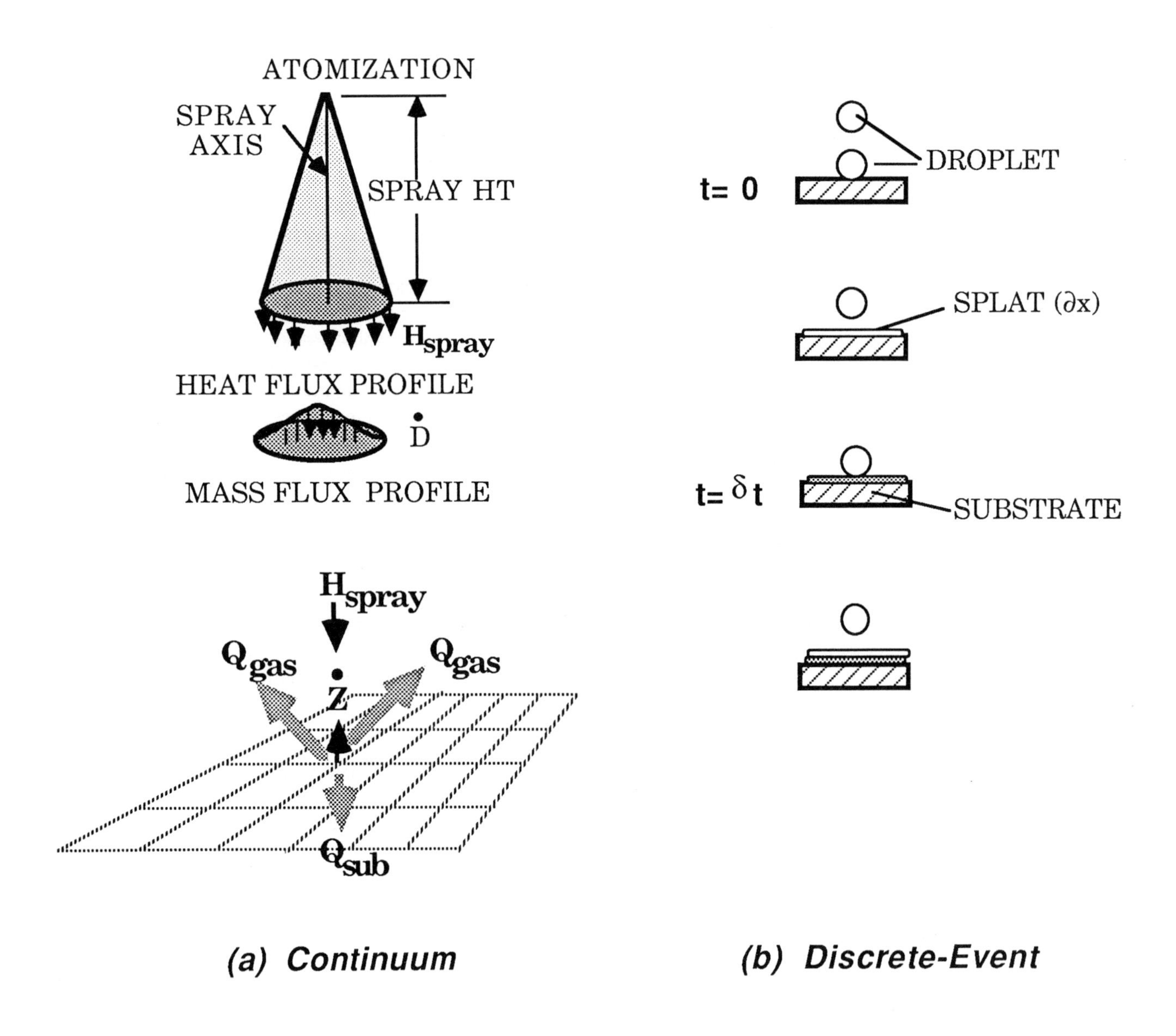
ATOMIZATION
SPRAY AXIS
SPRAY HT
H spray
HEAT FLUX PROFILE
D
MASS FLUX PROFILE
H spray
Q gas
Z
Q gas
Q sub
(a) Continuum
DROPLET
t= 0
SPLAT (∂x)
t= δt
SUBSTRATE
(b) Discrete-Event

Figure 11

of packing of the solid particles. Given a packing efficiency Π_p, the desired amount of liquid is the minimum quantity which will flow and fill the interstices between solid particles; any quantity less than this value will result in an amount of porosity given by:

$$\%P = [(100-\Pi_p) - \%L_{spray} (1 - \beta_s)] \cdot 100 \quad (11)$$

where β_s is the solidification shrinkage for the alloy and Π_p is governed by the state of the spray.

(e)Effect of Substrate Characteristics

The effect of physical and thermal properties of the substrate on preform characteristics is predominant only during the initial stages of deposit growth, and it is a strong function of the degree of contact at the interface. Smoothness of the substrate surface determines the degree of mechanical interlocking/adhesion between the deposit and the substrate; if the degree of contact is poor, thermal properties of the substrate exert *minimal influence* on preform solidification. For example, a metallic substrate (e.g. copper or steel) produces a high rate of heat extraction in the initial stages of deposition due to a good contact. This results in the formation of a pronounced initial chill layer (~1mm thick) with attendant porosity (~10-15%). However, the heat extraction decreases after 1-2 s due to the formation of an air gap between the deposit and the substrate. A refractory substrate (e.g. alumina) has been found to decrease the level of porosity in the initial stages; this effect is due to the lower thermal conductivity of the refractory. It is also possible to coat the surface of the substrate with non-conducting coatings (e.g. boron nitride) to control the heat transfer coefficient at the interface.

Heat flux into the substrate can be decreased by increasing the substrate temperature. Ideally, a high substrate temperature (>0.8 T_m) is desirable during the initial period of deposition (~2s) in order to reduce the level of porosity at the bottom surface of the deposit. Subsequently, the temperature should be maintained at or near room temperature (e.g. by water cooling) in order to sustain the heat flux and decrease the time required for deposit solidification. Preheating of the substrate is expected to promote bonding between the substrate and the deposit due to reduction in the cooling rate; this approach is being utilized in the manufacture of tubes on emplaced mandrels.

4. Summary

• Successful utilization of spray casting requires optimization of at least eight independent process parameters to achieve desired preform shape, microstructure and yield. The basis of the optimization is to achieve desired values of two critical dependent parameters of the process, namely the physical and thermal state of (i) the spray just prior to consolidation, and (ii) the state of the surface onto which the droplets impact.

• Velocity of droplets during flight is primarily a function of their mass. Droplet velocities predicted from fluid flow analyses are in the range 10-100 m/s under standard operating conditions of the Osprey process. Experimental determination of droplet size and velocity are currently limited due to the microscopic size of the droplets and the time scale of droplet flight (a few milliseconds).

• Density, melting point and heat contained within the freezing range of the alloy are the three significant material parameters which affect the state of the droplets/spray at impact. The diameter of the largest solidified droplet at X=400mm ranges from 30μm-125μm for the alloys investigated and increases with increasing melting point. The percentage of liquid in the spray at X=400mm ranges from ~5% for Cu to ~35% for aluminum and its alloys under identical spraying conditions.

• Both melt superheat and gas:metal ratio affect $\%L_{spray}$ to a similar extent (~10%) when their value is changed by 50%; however increase in the melt superheat increases $\%L_{spray}$ while a higher value of GMR produces a colder/more solidified spray at any flight distance. Changing the atomizing gas to argon (from nitrogen) does not significantly alter the droplet size distribution, however, even a four fold increase in the flow rate of argon is not as effective as nitrogen in cooling the spray.

• Preform yield can be represented by the product of target and consolidation efficiencies. Target efficiency for the production of a 120mm dia. disk preform was calculated as ~72% from the model to simulate preform shape/geometry. This model can be employed to predict the build up of thickness/shape of disks, billets, tubes and/or strip via spray deposition.

• Measured values of heat flux across the deposit-substrate interface indicate that the heat transfer coefficient is a maximum of ~10^5 W/m/K in the initial second but decreases to ~500 W/m/K during the bulk of the deposition period.

• Grain size in the preforms is determined by the spatial distribution of solid particles from the spray and cooling rates after deposition. The solidification process is two stage: a majority of the solidification (=100-$\%L_{spray}$) occurs in flight at cooling rates in the range 10^3-10^4 K/s while the remainder of the solidification occurs at <10°C/s.

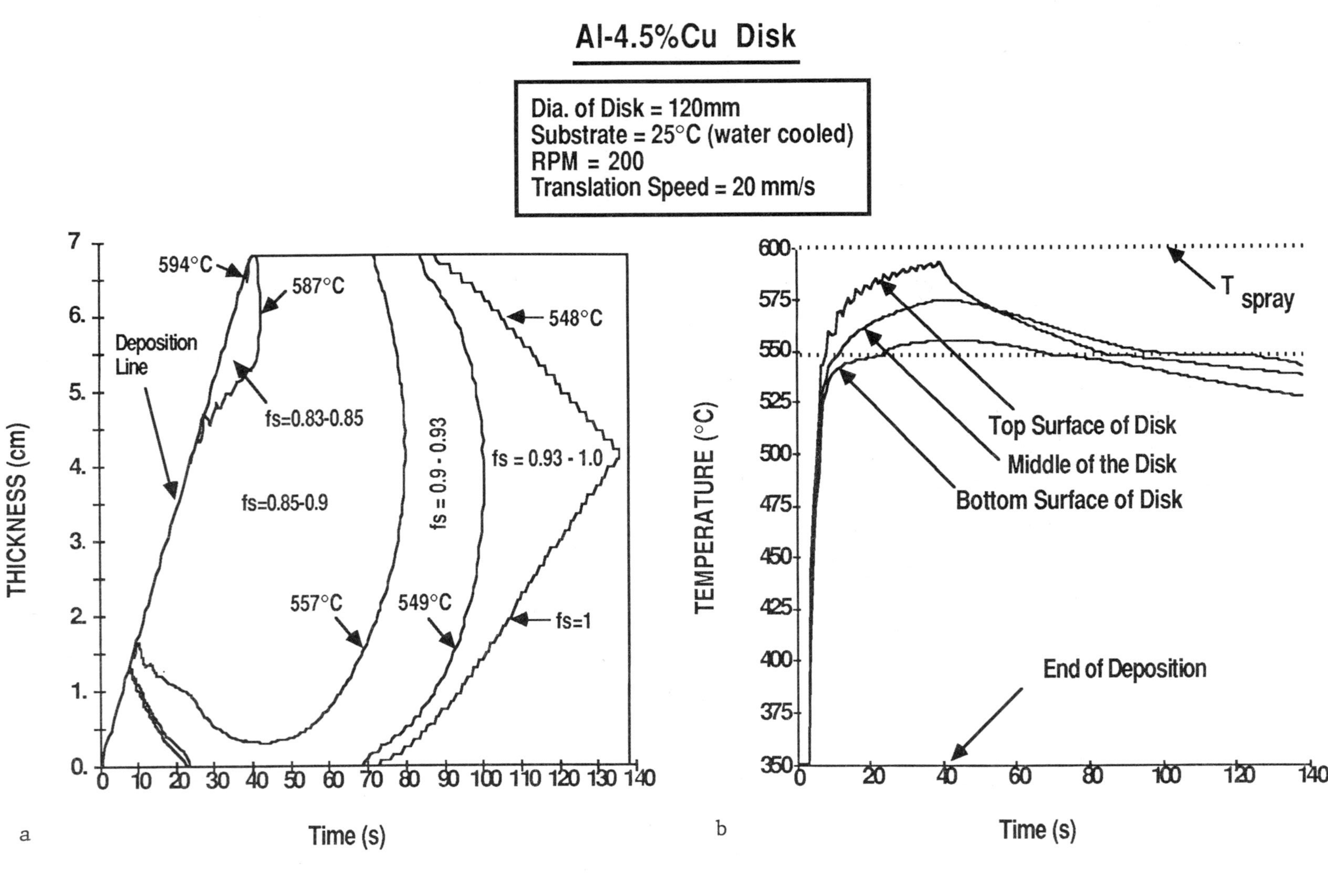
Al-4.5%Cu Disk
Dia. of Disk = 120mm
Substrate = 25°C (water cooled)
RPM = 200
Translation Speed = 20 mm/s
THICKNESS (cm)
594°C
587°C
548°C
Deposition Line
fs=0.83-0.85
fs=0.85-0.9
fs = 0.9 - 0.93
fs = 0.93 - 1.0
557°C
549°C
fs=1
Time (s)
a
TEMPERATURE (°C)
T spray
Top Surface of Disk
Middle of the Disk
Bottom Surface of Disk
End of Deposition
Time (s)
b

Figure 12

Acknowledgements

The authors wish to acknowledge discussions with and assistance from Dr Alan Lawley, Dr Dan Wei and Dr Suresh Annavarapu.

References

1. W.A. Tony, Iron & Steelmaker,12 p.11, December 1987.
2. P.W. Wright, Materials and Design, 8, 3, May/June 1987.
3. Net Shape Technology in Aerospace Structures, I-IV, National Academy Press,Washington, D.C., 1986.
4. A.G. Leatham, W. Reichelt, and Metelmann, Near Net Shape Manufacturing Processes, (eds.) P.W. Lee and B.L. Ferguson, ASM International, p.259, 1988.
5. R.W. Evans, A.G. Leatham and R.G. Brooks, Powder Metallurgy, 28, 1, p.13, 1985.
6. D. Apelian, G. Gillen and A. Leatham, Processing of Structural Metals by Rapid Solidification, (eds.) F.H. Froes and S.J. Savage, Amer. Soc. for Metals International, p.107, 1987.
7. E.J. Lavernia, G. Rai and N.J. Grant, Int. J. Powder Metallurgy, 22, 1, p.9, 1986.
8. E.J. Lavernia, and N.J. Grant, Metal Powder Rep., 4, p.255, 1986.
9. E. Gutierrez-Miravete, E.J. Lavernia, G.M. Trapaga, J. Szekely and N. J. Grant, Metall. Trans., 20A, 1, p.71, January 1989.
10. D. Apelian, M. Paliwal, R.W. Smith and W.F. Schilling, Int. Metals Rev., 28, 5, p.271, 1983.
11. A.R.E. Singer, J. Inst. Metals, 100, p.185, 1972.
12. A.R.E. Singer, Metals and Materials, 4, p.246-257, 1970.
13. A.R.E. Singer, in proc. of High Density P/M Consolidation Processes, P/M 84, Toronto, June 1984.
14. D.Apelian, B.H. Kear and H.W. Schadler, in Rapidly Solidified Crystalline Alloys, (ed.) S.K. Das, B.H. Kear and C.M. Adam, The Metallurgical Society, Warrendale, PA, p.93, 1985.
15. A.G. Leatham, A.J.W. Ogilvy, P.F. Chesney and O. Metelmann, Modern Developments in Powder Metallurgy, eds.: P.U. Gummeson and D.A. Gustafson, Metal Powder Industries Federation, Princeton, NJ, 18-21, in Press,1988.
16. Osprey Metals Ltd., private communication with R.G. Brooks and A.G. Leatham, 1988.
17. R.H. Bricknell, Met. Trans., 17A, 4, p.583, 1986.
18. H.C. Fiedler, T.F. Sawyer and R.W. Kopp, Spray Forming-An Evaluation Using IN718, General Electric Technical Information Series, 86CRD113, May 1986.
19. S. Annavarapu, A. Lawley and D. Apelian, Met. Trans, 19A, 12, p.3077, 1988.
20. A. Lawley and D. Apelian, A Fundamental Study of Thin Strip Casting of Plain Carbon Steel by Spray Deposition, NSF Report MSM-8519047, National Science Foundation, Washington D.C., 1988.
21. Mannesman Demag, W.Germany, private communication with W. Reichelt, 1989.
22. D. Apelian, A. Lawley, P. Mathur and X. Luo, Modern Developments in Powder Metallurgy, eds.: P.U. Gummeson and D.A. Gustafson, Metal Powder Industries Federation, Princeton, NJ, 19, p.397, 1988.
23. A. Moran and W.A. Palko, in Progress in Powder Metallurgy, (eds.) Freeby, C.L. and Hjort, H., Metal Powder Industries Federation, Princeton, NJ, 43, 1987.
24. R.P. Singh and A. Lawley, Modern Developments in Powder Metallurgy, (eds.) P.U. Gummeson and D.A. Gustafson, Metal Powder Industries Federation, Princeton, NJ, 19, p.489, 1988.
25. J. Duszczyk, J.L. Estrada, B.M. Korevarr, T.L.J. deHaan, D. Bialo, A.G. Leatham and A.J.W. Ogilvy, Modern Developments in Powder Metallurgy, eds.: P.U. Gummeson and D.A. Gustafson, Metal Powder Industries Federation, Princeton, NJ, 19, p.441, 1988.
26. J.F. Faure and L. Ackermann, Modern Developments in Powder Metallurgy, eds.: P.U. Gummeson and D.A. Gustafson, Metal Powder Industries Federation, Princeton, NJ, 19, p.425, 1988.
27. Y. Unigame, An Analysis of Oxide Dispersion Strengthening of Ferrous Alloys during Spray Casting, M.S. Thesis, Drexel University, Philadelphia, 1989.
28. A.R.E. Singer and R.W. Evans: Met. Tech., 10, p.61, 1983.
29. A. Lawley, J. Metals, 33,1, p.13, 1981.
30. A. Unal, Metall. Trans., 20B, 2, 1989.
31. A. Lawley and R.D. Doherty, in Rapidly Solidified Crystalline Alloys, (ed.) S.K. Das, B.H. Kear and C.M. Adam, The Metallurgical Society, Warrendale, PA, p.93, 1985.
32. D. Bradley, J. Phys. D., 6, p.1724, 1973.
33. H. Lubanska, J. Met., 22, 2, p.45, 1970.
34. R. Dunstan, A.G. Leatham, M.I. Negm and C. Moore, Opsrey Metals Ltd., presented at National P/M Conference, Philadelphia, May 3-6, 1981.
35. P.C. Mathur, Analysis of the Spray Deposition Process, Ph.D. Thesis, Drexel University, Philadelphia, 1988.
36. R.M. German, Powder Metallurgy Science, Metal Powder Institute Federation, Princeton, NJ, 1984.

37. P. Mathur, D. Apelian and A. Lawley, Acta Metall., 37,2, p.429, 1989.
38. P. Mathur, D. Wei and D. Apelian, Modeling and Control of Casting and Welding Processes IV, eds. A.F. Giamei and G.G. Abbaschian, The Minerals, Metals and Materials Soc., Warrendale, PA, p.275, 1988.
39. L.H. Kallien, P.N. Hansen and P.R. Sahm, Modeling and Control of Casting and Welding Processes IV, eds. A.F. Giamei and G.G. Abbaschian, The Minerals, Metals and Materials Soc., Warrendale, PA, p.543, 1988.
40. J. Liu, L. Arnberg, N. Backstrom, H. Klang and S. Savage, Mat. Sci, Eng., 98, p.43, 1988.
41. E.J. Lavernia, E.M. Gutirrez, J. Szekely and N.J. Grant, "Progress in Powder Metallurgy 1987", Metal Powder Industries Fed., Princeton, NJ, 43, p.683, 1987.
42. A.A. Ranger and J.A. Nicholls, AIAA Journal, 7, 1969.
43. S. Connelly, J.S. Coombs and J.O. Medwell, Metal Powder Report, 41, 9, p.653-661, 1986.
44. A.R. Anderson and F.R. Johns, A.S.M.E., 21, p.13, 1955.
45. S. Elghobashi, T. Abou-Arab, M. Rizk and A. Mostafa, Int. J. Multi-phase Flow, 10, 6, p.697, 1984.
46. B. Bewley and B. Cantor, in Int. Conf. on Rapidly Solidified Materials, eds. P. Lee and R. Carbonara, Amer. Soc. for Metals, Ohio, p.15, 1986.
47. Kurten, et. al., Bubbles, Drops and Particles, (ed.) R. Clift, J.R. Grace and M.E. Weber, Academic Press, N.Y., 1978.
48. K. Bauckhage, H. Flogel, U. Fritsching, U. Hiller and F. Schone, Simultaneous Size and Velocity Measurements in Multiphase Flow Systems/Some Extentions of the Phase Doppler Method, Proc. 3rd Intl. Symp. on Laser Anemometry, Amer. Soc. Mech. Eng., Winter Mtg., Boston, Mass. U.S.A, 1987.
49. P.C. Mathur, A Study of Droplet Flight during Spray Forming, M.S. Thesis, Drexel University, Philadelphia, 1986.
50. W.E. Ranz and W.R. Marshall, Chemical Engineering Progress, 48, 1952.
51. D. Apelian, A. Lawley, G. Gillen and P. Mathur, Spray Deposition: A Fundamental Study of Droplet Impingement, Spreading and Consolidation, ONR Technical Report 4, NR 650-025, Contract N 00014-84-K-0472, Office of Naval Research, Arlington, VA, 1988.
52. C.G. Levi and R. Mehrabian, Met. Trans., 13A, p.221, 1982.
53. W.J. Boettinger, S.R. Coriell and R.F. Sekerka, Mat. Sci. Eng., 65,1, p.27, 1984.
54. N.J. Grant, Casting of Near Net Shape Products, (eds.) Y. Sahai, J.E. Battles, R.S. Carbonara and C.E. Mobley, TMS, p.203, 1988.
55. Nippon Steel Corp., Japan - private communication with Y.Tomita, 1989.
56. S. Annavarapu, Ph.D. Thesis, Drexel University, Philadelphia, 1989.
57. E. Gutierrez-Miravete, G.M. Trapaga and J. Szekely: Casting of Near Net Shape Products, eds. Y. Sahai, J.E. Battles, R.S. Carbonara and C.E. Mobley, TMS, p.133, 1988.
58. D. Spencer, R. Mehrabian and M.C. Flemings, Met. Trans., 3A, p.1925, 1972.
59. H. Jones, Rapid Solidification Processing: Principles and Technologies, (eds.) R. Mehrabian, B.H. Kear and M. Cohen, Claitor's Publishers, Baton Rouge, LA, 1978.
60. R.D. Doherty, Met. Sci., 16, p.l, 1982.
61. J. Madjeski: Int. J. of Heat and Mass Transfer, 19, p.1009 , 1976.
62. H. Jones, Rapid Solidification of Metals and Alloys, The Institution of Metallurgists, p.43, 1982.
63. E. Garrity, Ph.D. Thesis, Drexel University, Philadelphia , 1989.
64. R.M. German, Liquid Phase Sintering, Plenum Press, NY, 1985.
65. P.E. Zovas, R.M. German, K.S. Hwang and C.J. Li, J. Metals, 35, 1, p.28, 1983.

3

Powder Production and Handling Electrolytic Powders

E. PEISSKER
Norddeutsche Affinerie Aktiengesellschaft, Hamburg, Germany

The first attempts to produce metal powder by electrolysis date back to the 19th century [1]. The interest in and demand for electrolytically deposited metal powders grew when the first sintered parts and sintered bearings came on the market. Information on the beginnings of the process are detailed by Jones and Mehl [1,2].

Systematic laboratory tests ran parallel to the development of commercial production methods with the aim of acquiring a better understanding of the various stages in depositing the powder (see in particular the works by Ibl and Trümpler [3]. Numerous authors (for example Kumar and Gaur [4]) have dealt at length with the regulation/optimization of the electrolytic processes during powder deposition. The electrolysis of aqueous solutions is chiefly used in the manufacture of powders, a limited account of which is given in the following.

Powders manufactured by electrolytic deposition are characterized by high purity as well as an extremely irregular particle shape and surface, features which guarantee very high green strength when the powders are processed to powder metallurgy parts. A continuous metal matrix which is very absorbent for a non-metallic filler results when processing to composite materials such as friction materials or carbon brushes. These advantages must be seen in relation to the disadvantage that in some cases a relatively high specific energy consumption goes hand in hand with the powder electrolysis.

Today, the powder electrolysis manufacturing process has tough competition from other methods of manufacturing powders which have been subsequently developed, such as atomization. However, electrolytically deposited metal powders have a good chance of standing up to this owing to the applications which are based on the typical property combinations for electrolytic powder.

The main powders now produced commercially by electrolysis are copper, iron and small amounts of silver [5]. In Western Europe, the Comecon countries and Japan more than half the copper powder production is covered by electrolytically deposited copper powder. In contrast, however, in North America electrolytic copper powder is hardly produced any more. In respect of iron powder the share of electrolytically deposited powder amounts to less than 1% of the entire powder production - for information on the manufacture and applications see Samal [6].

Special electrolytic processes have repeatedly been suggested and tested on a laboratory scale for alloy powders such as Cu-Sn, Cu-Pb, Cu-Ag, Cu-Ni, [Ref. 7, with numerous recommendations for further reading]. However, in commercial practice none of these processes was competitive up to now.

The manufacture of copper powder is elaborated in the following as a typical example of a metal powder electrolysis.

1. Production of Electrolytic Copper Powder

In suitable electrolysis conditions copper powder particles with a dendritic particle shape form direct on the cathode. Figure 1 shows the characteristic formations: either powder with an extremely dendritic particle shape, i.e. with low apparent density, or powder with a round dendritic particle shape and, as a result, a higher apparent density. Powder with a low apparent density is always deposited as fine particles, coarser powder exhibits a higher apparent density (Fig. 2). Figure 3 shows the possibilities of regulating the powder deposition. The respective apparent density/fineness required is mostly achieved by modifying the copper concentration in the electrolyte. Typical conditions for the tankhouse are:

- copper concentration in the electrolyte: 5 - 30g/l
- sulphuric acid concentration in the electrolyte: 150 - 250g/l
- anodic current density: 300 - 600A/m^2
- cathodic current density: 600 - 4000A/m^2
- electrolyte temperature: 40 - 60°C
- cell voltage: 1.0 - 2.0V

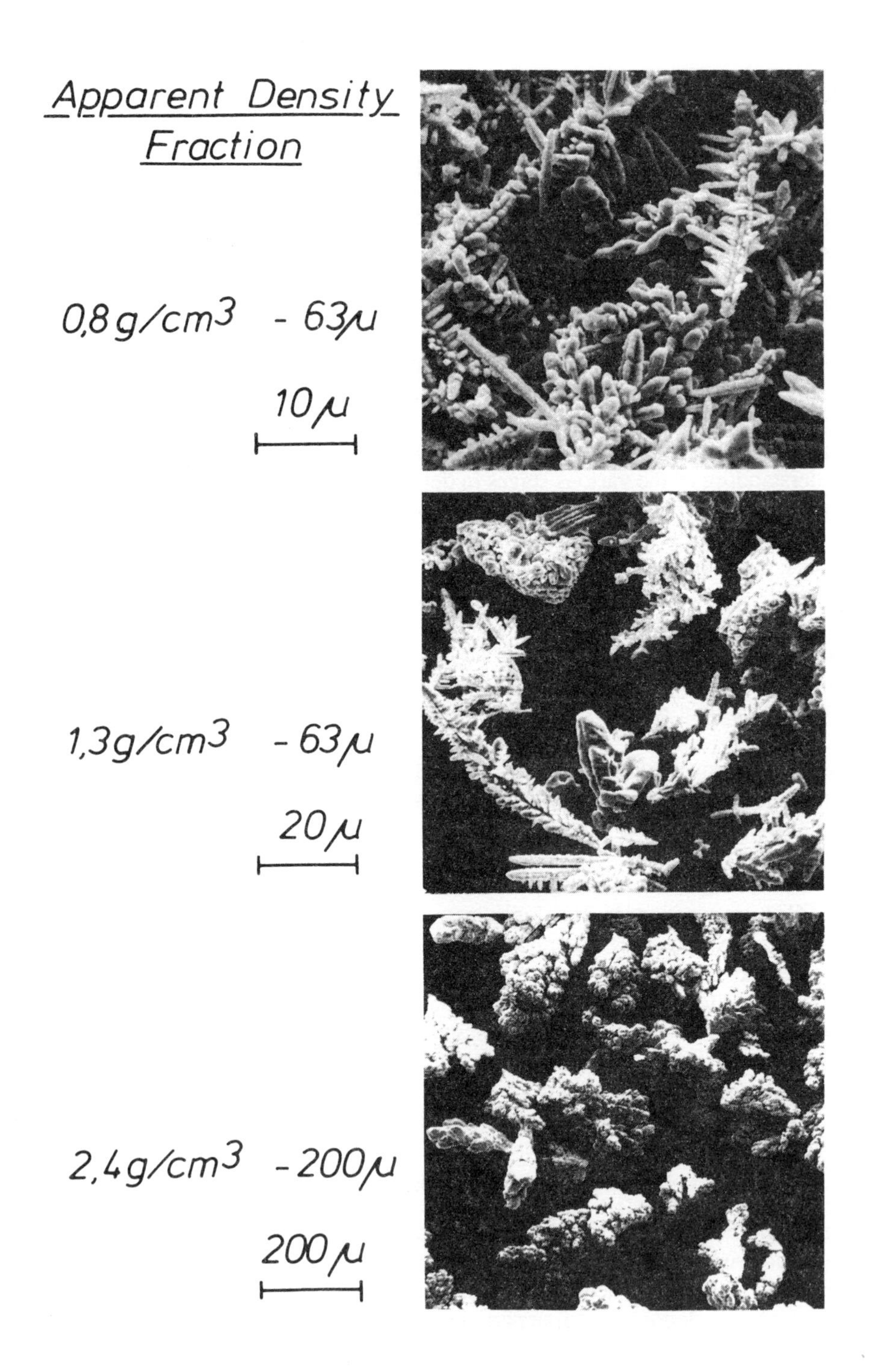

Figure 1 SEM micrographs of different dendritic copper powders. Source: ref. (8)

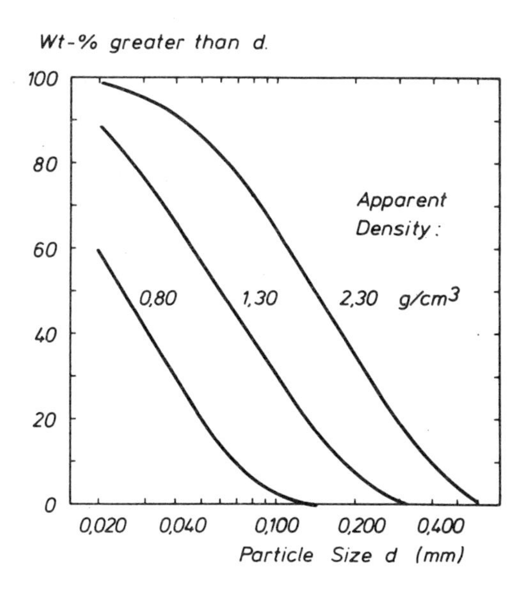

Figure 2 Particle size distributions in primary copper powder deposits. Source: ref. (8)

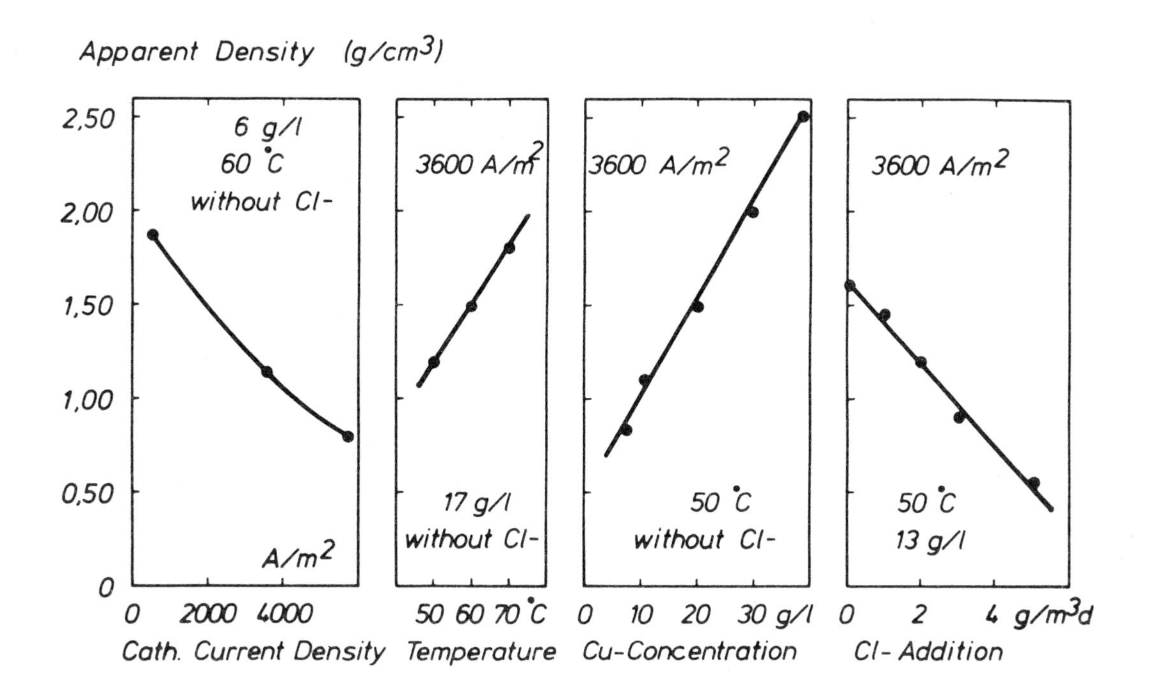

Figure 3 Apparent density in relation to four electrolysis parameters: cathode current density, temperature, copper concentration, chlorine additions. Source: ref. (8)

The primary deposited powder is, if necessary, separated from the cathode, removed from the tankhouse cell and washed to an acid-free state.

The further processing of the wet copper powder presents the following possibilities of actually influencing the properties of the copper powder being produced in addition to determining the deposition conditions. In practice, two main process variations have been developed:

1. Electrolysis with high cathodic current density, control of the apparent density/fineness as far as possible during the electrolytical deposition, hot air drying, see flow sheet (Fig. 4).

Advantages:
- very light dendritic powder with apparent densities up to 0.7 g/cm^3 produced without difficulties
- very little risk of reoxidation

Disadvantages:
- relatively complicated cathode bar/anode sheet geometry
- flowability is worse than with the following alternative

2. Electrolysis with low cathodic current density and to a great extent constant electrolysis conditions, drying and reduction in a belt furnace, subsequent milling of the resulting sintered cake. Control of the apparent density/fineness mostly by varying the processing conditions, see flow sheet - Fig. 5.

Advantages:
- simple cathode plate/anode plate geometry
- higher apparent density producible to 4 g/cm^3

Disadvantages:
- very light dendritic powder with apparent density less than 1.7 g/cm^3 not producible
- great risk of subsequent oxidation

A detailed description of the two production processes is not given here, since detailed information on the first alternative can be found in Peissker [8] and on the second alternative in Wills and Clugston [9] as well as in Taubenblat [10]. Instead some particular points are dealt with in the following.

2. Feed Material

A copper powder electrolysis always functions as a refining electrolysis as well so that, in principle, relatively highly impure, and as a result cheap, scrap should be suitable as the feed material. The latter is smelted to anode sheets or hung in lumps in the tankhouse cells in anode baskets. It is always to be expected that the electrolyte will be made impure by the scrap feed. This can, in turn, lead to intolerable changes in the deposition process. In practice, therefore, only slightly impure scrap or cathode copper which has been already electrolytically refined from the start is used.

On numerous occasions the attempt has been and will continue to be made to work with insoluble anodes and to feed copper in solution in the tankhouse cell. Corresponding solutions rich in copper are obtained for example from leaching ores low in copper with sulphuric acid. Occasionally, galvanic waste water rich in copper is used. Experience has shown that the copper powder resulting from this charge material is relatively impure, bad to reproduce, and only usable to a limited extent.

3. Electrolyte

Today, almost all commercial copper powder electrolysis operates with electrolyte based on H_2SO_4, although occasionally electrolyte with a $(NH_3)_2CO_3$ basis is also used. For electrolyte based on HCl see Ref.4 and based on NH_2HSO_3, see Kotovskaya *et al.* in Ref.10.

4. Electrode Material

Lead with up to 12% antimony as the standard material as well as specially prepared titanium sheets, graphite and stainless steel are used as insoluble anodes for the electrolysis in a sulphate aqueous solution.

Copper rod or sheet, more seldom nickel sheets, serve as cathode material. Coarser powder, in particular, sticks much too firmly to the cathodes in the deposition. Attempts to reduce adhesion by covering the cathodes with soft conductive caoutchouc or silicon caoutchouc were, unfortunately, to no avail. The deposited copper powder is brushed or hammered off the cathode sheets/rods.

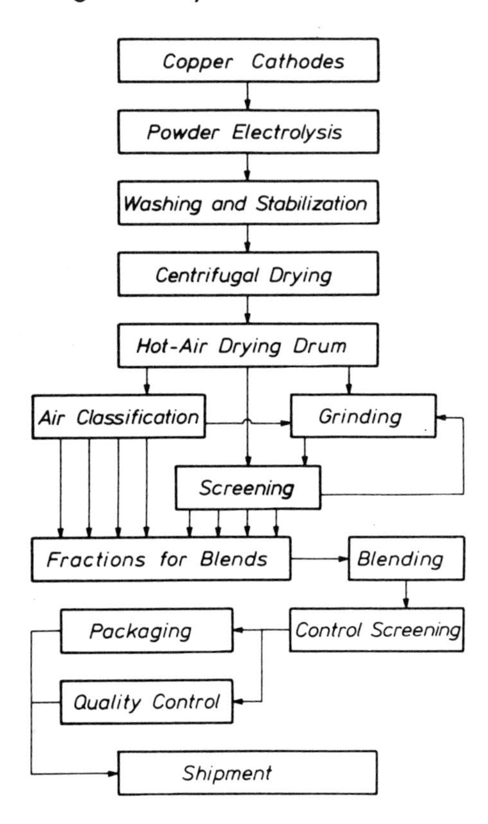

Figure 4 Flow sheet: electrolytic copper powder production, var. 1. Source: ref. (8)

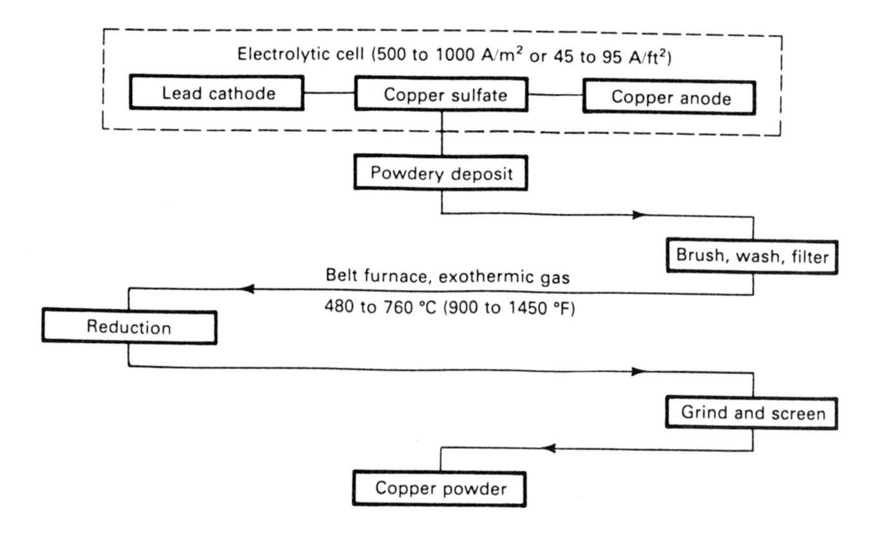

Figure 5 Flow sheet: electrolytic copper powder production, var. 2. Source: ref. (10)

	FFL	FL	FS	SSM	M	GF
Sieve Analysis: + 0,315 mm						./.
+ 0,200 mm				./.	./.	1
+ 0,080 mm	./.	./.	./.	26	40	94
+ 0,063 mm	1	1	2			
+ 0,040 mm	4	9	13	33	33	4
- 0,040 mm	95	90	85	41	27	1
Apparent Density (g/cm^3)	1,00	1,25	1,90	2,40	2,40	2,40
Flow Rate (sec/50 g)	∞	∞	∞	33	29	27
Specific Surface Fisher (cm^2/g)	1800	1400	850	500	450	300
Oxygen Content (wt %)	0,09	0,08	0,07	0,04	0,03	0,02

Figure 6 Typical data of commercial copper powders. Source: ref. (8)

5. Continuously Operated Electrolytic Cells

Electrolytic cells are mostly not operated continuously. As soon as the cathode sheet is covered with a corresponding powder layer or the bottom of the cell is filled with powder, the cell operation must be interrupted and the powder removed. The costs for halting the electrolysis and the labour intensive removal of the powder are relatively high. As a result there are numerous suggestions in the patent literature for continuously operating electrolytic cells. The more recent propose in the main fixed anodes and manoeuvrable, (e.g. rotating) cathodes, on which the powder is continuously deposited, brushed off and transported from the cell. In practice, such constructions have not as yet proved themselves practicable, in particular owing to the extensive mechanics.

6. Additions to the Electrolyte

Additions can help to influence the particle shape and size and/or to improve the current efficiency. For a long time it has been acknowledged that glue, glucose, glycerine and boric acid have the effect of reducing the particle size, partly in combination with increasing the apparent density. The influence of halogen ions and antimony salts was systematically examined by Russian authors [11]. Chlorine ion additions can be used with the object of controlling the apparent density, (Fig. 3). Walker and Duncan [12] managed to improve the current efficiency with additions of sodium sulphate, cupric chloride and benzotriacole. In many cases additions to the electrolyte have also influenced the tendency to oxidation in the resultant copper powder. In practice, therefore, additions are not used if at all possible.

7. Stabilization Against Reoxidation

The severe susceptibility of clean oxide-free copper surfaces to oxidation, especially in high relative humidity is generally acknowledged. Fine copper powder is particularly endangered owing to its large specific surfaces, which must also be protected against reoxidation by so-called stabilization.

7.1 Dry stabilization

Camphor, stearic acid, zinc stearate, corrosion inhibitors such as benzotriazole are, for example, added to dried or reduced copper powders. Occasionally substances dissolved in volatile mediums (alcohol, acetone) such as fatty acids or the salts of tartaric acid are also used. Dry stabilization which is the only possibility for reduced powders is mostly inferior to wet stabilization in its effectiveness.

7.2 Wet stabilization

Wet stabilization is carried out in connection with the acid free wash when a more or less continuous protective coating is formed on the whole of the copper powder surface, for example, by using solutions of chromate or potassium tartrate.

Most commercial copper powder manufacturers have special know-how at their disposal, not generally in the public domain. The last comprehensive relevant publication came from Pomosov *et al.* [13].

8. Integration of a Copper Powder Electrolysis in a Copper Smelter

The integration of the electrolytic process with a copper smelter, which there are many examples, has a number of advantages:

- Copper cathodes used as input material for powder production are always available in adequate quantities and their quality is invariably the same.

- The powder electrolysis is connected to the copper refinery tankhouse by a pipe system. Thus, for example, acid waste water from powder washing is absorbed by the electrolyte system of the refinery tankhouse.

- Since an excessive increase in anode current density must be avoided, copper cathodes serving as anodes in the powder electrolysis are not dissolved completely. The scrap copper sheets are returned to the refinery and cast into shapes, like any normal higher grade cathodes.

- The portion of the powder deposit which is too coarse for subsequent processing is also remelted.

Figure 6 gives a resumé of the properties of typical commercially produced copper powders. These powders were mainly used for the production of carbon brushes, friction materials and sintered bronze bearings.

References

1. W.D. Jones, Fundamental Principles of Powder Metallurgy, Edward Arnold (Publ.) Ltd., London, pp.142-176, 1960.
2. E. Mehl, Metal Treatment, 17, pp.118-128, 1950.
3. N. Ibl and G. Trümpler, in [1] pp.144.
4. D.K. Kumar and A.K. Gaur, J. Electrochem. Soc. India, 22-3, pp.211-216, 1973.
5. P.W. Taubenblat, Metals Handbook 9th Edition, 7, pp.71-72, 1984.
6. P.K. Samal, Metals Handbook 9th Edition, 7, pp.93-96, 1984.
7. A.T. Kuhn, P. Neufeld and K. Young, J. Appl, Electrochem. 14, pp.605-613, 1984.
8. E. Peißker, Int. J. Powder Met. 20, pp.87-101, 1984.
9. F. Wills and E.J. Clugston, J. Electrochem. Soc. 106, pp.362-366, 1959.
10. P.W. Taubenblat, Metals Handbook 9th Edition, 7, pp.110-116, 1984.
11. L.I. Gurevich and A.V. Pomosov, Soviet P/M Met. Cer., pp.10-15, Jan. 1969.
12. R. Walker and S.J. Duncan, Surface Techn., 23, pp.301-321, 1984.
13. A.V. Pomosov, M.I. Numberg and E.F. Krymakove, Soviet P/M Met. Cer., pp.175-177, March 1976.

4

Spark Erosion

J. L. WALTER
General Electric Corporate Research and Development, Schenectady, NY, USA

When two electrodes are in close proximity in a dielectric fluid, the application of a voltage pulse can produce a spark discharge between them. Some of the energy in the spark discharge is transferred to the electrodes and results in the heating of highly localized regions of the electrodes. If the regions are heated above their melting temperature, molten droplets or vaporized material may be ejected from the electrodes. Particles are produced by the freezing of the droplets or by condensation of the vaporized material in the dielectric fluid. This phenomenon was investigated many years ago, particularly by Svedberg [1] who used it for the preparation of colloidal suspensions. A number of more recent applications are described by Rudorff [2]. During the 1940s, spark erosion was developed into the technology of electric discharge machining (EDM) in the Soviet Union. EDM has become a widely used method for machining complex configurations in hard and tough materials [3].

Spark erosion has also been used as a technique for producing small quantities of very rapidly solidified micropowders [4-10], as a general micropowder method [11-17] and for applications such as ferrofluids [18]. Current investigations have the objectives of: extending the range of metals, alloys and compounds prepared by spark erosion; determining the influence of electrical, mechanical, and chemical operating parameters on particle production rate, size distribution, composition, microstructure, production and power efficiency.

Thus far, it has been demonstrated that spark erosion is a versatile and economical method for producing powders of a large variety of materials. The more noteworthy features of the process are the wide range of applicable materials and the small sizes of the particles obtained.

1. The Spark Erosion Process

Consider two electrodes closely spaced and immersed in a dielectric fluid. The electrodes are connected to a pulsed power source. When the field strength in the gap is sufficiently high, a spark is produced [19]. The spark results from the breakdown of the dielectric fluid. It is assumed that there is substantial electron emission from the cathode. These electrons and others in the gap gain energy from the electric field and ionize molecules of the dielectric fluid, producing more electrons and positive ions. A plasma channel of small diameter (perhaps 50μm) is established in approximately 10ns after the electric field is applied. The temperature in the plasma channel has been determined to be in excess of 100,000K [2]. The high temperature plasma column is surrounded by a sheath of vaporized dielectric which is confined by the relatively incompressible liquid; the pressure in the plasma column can be as high as 280MPa [20].

The temperature at the electrodes depends on the transfer of the kinetic energies of the electrons and ions to the anode and cathode, respectively. Power densities at the electrodes of 10^4-10^5Wmm^{-2} have been reported. These values lead to rates of temperature rise at the electrodes of 10^9-10^{10}Ks^{-1}; electrode melting occurs even with pulses shorter than 1us. The temperature of the locally heated regions of the electrodes is raised above the usual boiling point of the metal because of the pressure exerted by the plasma channel on this region. With the rapid decrease of pressure as the spark collapses at the end of the voltage pulse, the superheated regions boil violently, ejecting molten droplets and vapour into the dielectric liquid; the droplets are very rapidly quenched. Mechanisms, other than boiling of superheated regions, may become significant for pulse currents greater than 50A. Such large pulse currents are often encountered in spark erosion for particle production in contrast to EDM.

Most of the efforts to produce powder have used the tool and workpiece method in a manner analogous to EDM. However, the tool-workpiece arrangement, with its single sparking interface, produces powders at a low rate. To increase the number of sparking interfaces, a new operating unit was developed by Ishibashi [11-13] and by Berkowitz and Walter [21-23]. The unit may consist of a glass container for the dielectric fluid and the electrode-charge apparatus which is supported in the cell at some distance from the bottom. The electrodes are well separated and the space between the electrodes is filled with pieces of charge, 5-10mm in diameter, of the same material as the electrodes. The power source is connected to the two electrodes. In operation, the cell is shaken when power is applied to prevent welding of the charge and to continually provide gaps between the pieces of charge and between the charge and the electrodes. Sparks may occur simultaneously for a single voltage pulse. The particles drift to the bottom of the cell to be collected either later or during the process. With a cell about 10cm in diameter, powder production rates are about two orders of magnitude over that obtained using an EDM apparatus. The cell is easily adapted to the use of almost any type of dielectric liquid-organic, aqueous, or cryogenic. Both pulse generators and relaxation oscillators may be used to power spark erosion for powder production. Relaxation oscillators are capacitors which are charged from a D.C. source through a charging resistor and discharged across the load, i.e., the electrodes and the charge. The advantages of the relaxation oscillator are simplicity, economy, and the ability to deliver high peak currents.

Some examples of production rate and particle size distribution are presented for the case of an iron-base alloy which was spark eroded in the electrode-charge cell in distilled water using a relaxation oscillator. Less than 2% of the particles were greater than 75µm in diameter. Operating at 600V and 50uf, production rates of 400gm/hr of powder less than 75µm were obtained. Some 65% of the powder was less than 30µm in diameter and 45% was less than 20µm in diameter.

Spark erosion may also be utilized to produce alloys and compounds whose compositions are different from those of the electrodes/charge by reacting the eroded material with the dielectric liquid. A number of carbides [24] and oxides have been prepared because organic and aqueous dielectrics are easy first choices. If vaporized material is ejected, intimate atomic mixing can occur without considerations of diffusion lengths, concentration gradients, etc. [23].

References

1. T. Svedberg, "Colloid Chemistry", Part I, Chemical Catalog Co., New York, 1924.
2. D.W. Rudorff, Proc. Inst. Mech. Eng., 171, p.495, 1957.
3. B.R. Lazarenko and N.I. Lazarenko, Stanki Instrum., 17, p.8, 1946.
4. T. Yamaguchi and K. Narita, IEEE Trans. Magn., MAG-13, p.1621, 1977.
5. A.E. Berkowitz and J.L. Walter, Rapid Solidification Processing, II, (eds.) R. Mehrabian, B.H. Kear, and M. Cohen, Claitor's Publication Division, Baton Rouge, LA, p.294, 1980.
6. A.E. Berkowitz, J.L. Walter and K.F. Wall, Phys. Rev. Lett., 46, p.1484, 1981.
7. S. Aur, T. Egami, A.E. Berkowitz and J.L. Walter, Phys. Rev. B, 26, p.6355, 1982.
8. A.E. Berkowitz and J.L. Walter, Mater. Sci. Eng., 55, p.275, 1982.
9. J.L. Walter, A.E. Berkowitz, and E.F. Koch, Mater. Sci. Eng., 60, p.31, 1983.
10. J.L. Walter, and A.E. Berkowitz, Mater. Sci. Eng., 67, p.169, 1984.
11. W. Ishibashi, U.S. Patent #3355279, issued Nov. 28, 1967.
12. W. Ishibashi, Funtai Oyobi Funmatsuyakin, Jour. Jap. Soc. of Powder and Powder Metallurgy, 24, p.107, 1977, in Japanese.
13. W. Ishibashi, ibid 24, p.113, 1977, in Japanese.
14. H. Ruppersberg and H.J. Bold, Metall, 26, p.34, 1972.
15. S.F. Cogan, J.E. Rockwell III, F.H. Cocks, and M.L. Shepard, J. Phys. E.: Sci. Instrum. Vol. II., 1978.
16. M. Enokizono and K. Narita, Jpn. J. Appl. Phys., 20, p.2421, 1981.
17. 0. Pisarenko and M.A. Lunina, Russian Journal of Physical Chem., 51, p.1197, 1977.
18. A.E. Berkowitz and J.L. Walter, J. Magn. Magn. Mat., 39, p.75, 1983.
19. I.M. Crichton, J.A. McGeough, W. Munro, and C. White, Precis. Eng., 3, p.155, 1981.
20. H. Tsuchiya, T. Inoue, and Y. Mori, Proc. 7th Int. Conf. on Electro-Machining, ed., J.R. Crookall, IFS Publ. Ltd. and North Holland Pub. Co., p.107, 1983.
21. A.E. Berkowitz, J.D. Livingston, and J.L. Walter, J. Appl. Phys., 55, p.2106, 1984.
22. A.E. Berkowitz and J.L. Walter, J. Mater. Res., 2, p.277, 1987.
23. J.L. Walter and A.E. Berkowitz, Mat. Res. Soc. Symp., vol.80, p.179, 1987.
24. D. Ayers and K. Moore, Met. Trans., 15A, p.1117, 1984.

Section 2

Powder Characterization

5

Particle Characterization: Size and Morphology

M. GHADIRI, F. A. FARHADPOUR, R. CLIFT AND J. P. K. SEVILLE

Department of Chemical and Process Engineering, University of Surrey, Guildford, Surrey, UK

Abstract

The electrozone, sedimentation, image analysis and field scattering techniques for the characterization of size and shape of fine particulate solids are reviewed. For non-spherical particles, these techniques yield different measures of particle size. A method is presented for relating the various measures of particle size by taking account of particle shape, and by analysing the principles of operation of several most commonly used instruments.

For light scattering instruments, in addition to the particle shape, particle size is influenced by the optical properties of the particles. Some of the difficulties involved in the determination of particle size distribution by this technique are also reviewed.

1. Introduction

In recent years, interest in the characterization of fine particles has grown rapidly in fields which include environmental protection, industrial hygiene, and powder technology where fine particles have always been of primary concern. However, this interest now extends, for example, into manufacturing of advanced materials, where it is now well-known that particle size, shape and orientation profoundly influence the characteristics of the final products, while these properties in turn depend strongly on the processes to which the particles have been subjected.[1,2] For example, in fibre-reinforced composites the fibre size, aspect ratio and orientation govern not only the strength of the final composite product, but also the rheology of suspensions in processing and handling of the intermediate materials [1,3]; in the manufacturing of advanced ceramics, the number and size of agglomerates influence the ultimate strength and reliability of the ceramic products.[2]

The major techniques for the characterization of particle size and morphology are based on sieving, image analysis, sedimentation, inertial separation, electrical and optical sensing of single particles, light diffraction and photon correlation spectroscopy. The technique of light extinction is widely used, for example as a process diagnostic tool, and is discussed by various authors, e.g. Hinds [4]. However, this technique does not determine size distribution; at best, it gives a mean particle size if the concentration is known from an independent measurement. Light extinction is therefore not discussed in detail here.

It is not the intention here to review all the available instruments and techniques for particle characterization as there is a vast amount of work reported in the literature, and for which comprehensive and recent reviews are available.[4-10] Rather, our aim is to provide a basis for relating various measures of particle size by an analysis of the principles of operation of several commonly used instruments, and by taking account of the particle shape. The techniques which are examined in some detail are sizing by sedimentation, inertial motion, image analysis and light diffraction. A list of some of the instruments based on these techniques is given in Table 1. There has been a substantial increase in the number of devices in recent years and, although the list is extensive, it is not meant to be comprehensive. The emphasis here is on non-spherical particles and on the fine end of the particle size distribution, typically in the range 0.1 to 20μm which includes the "respirable" fraction.

For non-spherical particles, there is obviously no single definition of diameter and the most appropriate definition depends on the processes to which the particles are subjected.[8] However, quite often it may not be possible to measure this directly, and one of the problems in characterizing the size distribution of particulate materials lies in establishing how different "measures" of particle size obtained by different techniques are related. The comparison of various techniques is obviously the first step along this line, and there is a large body of work in which the results of size analyses of a particulate material by a variety of techniques have been reported, and in some cases accompanied by interpretations of the similarities or differences between the various measuring techniques.[11-13] The main difficulty with comparative work is that it is usually only applicable to the particular material under test, and the conclusions cannot be extended even to the same material but with different size distributions. This is illustrated by a recent and very extensive comparative

Table 1

Technique	Examples of Instruments	Diameter measured	Approximate Size Range/μm	Medium
Image Analysis: using optical or scanning electron microscopy	•Cambridge Instrument Quantimet 920 •Joyce-Loebl Magiscan •Kontron IBAS system	d_A	Depends on the resolution of the input peripheral	Liquid gas
Sedimentation:				
1. gravity	•Andreasen pipette •Micromeritics X-ray sedigraph •Quantachrome X-ray Microscan	d_s	2 - 100	liquid
2. centrifugal	•Ladal pipette centrifuge •Joyce-Loebl disc photocentrifuge •Brookhaven disc centrifugal sedimentometer •Horiba cuvet photocentrifuge •Shimadzu SA-CP3 centrifugal particle size analyser		0.01 - 60	liquid
Inertial separation	•Anderson cascade impactor •BCURA cyclone sampling train •TSI Aerodynamic particle sizer	d_I	0.3 - 30 0.3 - 30	gas gas gas
Electrical sensing zone method	•Coulter Counter •Particle Data Elzone 180 series	d_V	0.4 - 800	liquid
Light Scattering				
1. Single particle	•HIAC/Royco •Climet •PMS Airborne Optical Particle Counter •Brinkmann Particle Size Analyser •Polytec Particle Size Analyser	see text	0.2 - 20 0.7 - 1200 in several ranges	gas gas gas liquid
2. Multi-particle/ diffraction	•Malvern 3600E •Malvern Mastersizer •Sympatec Helos •Microtrac Small Particle Analyser •Microtrac Standard Range Analyser •Shimadzu SALD - 1100 •Leeds and Northrop		All instruments here are capable of measuring from around 0.1 μm to about 2000 μm in several ranges. The upper limit varies greatly between manufacturers.	gas/ liquid

Table 2 BCR reference materials

Material	Size Range, μm	dp, μm*	σ_g *	ρ_p, kg m^{-3}
BCR 66	0.35 - 2.5	1.14	1.75	2620
BCR 67	3 - 20	10.50	1.76	2650
BCR 69	12 - 90	38.50	1.82	2650
BCR 70	0.5 - 12	3.09	2.09	2640

* Estimated from size frequency plots by Allen and Davies [14]

work by Allen and Davies.[14] They carried out instrument evaluation tests using four BCR powdered quartz reference materials whose size range and density are given in Table 2. These powders are generally used for inter-laboratory control purposes and their size distributions are characterized by gravity sedimentation[15] so that the parameters reported in Table 2 are the Stokes diameters (see below).

The results of their evaluation tests are shown in Fig. 1. For each material there are two plots, one showing the mean deviation from the certified size distribution, and the other showing the reproducibility of the instrument, expressed in standard deviations. The mean deviation is defined as the average of the difference between the measured and standard size (in percentage) over the size distribution in the range of the standard percentiles 10-90%; because of lack of statistical reliability at the extremes of the distributions, the differences in these regions were not included. Where possible, Allen and Davies obtained a measure of reproducibility by analysing the same material six times and by calculating the average standard deviation over the full size range. These evaluation tests are the most comprehensive reported to date and are invaluable in elucidating the capabilities of the instruments most commonly used at present. The method of presentation of the results in the form of mean deviation and reproducibility is particularly useful for an overall performance index of the instruments.

Detailed features of the performance of each instrument with individual powders are given by Allen and Davies.[14] Some general conclusions are however worth mentioning here. It is expected that sedimentation techniques produce least deviation as they are based on the same principles as the BCR method for size certification, and this is generally the case but with a few exceptions. Of the Electrical Sensing Zone (ESZ) devices, the Coulter Counter performs generally better than the Elzone, and also produces very small deviations similar to the sedimentation techniques. This is probably a consequence of particle shape and is discussed further below. The reproducibility of sedimentation and ESZ techniques is generally very low. Instruments based on light diffraction show very large deviations from the standard size but also very high reproducibility. This is particularly the case with the Malvern Mastersizer. The mean deviation varies greatly between the instruments in this class for the same powder. This is an important feature which depends to a large extent on the method of calculation of size distribution of powder from the angular variation of the intensity of the scattered light, and is analysed in detail below in the section on light scattering.

The most salient feature of these results is that the performance of the instruments, i.e. both the deviation and reproducibility, varies greatly and in no predictable way between the four samples. As these samples are of the same material, but with different size distributions, it is clear that no generalization of the instrument performance can be made for other materials or even for other size distributions. This clearly highlights the difficulties in characterizing non-spherical particles.

A feature which has not been considered in the above analysis is particle shape. For any particular size range, the deviation from the standard size is essentially due to the particle shape. Because the shape can vary with particle size depending on the method of classification, the variation in the mean deviation between various size distributions could also be due to particle shape. It is therefore most desirable to determine a particle shape so as to relate different measures of particle size, and to provide a basis for tests of mutual consistency between different techniques.

A full analysis of the operation of individual instruments is outside the scope of this article. Moreover, the operation of commercial instruments currently in use is often controlled by microcomputers whose software has been developed by the manufacturers. This puts the user at the great disadvantage of not knowing exactly how the size distribution is determined. Therefore, in this article the principles of operation of a number of the most common techniques are reviewed, and methodologies are presented to relate various measures of particle size.

2. Definition and Measurement of Particle Diameter

For irregular particulate materials the definition of particle diameter must be specified. The following diameters are dealt with in this work.

2.1 Volume - Equivalent Diameter, d_v

This is defined as the diameter of the sphere with the same volume as the particle:

$$d_v = (6V/\pi)^{1/3} \tag{1}$$

The electrozone method, of which the Coulter Counter is a well-known example, measures essentially the volume-equivalent diameter.[16,17] It requires the particles to be suspended in an electrolyte at low concentrations. The suspension is drawn through a small orifice in an insulating wall having an immersed electrode on either side and across which a constant electric current passes. As each particle passes through the orifice it displaces its own volume of electrolyte, thus changing the resistance across the orifice. As the current across the orifice is kept constant, this produces a voltage pulse which is proportional to the particle volume. The pulses are measured electronically by a pulse height analyser, from which a distribution of particle number as a function of particle volume is calculated.

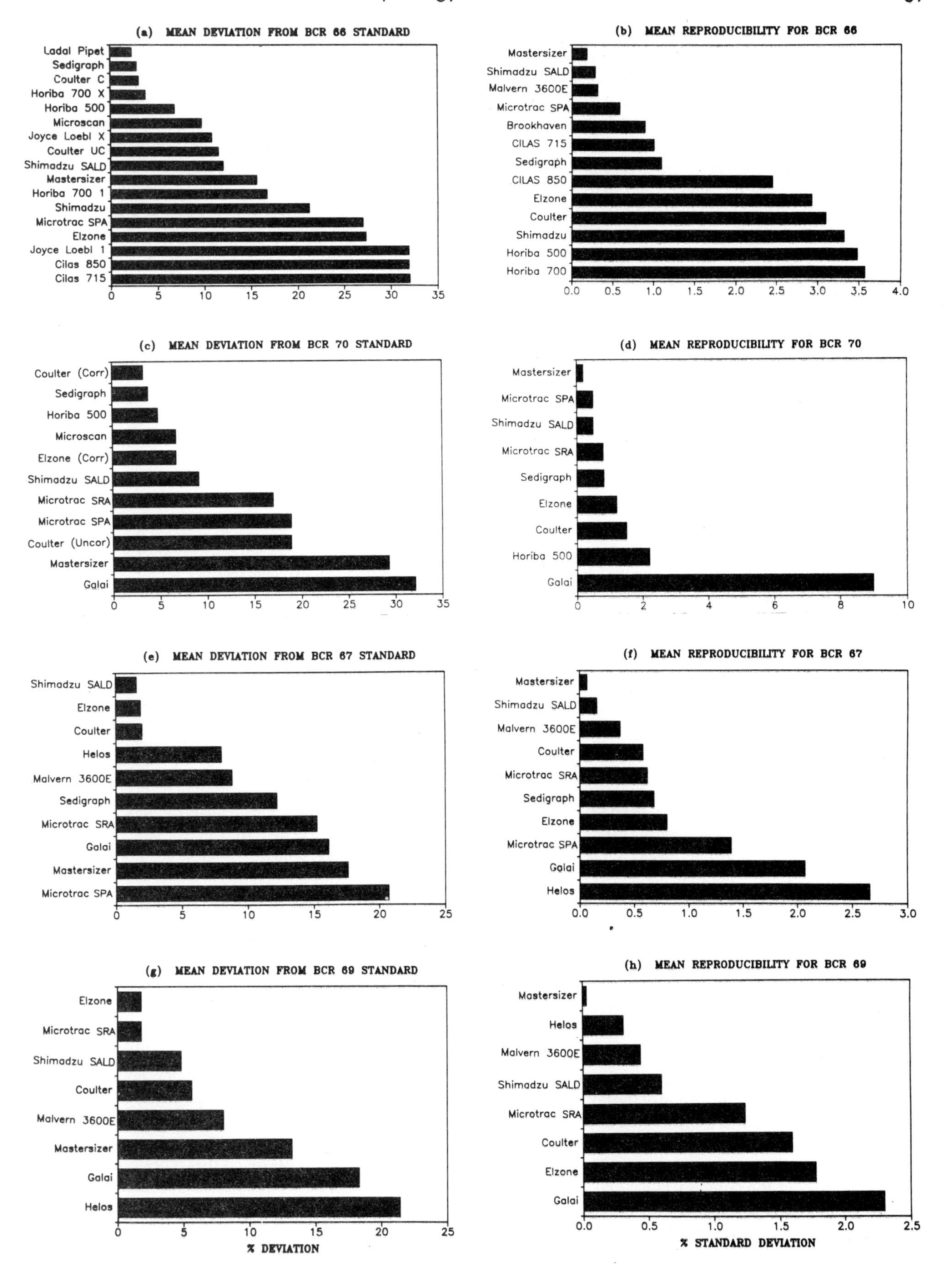

Figure 1 Allen and Davies' evaluation of several size analysers.

For moderately conductive particles, it may be necessary to increase the conductivity of the electrolyte to ensure a reliable measurement of particle size.[18-21] It is also important to select the appropriate orifice size for the size range to be measured; the response is generally linear over the size range of 1.5% to 60% of the orifice diameter. For powders with wide size distribution, it is necessary to make measurements with several different orifice sizes. Furthermore, the lower limit of measurable size is about 0.4μm, so that the applicability of the technique may be limited for powders with a substantial submicron fraction. Allen and Davies [14] recommend making a mass balance in all cases to reduce errors associated with resolution and inaccurate calibration. Some errors are also expected with particles of extreme shapes.[9] Particles in gases must be sampled and redispersed in an electrolyte solution first, and this process introduces its own problems which are discussed further below. However, provided reasonable care and attention are paid to the preparation of the suspension and to the principle of operation and limitation of the technique, reliable and reproducible measurements can be obtained for particles of arbitrary shape.[11-14]

2.2 Projected-Area Diameter, d_A

This is defined as the diameter of a circle with the same projected area as the particle:

$$d_A = (4A/\pi)^{1/2} \tag{2}$$

In general, d_A may refer to a preferred particle orientation, or to random orientation. In the latter case, and for convex particles only, the average projected area, A, is given by:

$$\bar{A} = S/4 \tag{3}$$

where S is the particle surface area.[22]

d_A is conveniently measured by image analysis equipment, and nowadays with the use of microcomputers a large number of particles can be measured automatically. In this technique, a two-dimensional projection of the particle is examined by optical or scanning electron microscopy. The information is usually analysed by an automatic image analyser, of which the Quantimet 920 is a good example. This technique is particularly suitable for shape characterization and a host of other parameters related to the shape can be determined (see below). Because the information obtained by this technique is based on a two-dimensional projection, the orientation of particles matters greatly, and this depends mainly on sample preparation. The analysis later in this article relies on estimates of d_A based on the maximum projected area of the particle, i.e. obtained for a particle lying in its plane of maximum stability on a flat surface and viewed normal to the surface. In general a particle in a moving fluid tends to orient itself to present its maximum cross-section to the flow.[23] Ghadiri [19] has shown that for particles dispersed in liquids it is possible to obtain the maximum d_A by depositing the particles on a flat membrane filter by drawing the suspending liquid through it. This procedure was found unsatisfactory for obtaining representative samples from a dispersion in a gas because of the non-uniform deposition of the particles on the filter. However, a slightly different approach has been shown to give deposits which are representative of the dust sampled from a gas stream and present their maximum projected area for image analysis: this is by the use of a pulse-charging pulse-precipitating electrostatic precipitator developed by Liu et al. [24], where a sample gas stream is passed through the device to produce a uniform dust deposit on the electrode which is suitable for image analysis. Sampling forms a very important and integral part of particle characterization,[8] but detailed discussion of sampling is outside the scope of this article.

In principle, light diffraction in the Fraunhofer range also gives the projected area for particles in random orientation and this is discussed further below. Similarly, light extinction can give the same measure of particle size provided the particle concentration is known. The measurement obtained in this way is, however, an average value and is therefore more representative for narrow size distributions. [For further details see Ref. 4.]

In addition to particle size, image analysis can be used for morphological characterization, and this can range from simple determination of circularity (the ratio of perimeter to projected area) to the more sophisticated Fourier [25], Delta [26] and Fractal [27] analyses.

For Fourier analysis, radii to the particle boundary are measured from a central point at preset angles. Fourier analysis of these data enables determination of the Fourier coefficients which are regarded on shape descriptors.[25] For example, the third harmonic is taken as a measure of particle triangularity and the fourth harmonic of squareness. This technique is really suited to rounded particles where subtle changes in particle shape can easily be detected, as it cannot deal readily with rugged or re-entrant particle shapes.[28] However, more recently Weichert [29] found that representation of the particle contour in the complex plane and use of the Digital Fourier Transform overcame these limitations. A new roundness parameter, computed from the complex Fourier coefficients, was found to describe the particle shape and to be independent of particle size, position and orientation.

Delta analysis is a relatively new development for measuring surface angularity on a fine scale. It is again based on the two-dimensional projection of the particle, and operates on a digitized outline of the particle image, where the angle between two vectors which traverse the particle boundary is measured.[26] A smooth curve gives a small angle, a rough one a large angle, so that a distribution of the angle size over the boundary gives an indication of surface roughness.

The Fractal analysis provides an indication of the surface roughness by taking a "structured walk" over the particle boundary at a preset length from which the perimeter is estimated from the total number of steps. The step length is reduced and the procedure repeated. An increase in the perimeter measured in this way as the step length is reduced is an indication of the extent of surface roughness. The calculation is continued until the perimeter reaches an asymptotic value from which the fractal dimension is calculated, representing the particle ruggedness.[27] The use of fractal analysis is spreading fast because of several features: it provides a measure of roughness over a range of scales, it does not have the limitation of the Fourier and Delta analyses, and under certain circumstances it can represent the roughness by a single parameter.

2.3 Stokes Diameter, d_s

The Stokes diameter, d_s, is defined as the diameter of a sphere with the same density and settling velocity as the particle in the same fluid, assuming Stokes' Law to be applicable to the sphere. For a particle of density ρ_p settling in a fluid of viscosity μ:

$$d_s = [18\mu\, U_t / C_s g(\rho_p - \rho_f)]^{1/2} \tag{4}$$

where U_t is the particle terminal velocity, C_s is the Cunningham slip correction factor for the sphere and ρ_p is the fluid density. Two closely related dimensions are the aerodynamic and inertial diameters. The aerodynamic diameter, d_a, is defined as the diameter of a sphere of standard density $\rho_{po} = 10^3 \text{kg m}^{-3}$, and slip correction factor C_a, which has the same settling velocity as the particle. In the Stokes regime:

$$d_a = [18\mu U_t/C_a g(\rho_p - \rho_f)]^{1/2} = d_s[C_s(\rho_p - \rho_f)/C_a(\rho_{po} - \rho_f)]^{1/2} \cong d_s[\rho_p/\rho_{po}]^{1/2} \tag{5}$$

The inertial diameter, d_I, is defined as the diameter of the sphere of the same density, ρ_p, and with the same inertial behaviour as the particle. Its relation to the Stokes diameter depends on particle and fluid flow conditions. It has been shown that d_I is equivalent to d_s under unsteady motion through a gas at low particle Reynolds number.[30] This connection however does not extend to particles in liquids, because the virtual mass and history terms in the equations of motion become significant or even dominant.[23,31] This is discussed further below.

For a non-spherical particle, the value of U_t and hence d_s depends on orientation. However, in simple sedimentation and in some inertial motions, Brownian motion "randomizes" orientation, so that the mean value of d_s is the mean over all possible orientations but measured values of d_s for non-spherical particles include a variance due to orientation. This is considered further below. Because d_s describes both sedimentation and inertial behaviour of particles (see below), it is frequently the most useful and commonly measured particle diameter. Consequently, there is a very wide range of instruments and methods available for its measurement. The simplest instrument for particles in suspension in a liquid is the manually operated Andreasen pipette in which the sedimentation rate is measured gravimetrically. More modern instruments use electromagnetic waves for sensing the concentration changes, and in order to speed up the operation the sensor may even be moved relative to the sedimentation vessel to scan the full length of the vessel. The lower size limit for gravity sedimentation is typically around 2μm and, to extend this limit, centrifugal sedimentation has to be used. However, in this method the particles are continuously accelerated in liquids, and, as will be seen below, this greatly complicates the procedure for deduction of particle size. In addition, measurement by sedimentation requires particles to be dispersed in a liquid, and this introduces problems concerning proper dispersion of the particles, especially in the high concentrations normally required for sedimentation measurements.

For particles dispersed in gases there are several devices currently available which are listed in Table 1. Inertial impaction is widely used, where the particles are classified by inertia and deposited on collector plates. An unavoidable problem with such measurements is that the impacting particles may rebound or be re-entrained from the collection plate. In general this leads to undersizing and low reproducibility.[30,32,33] This problem can be alleviated to some extent by coating the stage with an adhesive compound.[10] Cyclones with a sharp cut-off have been used and recently an instrument using an automated 3-cyclone train has been developed.[34] However, this remains a sampling device with relatively slow data acquisition. The TSI Aerodynamic Particle Sizer (APS) differs from the other instruments in measuring the inertial behaviour of individual particles by accelerating the gas through a nozzle and measuring the slip velocity of each particle relative to the gas as it issues from the nozzle.[35] This technique has been shown to be unreliable for non-spherical particles.[11,36] For plate-like particles, the APS undersizes significantly,[36] while for shapes approximated by cuboids with moderate aspect ratios, undersizing by up to 60% has been reported.[11] Seville et al.[11] argue that, in addition to possible effects of particle orientation, examined further below, the fluid flow conditions in the nozzle are such that particles can rotate and tumble; hence the diameter measured may be significantly different from d_s measured for steady motion with random orientation in the Stokes regime.

The above instruments all operate on sampled particles or particle-laden flows. Sampling and redispersion are therefore important steps prior to the size analysis. In addition to the usual difficulty of ensuring that the samples are representative, it is necessary to ensure that the state of agglomeration is unchanged. Some techniques, such as the ESZ, require very dilute hydrosols so that it is usually possible to disperse particles completely, for example with the aid of ultrasonic vibration and appropriate surface active agents. The danger is then that, for powders containing agglomerates, the agglomerates may be broken up in the liquid so that erroneously low sizes are measured. Sedimentation methods require more concentrated dispersions, and agglomeration in the liquid phase can occur leading to spuriously large apparent sizes.[30] It is therefore good practice to verify that the state of dispersion is the same in the original sample and the redispersed one. This is more critical for gas-borne particles, and may be done for example by collecting samples from a gas onto a microscope slide or SEM stubs using a sampling electrostatic precipitator,[24] or from a liquid onto a membrane filter, and comparing the state of dispersion by image analysis.

For particles entrained in gases, Tate *et al.*[37] propose a method in which d_s can be measured *in situ* and without sampling. The device is like a pitot tube, whose upstream tip forms a bluff body. It is inserted into the particle-laden gas stream with its axis parallel to the flow so that the stagnation streamline passes through the centre of the bluff tip. The particle velocity is measured very near the stagnation point by a fibre-optic laser Doppler anemometer. As the gas velocity drops to zero at the stagnation point, the difference between the particle velocity at the measurement point and the free-stream gas velocity approaching the bluff body is a function of the particle inertia, from which the particle size is determined. In addition to general applications, this device is potentially useful for detecting "surges" in particle size or concentration for use in gas cleaning equipment, protection of turbomachinery, and on-line automatic control of grinding and classification processes.

It is clear from the foregoing discussion that the measurement of d_s is not as straightforward as it first appears. This is supported by the experimental results of Allen and Davies [14] where greatest variations in reproducibility and in results between instruments are experienced with the sedimentation techniques. Ghadiri *et al.*[30] suggest that a more reliable estimate of the Stokes diameter may be obtained by approximating the particle to a well-defined shape whose hydrodynamic drag can be calculated. For this purpose, measurement of d_V and d_A is sufficient for shape characterization. This approach is based on an analysis of steady and unsteady motion of particles in fluids and is discussed further below.

3. Principles of Sizing by Particle Motion in Fluids

The Stokes diameter is based on steady settling of particles, but this condition is not ensured in centrifugal and inertial devices. In order to elucidate possible sources of error and to develop relationships between d_s and other diameters such as d_V, d_A and d_I, using some form of shape characterization, it is necessary to examine steady and unsteady motion of a non-spherical particle in a fluid.

3.1 Steady Settling

Consider a non-spherical particle moving through a fluid of density ρ_f and viscosity μ at velocity $\underline{U}$ in creeping flow, i.e. with $Re_p<1$ where Re_p is the particle Reynolds number:

$$Re_p = \rho_f d_S U/\mu \tag{6}$$

Many classes of particle shape possess three mutually orthogonal axes, known as the "principal axes of translation", such that the drag is parallel to $\underline{U}$ if $\underline{U}$ is steady and parallel to one of these axes.[38] For simple geometries, these axes correspond to obvious axes of symmetry.[23] Unit vectors parallel to the three principal axes will be denoted $\underline{i}, \underline{j}, \underline{k}$. If $\underline{U}$ is parallel to $\underline{i}$, then the drag is in the direction of $-\underline{U}$ and given by:

$$\underline{F}_d = -\mu c_1 U \underline{i} \tag{7}$$

where c_i is one of the "principal resistances" of the body. In general, the drag force can be written as:

$$\underline{F}_d = -\mu[c_1 U_1 \underline{i} + c_2 U_2 \underline{j} + c_3 U_3 \underline{k}] \tag{8}$$

where U_1, U_2, U_3 are the components of $\underline{U}$ parallel to $\underline{i}, \underline{j}, \underline{k}$. In general, the principal resistances c_1, c_2, c_3 are not equal, so that $\underline{F}_d$ is not in the direction $-\underline{U}$. It follows that a particle settling through a fluid in creeping flow does not move on a vertical path unless one of the principal axes of translation happens to be vertical. For example, if $\underline{i}$ happens to be vertical, then at the terminal velocity U_t the drag force counter balances the immersed weight and

$$F_d = -gV(\rho_p - \rho_f) = \mu c_1 U_t \tag{9}$$

so that

$$U_t = gV(\rho_p - \rho_f)/\mu c_1 \tag{10}$$

Otherwise, the terminal velocity contains a horizontal component. If a large number of particles are settling through the fluid, the orientations of the particles will usually be random (see above). Under these circumstances, the mean hydrodynamic resistance of the particles, $\bar{c}$, is given by the harmonic mean of the three principal resistances:[23,38]

$$\bar{c} = 3/(c_1^{-1} + c_2^{-1} + c_3^{-1}) \tag{11}$$

and the mean vertical component of the steady settling velocity is

$$U_t = gV(\rho_p - \rho_f)\mu\bar{c} \tag{12}$$

Equation 12 describes particles around which the fluid can be treated as a continuum, i.e. particles with diameter large by comparison with the mean free path of molecules of the fluid. Particles small enough for non-continuum (or "slip") effects to be significant experience reduced drag so that

$$U_t = CgV(\rho_p - \rho_f)/\mu\bar{c} \tag{13}$$

For a sphere, $c_1 = c_2 = c_3 = \bar{c} = 3\pi d$, $V = \pi d^3/6$, and $C = C_s$, so that equation 4 follows. For a non-spherical particle, we still have $V = \pi d_v^3/6$ from equation (1) and so

$$U_t = \pi C d^3{}_v g(\rho_p - \rho_f)/6\mu c \tag{14}$$

Therefore, from equations (4) and (13),

$$d_s = \left[\frac{C}{C_s}\,\frac{18V}{\bar{c}}\right]^{1/2} \tag{15}$$

This result is the basis of the "model shape" approach for relating d_s and dv, outlined below.

The slip correction factor is a function of particle size, and more weakly, shape and orientation.[23] Hence, in general, $C \neq C_s$. However, for particles in liquids, and for particles in gases under conditions of most practical interest, C is close to unity.[39] Therefore C/C_s may be taken as unity and equation (15) may be simplified to

$$d_s = \left[18V/\bar{c}\right]^{1/2} \tag{16}$$

and using equation (1)

$$d_s/d_v = \left[3\mu d_v/\bar{c}\right]^{1/2} \tag{17}$$

Even for extreme shapes such as a disc there is no significant variation of C with orientation,[31] so that C/C_s may still be taken as unity and equations (16) and (17) are good approximations.

3.2 Inertial Motion

Consider now a particle in unsteady motion through the fluid. For a sphere, the instantaneous drag in creeping continuum flow is given by the Basset-Boussinesq-Oseen equation:

$$-F_d = 3\pi\mu dU + \frac{\rho_f V}{2}\frac{dU}{dt} + \frac{3d^2}{2}\sqrt{\pi\rho_f\mu}\int_{-\infty}^{t}\left(\frac{dU}{dt}\right)_{t=s}\frac{ds}{\sqrt{t-s}} \tag{18}$$

where s is a time variable so that the last term contains an integral over past accelerations of the particle.[23] This term is usually known as the "Basset term" or the "history term". The penultimate term, dependent on the instantaneous particle acceleration, represents the "added mass" or "virtual mass" of fluid which, in effect, moves with the sphere. The first term on the right of equation 18 represents the familiar steady Stokes drag, equal to that in steady motion at the instantaneous velocity. For most motions of particles in gases, for which $\rho f \ll \rho_p$, the added mass and history terms in equation 18 can be neglected; in effect, the relaxation time of the fluid is short by comparison with that of the particle, so that the instantaneous drag is almost identical with the quasi-steady drag. However, this is not generally true for particles in liquids, where both virtual mass and history effects are significant. Analytically, the virtual mass term is reasonably tractable. Unfortunately, the less tractable and less familiar history term is normally the more significant of the two, [23] so that there is no point in including virtual mass but neglecting history. Calculating the unsteady motion of particles in liquids is therefore complex.

Returning to the simpler case of a particle in a gas under conditions such that the "quasi-steady drag" approximation can be made and taking the slip factor as unity, consider unsteady motion through a stagnant fluid. If the instantaneous velocity is $\underline{U}$ then equation 8 applies and the equation of motion for the particle is

$$\rho_p V \frac{d\underline{U}}{dt} = -\mu\left[c_1U_1\underline{i} + c_2U_2\underline{j} + c_3U_3\underline{k}\right] \qquad (19)$$

Because the principal resistances are not equal, the three components of $\underline{U}$ will in general decay at different rates. Therefore, in general, unsteady motion is not rectilinear. However, where $\underline{U}$ is in the direction of one of the principal axes, say $\underline{i}$, then the motion will be rectilinear for the same range of shapes as move rectilinearly in steady settling. Then

$$\rho_p V \frac{dU}{dt} = -\mu c_1 U \qquad (20)$$

For the specific case of a particle projected at velocity U_o into a stagnant gas, equation (20) leads to

$$U = U_o \exp\left[\frac{-\mu c_1 t}{\rho V}\right] \qquad (21)$$

Hence the distance travelled before coming to rest is

$$Z = \int_0^\infty U dt = \rho_p V U_o / c_1 = \pi d_V^3 \rho_p U_o / 6 c_1 \mu \qquad (22)$$

This "stopping distance" is a measure of the inertia of the particle in the fluid. Thus the inertial diameter of the particle is defined as the diameter of the sphere with the same stopping distance. For a sphere of diameter d_I, equation (22) becomes

$$Z = d_I^2 \rho_p U_o / 18\mu \qquad (23)$$

Thus, for equations (22) and (23), inertial equivalence demands

$$d_I / d_V = \left[3\pi d_V / c_1\right]^{1/2} \qquad (24)$$

or, averaged over all orientations using equation (11),

$$d_I / d_V = \left[3\pi d_V / \overline{c}\right]^{1/2} \qquad (25)$$

Comparing equations (17) and (25) shows that $d_S = d_I$; i.e. the diameters measured by sedimentation and inertial techniques should be identical.

This conclusion is subject to all the assumptions made in reaching it. In particular, inertial behaviour in liquids is quite different. For sedimentation instruments using centrifugal action, the acceleration of the particles is generally assumed to be instantaneous, but the effect of the drag has yet to be fully considered. Furthermore, equivalence of d_s and d_I only applies if the particle orientations are the same - either the same preferred orientation or distributed randomly - in the sedimenting and inertial flows and if Re_p remains in the creeping flow range. This probably explains why one particular inertial device, the TSI "Aerodynamic Particle Sizer" (discussed above), can give unexpected results for non-spherical particles. Particle Reynolds numbers in the device are typically above the creeping flow range, and it is likely that the nozzle flow used in the device gives the particles a preferred orientation.

4. Relationship Between Stokes, Volume and Area Diameters

To relate d_S, d_A and d_V, it is possible to use a "model shape" approach developed by Ghadiri *et al.*,[30] Griffiths *et al.*[36] and Dressel [40]. The basis of this approach is a useful general result due to Hill and Power[41]:

The drag on any particle (in creeping flow) is less than or equal to that on any body which encloses it, and greater than or equal to the drag on any body contained within it.

The general result implies that the drag on an irregular particle can be approximated by that on a body with roughly the same overall proportions. The "model shape" can then be selected as one for which the drag in creeping flow is known. A particularly convenient class of shapes is provided by the spheroids, which are ellipsoids of revolution (Fig. 2). If the axis of symmetry is the minor axis then the spheroid is oblate, whereas prolate spheroids have their major dimension along the axis. The shape of a spheroid is conveniently described by its "aspect ratio", the ratio of the polar to the equatorial diameter

$$\text{i.e. } E = d_p / d_E \qquad (26)$$

Thus $E < 1$ for oblate and $E > 1$ for prolate spheroids. The symmetry of the shape is such that one principal axis of translation is the axis of symmetry, while the other two can be taken as any two orthogonal directions in the equatorial plane. Dressel [40] has extended the treatment to general ellipsoids with three different axes, but the simpler spheroidal shape is an adequate model for most purposes. Oblate spheroids can be used to model platey or lenticular particles, with prolate spheroids as models for elongated or fibrous particles. Thin plate-like particles can be modelled as oblate spheroids with $E \rightarrow 0$ or as thin sharp-edged discs, but the resulting values for the principal resistances are very similar, as expected from Hill and Power's general theorem.[23.30] Similarly fibres can be modelled as very prolate spheroids ($E>>l$) or as long cylinders or "slender bodies",[42] and again the details of the shape assumptions prove to be unimportant.

Returning to equation (17), the ratio of Stokes to volume-equivalent diameter can be calculated if the resistance of the non-spherical particle is known. Explicit analytical results are available for spheroids which give the principal resistances as functions of d_E and E.[23] Therefore, via equation (17), d_S/d_V can be evaluated as a function of E. The analytical results are given by Ghadiri *et al.*,[30] and numerical values for different orientations are shown in Fig. 3.

In random orientation, $d_S \leq d_V$ with equality applying for spheres. This is consistent with a general result that the sphere has the largest average settling velocity in creeping flow of all bodies of a given volume.[43] Furthermore, a spheroid must be very far from spherical for d_S/d_V to be smaller than about 0.8. The extreme simplification that $d_S/d_V < 1$ and is about 0.9 should thus be a reasonable approximation except for fibrous particles.

To relate d_A to d_S and d_V, we note that an oblate spheroid in its plane of greatest stability presents its equatorial area, $\pi d_E^2/4$ to view. Hence d_A and d_E are identical, and so

$$dV/d_A = E^{1/3} \qquad (E<1) \tag{27}$$

Similarly a prolate spheroid presents a projected area $\pi d_E d_p/4 = E\pi d_E^2/4$

Hence

$$d_V/d_A = E^{-1/6} \qquad (E>1) \tag{28}$$

Thus, by using the "model shape" approach and approximating the particles as spheroids, E can be determined from any two of the three diameter measures (d_V, d_S, d_A) and used to predict the third. Alternatively, if all three diameters have been measured, these results can be used to check the consistency of the results. Ghadiri *et al.*[30] have tested this approach on two industrial particulates: a devolatilized coal char ("gasifier fines") and fly ash from a fluidized bed coal tailings combustor. Seville *et al.*[11] used the same materials to evaluate the performance of several commercial particle size measurement devices. For each particulate, the distribution of d_V was measured by the conventional electrozone technique, using a Model ZM Coulter Counter. The distribution of d_A was also determined by image analysis using a Quantimet Image Analysis Computer to count particles deposited from liquid suspension onto a membrane filter, as outlined above. Both of these materials are platey. Therefore they were "modelled" as oblate spheroids, and the aspect ratio estimated from the measured d_V and d_A using equation (27). The distribution of d_S was then estimated for the particles in random orientation and compared with direct measurements obtained by the Andreasen pipette sedimentation method.[44]

Results for the gasifier fines are shown in Fig. 4. As expected, d_A is substantially larger than d_V. However, the cumulative distribution lines are roughly straight and parallel; i.e. the distributions of both d_A and d_V are roughly log-normal with different means but the same geometric standard deviation. This suggests that d_V/d_A is constant over the whole size range, so that a single aspect ratio can be used to characterize the whole distribution. The mean value is $E = 0.196$. Using this value, the distribution of d_S was predicted and is shown in Fig. 4, along with determinations from four replicate Andreasen pipette analyses using settling in water from initial concentrations of order 0.1% v/v. The agreement is satisfactory, supporting the validity of this approach. In this case, the comparison also served as an indication of the appropriate concentration to use in the Andreasen pipette: reproducibility and consistency were good at this low starting concentration, but higher concentrations gave erratic results due to agglomeration or incomplete dispersion.

Figure 4 also shows measurements made with an Anderson personnel dust sampler, which should in principle measure $d_I = d_S$. However, the measurements bear no relationship to the other three determinations. Because the relationship between d_S, d_V and d_A is known and consistent, it is possible to conclude that it is the Anderson results which are at fault. The likely explanation is "rebound" of particles from the impactor stages, as reported by other workers.[32,33]

Corresponding results for fly ash are shown in Fig. 5. From the measured distributions of d_A and d_V, the mean aspect ratio is 0.325. Microscopic examination confirmed that the difference from gasifier fines is at least qualitatively correct: the fly ash is much less platey. Attempts to measure d_S by settling through an aqueous medium gave irreproducible results for this material. By comparison with the expected distribution, it was clear that the fly ash was agglomerating in suspension. Eventually, after trying various organic liquids, more repeatable results were obtained by settling in ethanol,

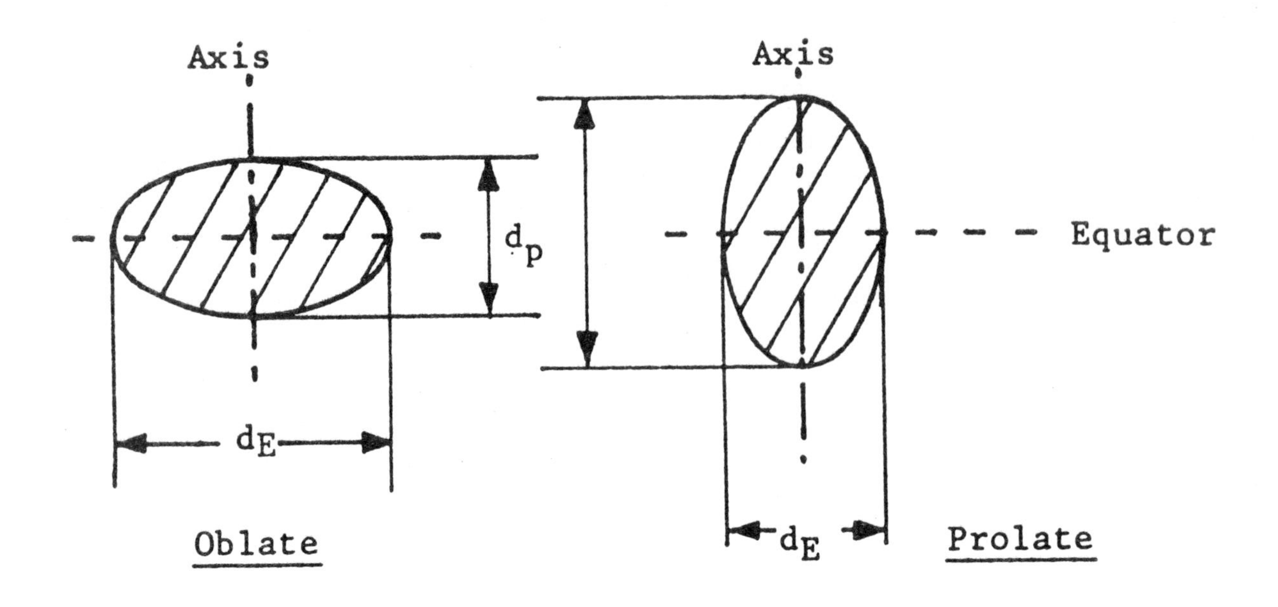

Figure 2 Spheroidal model particles.

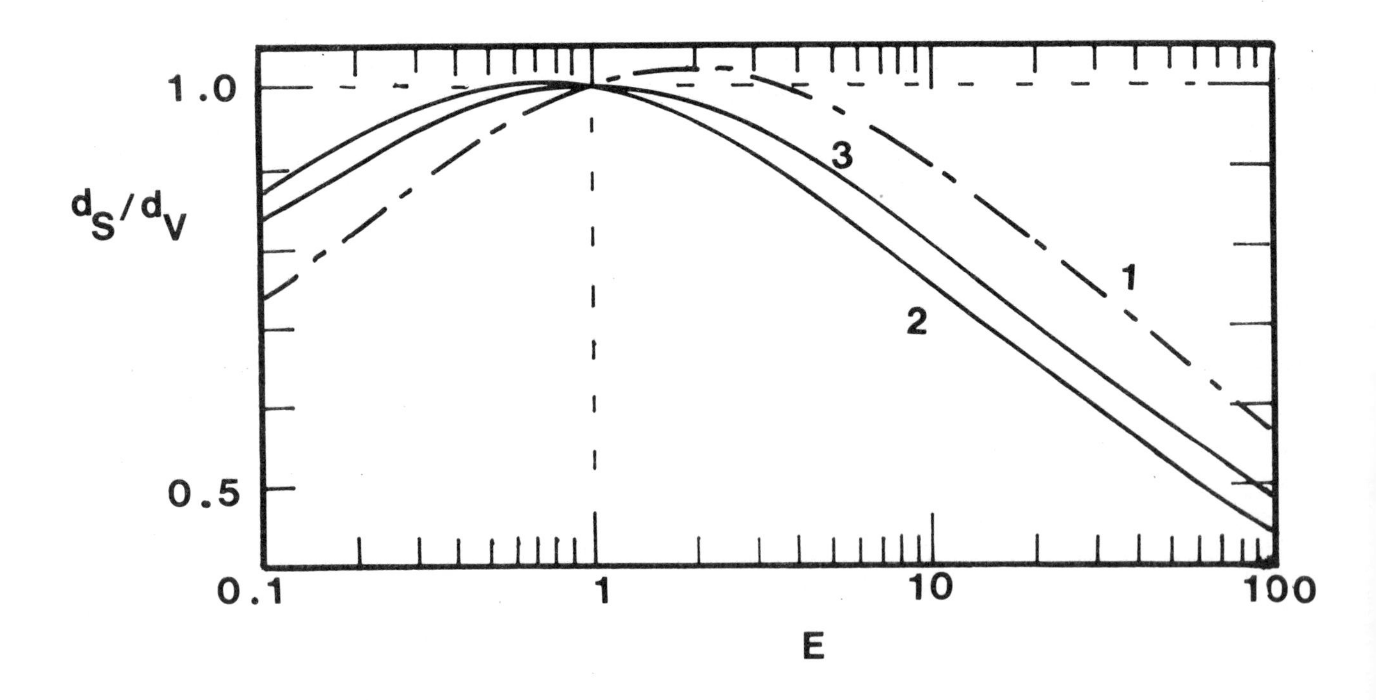

Figure 3 Ratio of Stokes to volume diameter for spheroids.

1. Motion parallel to axis. 2. Motion in plane of equator. 3. Random orientation.

and these are shown in Fig.5.[30] The results are now broadly compatible with the Coulter and Quantimet measurements, but there is evidently still some agglomeration. The difficulties experienced with sedimentation measurements for these particles, and in particular the wide variation obtained with different dispersants and liquid media, suggest that estimating d_S from d_A and d_V could be more reliable than direct sedimentation for particles liable to agglomeration.

5. Principles of Sizing by Light Scattering

Techniques based on light scattering have had the most rapid growth in instrument development in recent years, in spite of the inherent disadvantage that the response depends not only on the particle shape and surface characteristics but also on its optical properties. For this reason, the present section is included to highlight difficulties in determination of particle size distribution, particularly for field scattering instruments, and to outline conditions under which reliable data may be obtained.

The incentive for development of light scattering techniques is obviously their relative ease of use, automation, and capability for on-line size analysis. However, because particle sizing depends on the refractive index and absorbance of the particle,[11,12,45] it is in general not possible to obtain a unique "light scattering diameter" and to relate it to other particle diameters of interest. This can only be achieved for two special cases. In the Fraunhofer diffraction range, where the particle diameter is greater than the wavelength of the incident light, the projected area in random orientation is obtained (see below). In the Rayleigh range, where particle diameter is smaller than the wavelength of the incident light, the equivalent volume diameter is obtained.[46,47] It is clear that, because of these complexities, absolute size measurement by light scattering is not simple, and usually other techniques such as sedimentation or ESZ must be used. However, for on-line size analysis and particularly for detection of changes in particle size, it can be convenient to use light scattering techniques.

Light scattering instruments may be divided into two groups: single particle optical sensing zone in which light scattered by individual particles is measured, and field sensing zone where light scattered from an assemblage of particles is measured. Although both groups share common principles of operation, the method of measurement of light scattering differs between the two groups, and this causes discrepancies in the results for non-spherical particles.[11] In optical sensing zone instruments, the light scattered over all angles in the forward direction is collected and focussed onto a photomultiplier. As a single particle goes through the sensing zone, the scattered light produces an electric pulse whose height is related to the particle size. For non-spherical particles this also depends on the orientation, so that it may cause "spreading" of the observed size distribution whereby particles of the same size may be counted in different channels. An incandescent white light source is usually used to damp out angular oscillations of the scattered light. Furthermore, integration of light over the forward direction reduces the sensitivity to the refractive index, absorbance and departure from sphericity, and overcomes the problem of large angular variations of the scattered light intensity. However, in spite of this, optical sensing zone instruments do not give an accurate measure of particle size in practice because of shape effects and the influence of the optical properties on response of particles in the micron size range.[11]

These difficulties are also present in the field scattering instruments. However, in this case the scattered light energy is measured for an assemblage of particles where random orientation may be assumed. A particular feature of field scattering instruments is that, in order to obtain a size distribution, the angular distribution of the scattered light intensity has to be related to the particle size distribution by light scattering theory. This introduces difficulties due to particle shape, common with the optical sensing zone instruments, compounded by difficulties in the mathematical procedures required to extract the size distribution from the measured data. We concentrate here on field scattering instruments, because they represent a much more recent and popular development than optical sensing zone instruments. The latter is now well-established with a substantial body of literature, [see e.g. Refs. 46, 48 and 49], although the difficulties in predicting the scattered light intensity for non-spherical particles restricts its application.

5.1 Forward Lobe Light Scattering

Instruments based on laser light scattering of an assemblage of dispersed particles, of which the Malvern Mastersizer and Sympatec Helos are good examples, offer a non-intrusive technique particularly suited to on-line size analysis. In most applications the only sample preparation required is dilution of the particle laden stream to avoid multiple scattering effects. As seen above, light scattering instruments are capable of very high reproducibility but for particles in the micron size range usually yield a distribution which differs markedly from that obtained by other techniques.[11,13,14,50,51] These discrepancies can be traced to the theoretical assumptions and numerical procedures employed to extract the size distribution from the distribution of scattered light. The details of the procedures employed are closely guarded proprietary information. Therefore, this section can only provide an introduction to the major difficulties encountered in interpreting light scattering data for micron-sized particles. These include the arbitrary omission of the particle optical properties from the theoretical analysis, the assumption of spherical geometry for light scattered by irregular particles, and the mathematical limitations faced in recovering the size distribution from the measured data. The work described here enables identification of the experimental conditions under which the theoretical analysis can be simplified, and suggests a procedure whereby the size distribution recovered from light scattering instruments can be related to that obtained by other techniques.

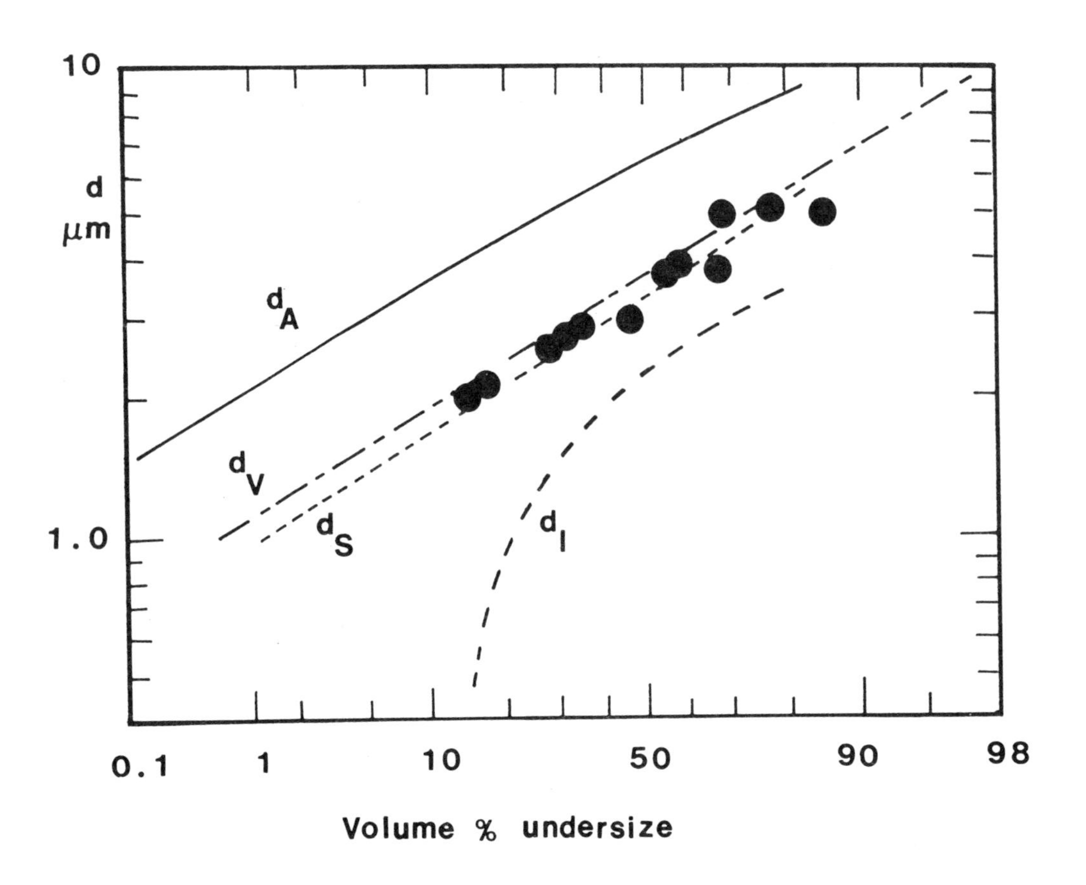

Figure 4 Size distribution of gasifier fines.
• Andreasen Pipette
d_I Anderson Impactor

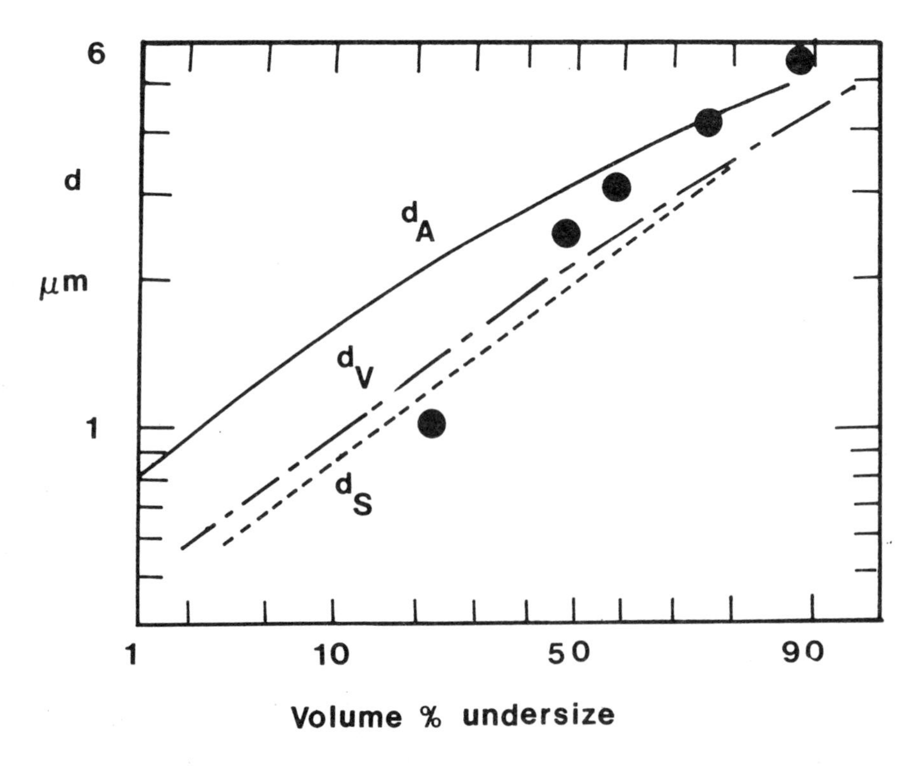

Figure 5 Size distribution of fly ash.
• Andreasen Pipette: sedimentation in ethanol.

A typical instrument consists of a collimated beam, usually from a He-Ne, laser light source ($\lambda = 0.6328\ \mu m$), which is scattered by the particles. The unscattered light is focussed onto the centre of a plane detector placed at the focal point of a suitable Fourier Transform lens, while the scattered light is usually focussed onto a set of (*semi*) circular photomultiplier rings placed at different radii in the detector plane. Each ring effectively captures the light scattered at a particular angle to the incident beam. Each particle scatters light to all radii within the focal plane, but the angular intensity diminishes faster for larger particles.[52] Let the ith ring have inner and outer radii r_1 and r_{i+1} respectively, corresponding to angles θ_i and θ_{i+1} to the incident beam. The total energy falling on this ring is given by the sum of contributions from each particle in the assembly:

$$S_i[\theta_i,\theta_{i+1},m] = \int_{\alpha_{min}}^{\alpha_{max}} \eta[\theta_i,\theta_{i+1},m,\alpha]\, n(\alpha)\, d\alpha + \delta S_i;\ i=1,2,...,N \tag{29}$$

Here m is the refractive index of the particle relative to that of the medium, $\alpha = pd/\lambda$ is a dimensionless size parameter, $n(\alpha)$ is the desired size distribution which is assumed bounded between given limits, $\eta[\theta_i,\theta_{i+1},m,\alpha]$ is the light energy contribution from particles of size α and δS_i is the measurement noise. Equation (29) is only valid in the absence of multiple scattering effects, but these can be neglected if less than 30% of the incident energy is scattered.[48]

The size distribution $n(\alpha)$ must be recovered by matching the theoretical and experimental energy distributions on the detector rings. The first step in the analysis is to adopt a theoretical expression to describe the single particle scattering function η. A rigourous expression which allows for reflection, refraction and diffraction is available in the form of the Mie solution of the Maxwell equations of electromagnetic radiation for spherical particles.[49,52] For other less symmetrical shapes the Maxwell equations must be solved numerically which demands an excessive computation time. The analysis of the scattered energy distribution is therefore normally based on the assumption of sphericity. Excluding particles with strong preferred orientation effects, this turns out to be a reasonable practical assumption for suspensions of irregular particles in random orientation.

The evaluation of the Mie solution for a sphere demands a knowledge of the refractive index of the particle. The refractive index is itself a complex number $m=m_R-jm_I$, where the real part reflects the phase shift and the complex part the light attenuation within the particle.[53] The real part of the refractive index can be measured by standard techniques but the measurement of the complex part for finely divided solids presents considerable difficulties. The majority of commercial instruments therefore adopt the Fraunhofer diffraction approximation to the rigourous Mie solution which does not demand a knowledge of the particle refractive index.

Particles much greater than the wavelength of the incident beam ($d >> \lambda$) can be assumed opaque; the scattering process is then dominated by diffraction of the incident light around the edges of the particle. For a three-dimensional particle, the edges as seen by the coherent beam correspond to the projected area in a direction normal to the incident beam. This is known as the Fraunhofer diffraction range and leads to simple analytical expressions.[48,52,54] However, as the particle size approaches the wavelength, λ, a fraction of the incident light is transmitted through the particle and this can interfere with the diffracted light to cause so-called "anomalous" diffraction patterns. The Fraunhofer approximation is then inadequate and its use can result in reproducible but "untrue" size distributions. The extent of deviation of the Fraunhofer approximation from the Mie theory has been the subject of several single-sphere error analyses at fixed scattering angles.[55,56] The deviation increases with decreasing size but decreases with increasing light absorption coefficient of the particle. For example, for a 5.67 μm sphere with $m_I = 0$, the calculated scattered energy falling on certain detector rings may be underestimated by as much as 100%. For the same sphere but with $m_I = 0.06$, indicating an absorbing particle, the scattered energies calculated by the Fraunhofer and Mie theories are in close agreement over all detector rings. For a 1.35μm sphere, on the other hand, even a value of m_I as high as 0.32 fails to provide close agreement between the two theories.

For particles approaching the wavelength of the incident beam the anomalous diffraction approximation provides a closer agreement with the Mie theory. The light transmitted through the particle is assumed to be unrefracted, leading to simple expressions to allow for the interaction between the diffracted and transmitted light.[12,45,52] This approximation is only valid when the refractive index of the particle, $m = m_R-jm_I$, is close to that of the medium in which the particles are suspended, i.e. $|m| \to 1$. Some commercial instruments claim to use the anomalous diffraction approximation to describe the single particle scattering function η. The exact details of the procedures used are not clear, but it is to be noted that no knowledge of the complex part of the particle refractive index is required. This suggests strongly that the analysis treats the particles as non-light absorbing spheres. This arbitrary omission of the complex part of the particle refractive index can lead to substantial errors.[12,45]

Farhadpour [45] calculated the scattered energy distribution on a commercial detector assembly (Malvern Particle Sizer) consisting of 31 semi-circular detector rings, according to Fraunhofer diffraction and anomalous diffraction for various levels of light absorbency. The results are compared in Fig.6. The particle refractive index relative to that of the medium was chosen as $m = 1.04-jm_I$, so that for all cases considered $1.04<|m|<1.09$, and hence in all cases anomalous

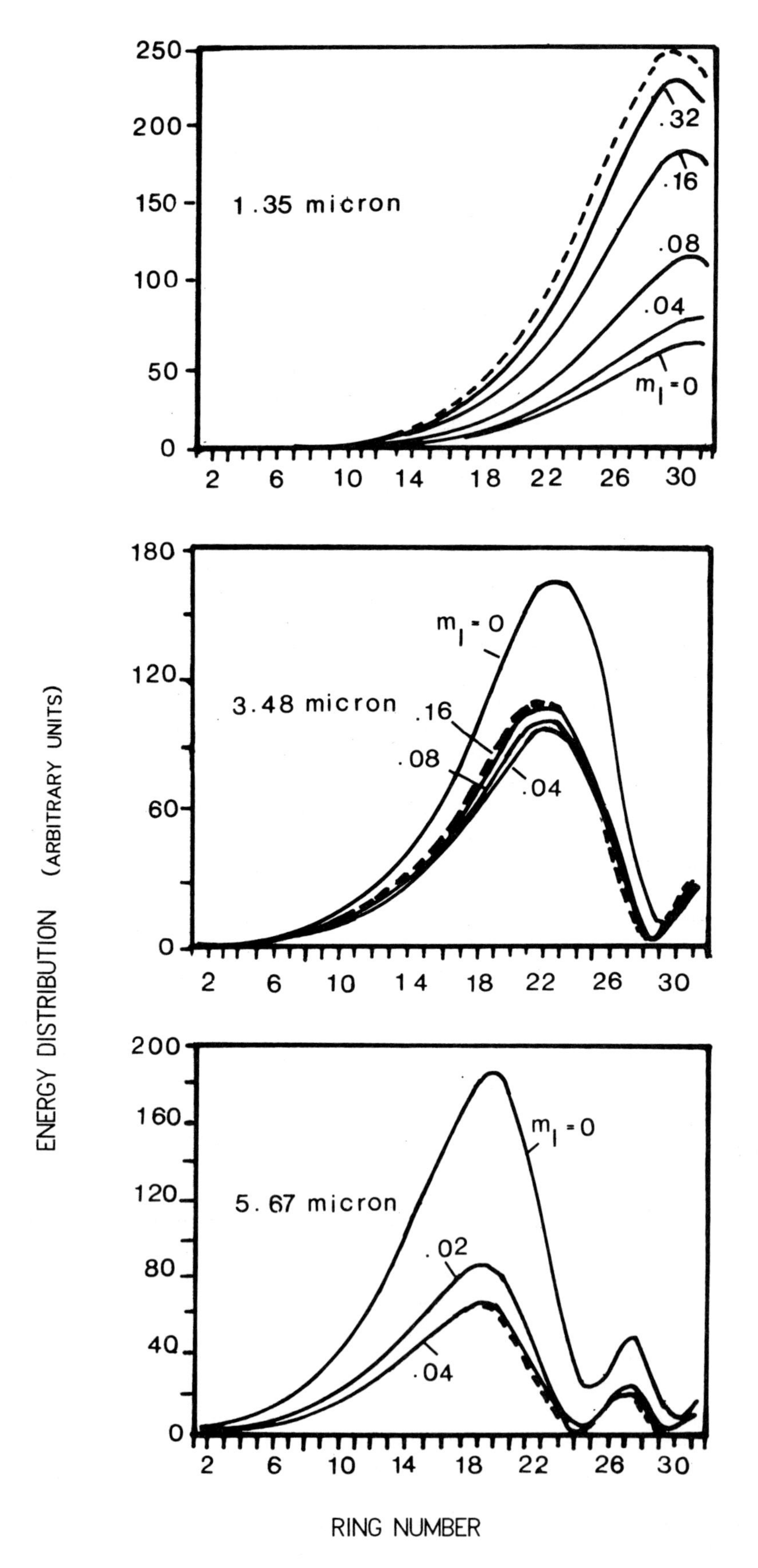

Figure 6 Comparison of energy distribution onto a commercial detector for Fraunhofer (---) and anomalous diffraction by uniform spheres.

diffraction provided a valid approximation. The very strong influence of light absorbency in the anomalous diffraction regime is due to the attenuation of the light penetrating the particles, which grows exponentially with particle size. Thus for a 5.67μm sphere a value of $m_I \cong 0.04$ reduces the transmitted light sufficiently to provide quantitative agreement with the Fraunhofer theory applicable to an opaque sphere. For a 1.35μm sphere, however, the influence of the transmitted light cannot be ignored even with a value of m_I as high as 0.32.

It is clear that for a polydispersed mixture of micron sized spheres the agreement between the two theories depends on the size distribution itself and on the level of light absorbency. For example, the under- and over-estimation for the larger and smaller particles indicated in Fig. 6 could lead to a close but fortuitous agreement. We have in fact observed a situation where, for a suspension in the 1-10μm size range, the invalid use of Fraunhofer diffraction software yields a distribution in close agreement with other techniques.[12] In contrast, the valid anomalous diffraction software seriously oversizes the particles. This has been traced to the fact that the anomalous diffraction software usually implemented is based on non-light absorbing spheres which, as seen in Fig. 6, can lead to substantial errors. Including light absorbency in the anomalous diffraction analysis has been shown to yield the correct size distribution.[45]

Given the complete refractive index of the particles, the anomalous diffraction theory provides a valid approximation for size analysis of particles in the micrometer size range. In many cases, however, the complex part of the particle refractive index is not known *a priori* and its measurement for finely divided solids is not a simple task. A practical solution in such cases is to determine the size distribution by an independent technique such as the electrical sensing zone method. The same suspension can then be measured by the light scattering instrument, and the level of light absorbency determined so that the two techniques provide reasonable agreement provided that the shape does not deviate significantly from a sphere.[45] This procedure is particularly suited to liquid suspensions of non-metallic particles, for which the relative refractive index is usually close to unity or can be brought sufficiently close to unity by a suitable choice of the medium. It is, however, less useful for micron sized aerosols where the relative refractive index is often greater than 1.3, and the anomalous diffraction theory is invalid. Such cases must be treated on the basis of the rigourous Mie theory. Because of high absorbance, anomalous diffraction is unlikely to be significant for metallic particles.

Size analysis in the anomalous diffraction regime also offers a number of other practical advantages. The data obtained will be free of many of the complications caused by reflection and edge effects, and will also be relatively insensitive to mild local variations in the particle refractive index. More significant, however, is the relative insensitivity of the anomalous diffraction results to particle shape in random orientation. Evidently this will not be the case for particles with distinct preferred orientation such as needles and long cylinders which demand individual attention.

The assumption of sphericity has been examined in a practical manner by Chylek and co-workers;[57-59] see also Welch and Cox [60] and Aquista.[61] Chylek *et al.*[57] argued that most of the individual features of the light scattered by randomly orientated irregular particles will be averaged out on summing over all possible particle orientations, summing over all possible particle shapes, and finally summing over the distribution of particle sizes. This would lead to a smooth scattering function which would only differ from that due to equivalent spheres in some general characteristic feature. It is then argued that spheres can support surface waves, which are responsible for the ripple structure in the extinction curve, and the well known glory effect which is a strong enhancement of the scattering in the backward direction.[52,58] Irregular particles on the other hand are unable to support surface waves. Chylek *et al.*[57] obtained a very good description of light scattering by irregular particles in the size range $\alpha<30$ and $m<2$ by a simple modification of the Mie solution for spheres, to suppress the elements responsible for surface waves. Particles, spherical or otherwise, cannot support surface waves in the anomalous diffraction regime. It could therefore be argued that the scattering function for randomly orientated irregular particles would show little difference from that for spheres; the assumption of sphericity in the anomalous diffraction regime is therefore likely to lead to less error. This obviously has implications for the appropriate choice of the suspension medium.

We now turn to the more difficult problem of recovering a size distribution from the measured scattered energy distributions. The inversion of the integral in equation (29) to recover the unknown distribution $n(\alpha)$ has attracted considerable attention in both the mathematics and light scattering literature.[62-67] The problem is known to be "ill-conditioned" since for smooth theoretical kernels $\eta(\theta_i,m,\alpha)$ the high frequency components of the distribution $n(\alpha)$ are severely damped out.[65] A direct attempt at inverting the integral can be attempted by dividing the size range into M intervals, α_j, $j = 1,2,\ldots,M$ with a uniform (or any other) distribution in each size range. The integral in equation (29) can then be replaced by a quadrature formula:

$$S_i = \sum_{j=1}^{M} \omega_j \varphi(\theta_i,\alpha_j)\, n_j \quad ; \quad i = 1,2,\ldots,N \tag{30}$$

where $\varphi(\theta_i,\alpha_j)$ is the contribution of particles in size range j to detector ring i, ω_j is a weighting function dependent on the quadrature formula used and n_j is the number of particles in the jth size range. Equation (30) can be rewritten in the compact matrix form:

$$\underline{S} = \begin{pmatrix} S_1 \\ \vdots \\ S_N \end{pmatrix} = \begin{pmatrix} \omega_1\varphi(\theta_1,\alpha_1) \ldots \omega_M\varphi(\theta_1,\alpha_M) \\ \vdots \qquad\qquad \vdots \\ \omega_1\varphi(\theta_N,\alpha_1) \ldots \omega_M\varphi(\theta_N,\alpha_M) \end{pmatrix} \begin{pmatrix} n_1 \\ \vdots \\ n_M \end{pmatrix} = [\varphi]\, \underline{n} \tag{31}$$

Direct inversion of equation (31) will in general fail and produce a wildly oscillating size distribution, because the elements of the matrix [φ] vary over several orders of magnitude, and therefore lead to a wide range of eigenvalues. The solution is also dominated by the eigenvectors corresponding to the smaller eigenvalues which are highly oscillatory.[65] The inversion problem is therefore inherently ill-conditioned.

Two approaches have been proposed to overcome the above difficulty. In one approach, the optimization route, the inversion problem is circumvented altogether. At the simplest level a parametric distribution, e.g. normal, log-normal, Rosin-Rammler, etc., is chosen to describe the form of the unknown distribution. The parameters can then be adjusted to give the best fit between the calculated and the measured energy distributions. This procedure is open to serious criticism on two grounds. First is the arbitrary assumption of a particular functional form for the unknown size distribution. Second and more serious is the observation by Twomey [64] that quite different distributions can lead to very similar calculated energy distributions. The results obtained by this procedure are therefore suspect and must be supported by other independent evidence.

The need for the assumption of a particular parametric form can be avoided by treating the problem as a multivariable mathematical programming problem. Here the number of particles in each size range is itself treated as an independent parameter and adjusted until the sum of the square of the error between the calculated and measured energy distributions is minimized. A variety of optimization techniques are available and can be used to solve this least square minimization problem; the Levenberg-Marquadt technique [68] is used most frequently. This so-called "model independent" procedure is implemented on most commercial instruments and is more satisfactory than the empirical curve fit approach. Its main drawbacks arise because, as with any nonlinear optimization problem, the search procedure can only provide a local minimum. Moreover, in many cases the final converged profile can depend on the initial guess used to start the search procedure. Some commercial software will allow starting the search from different user-defined initial conditions to check for such sensitivity. As a general rule, the mathematical programming approach tends to smear out any sharp features in the particle size distribution. A severe test of the software can be simply arranged by forming a bi-dispersed suspension of small and large particles. In many cases this bi-modal distribution will be recovered as a much broader and less sharp function with particles in the intermediate size range. Fortunately, this is an artificial test because most practical applications are concerned with a continuous distribution.

In a second and more direct approach to recovering the size distribution, the inversion problem is tackled under a set of smoothing constraints to avoid excessive oscillations. A procedure based on the curvature of the distribution function was first reported by Philips [63] and later improved by Twomey [64] who also examined various other criteria besides curvature. The application of the Twomey-Philips inversion technique in light scattering has been attempted by several workers, e.g. Yamamoto and Tanaka [69], King *et al.*[70], and Walters [71]. The original Twomey-Philips technique did not allow for non-negativity constraints on the size distribution, and also the degree of smoothing was chosen more or less arbitrarily. These deficiencies were tackled by Butler *et al.*[66] who developed an elegant procedure for including non-negativity constraints and an optimum choice of the smoothing level based on the signal/noise ratio obtained experimentally. Butler's technique was shown to be capable of accurate representation of uniform distributions, bi-modal uniform distributions and also gave a reasonable approximation to a Dirac delta function. At the present time, however, these more complex techniques are not available on commercial instruments, which largely rely on the model-independent mathematical programming approach.

6. Conclusions

Analysis of particle motion in fluids under steady motion, as in gravity sedimentation, and under non-steady motion, as in centrifugal sedimentation and inertial separators, indicates that the measurement of Stokes diameter and inertial diameter for non-spherical particles is not as straightforward as it first appears, and only under special circumstances are the two sizes equivalent. Some of the discrepancies between results from various instruments for measuring these diameters could emanate from differences in particle motion between these instruments. A method is therefore presented for obtaining explicit relationships between the volume-equivalent diameter, the maximum projected area-equivalent diameter and the hydrodynamic (Stokes) diameter by representing the particulate by a "model" shape. This approach is particularly useful in detecting erroneous measurements.

Light diffraction instruments show very large deviations from other measures of particle size such as the volume-equivalent diameter and Stokes diameter for micron sized particulates. This discrepancy is traced to the failure of commercial equipment to account for the light absorbency of the particles. Light absorbency can be taken into account readily by using the anomalous diffraction approximation, and it is shown to provide a better agreement. However, this requires the refractive index of the particle to be close to that of the medium in which the particles are suspended.

List of Symbols

A	Projected area of particle, m^2.
C_s, C	Slip correction factors for sphere and non-spherical particle.
c_1, c_2, c_3	Principal hydrodynamic resistances of particle, m.
c	Mean hydrodynamic resistance, m.
d	Particle diameter, m.
d_A	Area-equivalent diameter, m.
d_a	Aerodynamic diameter, m.
d_E	Equatorial diameter of spheroid, m.
d_I	Inertial diameter, m.
d_P	Polar diameter of spheroid, m.
d_S	Stokes-equivalent diameter, m.
d_V	Volume-equivalent diameter, m.
E	Aspect ratio of spheroid, d_p/d_E.
F_d	Drag of fluid on particle, N.
g	Acceleration due to gravity, m s^{-2}.
$\underline{i}$, $\underline{j}$, $\underline{k}$	Vectors defining principal axes of translation of particle.
M	No. of size classes.
m	(relative) refractive index of the particle.
m_I	Complex part of m.
m_R	real part of m.
N	no. of detector rings.
n (α)	Normalised particle size distribution.
Re_p	Particle Reynolds number.
r_i	Radius of detector ring i, m.
s	Dummy time variable.
S_i	light energy scattered on detector ring i.
t	Time, s.
$\underline{U}$	Particle velocity, m s^{-1}.
U_1,U_2,U_3	Components of $\underline{U}$ in directions i, j, k, m s^{-1}.
U_o	Initial particle velocity, m s^{-1}.
U_t	Terminal settling velocity, m s^{-1}.
V	Particle volume, m^3.
Z	Stopping distance of particle, m.
α	Dimensionless particle size parameter, πd/λ.
η	Light energy contributions for particles of size α.
θi	Angle of scattered light with respect to the incident beam, rad.
l	Wavelength of light, μm.
m	Fluid viscosity, kg m^{-1} s^{-1}.
rf	Fluid density, kg m^{-3}.
rp	Particle density, kg m^{-3}.
$\phi(\theta_i, \alpha_j)$	Energy scattered by particle size α_j at an angle θ_j.
ωj	quadrature weighting functions.

References

1. A. Kelly and S.T. Mileiko (ed.), Fabrication of Composites, Handbook of Composities, Oxford, North-Holland Pub. Co., 4, 1983.
2. K. Kendall, Powder Technology, 58, 151-161, 1989.
3. H.A. Barnes, J.F. Hatton and K. Walters, An Introduction to Rheology, Amsterdam, Elsevier, 1989.
4. W.C. Hinds, Aerosol Technology, New York, Wiley- Interscience, 1982.
5. H.G. Barth, S.-T. Sun and R.M. Nichol, Anal. Chem., 59, 142R-162R, 1987.
6. K. Leschonski, Part. Charact., 3, 99-103, 1986.
7. J.K. Beddow (ed.), Particle Characterization Technology, Boca Raton, Florida, CRC Press,1 & 2, 1984,

8. T. Allen, Particle Size Measurement, London, Chapman and Hall, 1981.
9. B.V. Miller and R.W. Lines, CRC Critical Reviews in Analytical Chemistry, Boca Roton, Florida, CRC Press, 20, 2, pp.75-116, 1988.
10. D.A. Lundgren, F.S. Harris, W.H. Marlow, M. Lippmann, W.E. Clark and M.D. Durham (ed.), Aerosol Measurement, Gainsville, University of Florida Press, 1979.
11. J.P.K. Seville, J.R. Coury, M. Ghadiri and R. Clift, Part. Charact., 1, 45-52, 1984.
12. U. Tuzun and F.A. Farhadpour, Part. Charact., 2, 104, 1985.
13. J.G. Harfield, A.W. Simmons, R.A. Wenman and R.A. Wharton, in Particle Size Analysis 1985, (ed. P.J. Lloyd), 211-221, 1987, Chichester, Wiley.
14. T. Allen and R. Davies, in 4th European Symposium on Particle Characterisation, Nuremberg, Preprints 1, 17-46, 19-21 April 1989.
15. BCR CRM 66, CRM 67, CRM 69 and CRM 70, Bureau Communité de Reference (BCR), 200 Rue de la Loi, B-1049, Brussels, Belgium.
16. J.G. Harfield, R.A. Wharton and R.W. Lines, Part. Charact., 1, 32, 1984.
17. B. Scarlett, 2nd European Symp. on Part. Charact., PARTEC, Nuremberg, 681, 1979.
18. B.A. Batch, J. Inst. Fuel, 455-461, 1964.
19. M. Ghadiri, Ph.D. Dissertation, Univ. of Cambridge, 1980.
20. J.P.K. Seville, Ph.D. Thesis, Univ. of Surrey, 1987.
21. J.G. Harfield, in Particle Size Analysis 1981, (ed. N.G. Stanley-Wood and T. Allen), Chichester, Wiley, 1982.
22. V. Vouk: Nature, 330-331, 1948.
23. R. Clift, J.R. Grace and M.L. Weber, Bubbles, Drops and Particles, New York Academic Press, 1978.
24. B.Y.H. Liu, A.C. Verma, Analyt. Chem., 40, 4, 843, 1968.
25. H. Bandemer, M. Albrecht and A. Kraut, Part. Charact., 2, 98, 1985.
26. T.P. Meloy and N. Clark, in Particulate and Multiphase Processes, 1: General Particulate Phenomena, (eds. T. Ariman and T.N. Veziroglu), 459-470, 1987, Washington D.C., Hemisphere.
27. B.H. Kaye, G.G. Clark, J.E. Leblanc and R.A. Trottier, in Part 1, 1st World Congr. Particle Technol., PARTEC, Nuremberg, 17-29, April 16-18 1986.
28. T.P. Meloy and N. Clark, in Particulate and Multiphase Processes, 1: General Particulate Phenomena, (eds. T. Ariman and T.N. Veziroglu), Washington D.C., Hemisphere, 437-445, 1987.
29. R. Weichert, in 4th European Symposium on Particle Characterisation, Nuremberg, Preprints 1, 409-420, 19-21 April 1989.
30. M. Ghadiri, J.P.K. Seville, J.A. Raper and R. Clift, in Part 1, 1st World Congress Particle Technol., PARTEC, Nuremberg, April 16-18, 203-220, 1986.
31. R. Clift, in Particle Size Analysis 1988, (ed. P.J. Lloyd), Chichester, Wiley-Interscience, 3-17, 1987.
32. Y.-S. Cheng and H.-C. Yeh, Environmental Sci. and Tech., 13, 1392, 1979.
33. N.A. Esmen and T.C. Lee, Am. Ind. Hygiene Ass. J., 41, 410, 1980.
34. J. C. F. Wang, Aerosol Sci. and Tech., 4, 301, 1985.
35. J.C. Wilson and B.Y.H. Liu, J. Aerosol Sci., 11, 139, 1980.
36. W.D. Griffiths, S. Patrick and A.P. Rood, J. Aerosol Sci., 15, 491, 1984.
37. A.H.J. Tate, J.P.K. Seville, A. Singh and R. Clift, I. Chem. E. Symp. Ser., 99, 89-99, 1986.
38. J. Happel and H. Brenner, Low Reynolds Number Hydrodynamics, The Hague, Martinus Nijhoff, 1983.
39. R. Clift, M. Ghadiri and K. V. Thamibimuthu, in Progress in Filtration and Separation, Vol. 2, (ed. R.J. Wakeman), Amsterdam, Elsevier, 1981.
40. M. Dressel, Part. Charact., 2, 62, 1985.
41. R. Hill and G. Power, Quart. J. Mech. Appl. Math., 9, 313, 1956.
42. R.G. Cox, J. Fluid Mech., 44, 791, 1970.
43. H.F. Weinberger, J. Fluid Mech., 52, 32, 1972.
44. British Standard Institution, BS 3406, 1963.
45. F.A. Farhadpour, in Particle Size Analysis 1988, (ed. P.J.Lloyd), Chichester, Wiley-Interscience, 323-334, 1987.
46. J.R. Hodkinson, in Aerosol Science, (ed. C.N. Davies), London, Academic Press, 1966.
47. J. Gebhart, in Preprints Fine Particle Society Annual Meeting, 7-5, April 22-26, 1987, Miami Beach.
48. L.P. Bayvel and A.R. Jones, Electromagnetic Scattering and its Application, London, Elselvier, Applied Sci. Pub., 1981.
49. C.E. Bohren and D.R. Huffman, Absorption and Scattering of Light by Small Particles, New York, John Wiley and Sons, 1983.
50. L. Brecivic and J. Garside, Chem. Eng. Sci., 36, 867, 1981.
51. T.A. Shippey and J. Garside, I. Chem. Eng. Jubilee Symp., London, 1981.

52. H.C. van de Hulst, Light Scattering by Small Particles, New York, Dover Publication Inc., 1981.
53. R.P. Feynman, R.B. Leighton and M. Sands, The Feynman Lectures on Physics (5th Ed.), Massachusetts, Addison-Wesley, 1975.
54. J. Swithenbank, J.M. Beer, D.S. Taylor, D. Abbot and C.G. McCreeath, Prog. Astronautics and Aeronautics, 53, pp.421-447, 1977.
55. A. R. Jones, J. Phys. D.: Appl. Phys., 10, L163, 1977.
56. S. Boron and B. Waldie, Applied Optics, 17, 1644, 1978.
57. P. Chylek, G.W. Grams and R.G. Pinnick, Science,193, 480,1976.
58. P. Chylek, J. Opt. Soc. Am., 66, 285, 1976.
59. P. Chylek and R.G. Pinnick, Applied Optics, 18, 1123, 1979.
60. R.M. Welch and S.K. Cox, Applied Optics, 17, 3159, 1978.
61. C. Aquista, Applied Optics, 17, 3851, 1978.
62. L. Fox, Numerical Solution of Ordinary and Partial Differential Equations, 1962, Oxford, Pergamon Press.
63. D.L. Phillips, J. Assoc. Computing Machinery, 9, 84, 1962.
64. S. Twomey, J. Assoc. Computing Machinery, 10, 97, 1963.
65. S. Twomey, Introduction to the Mathematics of Inversion in Remote Sensing and Indirect Measurement, New York, Elsevier Scientific Publishing Co., 1977.
66. J.P. Butler, J.A. Reeds and S.V. Dawson, SIAM J.Numer. Anal., 18 (3), 381, 1981.
67. M. Heuer and K. Leschonski, Part. Charact., 2, 7, 1985.
68. W.H. Press, B.P. Flannery, S.A. Teukolsky and W.T. Vetterling, Numerical Recipes, 14, 1988 Cambridge, Cambridge University Press.
69. G. Yamamoto and M. Tanaka, Applied Optics, 12, 1340, 1973.
70. M.D. King, D.M. Byrne, B.M. Herman and J.A. Reagon, J. Atmos. Sci., 35, 2153, 1978.
71. P.J. Walters, Applied Optics, 19, 2353, 1980.

6

Surface Chemical Characterization of Powders

J. F. WATTS AND T. J. CARNEY

Department of Materials Science and Engineering, University of Surrey, Guildford, Surrey, UK

Abstract

The surface analysis methods of X-ray photoelectron spectroscopy (XPS), Auger electron spectroscopy (AES), and scanning Auger microscopy (SAM) are described and the manner in which the data obtained may be interpreted is discussed. Particular problems associated with the analysis of powder samples are highlighted, and methods of sample mounting, and spectral and image acquisition are suggested that will mitigate such difficulties. A series of examples are used to illustrate the different levels of information attainable with the three techniques, showing the provision of analytical information with high spectral, depth and spatial resolution.

1. Introduction

A characteristic of all fine powders is their high surface to volume ratio, consequently a knowledge of the composition of the surface and near-surface regions of powders can be of critical importance as far as their subsequent processing is concerned. By near surface we are generally concerned with the outer 5nm or so of a material's surface and the departure of this outer region from the bulk composition. Such changes in composition generally result from one of two processes; they may be a result of the material interacting with the environment; for example, metal powders will always be covered with a thin layer of oxide; or they may result from segregation or other processes that have occurred during powder production or processing, for example, an Al-Mg powder will invariably show a certain degree of magnesium segregation to the surface. In order to be able to detect such phenomena, methods of analysis with a very small analysis depth are needed. The surface analysis methods based on emitted electron spectroscopy have shown themselves to be very versatile in this respect. In this chapter we shall consider the methods of X-ray photoelectron spectroscopy (XPS) and Auger electron spectroscopy (AES) together with its imaging counterpart scanning Auger microscopy (SAM). Before considering, in some detail, the special problems relating to the analysis of powders and looking at a series of case histories we shall first examine the basic principles of electron spectroscopy.

2. Surface Analysis by Electron Spectroscopy

XPS and AES have several features in common as they both rely on the analysis of low energy (<2000eV) electrons. Consequently it is possible to carry out both methods on the same analytical system. However, the source of primary radiation used is different; X-rays for XPS and a finely focussed electron beam for AES. The depth of analysis is comparable (2-5nm) and is determined by the energy of the outgoing electrons and matrix factors. This Chapter will first review the basic principles of the two techniques, for a detailed treatment the reader is referred to more advanced texts [1-5].

2.1 X-Ray Photoelectron Spectroscopy

In the XPS experiment the sample under investigation is irradiated with soft X-rays (usually AlKα $h\nu$=1486.6eV or MgKα $h\nu$=1253.6eV) this brings about the photoemission of core electrons whose binding energy is less than that of the photon source; for instance the Mg1s electron has a binding energy of 1305eV and this is accessible with AlKα radiation but not MgKα, where the Mg2p electron spectrum (binding energy 50eV) would be recorded. The photoemission process is shown schematically in Fig. 1, the kinetic energy (E_K) of the emitted electron is measured by the electron spectrometer and is a function of photon source and electron binding energy (E_B):

$$E_K = h\nu - E_B - \omega \tag{1}$$

where:
ω - spectrometer work function.

Thus the kinetic energy of the outgoing electron will vary with X-ray source and it is the binding energy of the electron that is characteristic of both atomic orbital and valence state. A photoelectron spectrum of a fine (sub 45μm) aluminium powder is shown in Fig. 2, peaks arising from Al2p, Al2s, C1s, and O1s can be identified. In addition a strong peak at approximately 750eV is seen, this is one of the *OKLL* Auger transitions resulting from the radiationless de-excitation of the ion following photoemission. This process is shown schematically in Fig. 3. The core hole created by the ejected 1s (K) electron is filled by a 2p ($L_{2,3}$) electron, and in order to conform with conservation of energy principles an electron must also be ejected from the atom, e.g. another 2p electron in Fig. 3. This electron is termed the $KL_{2,3}L_{2,3}$, Auger electron and it is the *kinetic* energy of this electron that is characteristic of its parent atom. Auger spectra of this type, i.e. convoluted within the photoelectron spectrum are sometimes termed X-AES indicating they are X-ray induced. One very important use of such transitions is in determining the Auger parameter of a material, this is the separation of the XPS and X-AES peaks and can be particularly useful for materials which show a small XPS chemical shift, this aspect of XPS will be discussed further in the next section. The inset of Fig. 2 shows the Al2p spectrum recorded at higher resolution and components attributable to metallic and oxidized aluminium can be seen. By the use of the appropriate computer peak-fitting methods the relative proportions can be calculated. Such chemical resolution is obtained only at the expense of spatial resolution and XPS is most frequently encountered in its large area ($10mm^2$) form, although small area XPS (100μm) has been available for some years and imaging XPS (5μm) has recently become commercially available. Thus as far as powders are concerned special preparation methods are necessary so that a uniform array of particles are presented for analysis.

Having obtained a spectrum of the type illustrated in Fig. 2 and assigned all the major peaks to the appropriate Auger or photoelectron transitions it is often advantageous to be able to produce a quantitative surface analysis and for XPS this is a relatively straightforward matter. The intensity (I) of a photoelectron peak from a homogeneous solid is given, (in a very simplified form), by:

$$I = J\rho\sigma K\lambda \tag{2}$$

where:
J - photon flux,
ρ - concentration of a particular species in the solid,
σ - photoelectric cross-section, (and depends on the atom, energy level, geometry, and mode of operation of the spectrometer),
K - a term which covers instrumental factors,
λ - electron inelastic mean free path or attenuation length within the solid.

In this way it is possible to calculate the concentration of surface species, but as several of the quantities in the above equation depend only on spectrometer design it is usual to employ relative sensitivity factors experimentally determined from standard compounds such as fluorides. Each measured peak intensity (preferably peak area) is simply divided by the appropriate atomic sensitivity factor and this normalized peak area represents the relative abundance of that element.

In summary, XPS is able to readily produce a quantitative surface analysis from metals, inorganic, and organic materials, and determine the chemical environment of the element in question. This is achieved at the expense of spatial resolution, as XPS is normally operated in an area averaging mode with some $10mm^2$ of the sample contributing to the analytical signal.

2.2 Auger Electron Spectroscopy

In Auger electron spectroscopy it is usual practice to use a finely focussed beam of electrons as the excitation source. Consequently it is possible to achieve a surface analysis at very high spatial resolution (<50nm) and it becomes possible to operate a scanning Auger microscope in the same modes as a SEM equipped with energy or wavelength dispersive X-ray analysis (often referred to as an electron probe micro-analyser), that is, data can be acquired as full spectra from a point, line-scans of a particular element, or a two-dimensional chemical map. The spectra of Fig. 4a are taken from an aluminium particle, and the elements carbon, oxygen, and aluminium can be seen (carbon is present merely as a thin contamination layer). However, the Auger spectrum is always associated with a very intense background and the Auger transitions themselves are seen as peaks superimposed on this intense background.

This has important ramifications in both the quantification of Auger spectra and the recording of Auger chemical maps. The intensity of the Auger spectrum, both peaks and background, is strongly dependent on matrix and geometrical parameters, it is the former effect that makes quantification in AES rather more involved than in XPS. It is for this reason (the very intense background) that Auger spectra have traditionally been recorded in the differential mode as shown in Fig. 4a, but with the advent of high resolution Auger microscopes the direct (or pulse counted) form of the spectrum, as shown in Fig. 4b, is becoming more popular.

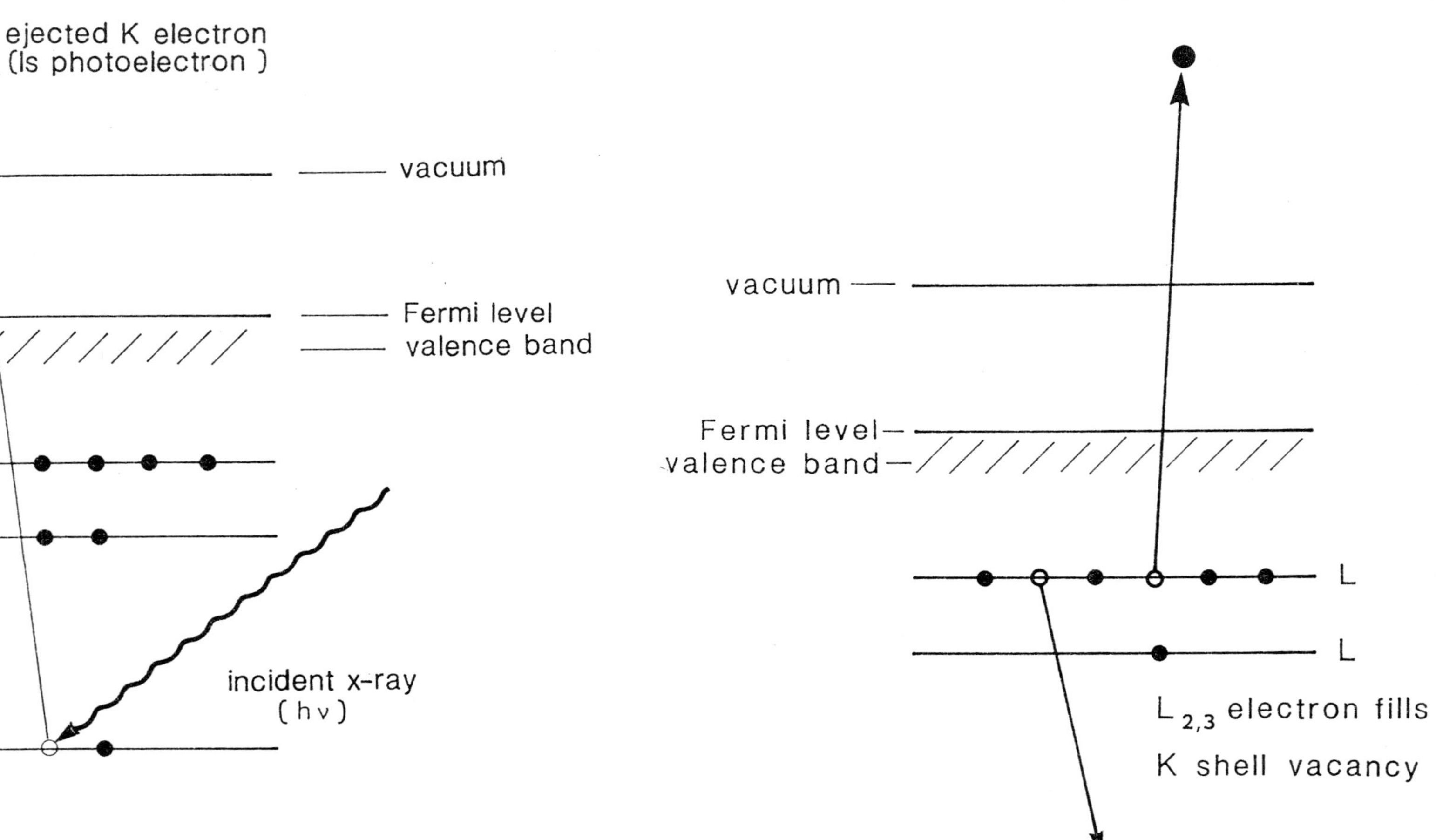

Figure 1 Schematic of the XPS process, showing photoionization of an atom by the ejection of a 1s electron.

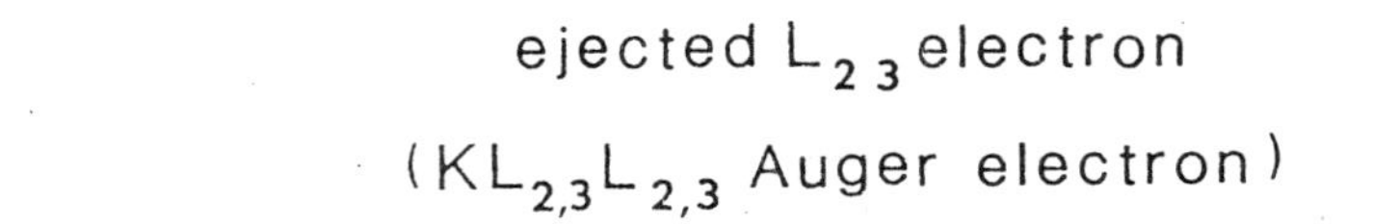

Figure 3 Relaxation of the ionized atom of Fig. 1 by the emission of a $KL_{2,3}L_{2,3}$ Auger electron.

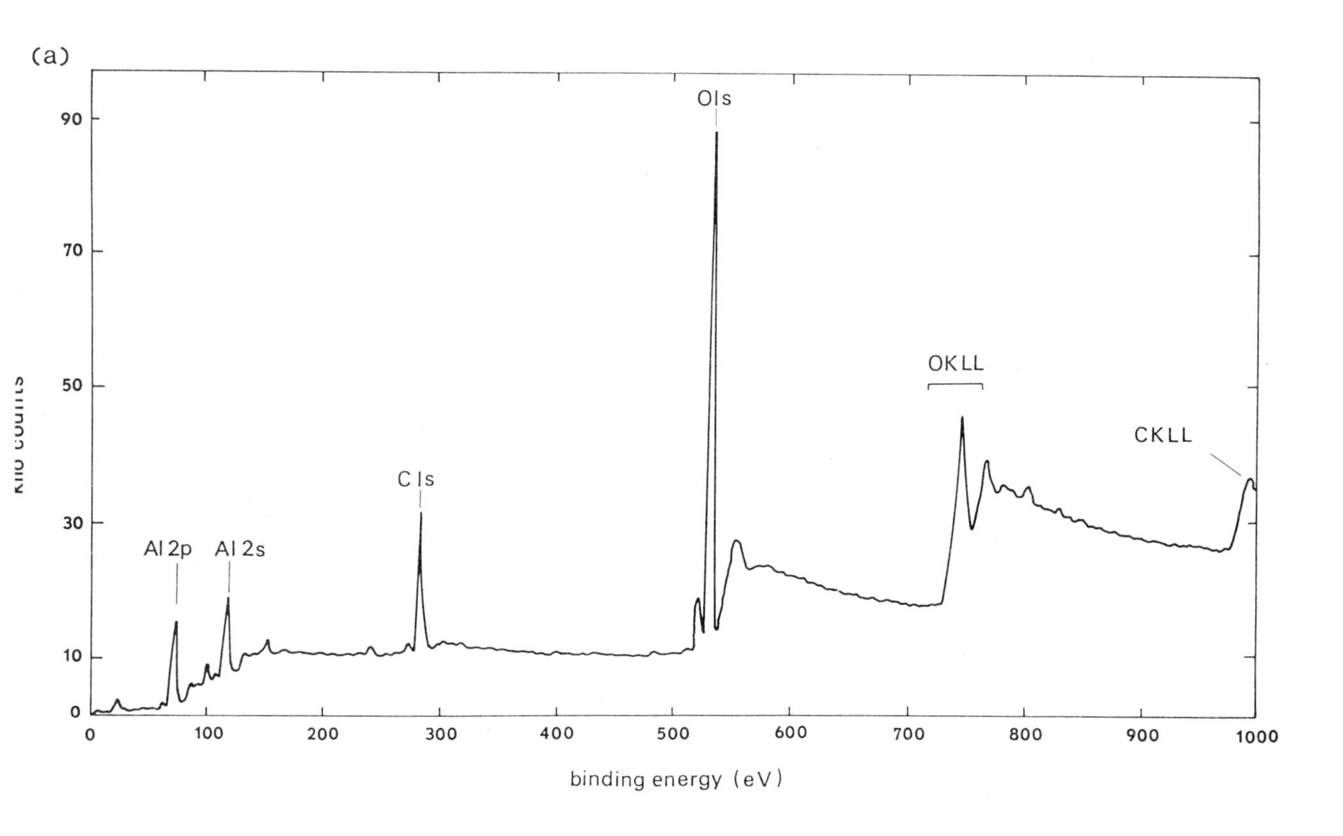

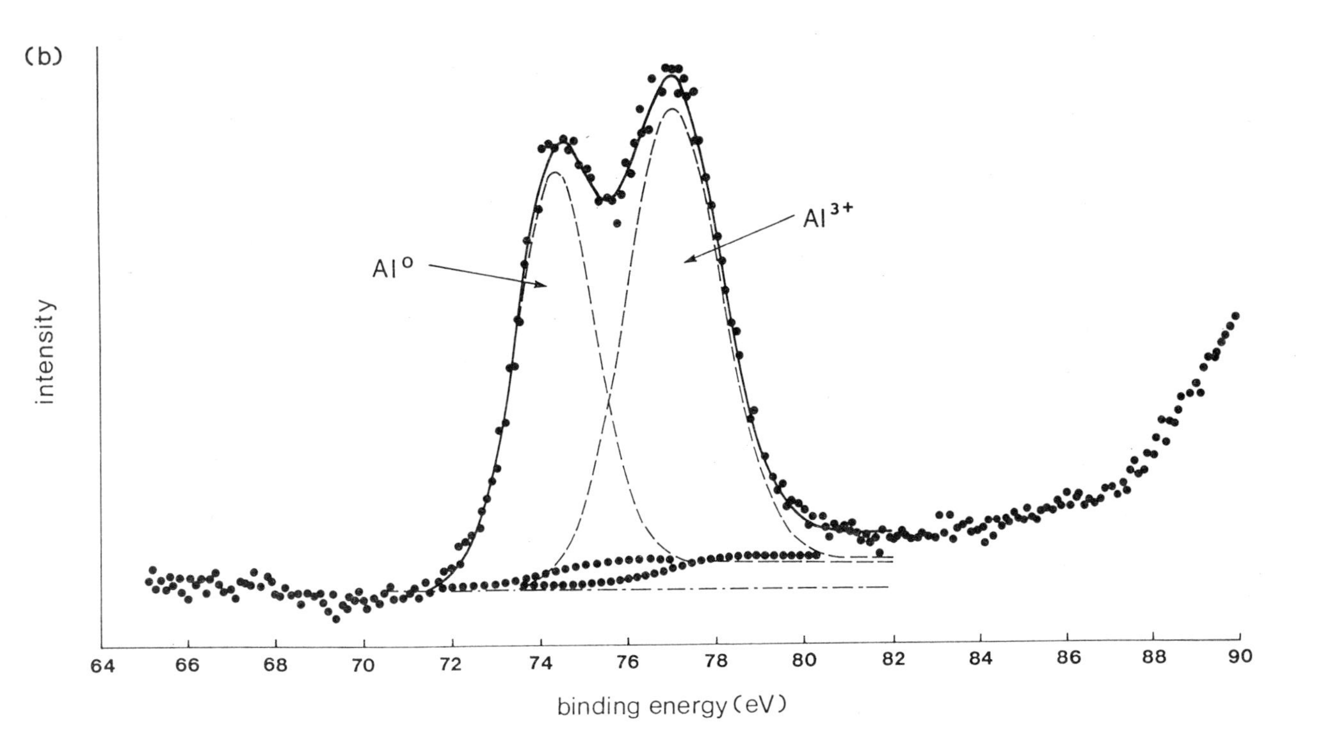

Figure 2 (a) XPS survey spectrum of fine aluminium powder using MgKα radiation.
(b) High resolution spectrum of the Al2p region showing metallic and oxide components resolved with the use of a computer curve fitting method.

In the quantification of AES data it is necessary to make use of experimental sensitivity factors obtained from binary or ternary alloys of the elements of interest obtained from spectrometers operating with the same experimental conditions. Auger spectra can be quantified quite successfully providing such standard data are available, if they are not (and often the instruments time will not be available or warranted), comparison of spectra will provide a qualitative guide as far as relative changes in surface composition are concerned.

In obtaining an Auger map of a sample such as spherical particles, surface roughness as well as matrix effects must be considered. The images of Fig. 5 (recorded at X5000) indicate the topography problem rather well. The SEM image (upper left) is very similar to that obtained by recording the intensity of any of the electrons of Fig. 4b irrespective of their energy, or whether they contribute to a characteristic peak (Al*KLL*, O*KLL*, etc.) or the background. The modulation in intensity of signal is merely a result of geometrical considerations. It is possible to go some way towards correcting for surface topography by subtracting the electron background (recorded at a kinetic energy some 20eV above the peak of interest), from the characteristic peak intensity, the (peak-background) map, as shown in Fig. 5, upper right. However, this procedure is not very successful and the usual algorithms employed normalize this quantity to provide (P-B)/B or (P-B)/(P+B) maps (where P=peak and B=background). The corrected image of Fig. 5 (lower left) was obtained using the (P-B)/B method and the spherical appearance of the powders has now been virtually removed from the image, the chemical information now appearing as two dimensional discs. There is a small degree of charging between the aluminium powder and the indium foil mount, and this allows us to record an oxygen image from the powder and from the substrate; the lower right quadrant of Fig. 5 shows an oxygen map from the indium oxide. However, as a result of the width of the Auger lines there appear to be certain "high spots" on the powder particles which are clearly not from the indium oxide, they are merely regions of high Auger yield, (some distance removed from the characteristic peak), that have not been successfully accounted for using the (P-B)/B algorithm. The shift in Auger peak position is, in this example, a result of differing surface potentials, not a chemical shift.

Thus in XPS and AES we have two techniques admirably suited to the study of fine powder surfaces. XPS is able to provide chemical information which can also be used to calculate oxide thicknesses from an array of powders whilst AES/SAM may be used to analyse individual particles. XPS can be used for the analysis of powders in a very straightforward manner but in the case of electron beam induced AES there needs to be a pathway for the "bleeding" of surface charge that may build up. This can even be a problem on metal powders if the surface oxide layer is more than a few Angstroms thick. There are many special considerations that must be taken into account when preparing fine powders for analysis by electron spectroscopy and in the next section these will be described and the interpretation of the resultant spectra discussed.

3. Special Considerations When Analysing Powders by Electron Spectroscopy

The difficulties of handling fine particulate material are legion and presenting them for analysis in an electron spectrometer operating at ultra-high vacuum presents its own special problems.

The most important requirements are that the surface under investigation is presented for analysis in the same condition that it was produced; this effectively precludes high shear methods such as compaction into a pellet as this would deform the particles and lead to contamination from the die. In addition the powder under investigation must not be allowed to contaminate the UHV system components such as valves and pumping lines, green compacts are at risk here as they can disintegrate during the pump-down stage. The most acceptable method is to mount powders on either indium foil (for XPS and AES [6]) or double sided adhesive tape (XPS only). The sticking coefficients will vary with powder size distributions but, in general, the signal from the mounting medium can be reduced to an acceptably low level. There can however be problems when peaks from the mounting material are coincident with those from the specimen as illustrated in Fig. 6. These spectra are from a rapidly solidified (RS) Al-8Fe alloy of sub 45μm particle size mounted by pressing into indium foil and dusting onto double-sided tape. In the former example the In3d peaks are apparent at 450eV but of more concern is the In3p1/2 peak which is exactly coincident with the Fe2p3/2 photoelectron peak. In this case the problem was resolved by the use of adhesive tape and the other spectrum (Fig. 6b) shows the Fe2p region, where the Fe2p3/2 is clearly seen as a peak in its own right indicating the presence of iron in the surface layers.

The other problem associated with surface analysis of powders is the quality of information attainable in XPS from a raft of powders compared with that from a planar sample. The situation is shown schematically in Fig. 7 and the two problems encountered are the shadowing of the powder surface from the X-ray beam and the inelastic scattering of electrons emitted from the specimen by other particles. The former reduces the net count rate whilst the latter has the effect of increasing the energy loss background of the spectrum which brings about a deterioration in the counting statistics. The usual way of overcoming these problems is simply to increase the acquisition time for each spectrum.

Although metal powders are regarded as conducting they will invariably be covered in a thin oxide film as illustrated in Fig. 5, and this may make Auger spectroscopy rather difficult.

The use of indium foil for mounting is essential in these cases and sparse coverage of the powder enables exposed areas of metal foil to bleed off the surface charge (for XPS a high coverage of powder is required to reduce the foil signal to a minimum). It is also possible to carry out AES on ceramic powders in this way. XPS and AES/SAM provide

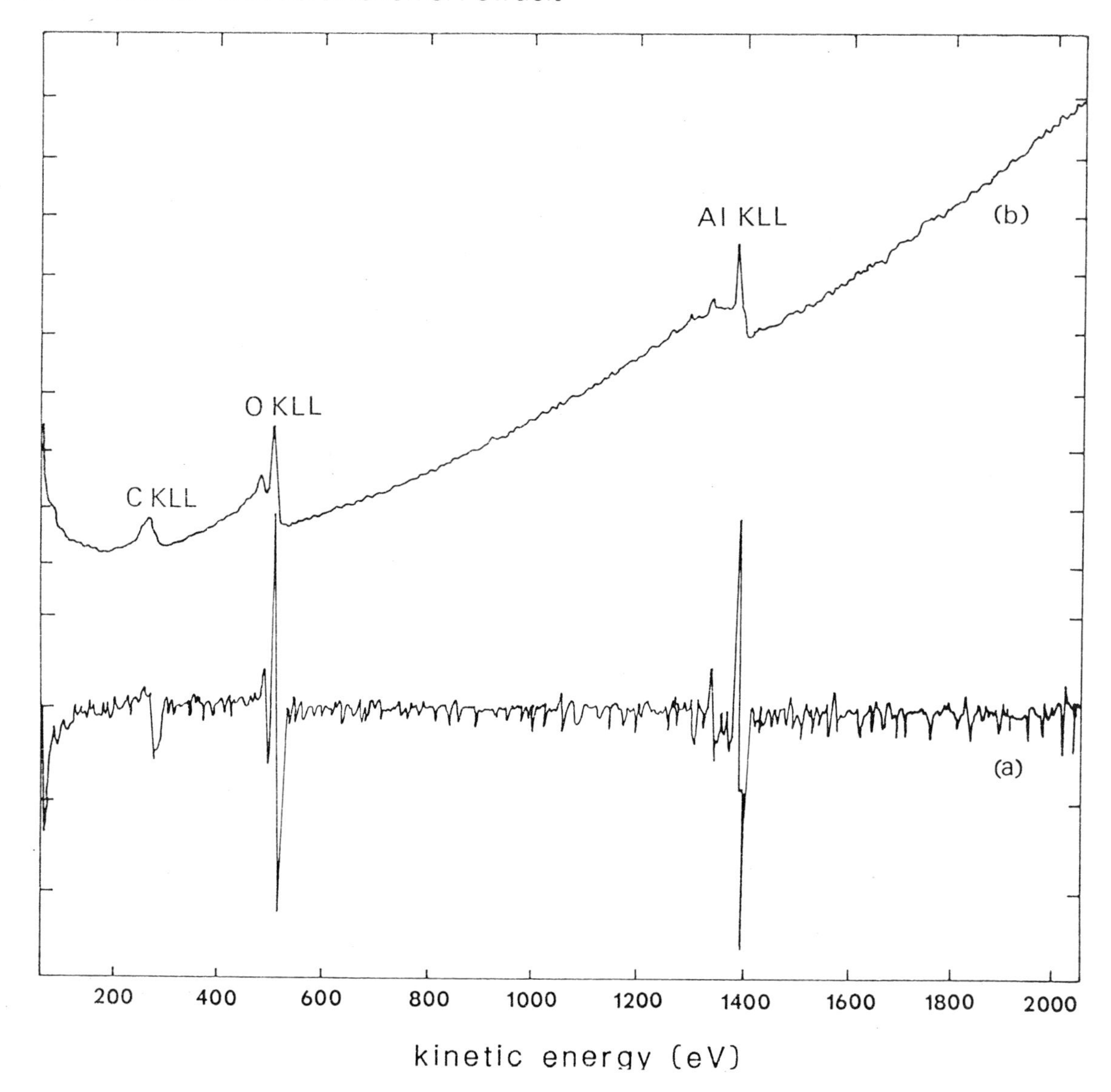

Figure 4 Auger spectra from an aluminium powder particle in the (a) differential, and (b) the direct, modes of acquisition.

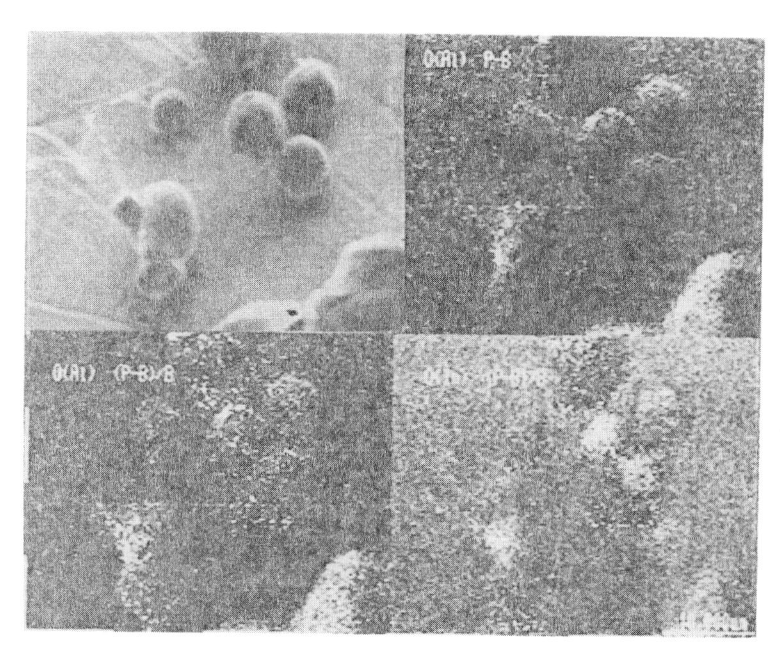

Figure 5 Scanning Auger microscopy of an aluminium powder, magnification X5000. Upper left: secondary electron image, Upper right: (P-B) oxygen (aluminium particles) map, Lower left: (P-B)/B oxygen (aluminium particles) map. Lower right: (P-B)/ B oxygen (indium mount) map.

complementary techniques for elucidating surface chemical characterization of powders. The former provides information on the chemical nature of the elements present and SAM indicates their distribution on the individual powder particle. This is not to say that Auger spectra do not yield chemical information and the X-ray excited Auger transitions are often combined with photoelectron data to determine the Auger Parameter of a material. This can be particularly useful for the chemical interrogation of alumosilicate materials.

3.1 The Auger Parameter

Although the problem of electrostatic charging does not preclude analysis in XPS as it would in electron beam techniques, it can sometimes lead to problems in interpretation. The effect of such charging on the electron spectrum is to slightly reduce the kinetic energy of all emitted electrons (i.e. an apparent increase in binding energy). The usual way of correcting for such charging is to correct the position of all peaks present in the spectrum by choosing one particular line and assuming an energy value. The Cls line at a binding energy of 285eV is often used for this purpose but the value will depend on both the chemical and physical form (C1s values varying between 284.6-285.0 are reported for adventitious carbon !). Thus the uncertainty associated with the corrected binding energy value may be greater than the chemical shift being studied. By considering the separation of a photoelectron peak and its associated Auger transition it is possible to eliminate this uncertainty by the use of the Auger Parameter, (α^*), which is conventionally defined as;

$$\alpha^* = E_K + E_B - h\nu \tag{3}$$

where:
E_K - kinetic energy of the Auger transition,
E_B - binding energy of the photoelectron,
$h\nu$ - photon energy.

This methodology is only useful where an energy shift can be observed in the Auger peak i.e. where the core vacancy is filled by an electron from another core-like energy level e.g. CuLMM Auger transitions. If such holes are filled by an electron from the band structure of the solid the peaks will be broad and contain little chemical state information (e.g the *KLL* lines of carbon and oxygen). In some cases the Auger chemical shift can be larger than that exhibited by the XPS peak, and it may be essential to consider both to obtain reliable information.

Thus valuable chemical information can be obtained from certain Auger parameters and a compilation of data is given in Table 1. For elements such as aluminium and silicon the *KLL* Auger peak is not directly accessible using conventional radiation, however the Bremstrahlung continuum from an achromatic X-ray source will provide sufficient intensity to excite such a line [7]. If we consider the Si*KLL* transition at a kinetic energy of approximately 1610eV this will occur in the negative binding of the photoelectron spectrum as indicated in Fig. 8, the intensity of the Auger peak is very small compared with the Si2p but as the background is fairly low it is possible to obtain good counting statistics. This Bremstrahlung induced Auger parameter is especially informative in the case of silicon where it has been shown to be particularly sensitive to molecular and crystalline structure as well as independent of electrostatic sample charging, as shown by the data of Table 2. Auger parameters can be obtained in a similar manner for Al, P, S, and Cl. Nonetheless the major analytical use of Auger electrons is in recording a surface analysis at high spatial resolution as described earlier and we shall now proceed to examine possible pitfalls in the SAM analysis of fine powders.

3.2 Scanning Auger Microscopy of Powders

A potential difficulty, for the electron microscopist, of any analysis using the surface sensitive techniques described above is their inability to allow charge compensating coatings to be used. In the scanning electron microscopy of powders which are coated in a very thin (sub - 2nm), insulating oxide the specimen can simply be coated in a thin layer of a conducting medium such as gold to a thickness of 10nm or so. However, if this coated specimen is then analysed by AES or XPS only the gold signal would be detected. Considering AES with an electron gun, without this charge compensating layer the specimens will tend to charge as there is a disparity between primary and secondary electron flux, and no electrical connection to earth to allow a compensating (charge neutralizing) flow of electrons. This situation is illustrated in Fig. 9a, and will cause a shifting of the peak position as illustrated below. The simplified expression for the kinetic energy of an Auger $KL_{2,3}L_{2,3}$ transition can be expressed approximately as:

$$E_{KLL} = E_K - E_{L2,3} - E_{L2,3} + \delta \tag{4}$$

where:
E_{KLL} - kinetic energy of an electron as detected from a $KL_{2,3}L_{2,3}$ transition,
E_K, $E_{L2,3}$, and $E_{L2,3}$ - binding energies of the specific energy levels,
δ - extra energy gained by the electron from the electric field established on the charged surface.

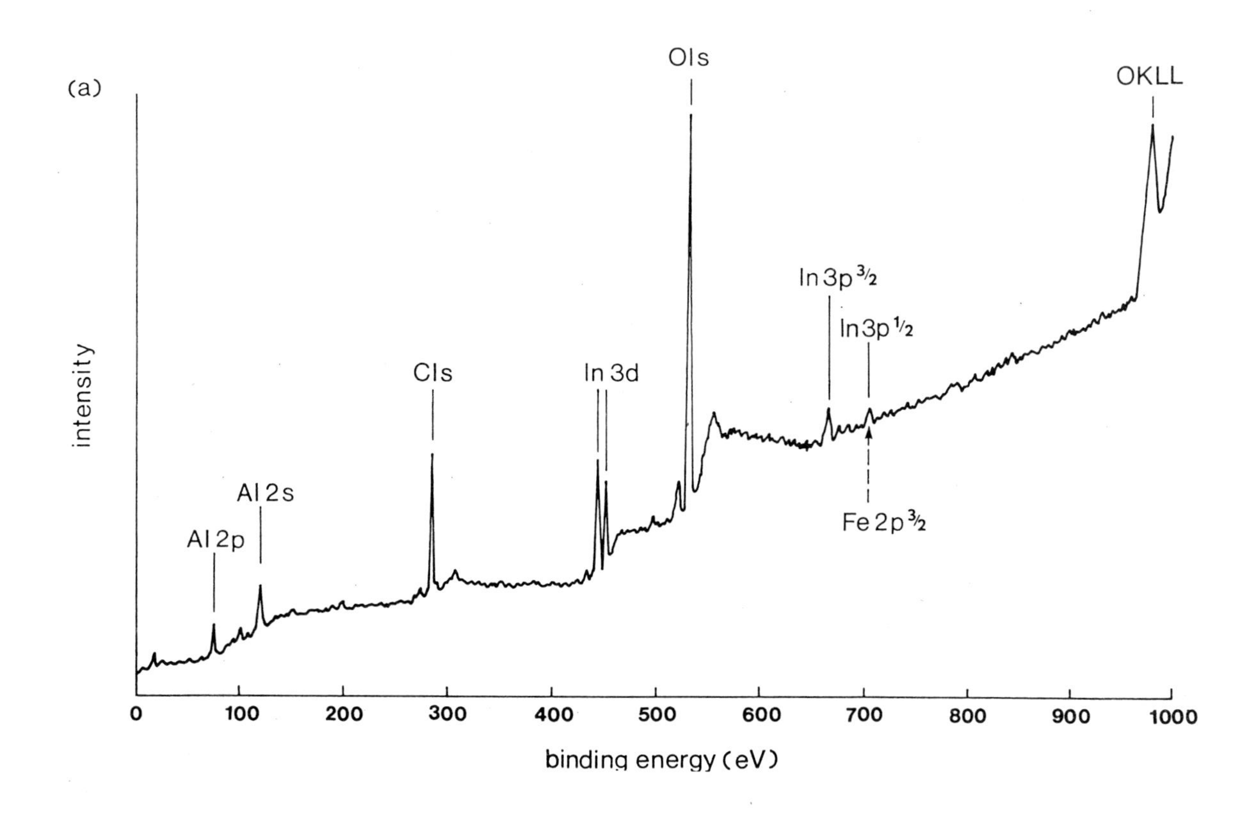

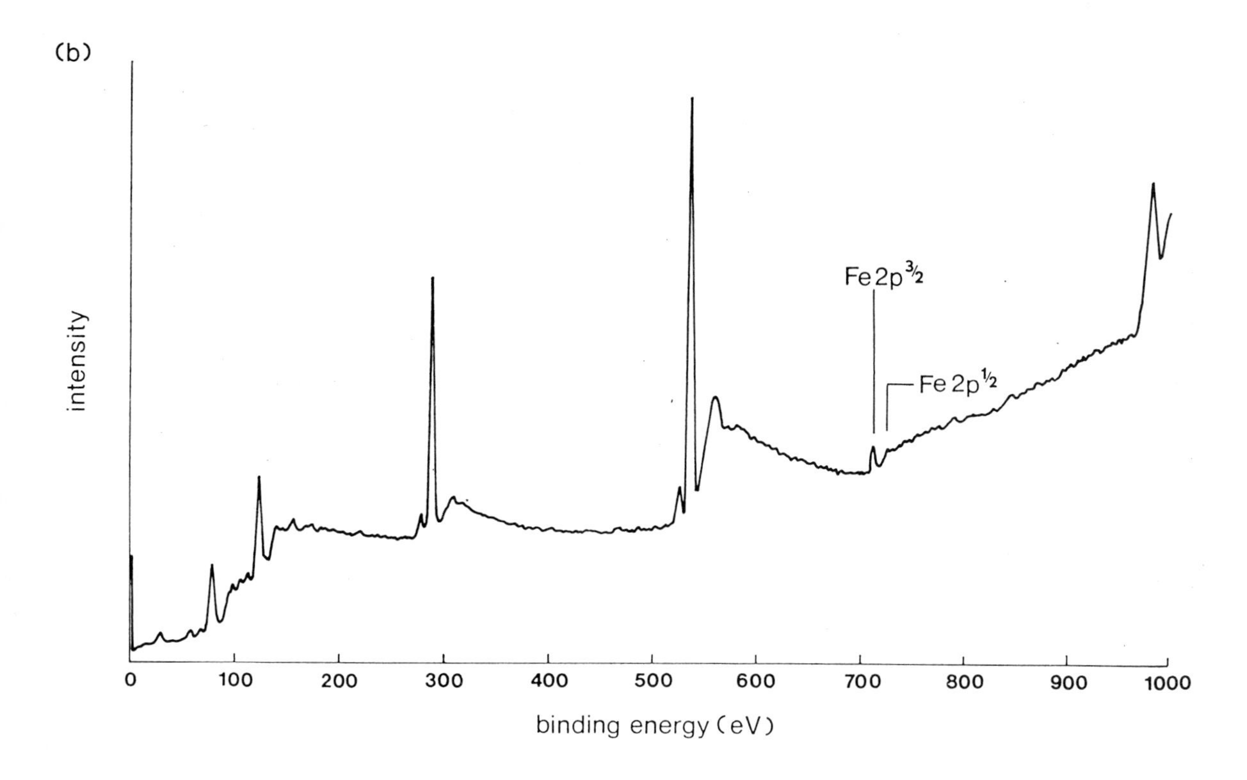

Figure 6 XPS spectra of an Al-8Fe powder. (a) Mounted by embedding in indium foil, the Fe2p3/2 peak is obscured by the In3p1/2. (b) Mounted on adhesive tape the Fe2p3/2 is clearly resolved as a peak in its own right.

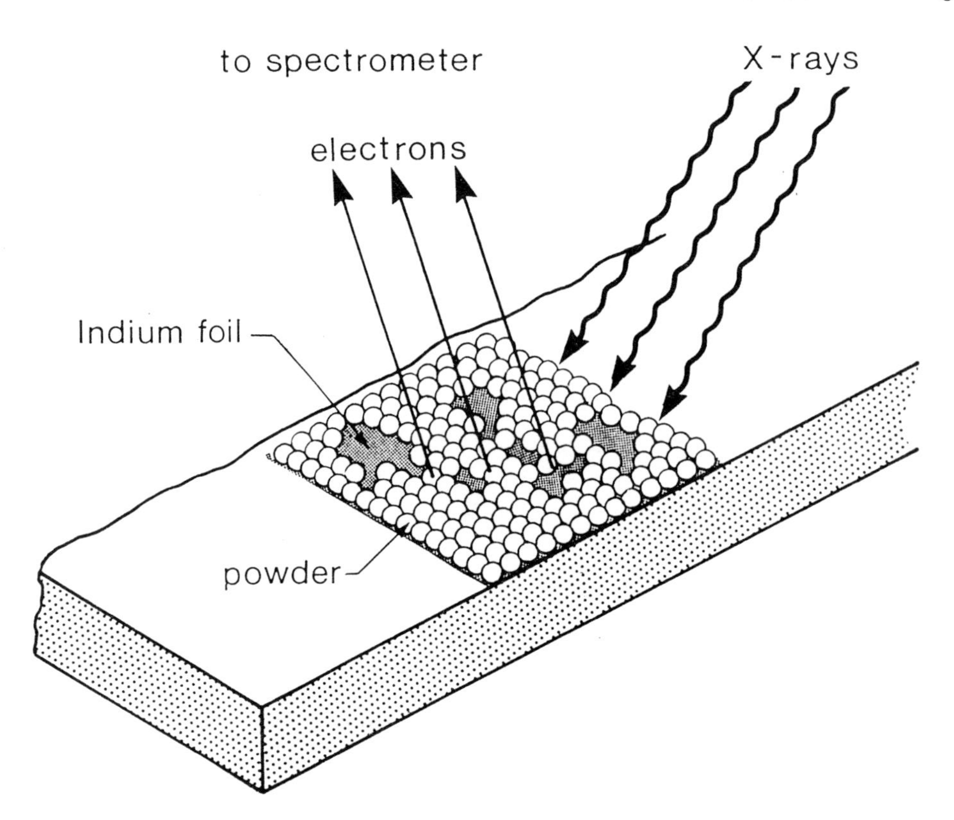

Figure 7 Powder particles mounted as a raft on indium foil or adhesive tape. In practice there will be a distribution of sizes to ensure good coverage of the mounting medium.

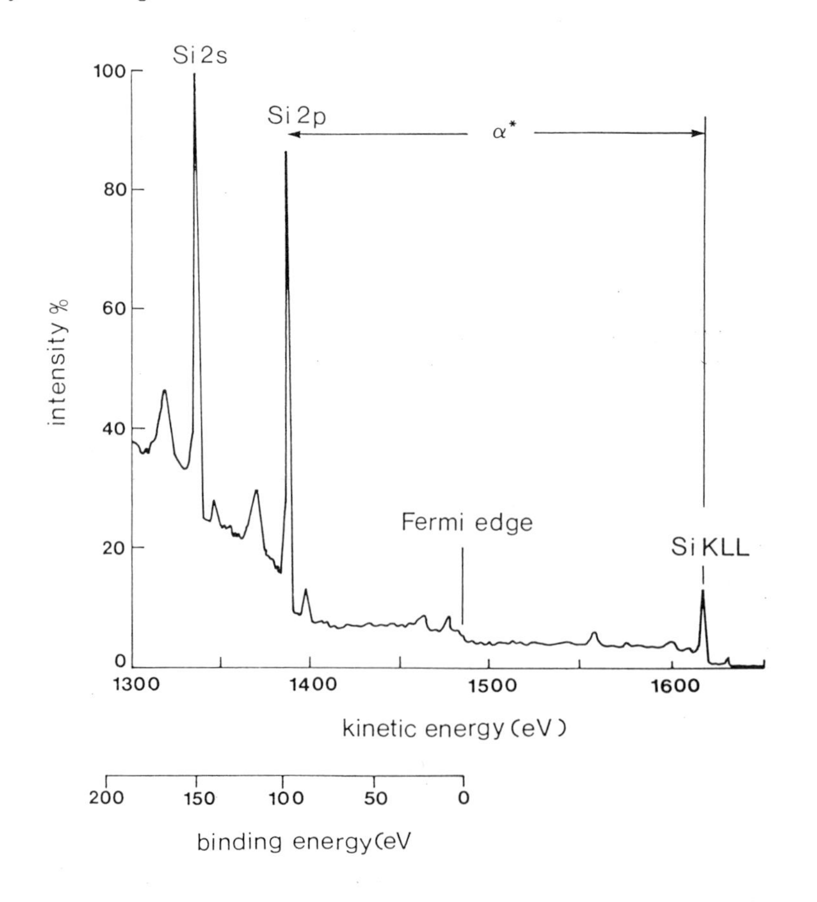

Figure 8 The Auger parameter (α) for silicon, showing the Si2p photoelectron peak and the Bremstrahlung induced SiKLL Auger peak.*

The net charge on the surface may be positive or negative depending on whether the secondary electron emission coefficient of the specimen is greater than unity (high beam energies) or less than unity (low energies), this will lead to either a slight retardation or acceleration of the electrons as they leave the specimen surface, and a consequent shift of the Auger peak to a lower or higher kinetic energy. The magnitude of the term δ can vary according the degree of charging across a surface, which in the case of powders can vary from particle to particle. This is illustrated quite dramatically in Fig. 5 where the indium mount and the powder particles are at a slightly different electrical potential. The subsequent shift of the *OKLL* peaks relative to each other allow individual oxygen maps to be recorded for the mount and the powder.

The net result of this is a shifting (as described above), and broadening of the Auger peaks, in some cases making them indistinct. In order to map regions for specific elements on the surface of the powders (such as magnesium, iron etc.) it is necessary to establish windows for not only the peak position for the specific element but also the background level. With the possibility of differential charging between individual powder particles it is possible that the enrichment of certain elements may be lost as the charging effect shifts the Auger peak position outside the window of energy analysis. With XPS to a certain extent the charging can be compensated by the addition of a flood gun [8] or the flood gun effect of certain X-ray guns. However, with the use of a finely focussed electron beam in AES or SAM the removal of excess charge is more difficult. An obvious method of reducing charging is to gain the best possible earthing for the specimen, which involves embedding the powder particles at a fine dispersion in a conducting medium such as indium foil as illustrated in Fig. 9b.

Reducing the kinetic energy of the electron beam, and operating at low specimen currents (<10nA) will also be beneficial, as it may be possible to operate in a regime in which the primary electron flux is approximately equalled by the outgoing secondary electrons. However, this action will reduce the Auger electron yield, and hence involve longer counting times in order to gain reasonable counting statistics, which dictates the need for high stability electron sources and specimen stages.

In conclusion AES of powders is often difficult due to charging, but this problem can usually be circumvented by the use of one or other of the measures described above, and a skilful Auger microscopist.

4. Technological Importance of the Characterization of Powder Surfaces

The production and storage of a fine particulate material is, in the majority of cases where powder metallurgy processing routes are involved, an intermediate stage between the starting alloy and a fully dense or near to fully dense monolithic structure. The condition of the powder prior to processing may have grave implications on the physical and chemical properties of the end product, and it is for this reason that a detailed knowledge of the surface composition of the powder is desirable. In considering changes that may occur there is the possibility of elemental segregation together with oxide growth occurring during powder production, which is particularly likely if an atomization route is involved. Once the powder has been manufactured careful attention to storage must be paid if gross degradation is to be avoided. As an example of the dramatic oxide growth that may occur on a powder particle, Fig. 10 shows SEM micrographs of RS Lital A (Al-2.5Li-1.3Cu-0.8Mg-0.12Zr) immediately after gas atomization (Fig. 10a) and following prolonged exposure to the atmosphere (Fig. 10b). The problems of attempting to consolidate powders of this type if they have been improperly stored are readily appreciated as the resultant compact, even if fully dense, will contain fragments of oxide with a consequential deleterious effect on mechanical properties. However most oxide films will be less spectacular than this and often take the form of very thin highly adherent films that defy identification by visual methods, but are readily characterized by surface analysis techniques. Let us first address the question of oxide thickness. If we consider electron emission from a planar surface it can be shown that the intensity of electrons from a depth d, I_d, is given by:

$$I_d = I_\infty \exp(-d/\lambda\sin\theta) \qquad (5)$$

where:
I_∞ - the intensity from an infinitely thick clean substrate,
λ - the electron inelastic mean free path,
θ - the electron take-off angle relative to the sample surface.

Similarly the intensity from a thin overlayer (an oxide for example), I_o, is given by:

$$I_o = I\,[1 - \exp(-d/\lambda\sin\theta)] \qquad (6)$$

By arranging these two expressions it is possible to derive an expression for oxide thickness, d, where the subscripts d and o represent the photoelectron intensities from the metal and oxide respectively:

$$d = \lambda\sin\theta\,\ln(I_o/I_d + 1) \qquad (7)$$

Table 1

Compound	Peaks Employed*	Photoelectron Binding Energy E_B (eV)	Auger Electron Kinetic Energy E_k (eV)	Auger Parameter• ($\alpha^*=E_k+E_B-h\nu$) (eV)
LaF_3	1s	684.3	658.2	88.9
InF_3	1s	685.0	656.6	88.0
LiF	1s	685.0	655.0	86.4
Na	1s	1071.6	994.5	812.5
NaI	1s	1071.6	991.4	809.4
NaF	1s	1071.0	988.8	806.2
Cu	2p3/2	932.7	918.6	597.7
Cu_2O	2p3/2	932.2	917.4	596.0
CuCl	2p3/2	932.2	915.2	593.8
Zn	2p3/2	1021.3	992.4	760.4
ZnO	2p3/2	1021.6	988.2	756.2
Ag	3d5/2	368.3	357.6	725.9
Ag_2O	3d5/2	367.4	357.1	724.5
AgO	3d5/2	443.6	355.6	723.4
In	3d5/2	443.6	410.8	854.4
InS	3d5/2	444.3	408.5	852.8
In_2O	3d5/2	444.0	407.0	851.0

*1s indicates 1s photoelectron and $KL_{2,3}L_{2,3}$ Auger peak positions were used for determination of α^*; 2p3/2 indicates 2p3/2 and $L_3M_{4,5}M_{4,5}$; 3d5/2 indicates 3d5/2 and $M_4N_{4,5}N_{4,5}$.

•If the value of α^* is negative (ie Ag or In) the photon energy, by convention, is added to the values of α^* so the expression becomes $\alpha^* = E_k + E_B$.

Table 2

Compound	Formula	Si2p E_B (eV)	Si *KLL* E_k (eV)	α^*_{Si} (eV)
Silicon	Si	99.6	1616.4	462.4
Silicon Carbide	SiC	103.9	1610.2	460.5
Silicon Nitride	Si_3N_4	102.0	1611.7	460.1
Silicon Dioxide	SiO_2	102.6	1609.6	458.6
Zinc Silicate	$ZnSiO_3$	108.1	1603.7	458.2
Silica Gel	$SiO_2.xH_2O$	107.0	1604.3	457.7

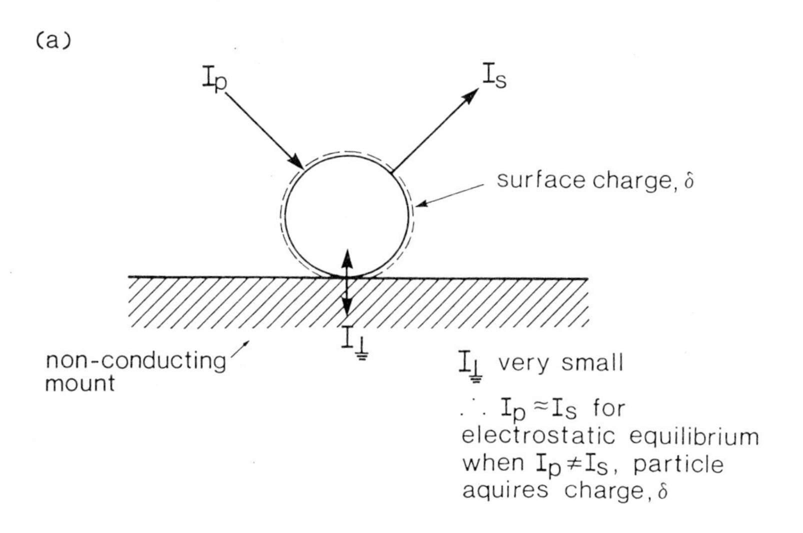

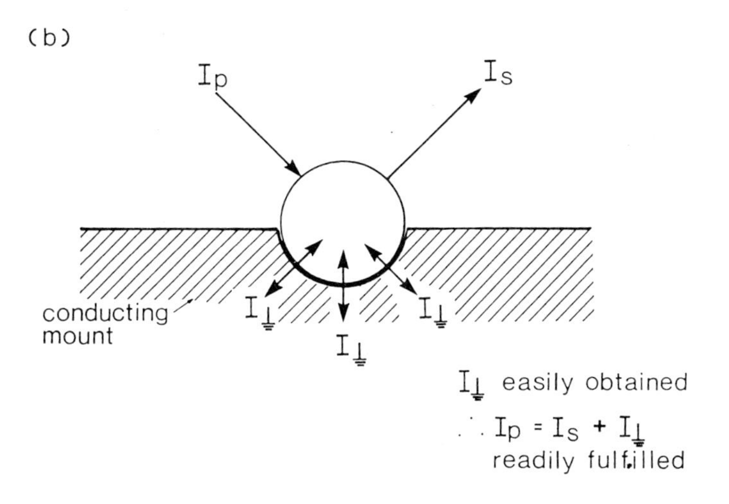

Figure 9 (a) Powder mounting for AES using a non-conducting medium prevents the dissipation of surface charge to earth, consequently the particle acquires a net charge that will effect the energy of the outgoing electrons and perhaps distort the spectrum and/or image. (b) By using a conducting mount the drain current to earth, or the return current to the specimen, is easily accommodated preventing charge build-up.

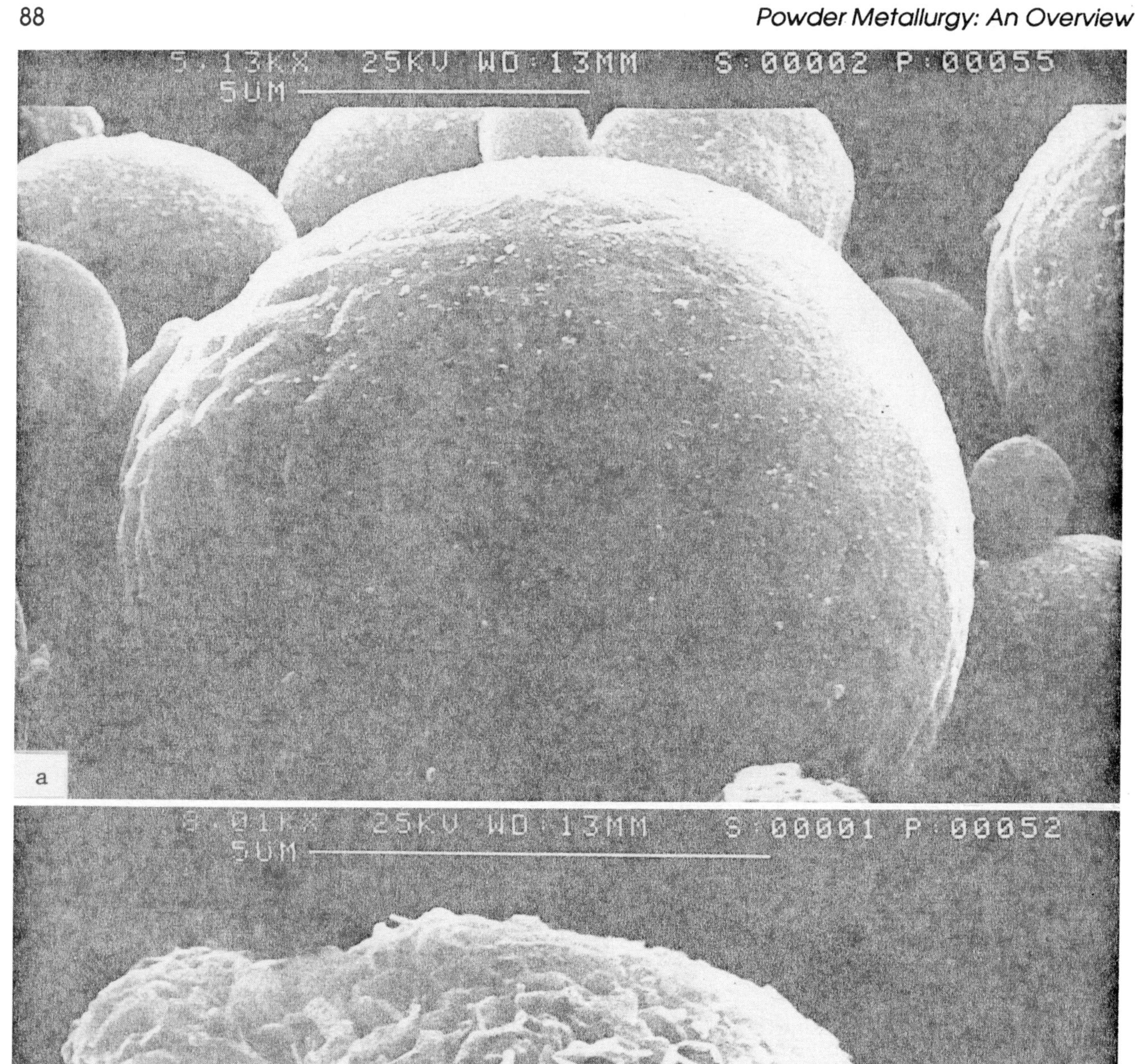

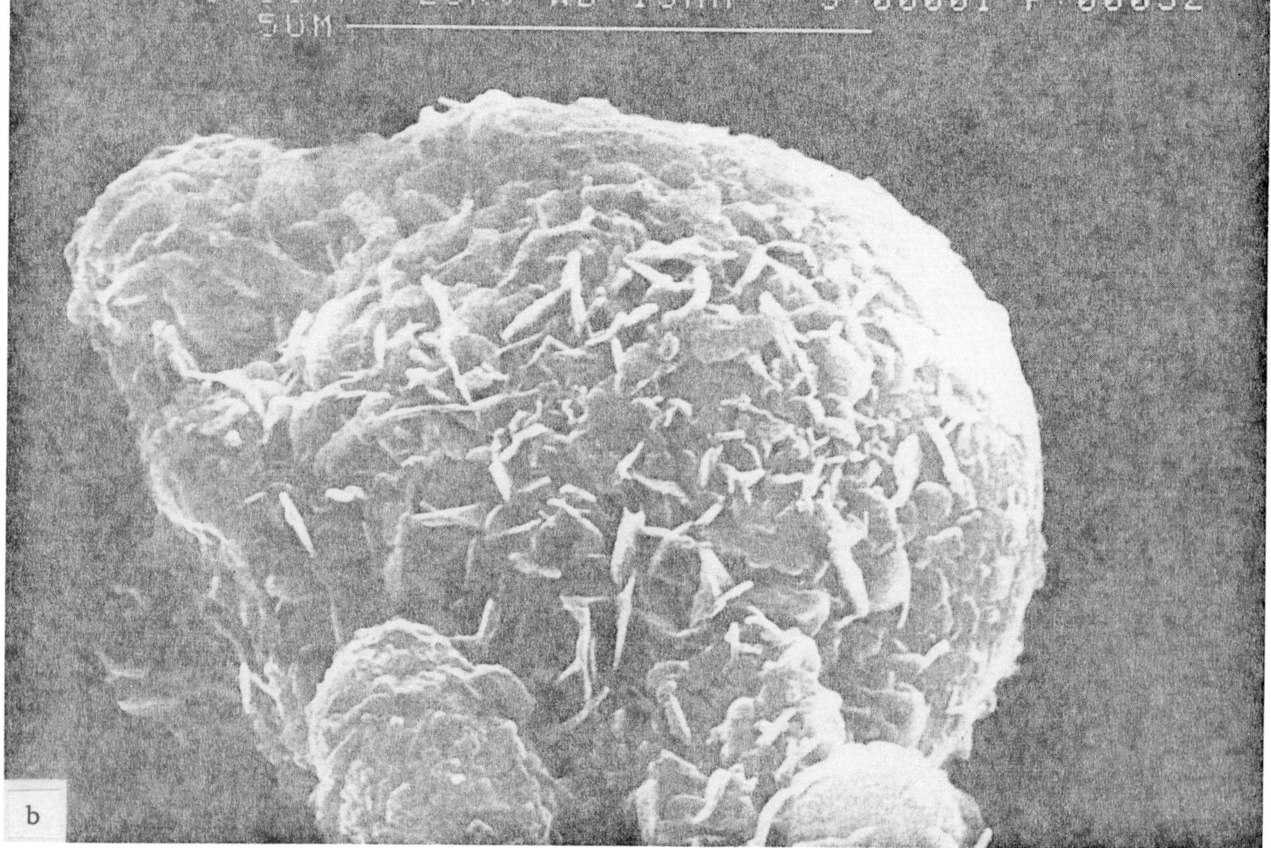

Figure 10 SEM images of a sample of Lital A powder in the (a) as-atomized condition, and (b) following exposure to humid air for several months.

However, this expression is only valid for a planar surface, in the case of spherical or near spherical particles we must make a geometrical modification to account for the range of θ values encountered (at differing points on the surface of the sphere), this has been described in detail by Cross and Dewing [9], and Carney *et al.* [10] and the expression now becomes:

$$d = 0.5\lambda \ln(I_o/I_d + 1) \quad (8)$$

In this manner the determination of oxide thickness is a simple matter, Fig. 11 shows the Pb4f spectrum of a fine lead powder, the various spectral components have been fitted using a computer routine, and use of the above equation indicates the oxide thickness to be of the order of 2.0nm.

Using this approach it has been possible to determine the thickness of the initial oxides formed on rapidly solidified powders. Values substantially less than the Mott limit of 2nm were recorded immediately after the production of various aluminium alloy powders and remain constant upon several hours air exposure [10].

A large number of RS aluminium based alloys produced by inert gas atomization at the University of Surrey have been analysed using XPS. Information that is required for a full understanding of the oxidation processes, solidification processes and corrosion processes of the powders are summarized in Fig. 12 and Table 3 below.

PARAMETER	INFORMATION
Oxide thickness	Investigation of primary oxidation processes (e.g. high or low temperature oxidation), effect of storage under varying conditions (e.g. hydration, corrosion, etc), the effect of oxide on the consolidated material (e.g. stringers).
Surface Chemistry	Enrichment or depletion of alloying or impurity elements in the oxide film or the underlying metallic region. Effect of enrichment of certain elements on the corrosive behaviour of the powder (e.g. magnesium or lithium), and the effect of these elements on the final consolidated product. Understanding of the surface segregation phenomena present in these materials. Effect of storage on surface oxide film (e.g. hydration of the film) and effect of adsorbed species such as CO_2 (e.g. the formation of Li_2CO_3).

Table 3 Summary of Information Required from the Surface of Rapidly Solidified Aluminium Powders.

The oxide thickness can be calculated using the expression indicated above, and the varying thicknesses noted in varying conditions. One major limitation of the approximation is the assumption that the aluminium oxide is present as a surface layer (save for a thin carbon contamination layer). In certain cases, such as with the presence of relatively large concentrations of magnesium, it is possible that a layered structure could exist, with magnesium oxide being the outermost film, followed by aluminium oxide then the metallic region as illustrated in Fig. 13. By merely reporting the ratio of the metallic to ionized species of aluminium the approximation would not account for a secondary overlayer. However, XPS in this case can still produce useful information regarding oxide thicknesses as a knowledge of the electron mean free path (λ) can be used to provide different depth resolutions in different X-radiations. For example, in AlKα radiation if the metallic portion of the Al2p peak can be resolved in the spectrum the oxide film is less than about 8nm, in MgKα radiation the oxide film is less than 7nm.

The presence of alloying or impurity elements in the surface region can easily be detected (as can their absence!). In order to quantify this enrichment or depletion, Carney *et al.* [10] introduced the enrichment factor term:

$$Ef^{\tau} = \frac{\{[M]/[M]+[Al]\}_{surface}}{\{[M]/[M]+[Al]\}_{bulk}} \quad (9)$$

Where the square brackets refer to the concentrations of element M and aluminium, in either the surface region detected by XPS or in the bulk region detected by, for example, atomic adsorption. Ef^{τ} is the enrichment factor for all the analysed region, i.e. elements in their metallic and ionized forms. A family of enrichment factors can be derived to quantify the enrichment or depletion in set regions of the powder e.g. in the oxide film or the underlying metallic region hence:

$$Ef^{\phi} = \frac{\{[M^{\phi}]/[M^{\phi}]+[Al^{\phi}]\}_{surface}}{\{[M]/[M]+[Al]\}_{bulk}} \quad (10)$$

$$Ef^{\tau} = \frac{\{[M^{\tau+}]/[M^{\tau+}]+[Al^{3+}]\}_{surface}}{\{[M]/[M]+[Al]\}_{bulk}} \quad (11)$$

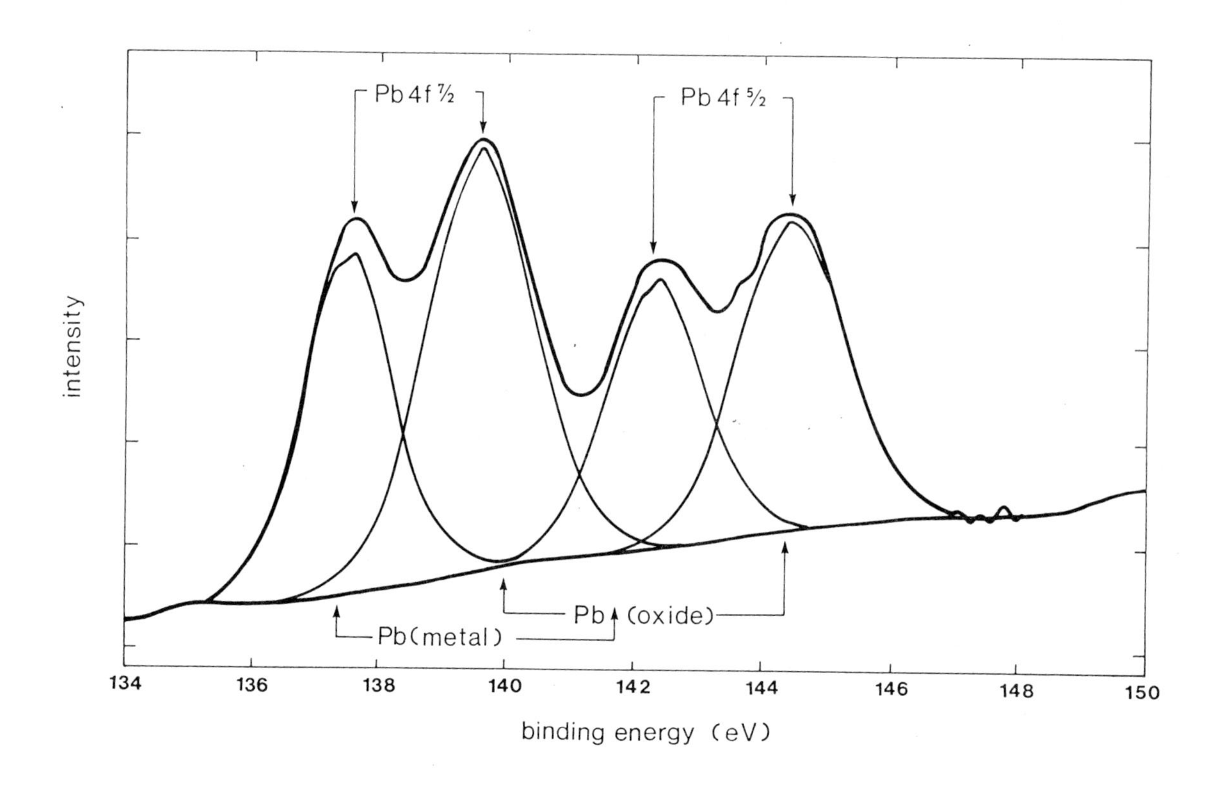

Figure 11 The Pb4f spectrum from a lead powder showing metal and oxide components of the Pb4f7/2 and Pb4f5/2.

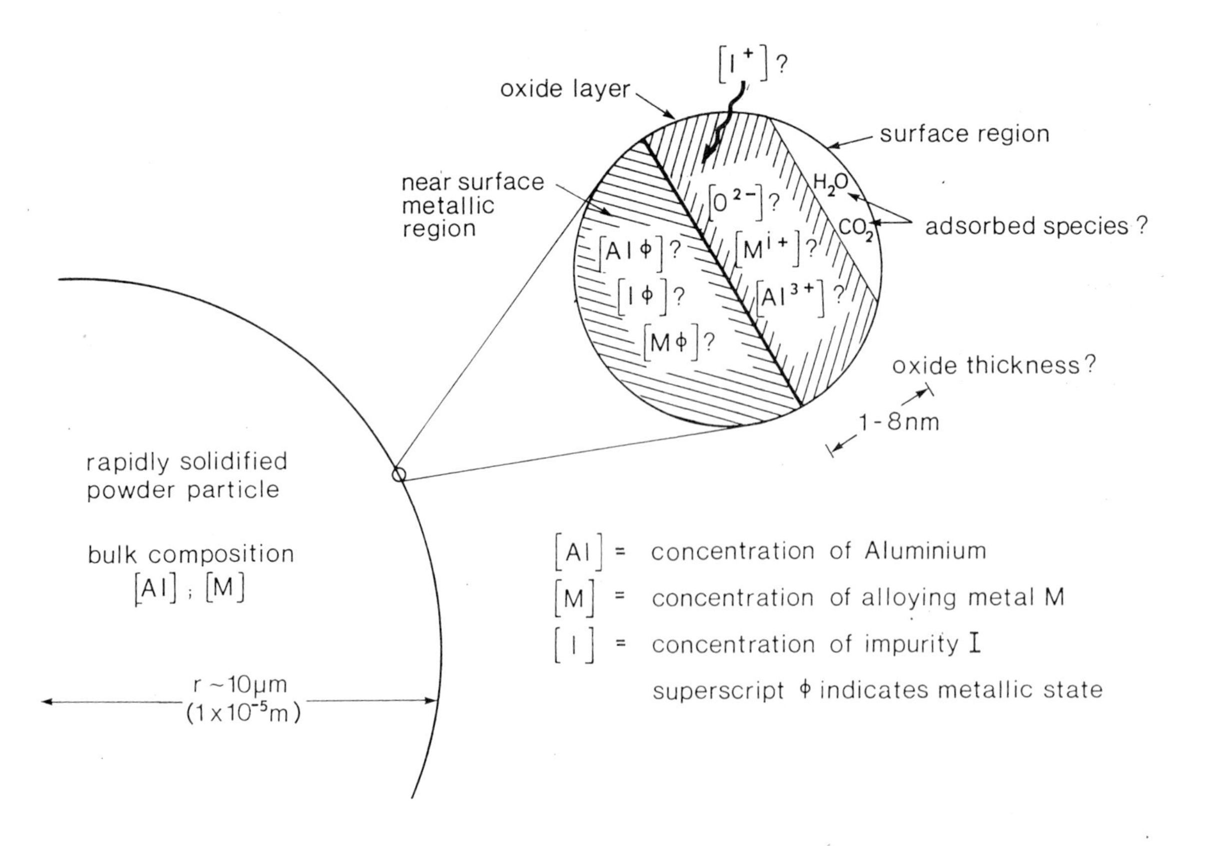

Figure 12 Information available from powder surfaces by the use of XPS and AES.

Here ϕ refers to elements in their metallic state and τ to their cationic state. Again the use of differing X-ray radiation can gain valuable information of the chemical structure of the powder's surface. For example, referring to Fig. 14, in a rapidly solidified Al-2Ni alloy probed by MgKα radiation, (Fig. 14a), the Ni3/2p peaks cannot be detected, however by using AlKα radiation the peaks are clearly defined, (Fig. 14b). The conclusion that may be drawn from this result is that nickel is depleted from the metallic region immediately below the oxide as shown in Fig. 14c. So once again a layered structure may be inferred from the XPS data.

An alternative method of assessing oxide thickness is the use of an ion beam to remove material within the spectrometer followed by sequential surface analysis; compositional depth profiling by XPS/AES in conjunction with ion sputtering. In the case of AES the electron probe will generally be smaller than the powder particles and a depth profile can be produced from an individual particle. With XPS the situation is rather problematic because of the geometrical considerations outlined above, as a large raft of powders is analysed. In addition there may be difficulties associated with the geometry of the apparatus, regions of the surface being shadowed from either the X-ray or the ion beam.

An Auger spectrum from an as-atomized powder of nickel superalloy (B-1950) is shown in Fig. 15 together with the depth profile obtained by sequential sputtering and AES, the oxide thickness can be estimated at approximately 3nm [11]. Although the accuracy of such a figure is not as reliable as that obtained from non-destructive methods outlined for XPS above, it does become the only method when considering thick oxide layers. This approach is illustrated in Fig. 16 which shows the oxygen concentration profiles for a series of martensitic chromium steels, nitrogen atomized and then heat treated in various oxidation conditions. Figure 16b shows the variation in oxide thickness as a function of bulk oxygen concentration of the powder. Thus, an alternative method of determining oxide thickness would seem to be measurement

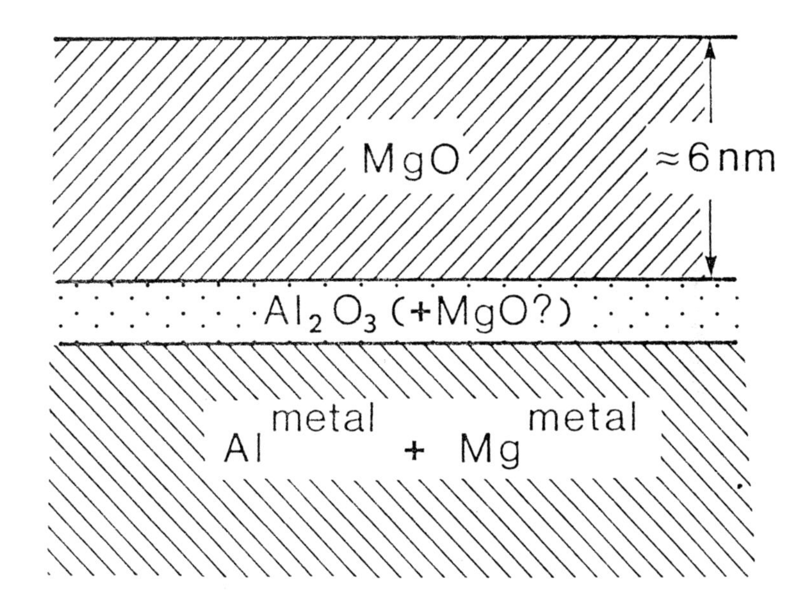

Figure 13 Heirarchy of oxide layers formed on a rapidly solidified Al-20Mg powder.

of bulk oxygen concentration once a calibration curve for a particular particle shape and alloy has been established. Although the oxide thickness seemed to have little effect on the strength of the powders after consolidation by HIPing, the ductility and impact properties were dramatically reduced. The latter parameter being reduced from 480 to 59kJm^2 as the oxide thickness increased from 8 to 38nm [12].

The XPS chemical shift can be employed in sputter depth profiling although corrections must be made for specimen and system geometry. This has been attempted by Nyborg *et al.* [13] who have produced calibration curves of the type shown in Fig. 17; sputter depth versus normalized intensity of the metallic component of the Al2p peak, calculated for various thicknesses of oxide on powder particles. The dashed line connects points where the oxide thickness and etch depth are judged to be equivalent. In this manner an experimental normalized intensity plot can be superimposed on that of equivalent depth/thickness and the intercept provides an estimation of oxide thickness [14]. The accuracy of this method depends on two factors, the spheroidicity of the particles and the accuracy with which the etch rate of the material is known.

Although oxide growth is the main concern in the analysis of environmentally formed products, other inorganic compounds can sometimes be present. A recent investigation of an Al-2.5Li-0.12Zr RS powder showed that the green compacts had very little strength and tended to crumble on handling. Analysis by XPS indicated surface segregation of lithium (seen as a Li1s peak at 55eV) whilst the C1s portion of the spectrum showed a carbonate species to be present on the surface [15]. The lithium and carbonate components of the spectrum are readily seen on the as-manufactured sample but following heat treatment they become very strong indeed as shown in Fig. 18. The free energy of formation of the reaction:

$$Li_2O + CO_2 = Li_2CO_3$$

is -170kJ mol^{-1} at 25°C and a pCO_2 value of 10^{-30} atmospheres is required for the reaction to proceed. Thus the formation of lithium carbonate will occur at least to some extent if lithium surface segregates, the oxide being an intermediate compound. The problems in compaction in this case were due to the presence of lithium carbonate as a surface phase.

5. Analysis Following Consolidation

The fate of surface phases during consolidation can be followed by studying metallographic cross-sections in the conventional way and this allows the ready identification of gross oxide inclusions. However, the fate of the thin oxide layers at prior particle boundaries cannot be followed in this way unless oxide aggregates are formed. A potentially elegant way of studying this phenomenon is to carry out consolidation by the normal route using a range of temperatures, the consolidated material can be fabricated into test pieces suitable for fracture within the Auger microscope and analysis of the fracture surface can be achieved without atmospheric contamination and oxidation. In this way it is possible to follow interfacial reactions that occur at prior particle boundaries as a function of consolidation temperature. In a study of this type Nyborg and Olefjord [16,17] investigated the HIPing of martensitic chromium steel powder at temperatures of 770°C, 1000°C, and 1100°C. The former temperatures gave rise to fracture at the prior particle boundaries whilst HIPing at 1100°C increases densification to an extent that mechanical properties are much greater. Auger spectra from the fracture surfaces of the material consolidated at the three temperatures are presented in Fig. 19, the surface obtained at the lowest temperature, when compared with the powder surface prior to HIPing is characterized by coarsening of the carbide and nitride particles and manganese is enhanced at the surface. At the intermediate temperature metallic bonding between powder particles occurs leading to a reduction in non-metallic elements at the fracture surface, together with a concomitant increase in the intensity of the iron Auger peaks.

6. Concluding Remarks

In this Chapter the basic principles of XPS and AES have been outlined with a view to providing the reader with a guide to the relative strengths and weaknesses of the two methods. XPS provides chemical information from conducting and non-conducting specimens which is readily amenable to quantification, but the technique is essentially an area averaging spectroscopy with approximately 10mm^2 contributing to the analysis. AES on the other hand is a true microanalysis method, and a spatial resolution as good as 20nm is attainable on current state of the art scanning Auger microscopes, but chemical information is rather limited. As far as quantification is concerned XPS is more advanced in this respect than AES, but with sufficient care both can provide an accuracy of ± 5%.

However, the weaknesses of the two methods are gradually being eroded; XPS is now routinely available in its small area form with an ability to analyse surface features as small as 100μm, and imaging XPS with a spatial resolution of around 5μm has recently become available. With these advances a stage is being reached in which the analysis of

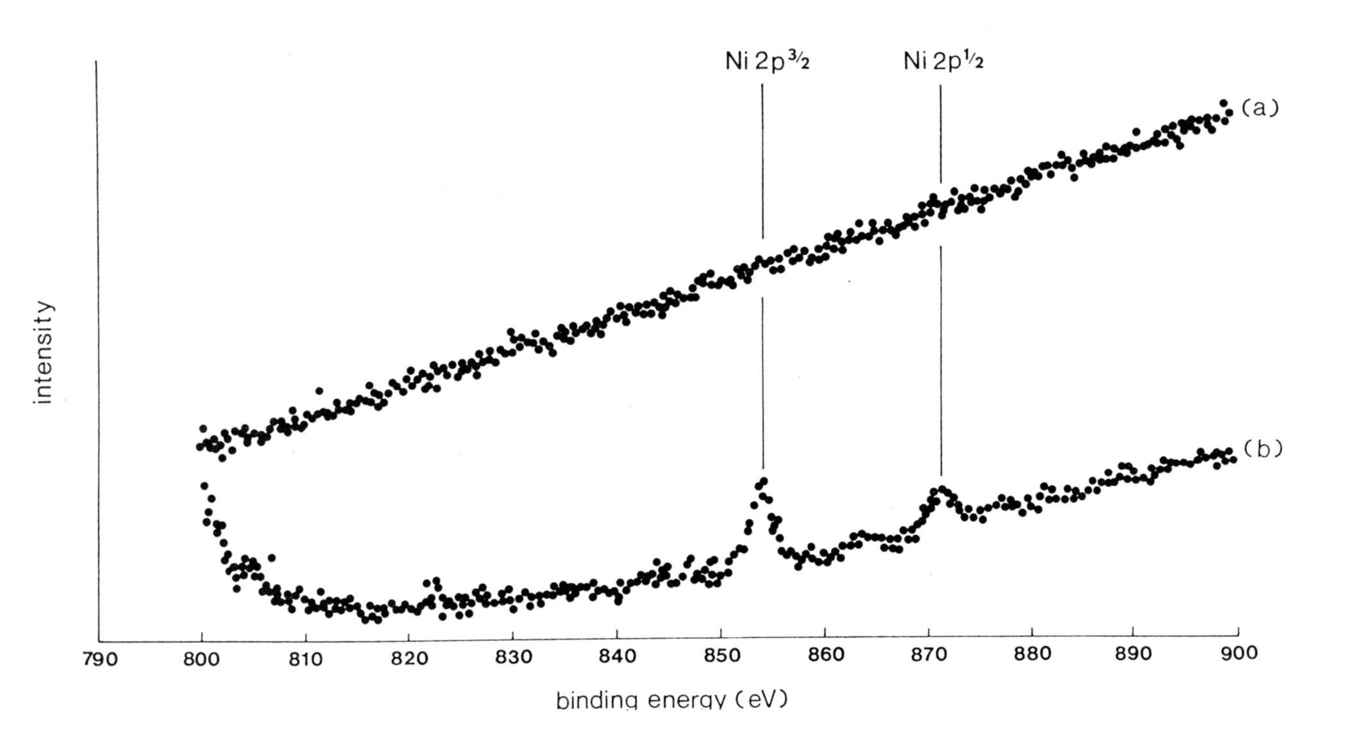

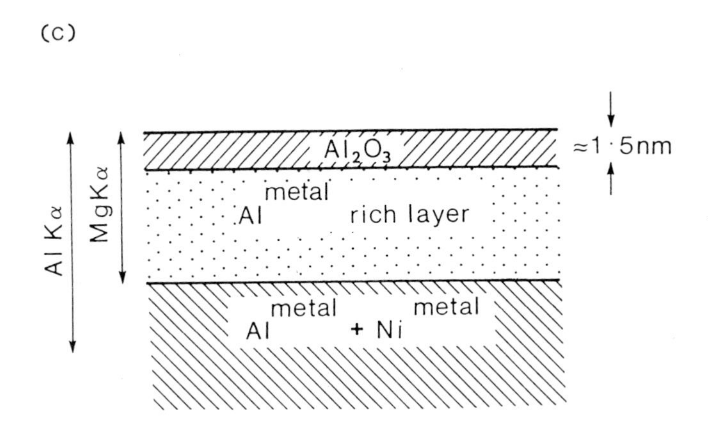

Figure 14 Ni2p region of the XPS spectrum of a Al-2Ni powder analysed using (a) MgKa radiation in which the nickel is not detected, and (b) AlKα radiation where the slight increase in analysis depth allows the identification of nickel. (c) Schematic representation of the layered structure existing at the surface of this powder.

(a)

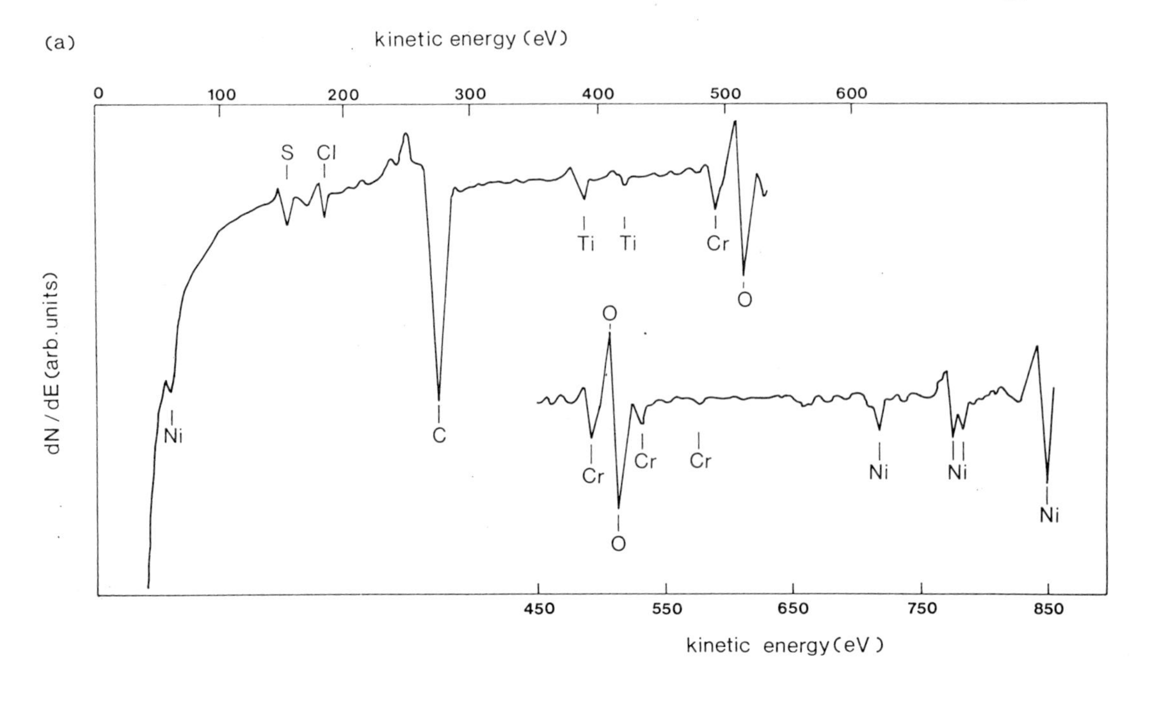

(b)

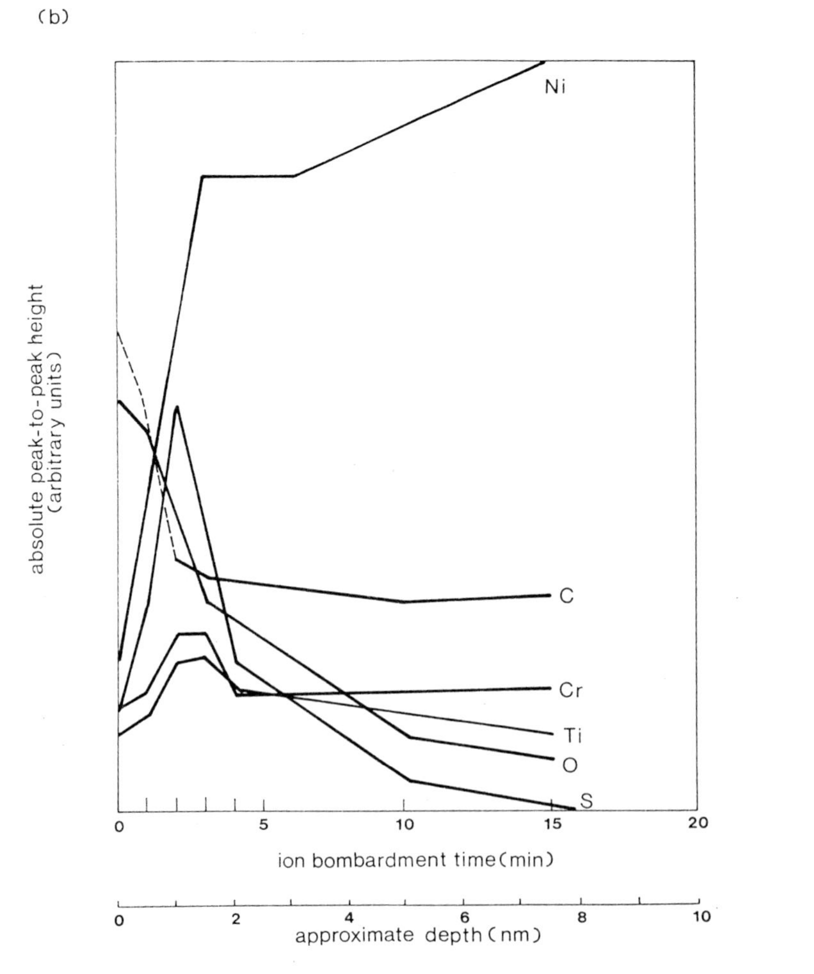

Figure 15 (a) AES spectrum from rapidly solidified superalloy (B-1950) powder. (b) AES depth profile from the same powder. (Reproduced, with permission, from reference [11].)

(a)

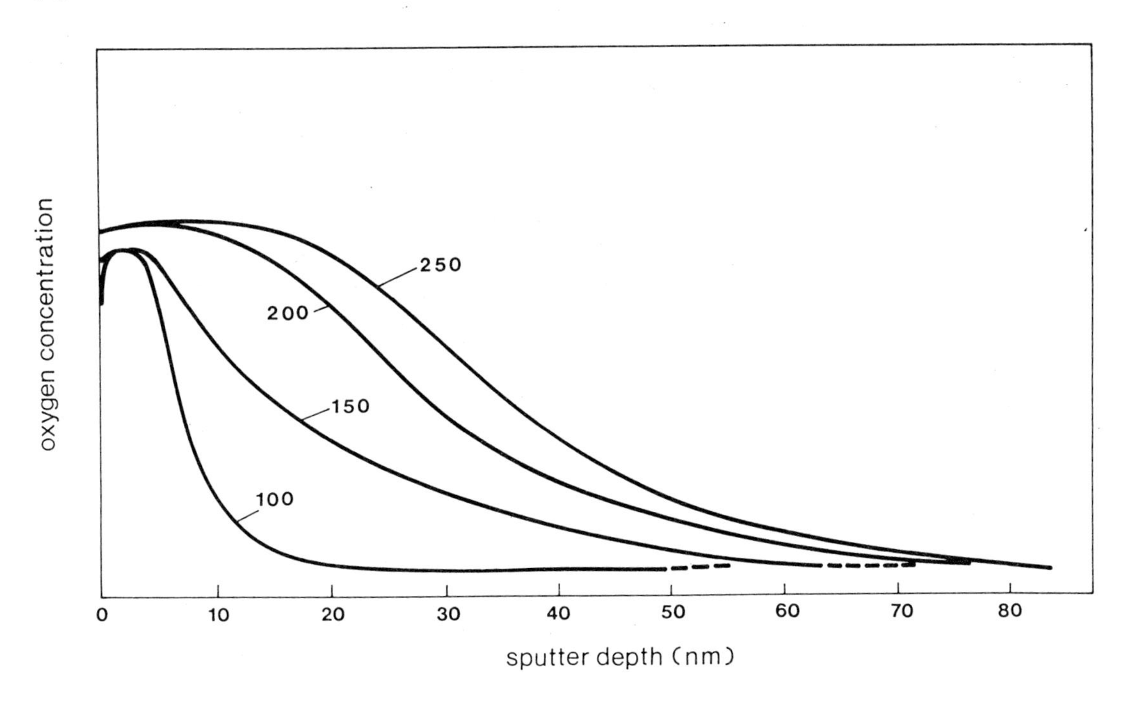

(b)

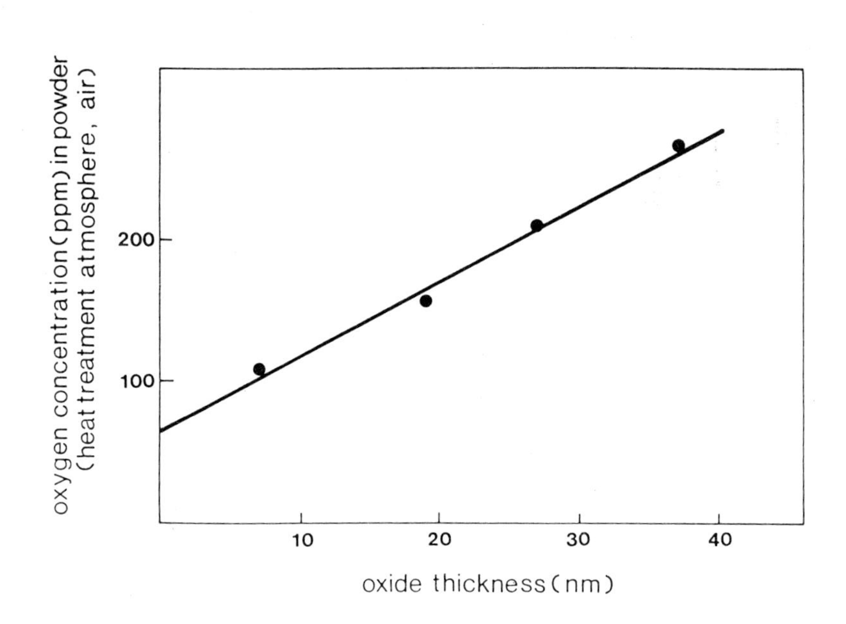

Figure 16 (a) AES oxygen concentration profiles from nitrogen atomized martensitic chromium steel. Numbers on the curves refer to the nominal oxygen concentration in the powders following subsequent heat treatment processes in air. (b) Oxide thickness as a function of bulk oxygen concentration of the oxidized powder. (Reproduced, with permission, from reference [12].)

individual powder particles by XPS becomes a reality with the attendant possibilities of mapping the chemical environment (oxidation state) of the constituent elements. In AES the emphasis will probably be on the extraction of chemical information from Auger spectra, which requires the use of analysers with high spectral resolution, until recently the preserve of XPS. Although the level of chemical information will never be as wide ranging as that available from XPS one can foresee a scenario in which AES/SAM are used for sub-micron surface studies and XPS in its various guises provides elemental and chemical information from analysis areas ranging from several square microns to tens of square millimetres.

In the application of surface analysis methods to powders, the emphasis has so far been on the use of AES because of the ease with which single powder particles and fracture surfaces can be analysed. However, XPS can be used to great effect on powders by preparing a raft of particles on double sided adhesive tape or embedded in soft metal foil. In this manner insulating powders such as minerals and polymers can be analysed without trouble. The ability of XPS to provide a non-destructive estimation of oxide thickness on metallic powders where the oxide is fairly thin, requires careful curve fitting of the metallic spectra, but these facilities are routinely available on spectrometer data sytems nowadays. An added level of sophistication can be introduced to this approach by defining metallic and cationic enrichment factors by combining XPS data with that from bulk analyses. In short it would appear that surface analysis methods have much to commend them to those in the powder metallurgy field, whether their interests be in trouble-shooting and quality assurance or basic research.

7. References

1. D. Briggs and M.P. Seah, (eds) Practical Surface Analysis by Auger and X-ray Photoelectron Spectroscopy, J. Wiley and Sons, Chichester, UK, 1983.
2. V.I. Nefedov, X-ray Photoelectron Spectroscopy of Solid Surfaces, VSP BV, Utrecht, The Netherlands, 1988.
3. J.E. Castle, in Analysis of High Temperature Materials, (ed.) O. Van der Biest, Applied Science Publishers, London, pp.141-188, 1983.
4. J.C. Riviere, The Analyst, 108, pp.649-684, 1982.
5. J.F. Watts, J. Microscopy, 140, pp.243-260, 1985.

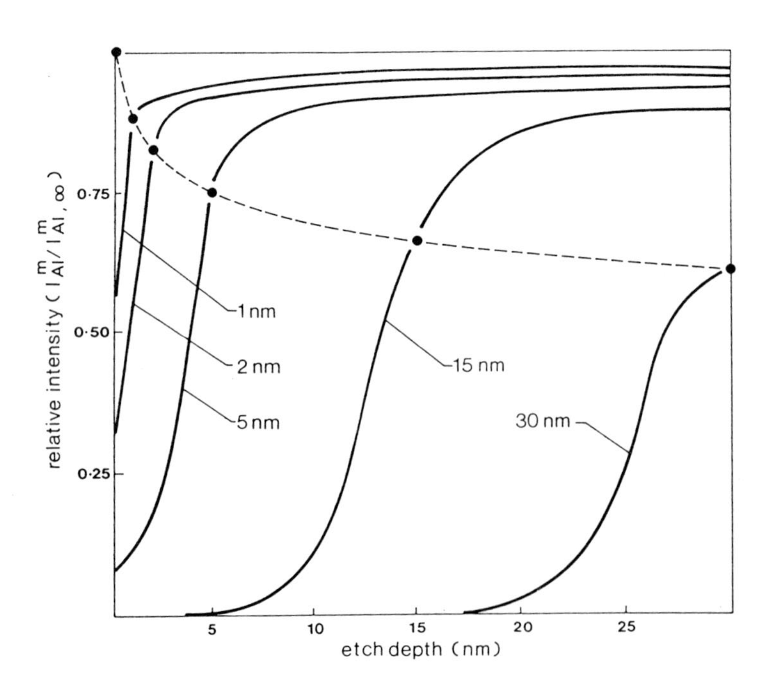

Figure 17 Calibration curve for the use of XPS in conjunction with ion sputtering for the determination of oxide thickness, calculated for various thicknesses of oxide on aluminium particles. The dashed line indicates equality of oxide thickness and etch depth. (Reproduced, with permission, from reference [13]).

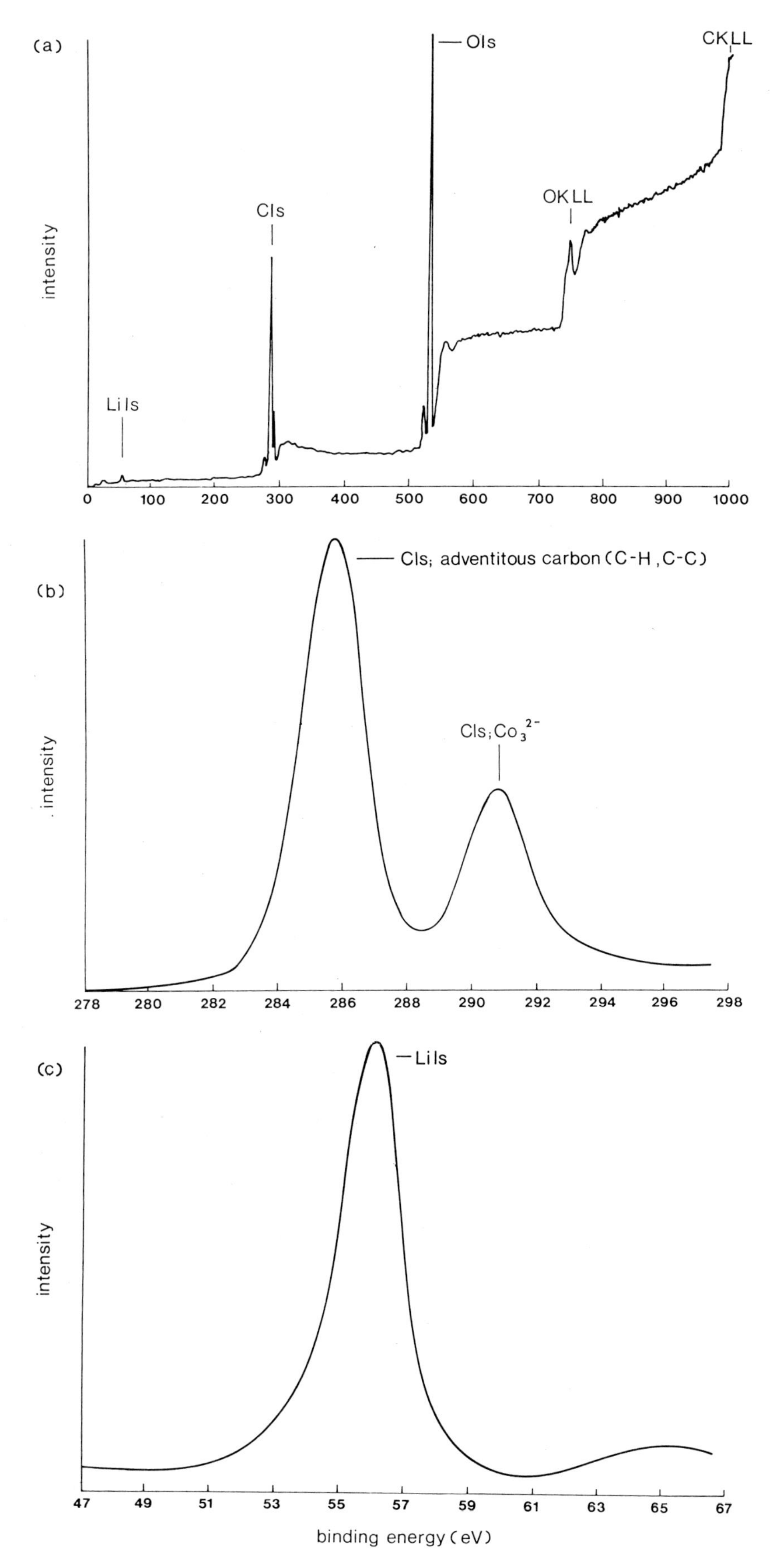

Figure 18 XPS spectra of Al-2.5Li-0.12Zr powder showing the formation of Li_2CO_3 as the surface phase. (a) Survey spectrum, (b) C1s spectrum, (c) Li1s spectrum.

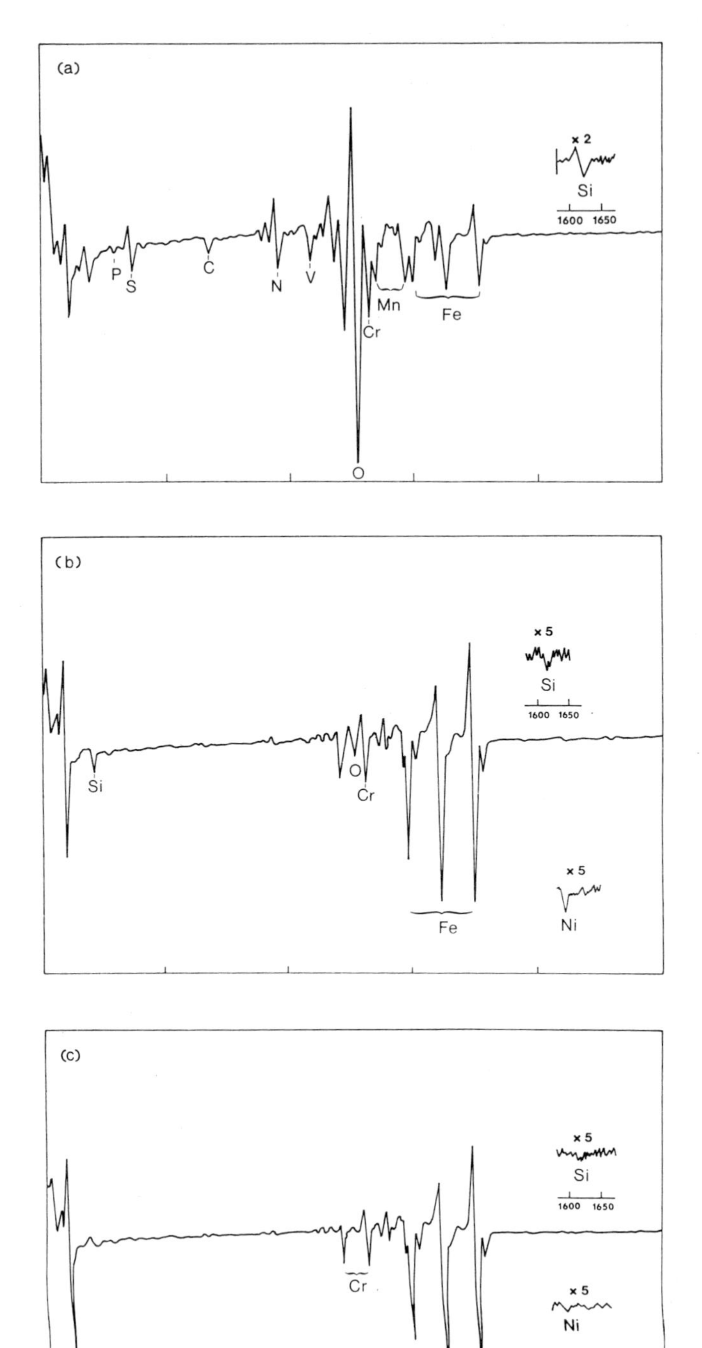

Figure 19 AES spectra recorded from the fracture surfaces of HIPed martensitic (Fe-12Cr) steel. HIPing temperatures were (a) 770°C, (b) 1000°C, (c) 1150°C. (Reproduced, with permission, from reference [17]).

6. J.F. Watts, An Introduction to Surface Analysis by Electron Spectroscopy, RMS Handbook, Oxford University Press, 1989.
7. J.E. Castle and R.H. West, J. Elec. Spec., 18, pp.355-358, 1980.
8. M.J. Edgell, R.W. Paynter and J.E. Castle, Surf. Interf. Anal., 8, pp.113-119, 1986.
9. Y.M. Cross and J. Dewing, Surf. Interf. Anal., 1, pp.26-31, 1979.
10. T.J. Carney, J.E. Castle, P. Tsakiropoulos, M.B. Waldron and J.F. Watts, Proceedings of Int. Conf. on Powder Metallurgy and Related High Temperature Materials: PM 87, India.
11. P.N Ross and B.H. Kear, in Rapid Solidification Processing and Technologies, (eds) R. Mehrabian, B.H. Kear, M. Cohen, Claitor's Publishing Division, Baton Rouge, LA, USA, pp.278-284, 1978.
12. L. Arnberg and A. Karlsson, Int. J. Powder Met., 24, pp.107-112, 1988.
13. L. Nyborg, A. Nylund and I. Olefjord, Surf. Interf. Anal., 12, pp.110-114, 1988.
14. I. Olefjord and A. Nylund, Surf. Interf. Anal., 12, pp.401-406, 1988.
15. J.F. Watts and M. Mahmoud, unpublished work.
16. L. Nyborg and I. Olefjord, Powder Met., 31, pp.33-39, 1988.
17. L. Nyborg and I. Olefjord, Powder Met., 31, pp.40-44, 1988.

Section 3

Consolidation

7

Metal Powder Injection Moulding

R.M. GERMAN
Materials Engineering Department, Rensselaer Polytechnic Institute, Troy, New York, USA

Abstract

This presentation covers the process of powder injection moulding (PIM) and its application to forming complex shaped, high performance metallic components. The PIM approach allows net shaping of a variety of materials, including metals, ceramics, cermets, intermetallics, and composites. In this process a high concentration of powder is mixed with a thermoplastic binder to form a low viscosity slurry when heated. This slurry is shaped using conventional injection moulding practices. Subsequently, the thermoplastic is removed from the compact and the powder sintered to near full density. The process has advantages for the fabrication of complicated shapes for demanding applications.

1. Basic Concepts

There is a multitude of low cost, intricate plastic parts around us everywhere, often formed by injection moulding. Even so, thermoplastic polymers are structurally inferior in comparison with other engineering materials like steel. In spite of the property differences, plastics formed by injection moulding are used because of the desirable combination of low cost and high complexity. In its simplest form, injection moulding involves heating a plastic to a temperature where flow is possible, then forcing the plastic into a shaped cavity where it is cooled before ejection. The process has similarities to metal casting, but results in greater precision and much finer surface detail.

In turn, injection moulding of polymers filled with powders is a recognized route for improving the polymer strength. Filled polymers are routinely used for applications requiring electrical conductivity. The filled polymer is a composite incorporating the fabricability of the polymer and the mechanical, thermal, magnetic, or electrical properties of the filler [1].

The recent evolution has been to maximize the content of solid particles and to remove the binder during sintering. As a consequence, a new powder forming process has evolved, providing shape complexity, low cost forming, and high performance properties [2]. The key steps in this process are outlined in Fig. 1. This new shaping process, termed powder injection moulding (PIM), is begun by mixing selected powders and binders. The mixture is granulated and injection moulded into the desired shape. The polymer imparts viscous flow characteristics to the mixture to aid forming, die filling, and uniform packing. After moulding, the binder is removed and the remaining powder structure sintered. The product may then be further densified, heat treated, or machined to complete the fabrication process. The sintered compact has the desirable complex shape and high precision of plastic injection moulding, but from materials capable of performance levels unattainable with filled polymers.

The process of powder injection moulding has been in incubation since the late 1920s. Early applications were to forming ceramic spark plugs, with more recent emphasis on metallic, composite, and cermet materials with complex shapes for high performance applications [3-5]. This interest accelerated as the property and shape limitations of conventional powder processing routes were realized by design engineers and parts fabricators. In the past ten years major progress has been made with forming heat engine components by PIM techniques [6-11]. Additional recognition came in 1979 with two awards of distinction given to PIM products by the Metal Powder Industries Federation [12]. One component was a screw seal used in commercial jetliners. The second award was given to a PIM niobium alloy thrust chamber and injector for a liquid propellant rocket engine. Thus, in the recent past PIM has moved into the forefront of advanced materials manufacturing.

2. Process Outline

As noted in Figs. 1 and 2, the steps involved in forming a component by powder injection moulding include the following:

1. Selecting and tailoring a powder for the process.
2. Mixing the powder with a suitable binder.

3. Producing homogeneous granular pellets of mixed powder and binder.
4. Forming the part by injection moulding in a closed die.
5. Processing the formed part to remove the binder (termed debinding).
6. Densifying the compact by high temperature sintering.
7. Post-sintering processing as appropriate, including heat treatment or further densification.

The equipment used in shaping the compact is the same as used for polymer injection moulding. A typical moulding machine is sketched in Fig. 3. It consists of a hydraulically clamped die that is filled through a gate from a pressurized and heated barrel. A motor driven reciprocating screw is employed to maintain mixture homogeneity and to generate the pressure needed to fill the die. The feedstock enters as granules from the loading hopper. Proper control of the moulding operation is attained using a closed-loop feedback system. A schematic cross-section through the die area is given in Fig. 4 with the sequence of moulding steps outlined in Fig. 5. The first steps are die clamping and filling. A high pressure is attained rapidly and maintained until the compact becomes rigid. After sufficient cooling in the die, the compact is ejected and the cycle repeated. Figure 6 is a sketch of the pressure versus time during the actual moulding cycle (approximately 20s). The initial pressure rise causes mould filling in a very short time. The maximum pressure (usually less than 15MPa) is held until the gate has chilled.

This technology is in practice with many variations, yet the basic facts are similar. Often the binder is a thermoplastic polymeric material, but water and various inorganic substances are used successfully. A typical binder is composed of waxes and plastics, such as paraffin and polypropylene, with appropriate wetting agents to provide binder adhesion to the powder. The amount of binder ranges from 15 to 50 volume per cent of the mixture, depending on the powder packing characteristics. Since monosized spheres exhibit a random packing density of 64% of theoretical, a PIM mixture using spheres will have near 36% by volume of binder. It is desirable to attain a high packing density of particles in the binder system while maintaining a low mixture viscosity; this dictates that there is sufficient binder to fill all interparticle voids and to lubricate particle sliding during moulding. Best moulding success is attained with mixture viscosities below 100Pas. This viscosity is dependent on the inherent binder viscosity, mixture temperature, shear rate in moulding, packing density of the powder, polar dispersants, and concentration of powder in the binder [13,14].

Small particles are used in PIM to aid sintering densification. Carbonyl iron and gas atomized stainless steel powders with spherical shapes and mean particle sizes of 4 to 15 μm are typical. Some progress has been made using particles as large as 100 μm, but compact shape distortion is a greater problem. The loading of solid particles in the feedstock is an important moulding consideration [15]. This parameter determines the amount of binder as well as the mixture viscosity and subsequent compact shrinkage in sintering.

The injection moulding step involves concurrent heating and pressurization of the feedstock. Because of the high thermal conductivity of PIM mixtures the moulding process requires careful attention to several variables. Much progress in computer modelling and control is occurring to better control defects that might arise in moulding [16].

After moulding the binder is removed from the compact by a process termed debinding. As illustrated in Fig. 7 there are several options for debinding based on thermal, solvent, and capillary extraction approaches [17,18]. For metals, thermal debinding is used most frequently, where the compact is heated slowly to 600°C in air to decompose the binder and oxidize the particles. This oxidation process provides some handling strength to the compact after the binder is removed. Several defects can arise in the debinding process, especially if the heating rate is more than a few K/h. A multiple component binder helps overcome such problems by leaving one component in the compact to hold the particles in place while a lower stability component is removed by evaporation. Alternatively, use of capillary wicking and solvent extraction processes decrease the debinding time, but are less widely applied as yet. These also rely on multiple component binder systems to progressively open the pore structure while holding the particles in place.

At the end of debinding some binder may remain in the pores to provide handling strength. The next step is sintering, which can be incorporated directly into the debinding cycle. Sintering provides strong interparticle bonds and removes the void space by densification. Isotropic powder packing allows for predictable and uniform shrinkage, so tooling is oversized as appropriate for the final compact dimensions and powder packing density. Computer techniques are being developed to handle tool design to account for the sintering shrinkage [19]. After sintering the compact has good strength and microstructural homogeneity with isotropic properties superior to those available with many other processing routes.

3. Basic Attributes

Powder injection moulding is capable of producing a new range of components from powders. A main attraction is the economical production of complex parts from high performance engineering materials. The Venn diagram shown in Fig. 8 reinforces this concept. The three basic considerations are shown as overlapping circles. The intersection of these three circles is an attractive area for the application of PIM. Because of the high final density, the PIM products are suitable for high performance levels. Examples of these attributes for PIM are illustrated in Fig. 9, which provides photographs of a few compacts formed by the PIM route.

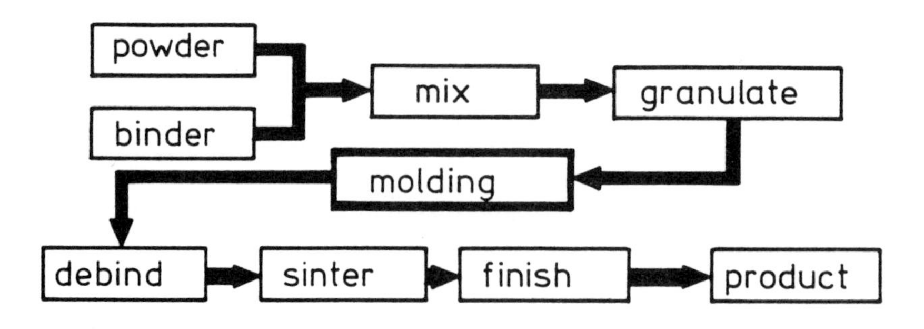

Figure 1 A flow chart identifying the key steps in the PIM process.

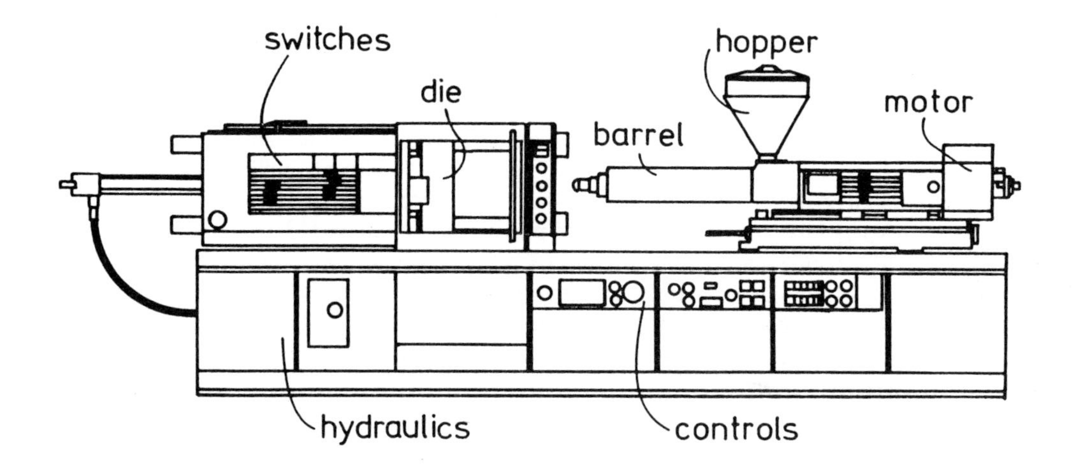

Figure 3 A sketch of an injection moulding machine as used for PIM processing showing the key components.

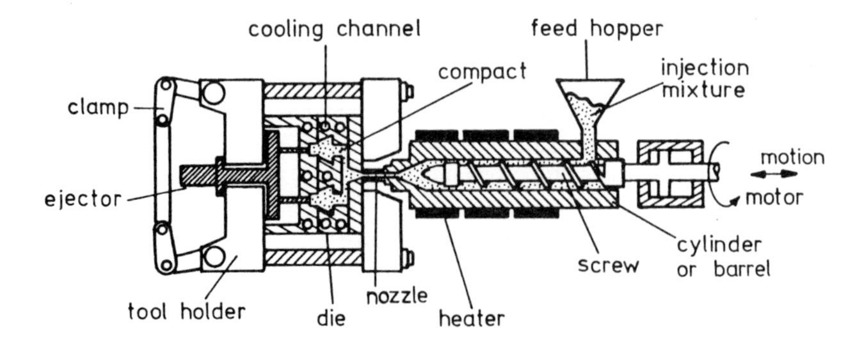

Figure 4 A cross sectional view of the die and barrel areas during moulding.

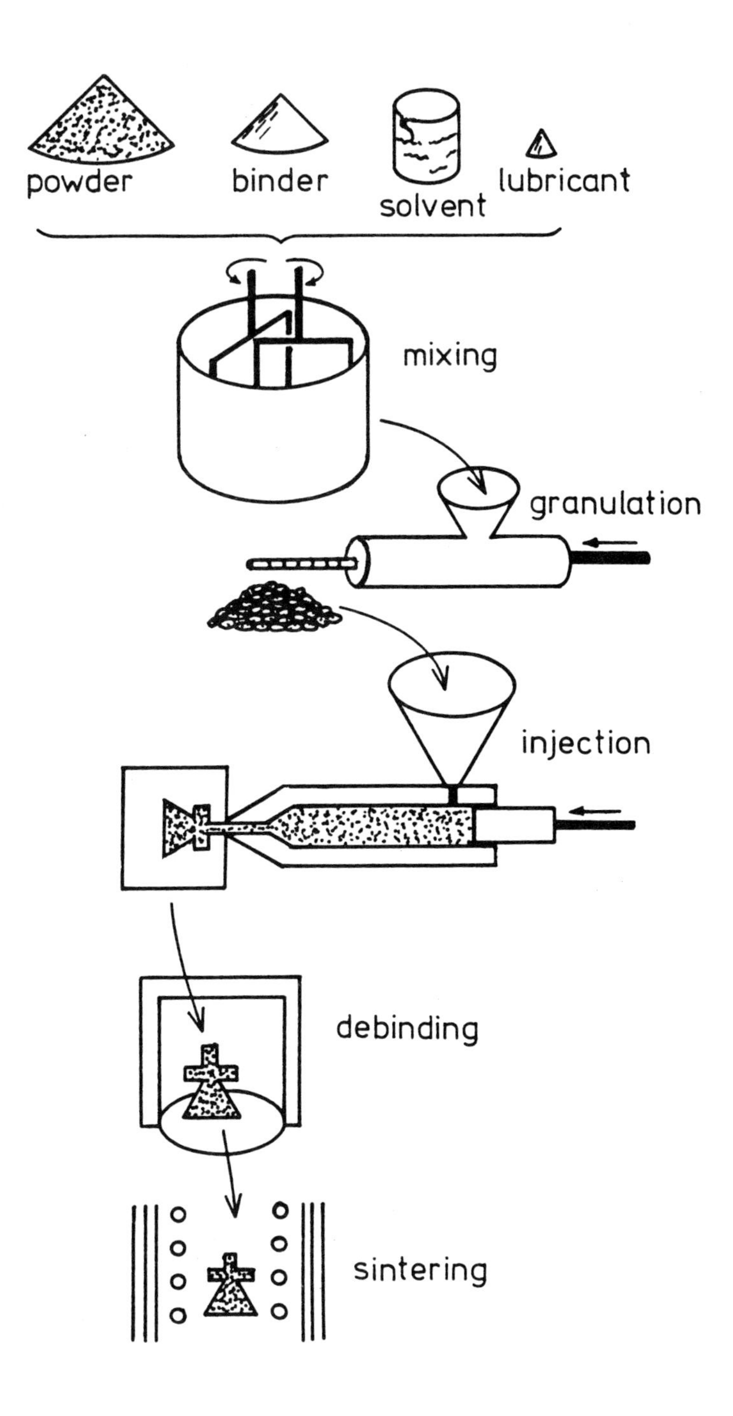

Figure 2 A schematic diagram of the processing steps and equipment involved in powder injection moulding.

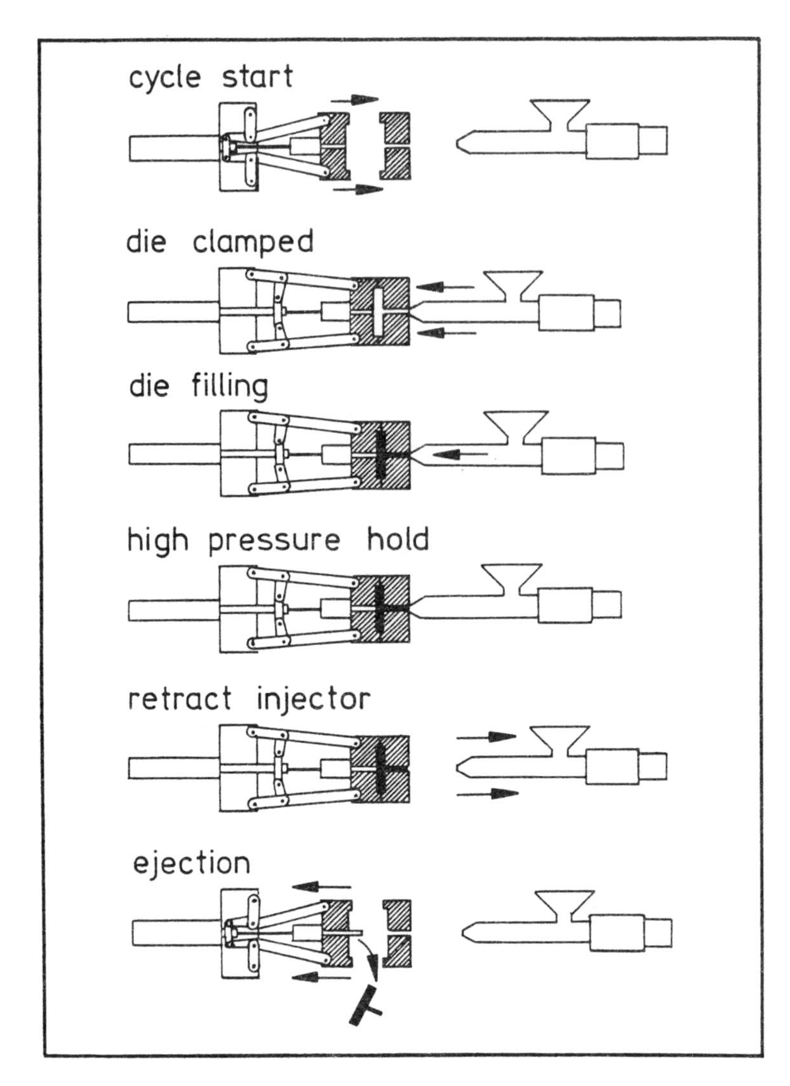

Figure 5 The moulding sequence showing the key points in the PIM cycle.

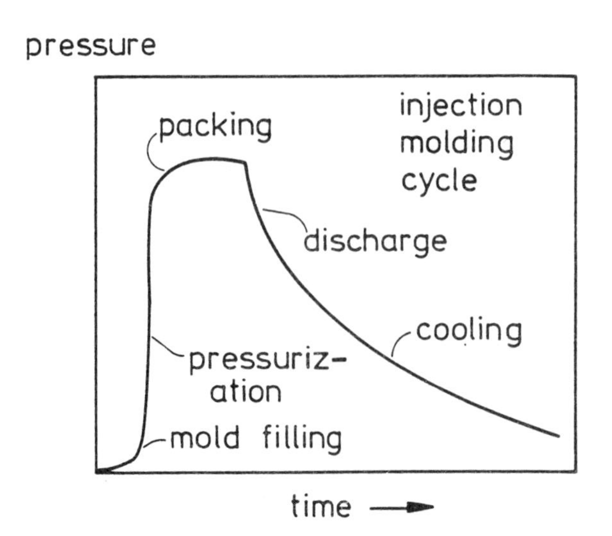

Figure 6 The pressure versus time during the PIM molding cycle; a typical peak pressure might be 15 MPa and the cycle time might be 20 s.

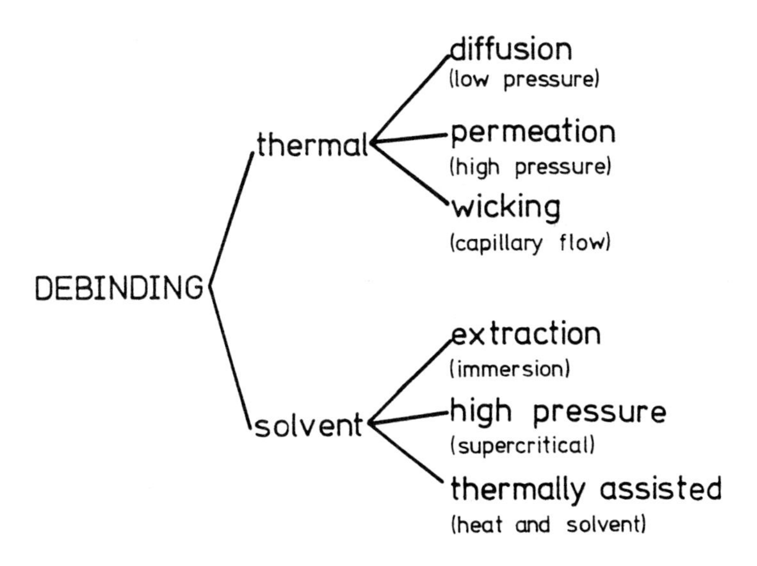

Figure 7 The options available for debinding PIM components include both solvent and thermal decomposition steps.

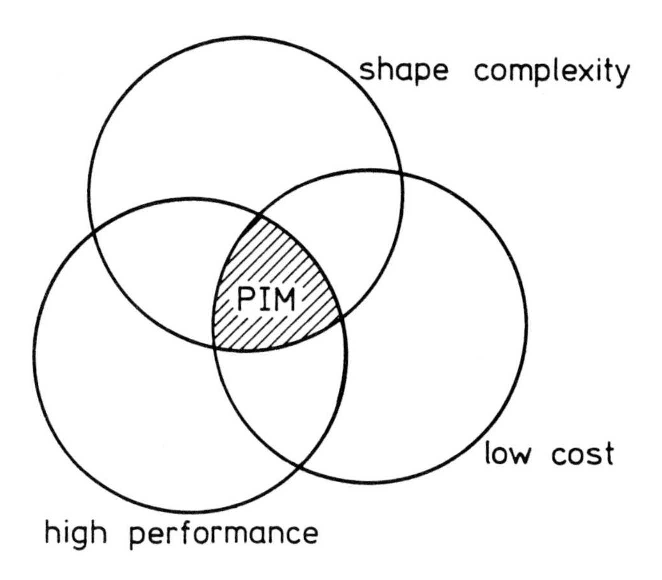

Figure 8 This Venn diagram identifies the optimal application of PIM with the combination of concerns over processing costs, performance levels, and component shape complexity.

Besides the primary advantages of shape complexity, low cost, and high performance, several secondary attributes are worth notice. These include near-net shape geometries with concomitant material and processing savings through the recycling of wastes. The optimum part and mould design may be modelled by methods already established for thermoplastics [20]. Also, capital equipment costs are relatively low in contrast with other shaping technologies. Finally, the process has a high level of automation, yet is applicable to short production runs.

The high final density and good properties make PIM a natural choice for the fabrication of high performance components. Table 1 provides a summary of a few systems and the measured mechanical properties. These properties are competitive with alternative fabrication approaches. The density and property advantages result from the uniform powder packing attained in the PIM process. For comparison, consider the pressure gradients in die compaction [21,22]. The wall friction between the powder and the die induces pressure gradients during compaction which lead to a non-uniform powder density. Figure 10 shows the density gradients in a copper compact die pressed to a mean density of 5.3 g/cm^3. Dimensional change (typically shrinkage) during sintering depends on the powder packing density; low density regions shrink more than high density regions. Because die compacted powders exhibit non-uniform shrinkage, they are often sintered at lower temperatures where densification is minimized, necessarily giving inferior sintered properties.

For the powder compact shown in Fig. 10, the centre portion will shrink more than the top and bottom, resulting in a hour-glass shape after sintering. Figure 11 plots the standard deviation in diameter for green and sintered cylindrical compacts formed by traditional die compaction and PIM. In the green condition the die compacted specimen has the smallest standard deviation in diameter, nearly half that of the PIM specimen. However, after sintering the non-uniform shrinkage of the die compacted powder leads to a much larger standard deviation. In most situations, such non-uniformity of final dimensions is not acceptable. Thus, low sintering temperatures are employed after die compaction to avoid dimensional warpage, with a sacrifice in final density.

The porosity of the final PIM product is the primary determinant of its properties; a lower porosity leads to improvements in such attributes as strength, toughness, ductility, conductivity, and magnetic response [23,24]. The hydrostatic forming character of PIM minimizes density gradients, which allows more predictable sintering densification and higher properties. Generally the sintered density exceeds 93% of theoretical, as compared with the 85% density attained in die compaction. Furthermore, the pores in PIM compacts are not interconnected and are less detrimental to properties. Consequently, PIM compacts have very attractive properties [6-11,25-30].

PIM is applicable to a wide range of materials including steels, nickel, most ceramics, cemented carbides, tungsten heavy alloys, niobium, silicon nitride, silicon carbide, superalloys, aluminide intermetallics, cobalt alloys, and fibre reinforced metals and ceramics. One area of current research is the fabrication of composite materials by PIM techniques. As Fig. 12 illustrates, PIM applicability falls naturally into that range termed "composite". From this simple ternary diagram, with polymer, metal, and ceramic as the terminal points, it is evident that injection moulding can be used to form a variety of composites. This includes metal matrix, ceramic matrix, and polymer matrix materials with dispersed fibres or particles of ceramic or metal. Figure 13 sketches how the flow pattern in moulding can be adjusted to place fibres into specific orientations in the compacts. This capability opens a new area of tailored composite microstructures that is yet to be exploited.

One important attribute is the application of PIM to the emerging high performance materials. These materials have a strong dependence on processing. Although several routes are available for producing the new materials, PIM approaches are attractive because of the combination of technology, cost, and feasibility [31]. Another advantage of PIM is the co-moulding of different materials, i.e. forming part of a component from one material and then another portion from a second material. Such components can be joined together in the green condition [32]. This option has merit for forming corrosion barriers, wear surfaces, and electrical interconnections in ceramics.

For the producer of PIM parts, some important attributes relate to manufacturing ease, including process control, flexibility, and automation. Small production runs are possible with as few as 5000 parts being economical. This flexibility fits well with the current direction in manufacturing, especially for the advanced ceramics [33]. There is always a desire to make rapid shifts in production with minimum expense. In this regard, PIM is attractive since much of the processing equipment remains invariant while the powder composition and tooling change. The process is an alternative to investment casting and die casting for metals, and slip casting and cold isostatic compaction for ceramics, with shape being less of a cost factor in manufacturing.

As with all technologies, the bottom line is the economics. Here, PIM can be defended as cost advantageous for the more complex shapes and specialist materials. One advantage is PIM's reduced number of production steps as compared with other techniques [26,34]. This is partly due to the elimination of secondary operations like grinding, drilling, and boring that typically are required for precision components. Also, since the feedstock material can be recycled, material use is nearly 100%. This is especially important for costly raw materials such as refractory metals, specialist ceramics, and precious metals [11,25,35,36].

Figure 9 Some examples of the complex shapes being formed by the PIM process.

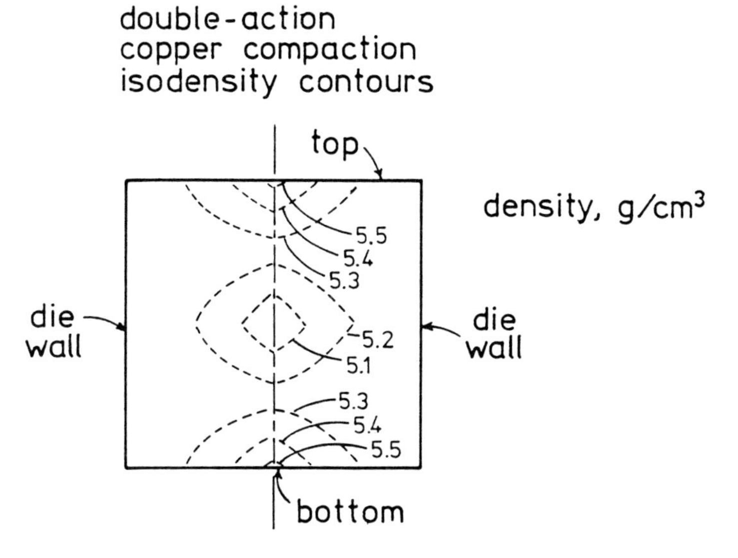

Figure 10 Density gradients in a cylindrical powder compact made using double action die compaction. The isodensity lines are based on the analysis of Thompson [22].

Table 1 Example room temperature mechanical properties for PIM materials

material	fractional density, %	yield strength, MPa	failure elongation, %
alumina	99	510	---
304L stainless	96	185	35
316L stainless	98	205	68
17-4 PH stainless	96	900	---
silicon carbide	95	300	---
silicon nitride	98	740	---
Fe	95	100	28
Fe 2% Ni steel	93	185	33
Fe 2% Ni steel HT	93	1090	4
Fe 8% Ni 1% C steel	96	970	5
Fe 50% Ni	96	170	21
Fe 2% Cu steel	95	335	23
Fe 3% Si	99	345	25
90% W heavy alloy	99	600	32
97% W heavy alloy	99	710	8
WC-10% Co	99	1140	---

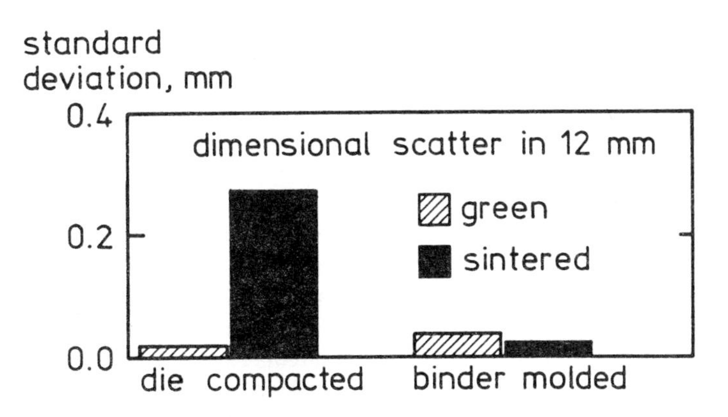

Figure 11 The standard deviation in diameter for two stages of each process; die compacted values are shown on the left and binder moulded values are shown on the right.

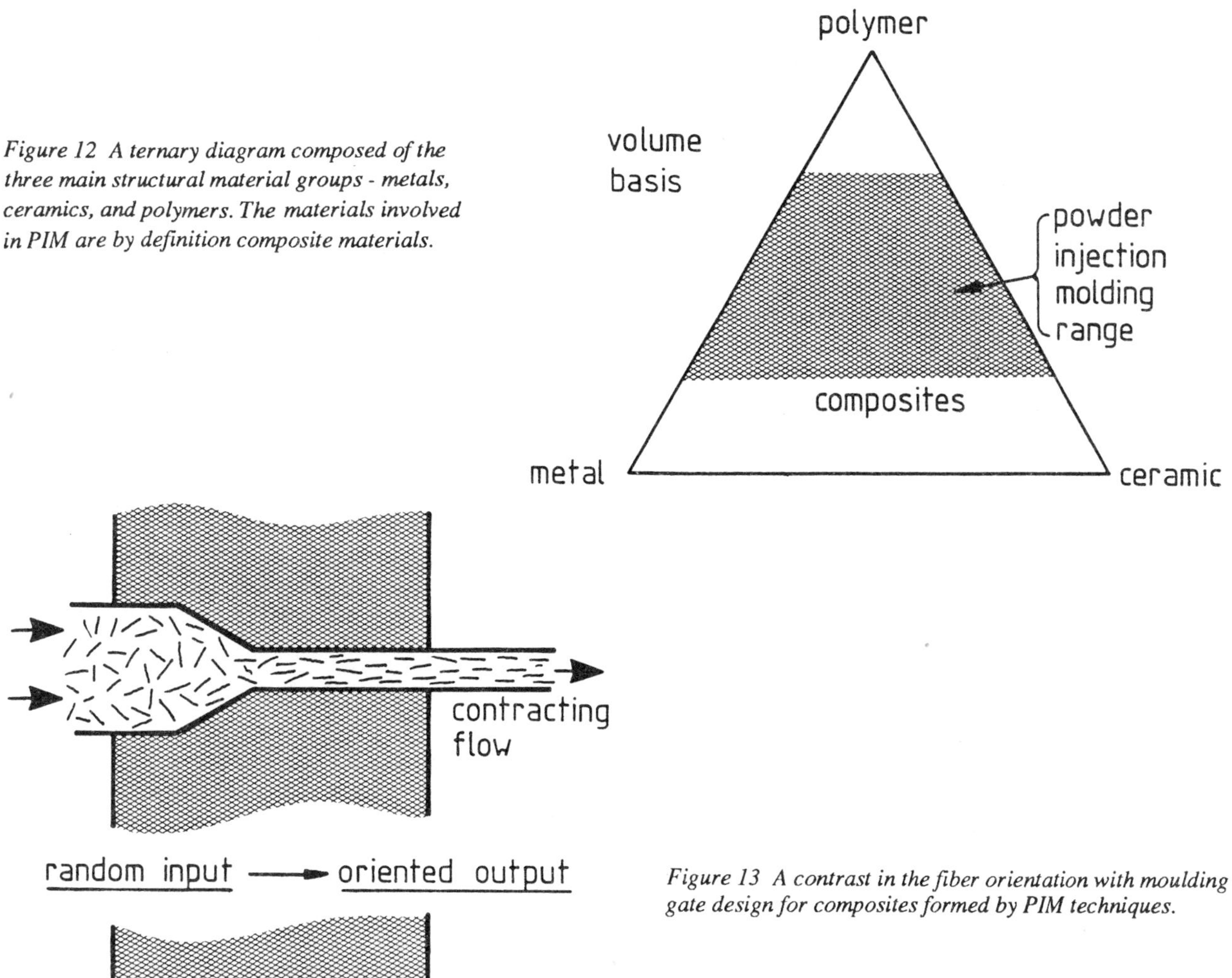

Figure 12 A ternary diagram composed of the three main structural material groups - metals, ceramics, and polymers. The materials involved in PIM are by definition composite materials.

Figure 13 A contrast in the fiber orientation with moulding gate design for composites formed by PIM techniques.

4. Process Limitations

The previous section outlined several attributes of PIM. Generally, it is viable for all shapes that can be formed by plastic injection moulding. Still, it is not cost competitive with traditional die compaction for shapes with relatively simple or axial-symmetric geometries. Also, the availability of suitable powders limits the current materials selection. Other limitations trace to equipment size and sophistication. Large components require larger moulding and sintering devices which prove more difficult to control. Accordingly, PIM is best applied to complex, small shapes.

The formation of thick cross-sections remains a problem with PIM because binder removal times scale with the square of the section thickness [17]. The current upper limit on section thickness is approximately 100 mm with a total component volume below 100cm^3. Section thicknesses as low as 1mm have been formed by PIM. Dimensional tolerances are typically in the range from 0.1 to 0.5% of the section thickness. For better dimensional control, machining or coining is required after sintering, which adds to the cost. Also, density gradients can arise during mould filling from uneven flow velocities. This is especially a problem when the feedstock sees a series of thickness variations during mould filling. These density gradients cause subsequent component warpage during sintering. Thus, rapid changes in section thickness are avoided and if possible the variation in thickness is held within a factor of two. Powder expense and availability are also cited as limitations to the growth of the PIM process.

Finally, some limitations reflect an overall poor understanding of the process. These problems are most acute with the ceramic materials. Microstructural flaws induced during moulding limit the mechanical properties of the final product [9,37-39]. Much of PIM's future growth is linked to solving problems relating to process control, powder availability, powder cost, dimensional control, batch processing, variable thicknesses, large dimensions, moulding defects, and distortion.

5. Growth Patterns

Early experience with PIM established high production yields with good tolerances. As the engineering community better appreciates these attributes, PIM will continue to grow. Although the current industry is relatively small, the anticipated growth is quite impressive [5]. Recent reports suggest growth rates as high as 50% per year, or a doubling of the sales approximately every 1.75 years. For this reason, many new ventures and development efforts are under way. Although the knowledge concerning PIM is restricted to a relatively small number of companies, there are similarities between established PIM processes. These similarities allow careful study of the underlying science and should lead to a wider understanding of PIM.

Many of the markets for PIM products have been developed. They include medical and dental devices, office equipment and instrumentation parts, high temperature aerospace parts, printed circuit substates, metalworking tools, electrical and electronic materials, military hardware, aircraft components, household appliance parts, machine tools, computer peripheral components, camera parts, firearm components, and high temperature ceramic turbines [5].

One of the critical issues facing those involved with PIM is the lack of adequately trained personnel. Indeed, all appearances indicate that the growth rate will be dependent on the ability to engineer specific systems and solutions to the many identified applications for PIM. This translates into a need for engineers who can work comfortably with this new technology. Likewise, there is a need for more research on the process basics associated with moulding, powder characteristics, binder formulation, automation, and sintering.

6. Acknowledgements

Much of the information in this article is based on the current research at Rensselaer being performed as part of the Advanced Powder Processing program in the Center for Metal Powder Injection Moulding Manufacturing Productivity and Technology Transfer.

7. References

1. S.K. Bhattacharya (ed.), Metal-Filled Polymers, Marcel Dekker, New York, NY, 1986.
2. R.M. German, Powder Injection Moulding, Metal Powder Industries Federation, Princeton, NJ, 1990.
3. K. Schwartzwalder, Ceramic Bull., 28, pp.459-461, 1949.
4. B. Haworth and P.J. James, Metal Powder Rept., 41, pp.146-149, 1986.
5. L.F. Pease, Inter. J. Powder Met., 22, pp.177-184, 1986.
6. T.J. Whalen and C.F. Johnson, Ceramic Bull., 60, pp.216-220, 1981.
7. P.A. Willermet, R.A. Pett and T.J. Whalen, Ceramic Bull., 57, pp.744-747, 1978.
8. W. Engel, E. Lange and N. Muller, Ceramics for High Performance Applications - II, J.J. Burke, E.N. Lenoe and R.N. Katz (eds.), Brook Hill Publishing Co., Chestnut Hill, MA, pp.527-538, 1978.

9. C.F. Johnson and T.G. Mohr, Ceramics for High Performance Applications - II, J.J. Burke, E.N. Lenoe and R.N. Katz (eds.), Brook Hill Publishing, Chestnut Hill, MA, pp.193-205, 1978.
10. B.C. Mutsuddy and D.K. Shetty, Technical Aspects of Critical Materials Use by the Steel Industry, Report NBSIR 83-2679-2, National Bureau of Standards, Washington, DC, vol.IIB, pp.43.1-43.28, 1983.
11. R.S. Storm, R.W. Ohnsorg and F.J. Frechette, J. Eng. Power, 104, pp.601-606, 1982.
12. P.K. Johnson, Inter. J. Powder Met. Powder Tech., 15, pp.323-329, 1979.
13. R.L. Mackey and B.R. Patterson, Prog. Powder Met., 43, pp.843-857, 1987.
14. J.A. Mangels and W. Trela, Forming of Ceramics, J.A. Mangels and G.L. Messing (eds.), American Ceramic Society, Columbus, OH, pp.220-233, 1984.
15. J. Warren and R.M. German, Modern Developments in Powder Metallurgy, P.U. Gummeson and D.A. Gustafson (eds.), Metal Powder Industries Federation, Princeton, NJ, 18, pp.391-402, 1988.
16. D. Lee, K.F. Hens, B.O. Rhee and C.M. Sierra, Modern Developments in Powder Metallurgy, P.U. Gummeson and D.A. Gustafson (eds.), Metal Powder Industries Federation, Princeton, NJ, 18, pp.417-429, 1988.
17. R.M. German, Inter. J. Powder Met., 22, pp.237-245, 1987.
18. M. Kimoto and S. Uchida, J. Japan Soc. Powder Powder Met., 34, pp.369-372, 1987.
19. C.M. Sierra and D. Lee, Powder Met. Inter., 20, pp.28-33, 1988.
20. E.C. Bernhardt (ed.), Computer Aided Engineering for Injection Moulding, Hanser Publ., Munich, Federal Republic of Germany, 1983.
21. R.M. German, Powder Metallurgy Science, Metal Powder Industries Federation, Princeton, NJ, pp.119-122, 1984.
22. R.A. Thompson, Ceramic Bull., 60, pp.237-243, 1981.
23. R. Haynes, Rev. Deform. Behav. Mater., 3, pp.1-101, 1981.
24. G.F. Bocchini, Rev. Powder Met. Phys. Ceram., 2, pp.313-359, 1985.
25. E. Lang and N. Muller, Powder Met. Inter., 18, pp.416-419, 1986.
26. R. Billiet, Proceedings PIM-82, Associazione Italiana die Metallurgia, Milano, Italy, pp.603-610, 1982.
27. P.C. Chen and C.K. Lim, Prog. Powder Met., 39, pp.153-161, 1983.
28. T.S. Wei and R.M. German, Inter. J. Powder Met., 24, pp.327-335, 1988.
29. G.D. Schnittgrund, SAMPE Quart. July, pp.8-13, 1981.
30. E. Lange and N. Muller, Powder Met. Inter., 18, pp.416-421, 1986.
31. B. Mutsuddy, Indust. Res. Dev. July, pp.76-80, 1983.
32. A.R. Erickson and R.E. Wiech, Jr., Metals Handbook, ninth edition, American Society for Metals, Metals Park, OH, 7, pp.495-500, 1984.
33. T.L. Francis, Powder Met. Inter., 17, pp.185-188, 1985.
34. R.J. Waikar and B.R. Patterson, Horizons of Powder Metallurgy, Part II, W.A. Kaysser and W.J. Huppmann (eds.), Verlag Schmid, Freiburg, Federal Republic of Germany, pp.661-665, 1986.
35. R. Billiet, Prog. Powder Met., 38, pp.45-52, 1982.
36. R. Billiet, Inter. J. Powder Met. Powder Tech., 2, pp.119-129, 1985.
37. D.W. Richerson, J.R. Smyth and K.H. Styhr, Ceramic Eng. Sci. Proc., 4, pp.841-852, 1983.
38. J.A. Mangels, Ceramics for High Performance Applications - II, J. J. Burke, E. N. Lenoe and R. N. Katz (eds.), Brook Hill Publishing Co., Chestnut Hill, MA, pp.113-130, 1978.
39. A. Nagel, G. Wingefeld, D. Agranov and G. Petzow, Horizons of Powder Metallurgy, Part II, W.A. Kaysser and W.J. Huppmann (eds.), Verlag Schmid, Freiburg, Federal Republic of Germany, pp.636-640, 1986.

8

Cold Isostatic Pressing of Metal Powders

J. TENGZELIUS AND O. PETTERSSON
Höganäs AB, S-263 83 Höganäs, Sweden

1. Introduction

Cold isostatic pressing has been used for more than 50 years for forming powders, especially ceramic powders. During recent years the pressing of metal powders has increased primarily due to new presses and tools which are suitable for automatic production of precision components. Cold isostatic pressing (CIP) is well described in the literature [1,2,3,4] and also compared with other methods [5]. The pressing of different powders has also been described [6,7,8,9,10]. This short description of CIP gives the state of the art and indicates some future possibilities for the utilization of this technology for production of PM parts. → p121

2. Technical Aspects

2.1 Powder

Isostatic pressing can be performed on most types of powders independent of flowability, compressibility, particle shape, etc. However, the result after pressing is largely dependent on the powder's characteristics. It is also necessary to use different pressing and tooling concepts for different powders. As in all PM processes, for CIP it is of the utmost importance that the powder properties are well known and stable. The following powder characteristics influence the pressing procedure and properties of the finished product.

Flow

Most tools for CIP are filled outside the press and therefore the filling of powder can be carried out without the common limitations such as lack of space, uncontrolled vibrations from the press, etc. For this reason, powders which are not free-flowing can also be used. For complex shapes and narrow tool cavities, however, a good flow of the powder is often necessary and at least results in closer dimensional tolerances.

Apparent density

This powder property strongly influences the dimensions of the compacted part. A low scatter of apparent density is important because the volume of the tool during filling is always constant. In order to improve tolerances it is possible to vibrate the powder to a predetermined height in the tool thereby controlling the total volume of tool cavity filled with powder. Instead of using a constant volume for filling, the powder can be weighed before filling. In the latter case a larger scatter of the apparent density can be accepted.

Particle size

Fine powders are, for tool wear reasons, better than powders with large particle size. During compaction the flexible tool material is extruded into the voids between the particles causing a tearing effect. Larger particles give larger voids which increase the tearing. Tool wear when compacting powders with particle sizes below 200μm is normally low.

Green strength

Green strength after cold isostatic pressing is higher than that after uniaxial pressing. Besides the positive effect of the isotropic stresses the possibility to press without lubricant has a large beneficial effect on green strength. The diagram in Fig. 1 compares the green strength of iron powder compacted respectively under the uniaxial and isostatic stresses.

Compressibility

A number of factors influence the density achieved after compaction. When comparing uniaxial and isostatic pressing at high pressures the most common result is that isostatic pressing gives a somewhat higher density. This is primarily due to the following three reasons:

- More efficient transfer of pressure within the compact.
- Elimination of die wall friction.
- Elimination or reduction of lubricant.

Figure 2 shows compressibility curves for iron powder compacted with the two types of pressures.

2.2 Tools

Tool materials

Different types of rubber or plastic often in combination with steel are used for tools [11]. The most common tool materials are described in Table 1.

Besides these flexible materials, steel also is used for some parts of the tools. The tolerances of the component are better on better on surfaces compacted against rubber. For this reason tool parts of steel will preferably be used for compacting areas with critical dimenions, and can also act as support for the plastic or rubber parts.

Tool design

The following parameters are of great importance for the design of a tool for isostatic compaction:

1. Powder properties.
2. Compacting pressure.
3. Tolerances of component.
4. Geometry of component.
5. Powder filling method.
6. Number of components to be manufactured.

The tolerances of the component are better on surfaces compacted against steel than on those compacted against rubber. For this reason, tool parts of steel will preferably be used for compacting areas with critical dimensions.

Components for CIP can be designed with threads, undercuts, conical shape, etc. However, the need to discharge the part from the tool must be considered. A powder with a large difference between powder density and compressed density can accept larger flanges or deeper threads than a powder which is not densified so much during compaction.

Some tools for cold isostatic pressing are shown in Fig. 3.

2.3 Tolerances

The tolerances of cold isostatically pressed components are normally wider than those of uniaxially pressed components. However, during recent years, tools and tool materials have been developed which can be used for the manufacture of precision parts [12]. Besides tools, dry-bag systems with automated filling have contributed to this development. The closest tolerances are achieved if very hard polyurethane rubber is used as a tool material. A hard rubber or polyurethane is not deformed during filling and therefore final dimensions will not vary to any large extent. These types of elastomers are also stable with time, and tool dimensions are not changed even after compaction of thousands of cycles.

The largest influence on tolerances is of the powder and filling procedure. If the powder density of the filled tool varies, the dimensions of the compacted part will also vary. It is possible to achieve a tolerance of about 1% of the thickness of the component if both powder and filling procedure is well controlled. Often, however, it is necessary to calculate on a 2-4% scatter.

The surfaces compacted against steel parts of the tools achieve extremely close tolerances - as good as those achieved by uniaxial pressing. The surface finish of a component is either very smooth or rather rough. The smooth surfaces are those pressed against steel and the rough ones are those pressed against the flexible material. The micrographs in Fig. 4 show cross-sections of these two types of surfaces. The roughness of the surface compacted by the flexible tool material is related to the particle size of the powder. A powder with a fine particle size also achieves a fine surface.

3. Wet-Bag/Dry-Bag

Cold isostatic presses are designed either according to the wet-bag or dry-bag principle (Fig. 5). The main difference between the two types is the tool handling systems. In a wet-bag press the tool is placed directly in the liquid in the high pressure vessel whereas the tool in a dry-bag press is separated from the liquid by a pressure transmitting membrane. Due to these differences wet-bag presses are normally suitable for production of a short series of complex parts where manual handling can be accepted. Dry-bag presses are suitable for mass production with a high degree of automation.

3.1 Wet bag presses

Wet-bag presses consist of a pressure vessel which during pressing is mounted in a frame. Both pressure vessel and frame have to withstand extremely large forces and are therefore often manufactured with pre-stressed wires. The axial force of a CIP press with an inner vessel diameter of 300mm at a pressure of 400MPa is more than 2500 tons.

A pressing cycle in a wet-bag press is performed with the following steps:

1. Filling the tool and closing it in order to avoid liquid penetration into the powder.
2. Charging the pressure vessel with tools.
3. Closing the vessel into the frame.

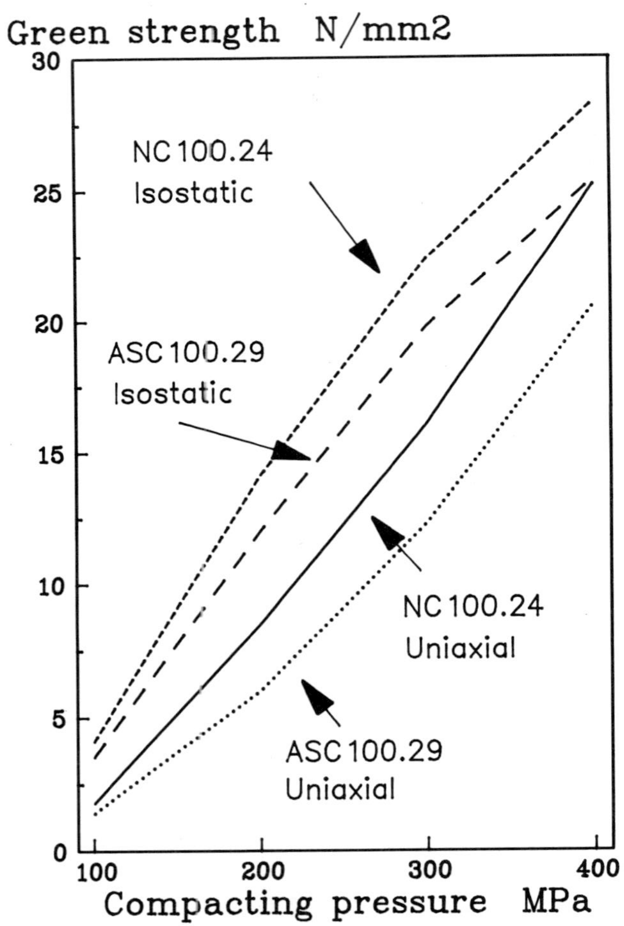

Figure 1 Green strength after isostatic resp. uniaxial compaction for sponge-based (NC100.24) and water atomized (ASC100.29) iron powder. Unlubricated powder.

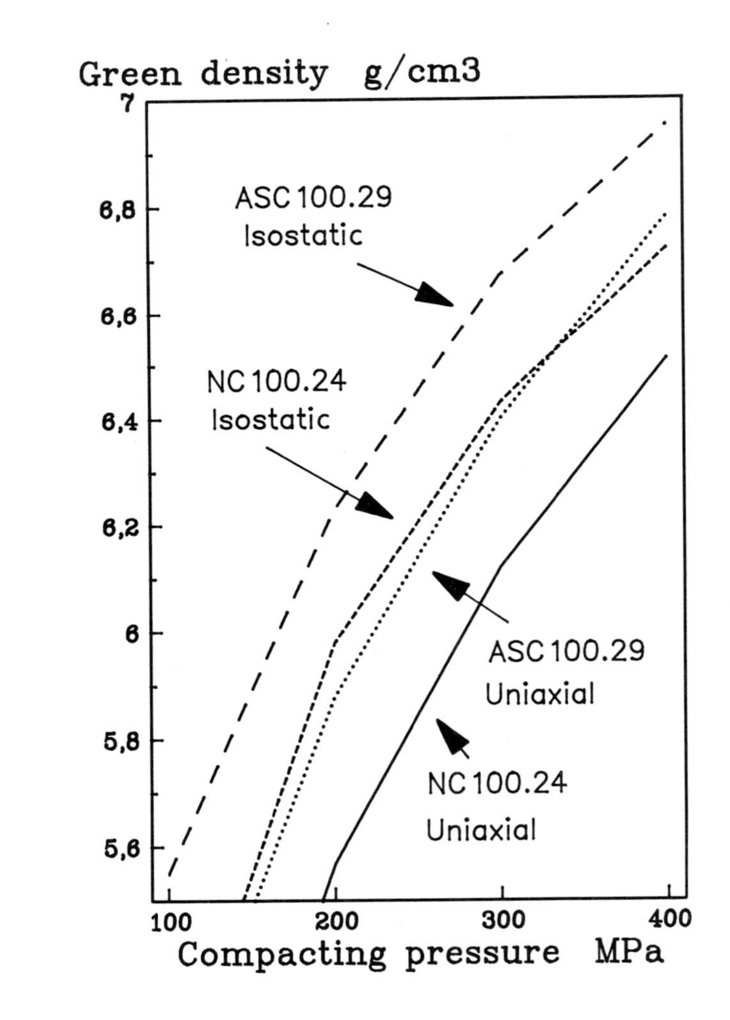

Figure 2 Green density after isostatic resp. uniaxial compaction for sponge-based (NC100.24) and water atomized (ASC100.20) iron powder. Unlubricated powder.

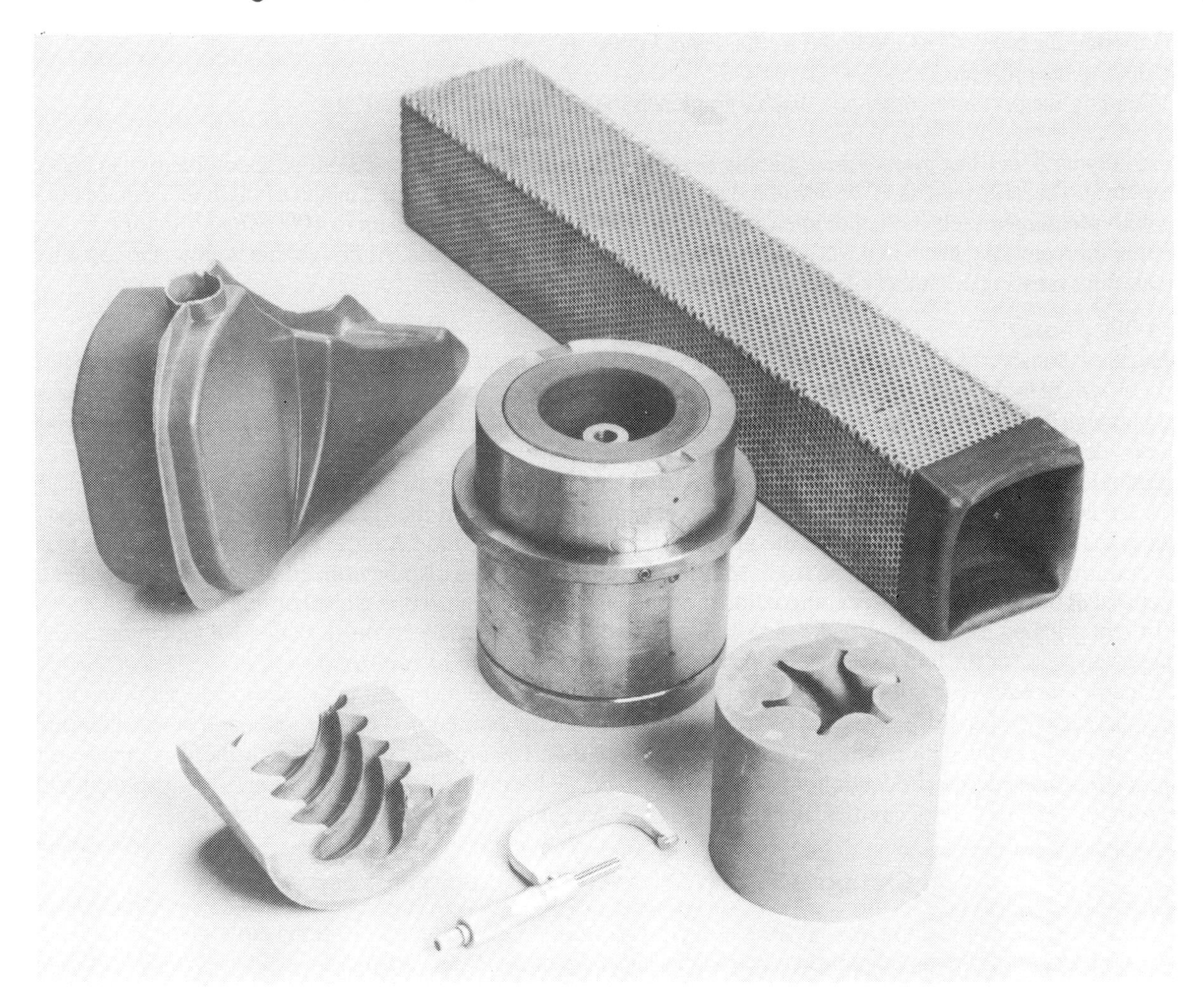

Figure 3 Examples of tools for CIP.

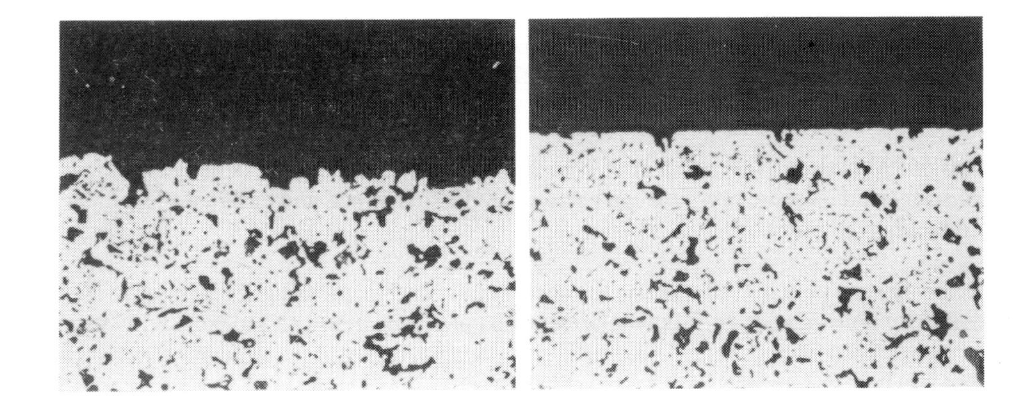

Figure 4 Finish of surfaces compacted against elastomer (left) and steel (right).

4. Increasing the pressure of the liquid to the desired pressure.
5. Reducing the pressure.
6. Unloading the pressure vessel and discharging each tool from the compacted part.

Industrially used wet-bag presses are typically designed for pressures between 200MPa (about 2t/cm^2) and 600MPa (about 6t/cm^2). The largest press in the world today has an inner vessel diameter of 2 m and a height of 3.5m. Laboratory presses with pressure vessels having diameters between 20-40mm use pressures up to 1000MPa (10t/cm^2).

The time for a pressing cycle in a wet-bag press is between 5 and 45 minutes. Although this is slow, the capacity can be increased by pressing a number of tools simultaneously.

3.2 Dry-bag presses

A dry-bag press consists of substantially the same parts as a wet-bag press [13]. The only major difference is the flexible barrier between the tool and the pressing liquid. Besides the advantage of the increased degree of automation and capacity it is an advantage that the tool is dry. With a dry tool there is no risk of contact between the pressing liquid (normally water) and the powder or compacted part.

An example of a layout for a modern dry-bag press and surrounding equipment is shown in Fig. 6. A number of tool cassettes are activated simultaneously in order to achieve high capacity. In the filling station tools can be filled with different types of powders depending on the product being produced. If the filling is more time-consuming than the pressing, multiple filling stations can be used. Multiple filling stations can also be utilized if different components are manufactured at the same time. After compaction the tool cassette is transferred to a discharging station which also can have different gripping mechanisms depending on the geometry of the part. By having a station for tool maintenance on the transport system for the tool cassettes the productivity of the plant can be improved.

The capacity for a dry-bag cold isostatic press is strongly related to the compacting pressure. The diagrams in Fig. 7 compare the capacity for different systems at various compacting pressures. New dry-bag systems combine high pressures, above 400MPa, with production capacities of more than 100 pressing cycles/hour. If these systems use tools with a number of cavities, the production capacity is close to that achieved with more conventional compacting methods. If, for example, each tool has 5 cavities, the capacity is 500 parts/hour or about 2 million parts/year.

4. Comparison With Other Manufacturing Processes

When is cold isostatic compaction a suitable manufacturing process? The brief description below gives some guidance and also comparison with uniaxial pressing, machining of wrought steel and casting.

Material yield is very high - between 90-100% - due to the close tolerances, design possibilities and re-use of powder after machining in the green condition. This is about the same yield as for conventional PM but much higher than that of machined wrought steel or casting.

Shape complexity can be achieved with all of these processes but uniaxial compaction is less flexible than isostatic compaction. For a large production series of complex-shaped parts both PM processes have cost advantages as compared with the other two.

Tolerances of cold isostatically pressed parts are similar to uniaxially pressed parts on surfaces compacted against steel but wider on surfaces compacted by the flexible part of the tool. Compared to casting, cold isostatic compaction results in much closer tolerances.

Mechanical properties of PM parts are independent of compacting method if the same density is achieved. Therefore, PM components formed by cold isostatic pressing can be compared to machined and cast materials in the same way as PM materials formed by uniaxial pressing.

Specially adjusted properties can be achieved on parts formed by cold isostatic pressing. This is due to the independence of powder properties and flexible tool design. Both materials with composite matrices and parts with composite designs can be produced. No interference from pressing aids, such as lubricants, makes it possible to carry out the sintering or heat treatment without any burn-off.

Table 2 indicates some typical characteristics for the different manufacturing processes.

5. Case Study: Cylinder Liners by CIP

The shape of cylinder liners is very suitable for isostatic compaction (Fig. 8 shows some typical cylinder liners) and the properties of PM steel can be adjusted to meet the stringent requirement demanded for such an engine component [14,

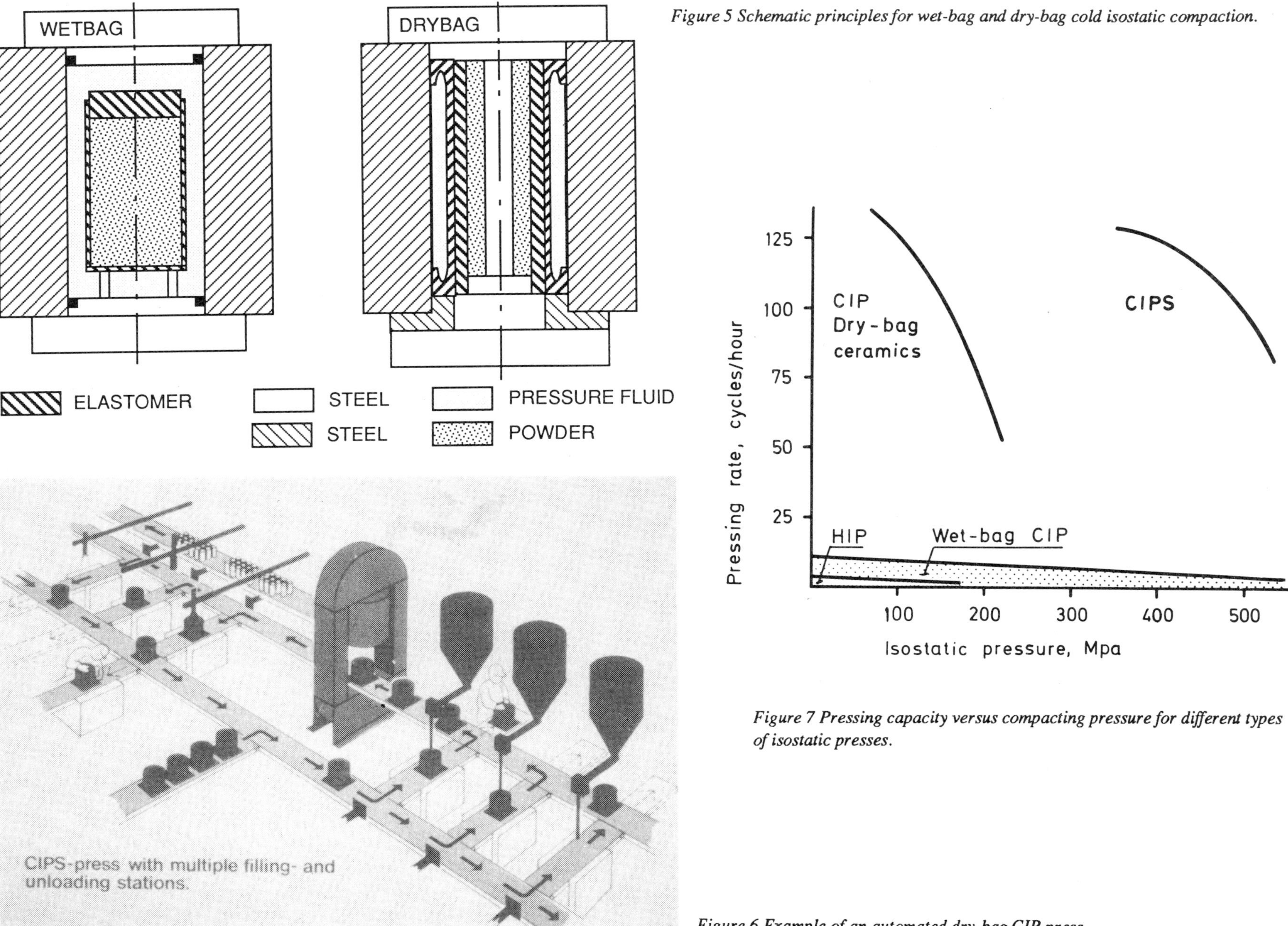

Figure 5 Schematic principles for wet-bag and dry-bag cold isostatic compaction.

Figure 7 Pressing capacity versus compacting pressure for different types of isostatic presses.

Figure 6 Example of an automated dry-bag CIP press.

Table 1 Tool materials for cold isostatic pressing

Material	Hardness	Elasticity	Life	Suitable pressure MPa	Cost	Wet-bag / dry-bag
Latex	Low	High	Short	<200	Low	Wet-bag
PVC	Medium	Low	Medium	<200	Low	Wet-bag
Silicon	Low	Medium	Short	<200	High	Wet-bag
Thermo-plastic	Medium	Low	One cycle	$\leq$400	Low	Wet-bag
Natural rubber	Medium/ high	Medium	Long	$\leq$200	High	Wet-bag / dry-bag
Polyurethane	Medium/ high	Medium/ high	Long	$\leq$600	High	Wet-bag / dry-bag

Table 2 Comparison between different manufacturing methods for mass production of steel components

	Powder metallurgy		Machining	Casting
	Uniaxial pressing	CIP	Wrought steel	
Material yield	High	High	Low	Medium
Shape complexity	High	High	High	High
Tolerances	Narrow	Narrow/ average	Narrow	Coarse
Mechanical properties	High	High	Very high	Medium
Number of manu-facturing steps	Few	Few	Many	Medium
Energy consumption	Low	Low	Medium	Medium
Environmental influence	Low	Low	Low	High

Figure 8 Typical PM cylinder liners.

15].

5.1 Background
A diesel engine for a small truck uses cast iron cylinder liners. These cast iron cylinder liners have drawbacks regarding wear resistance, rigidity, machining cost and material yield during production. Cold isostatically pressed cylinder liners were evaluated and showed improvements on all aspects.

The wear resistance was analysed by measuring the wear of the cylinder liners after engine tests. By using an iron-based PM material alloyed with phosphorus and carbon, the wear was reduced to about half that of cast iron.

The modulus of elasticity of grey cast iron is about 100,000N/mm^2. A PM steel alloyed with phosphorus and carbon has at a density of 6.5g/cm^3, a modulus of elasticity of about 120,000N/mm^2, and at 7.0g/cm^3 one of about 150,000N/mm^2. This increased rigidity improves the engine performance and makes it possible to design the engine with weight reduction.

The final thickness of this cylinder liner is about 1.5mm which is very difficult to attain when machining a cast iron blank. A number of machining steps must be carried out and the large pores and graphite flakes often result in fractures. The PM blank can, after pressing, sintering and sizing, have a wall thickness of about 1.6mm thereby making only a final grinding necessary.

The improved tolerances of the PM blank as compared to the cast iron blank increases the material yield from about 30% to 95%.

Specifications

Dimensions:	Outer diameter: 105 mm
	Inner diameter: 102 mm
	Height: 190 mm
	Weight: 0.6 kg
Raw material:	Iron powder PNC45 (Fe+0.45% P) with 1.8% graphite
Structure:	Pearlite with 0.5% free graphite
Density:	6.5 g/cm^3

Processing
1. Compaction without lubricant at 400 MPa in a dry-bag cold isostatic press.
2. Turning outer surface in as-pressed condition and return of swarf powder to incoming raw material.
3. Sintering at 1050°C 30 min. in an atmosphere of 90% N_2/10% H_2.
4. Sizing.
5. Grinding (honing).

Tolerances
The tolerances of the inner surface after sintering is best described with the diagram shown in Fig. 9. The dimensional tolerance of the inner surface can also be defined by its cylindricity which defines the machining allowance to achieve a perfectly cylindrical surface. The cylindricity after sizing of this thin-walled liner was below 0.1 mm for a batch of 1000 liners.

Production Cost
The production cost for this cylinder liner based on a yearly production of 1 million liners has been calculated both for PM and cast iron. The bar diagram in Fig. 10 shows that, including both capital and direct costs, the cost for the PM liner is about 30% lower.

6. Future Possibilities

The future utilization of cold isostatic pressing for the manufacture of metallic components will probably increase substantially, particularly for the production of besides the possibility of producing complex parts. New automated dry-bag systems suitable for mass production will also contribute to the development of the process. An important factor for choosing the CIP route is also the special material properties which can be achieved. Some of those are:

- Wear resistant materials made of hard particles. (In a CIP tool the particle hardness has no influence on tool wear.)
- Combination of powders in composite designs. Powders can be combined even if their powder properties differ significantly.

Other advantages of CIP are:
1. No lubricant necessary.
2. Very high green strength.
3. Uniform density.
4. High density.
5. Easy filling for composite parts.

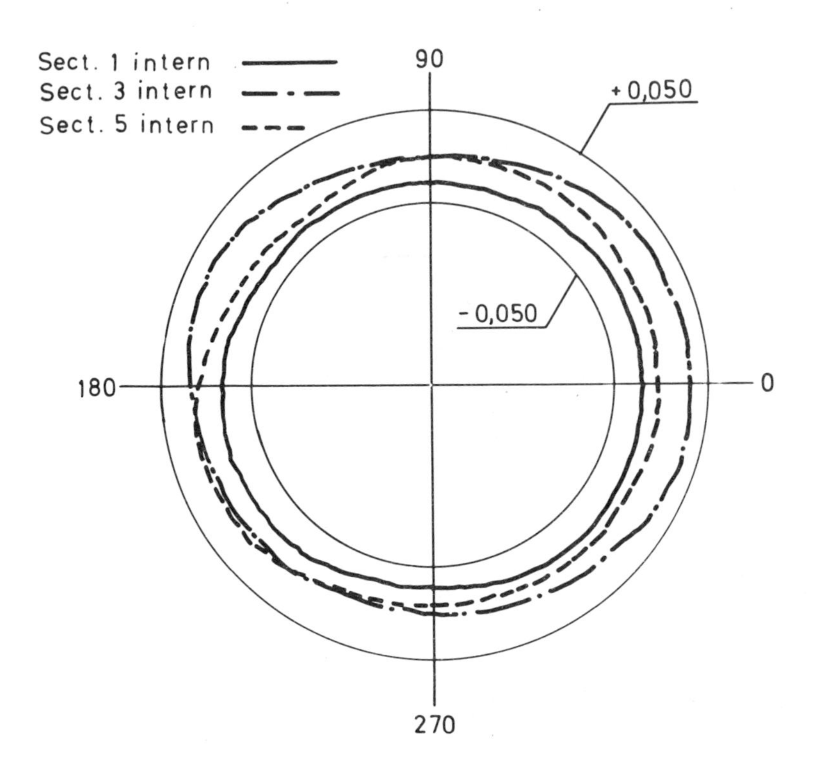

Figure 9 Measurement of the inner surface of a PM cylinder liner after sizing. The figures refer to the deviations from a perfect cylindrical surface. Measured at three different cross sections.

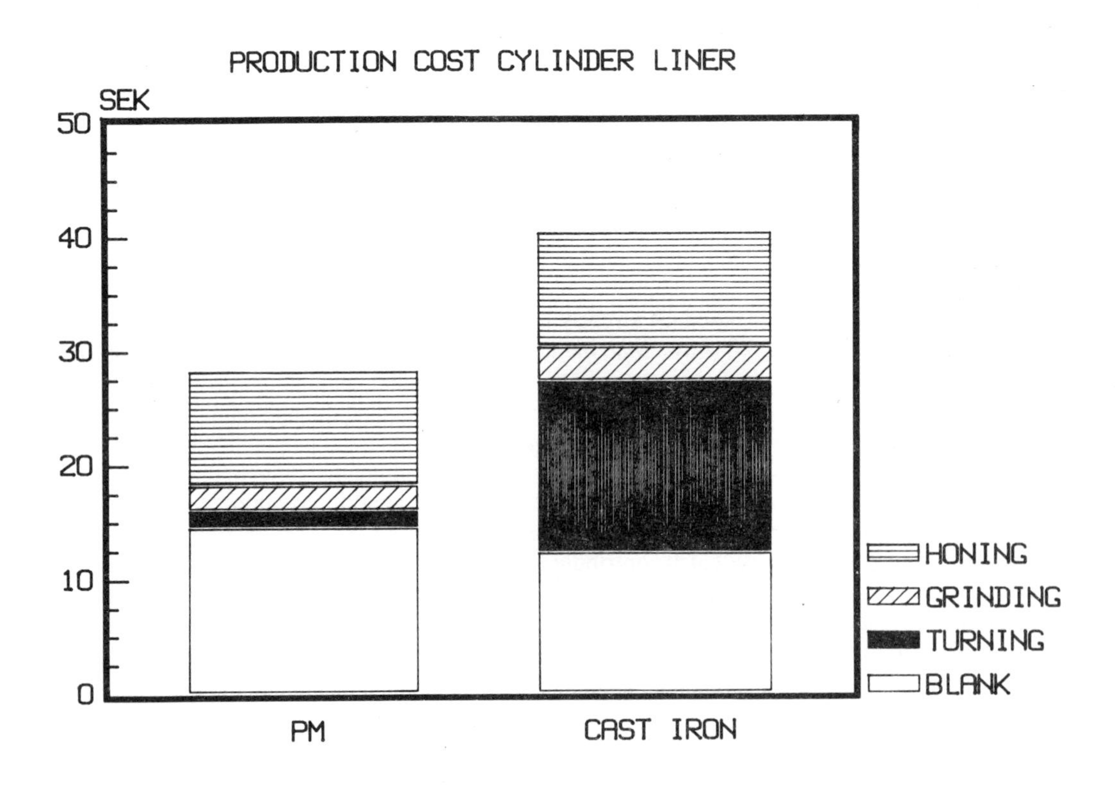

Figure 10 Comparison between the production costs for the described thin-walled cylinder liner when using different production methods.

6. Large components.
7. Low production cost.

There are a number of components suitable for CIP described in the literature [16,17,18,19,20]. Some components which are presently being evaluated for CIP are composite cylinder liners with an outer layer of solid steel [21], cylinder liners with an upper part with high wear and corrosion resistance, blanks for powder forging either with a very close weight control or with compound filling, moulds for production of glass bottles, moulds for production of plastic components, large bearings and bushes, valve guides, etc.

References

1. P.J. James, Isostatic Pressing Technology, Applied Science Publishers Ltd., 1983.
2. H. Ittner, Cold Isostatic Pressing with the R.T.S. System, Ceramic Forum Int. 63, 3, pp.126-128, March 1986.
3. J.T. Winship, Upsurge in isostatics, American Machinist, April 1984.
4. T. Nishimoto, Y. Kishi, T. Naoi, Recent Trends in Cold and Warm Isostatic Pressing Equipment, MPR, September 1987.
5. W.B. James, New Shaping Methods for Powder Metallurgy Components, Materials & Design, 8, 4, July/August 1987.
6. A.V. Krupin, L. Yu Maksinov, N.I. Sheftel, V.N. Bratenko, N.A. Maksinova, I. Ya Kondratov, E.N. Smirnova, The effect of cold isostatic compaction on the thermophysical properties of powder materials, High Temp. - High Pressures, 8, 6, pp.684-686, 1976.
7. P.J. James, Particle deformation during cold isostatic pressing of metal powders, Powder Metallurgy, 4, pp.199-204, 1977.
8. L.M. Barker, A.H. Jones, Fracture Toughness of CIP-HIP Beryllium at Elevated Temperatures, Theoretical and Applied Mechanics, 7, pp.45-49, 1987.
9. 0. Abe, S. Iwai, S. Kanzaki, M. Ohashi, H. Tabata, Influence of Size and Shape on Homogeneity of Powder Compacts Formed by Cold Isostatic Pressing (Part I) Press Forming of Thick Cylinders, J. Ceram. Soc. Jpn. 94, 10, pp.1092-1098, 1986.
10. P. Kostka, A. Issleib, Grundlegende Untersuchungen zum isostatischen Kaltpressen von Pulver aus Schnellarbeitsstählen, Wiss. Z. Techn. Hochsch. Magdeburg, 25, Heft 5, 1981.
11. K.J. Morris, Tooling Design for Cold Isostatic Pressing, Met Powder Rep, September 1981.
12. P.R.M. Skoglund, Rapid measurement and statistical evaluation leading to improved tolerances for PM cylinder liners manufactured by cipping, Powder Met., 30, 4, 1987.
13. J. Tengzelius, O. Pettersson, Design and Manufacturing Possibilities with Dry-bag Isostatic Pressing, Coll. on conv. forming processes in PM, Paris, 19-20 September 1984.
14. O. Pettersson, Cylinder liners made by P/M technique, Pow. Met. Int., 13, 1, 1981.
15. G. Wastenson, Production of iron powder cylinder liners by cold isostatic pressing, 2nd Int Conf on Isostatic Pressing, l, September 1982.
16. D.A. Van Cleave, Cold pressing opens the door on complex shapes, Iron Age, 9 May, 1977.
17. M. Niino, A. Kumakawa, R. Watanabe, Y. Doi, Fabrication of a High Pressure Thrust Chamber by the CIP Forming Method, MPR September 1986.
18. L. Albano-Muller, Filter Elements of Highly Porous Sintered Metals, Powd. Met. Int., 14, 2, 1982.
19. R. Smalley, Paper at the 1st International Conf. on Isostatic Pressing, University of Technology Loughborough, 19-21 September 1978.
20. J. Gestwa, R. Rohlig, P. Beiss, New applications of Cold Isostatic Pressing in Powder Metallurgy, MPR January 1984.
21. J. Tengzelius, E. Wirgarth, PM Cylinder Liners for Large Diesel Engines by CIP, Proc. of 3rd Int Conf on Isost Pressing, 1, London, 10-12 November 1986.

9

Dynamic Compaction Technology

D. RAYBOULD
Metals & Ceramics Laboratory, Research & Technology, Allied Signal Inc., Morristown, NJ, USA

Abstract

Dynamic compaction involves compacting powder by a shock wave which results in work hardened particles bonded by rapidly solidified welds; the technique also has the ability to produce bulk materials which can not be consolidated by other means. These unique properties and microstructures are reviewed and some of the unique materials such as amorphous/nano crystalline alloys, composites and high temperature superconductors, are briefly described. Emphasis is placed on potential applications and why some of these have not as yet materialized. Mention is made of others, which are showing promise.

The theory behind interparticle melting and shock wave interactions is described, as are the reasons why an understanding of them and the availability of calculations to predict them in a normal advanced metalworking environment is necessary for successful exploitation of the technique.

1. Introduction

Dynamic compaction offers a means of producing unique materials that in some cases can not be obtained by other techniques [1-4]. For conventional materials, the shock waves associated with the technique result in high dislocation densities, increased hardness and an associated increase in fatigue strength [5], while the good contact between particles results in activation of sintering [6,7]. Also, the interparticle melting that the technique can produce during cold compaction allows the production of many unique alloys of metals, ceramics and plastics which would loose their properties of interest if exposed to a high temperature and thus can be fabricated as 3-D objects only by dynamic compaction.

The reasons that these advantages have not resulted in wider commercialization are numerous. One important factor was the break up of the group holding the initial patents. It is, however, important to appreciate that the initial US patent [8] on interparticle bonding was granted in 1980 and that it is not unusual for a pioneering patent, such as this, not to achieve successful commercial exploitation in the life time of the patent. At present, groups in the USA, USSR, Japan, UK & Germany are carrying out interesting work on the technique.

Extensive work has been carried out in many areas, which could result in commercial applications. Often, however, dynamic compaction allows a cheap way of obtaining an estimate of the potential of a material, sometimes showing that the material is not of interest. If it is, then efforts are made to produce it by other techniques. The final benefit to dynamic compaction technology is small and tends to concentrate resources to the sophisticated high velocity impactors, as required for superalloys, rather than the low velocity/energy techniques, which are cheap and more easily commercialized, as required, for example, for aluminum metal matrix composites, plastics or rock drill bits.

An area that may be showing commercial potential at the present is the consolidation of superconducting powders to 3-D parts. Other techniques can consolidate these materials, but it appears from Japanese work [9] that dynamic compaction can result, after annealing, in a material with superior current carrying capacity at "high" temperatures. The reasons for this are not clear, but could be associated with the high dislocation density associated with shock loading. This is an area requiring further investigation. Also of interest is the direct synthesis of these materials during dynamic compaction, as has previously been carried out for other materials [1].

One essential requirement for the successful exploitation of dynamic compaction technology is an understanding of the shock waves and their interaction. This explains the understandable advantage of laboratories with such expertise, as a result of weapons related work. However, the pressures involved in dynamic compaction are lower than in many shock wave computer codes developed for such work. In addition, simple, easy-to-follow calculations must be available to allow the setting up of conditions easily, by the same type of personnel who would operate other advanced metal working facilities. Research to develop information on a variety of powders and computer calculations that can be run on readily available computers is being done, but this is one area in which extensive further work is required, in order to ensure successful commercialization in the future. This should, in particular, look at the shock wave reflections on a macro scale, so that multiple shock wave reflections can be predicted. Past experience indicates that such work will tend to be carried out by university-type organizations. This does not imply that successful commercialization is not possible at present, especially if impact velocities/energy are kept at modest levels. In addition, several parts/billets have already been produced successfully in significant quantities, demonstrating reproducibility under a given set of fixed conditions, once the optimum parameters have been established. These runs of 10 to 100 parts have shown that the technique is comparable in cost to other metalworking techniques for advanced materials.

2. Technique

Dynamic compaction, as described in this chapter, involves the consolidation of powder by discrete shock waves that are created by the impact of a light, high velocity punch (Fig.1), or by explosives placed close to the powder, but separated from it. There are, in fact, several high speed consolidation techniques with which dynamic compaction may be confused. Even the explosive and light punch techniques have significant differences, both between them, and also within each technique, depending on, in one case, the punch velocity and impedance; and in the other, the type quantity and location of the explosive. A simplified comparison of the compaction parameters for some high speed consolidation techniques is given in Table 1.

A shock wave is a discontinuity which moves through the material. Shock waves were first extensively studied in gases and fluids and equations developed for these materials are the basis for our understanding of shock waves in solids. Rankine and Hugoniot took a controlled volume around a shock wave in a fluid and assumed that mass, momentum and energy must be conserved. This allows three equations containing five variables to be drawn up. For a unique solution, a further equation is required. Often this is obtained experimentally, by measuring the shock front (wave) velocity and the velocity at which the material behind the shock (particle) moves. Several experiments are required to obtain an equation relating these two velocities [10].

Once obtained, the other equations may be solved and it is possible to draw the Hugoniot curve of the material. This is essentially a locus of all the pressure-density states obtainable by the propagation of a shock wave for a material with a given initial starting density and pressure, i.e. a dynamic compressibility curve. For a powder, if simplifying assumptions are made, such as limiting the maximum pressure to that required for full densification [11], then Raybould has shown that it is possible to use the quasistatic pressure-density curve, rather than determining a dynamic curve. Previously, Davidson [13] showed that at very high pressures, porous solids are insensitive to strain rates and at lower pressure and strain rates, Sheppard [14] reported a similar insensitivity for powders. This allows the pressure created on impact of a known projectile to be calculated from the quasistatic pressure density curve, which often is information readily available from the powder supplier. This approach has been shown to give good agreement between predicted and experimental pressures for a range of different powders. However, Page [15] showed this could be improved upon once some dynamic data was obtained.

Despite the close similarity of the quasistatic and dynamic pressure density curves during loading, until the shock pressure is relieved further densification may occur. This usually results in dynamic compaction giving a significantly higher density than quasistatic compaction for the same compaction pressure.

A unique aspect of dynamic compaction, resulting from the compaction being carried out by one or two discrete shock waves, is that it is possible to transform the powder to a solid. This is because the work of compaction is concentrated in the short time that it takes the shock front to pass a point, this results in the temperature rise caused by this work of compaction being sufficient to melt the material and gives strong interparticle bonding, (Fig. 2) [8,17]. Hence, the technique is of interest for materials which would degrade if exposed to normal sintering temperatures.

The work of compaction may be calculated from the area under a pressure volume (displacement) curve, (Fig. 3). For quasistatic compaction, it is the area under the curve connecting the start and finish pressure, but for dynamic compaction it is the area under the Rayleigh line, which is a straight line connecting these two points. This is because the shock wave does not allow the material to change gradually, but, instead, jumps from one pressure state to another, with the result that twice as much work may be carried out on the powder as compared to quasistatic compaction. This work is carried out in the time it takes a shock wave to pass a point (rise time), which, for powders, may be assumed to be the time for 3-5 shock wave reflections within a particle, usually of the order of a few μs [12,17]. Despite this, the overall temperature rise is still small unless very high pressures are employed. Raybould [8,12,17] realized that this work occurs primarily at the interfaces between particles in the form of friction and local deformation, while the short time during which it occurs does not allow heat to be conducted away to the centre of the particles. Calculating the rate of conduction of this heat away from the particles surface allowed the calculation of the particles surface temperature and the time during which the work of compaction must be carried out for interparticle melting to occur for a given compaction pressure. These analyses may be combined to give a single equation [8,12], which shows the importance of the different variables and allows the parameters to be optimized prior to carrying out any dynamic compaction.

Equation (1) shows that the possibility of interparticle bonding R, which should equal 1 for complete interparticle melting to commence, increases linearly with the powder particle diameter d and inversely with the melting temperature Ts of the material.

$$R = S\,[d(1-a)((a\,b)^{5/2}\,Ts\,Cp\,K\,p^{3/2})^{-1}]\,.\,(b\,P)^2\,.\,(1+bP)^{1/2} \qquad (1)$$

Cp - specific heat of the solid material.
K - thermal conductivity of the solid material.
Ts - melting point of the solid material.
p - density of the solid material.
P - compaction pressure behind the shock wave.
b - compaction constant representing powder stiffness and is obtained from the pressure/density relation [11,12,15].
a - initial density of the powder expressed as fraction of the solid material density. [for 60 % T.D., a = 0.60, [see also Refs.12,15]
d - average particle diameter, or the best estimate of the diameter of an "equivalent" spherical particle.
S - a shape factor describing the geometry and initial state of the particle surface. Some experimental values have been determined for different types of powders [8].

The above approach is simplified in order to allow all the key parameters to be expressed in one equation. One simplification was to ignore the melted material and to consider one dimensional heat flow at the particles surface. Detailed calculations of the heat flow from a spherical particles surface, which made no simplifying assumptions taking into account for instance the heat of melting and the temperature rise of the molten material, showed, Table 2, that the errors introduced by the simplified approach are negligible, until a large amount of interparticle melting occurs. The object of the simplified approach is, of course, to optimize the parameters, so that a small amount of interparticle melting occurs at the minimum pressure possible.

Area of Particle Heated	Melted Material	Volume Melted	Solidification Time Time to Melt
Total	Remains	2.1%	1.2
Half	Remains	9.6%	3.5
Total	Removed	1.9%	1.2
Half	Removed	14.0%	2.9

Table 2 Detailed heat flow calculations for a 200 micron diameter aluminum powder compacted by one shock wave to 800 MN/m². Showing the effect of removing the molten material once it has melted, and also of depositing the heat on only 50% of the particle surface to illustrate the importance of the irregular surface of a real powder.

An alternative simple approach to calculate the fraction of material melted (L), giving a single equation that represents an upper bound, was subsequently developed by Schwartz [18].

$$L = P\,p^{-1}\,(m-1)\,.\,[2(Cp\,Ts + Hs)]^{-1} \qquad (2)$$

Hs - heat of fusion of the solid.
m - distention, as for the constant a, m represents the powder's starting density as a fraction of the solid density. $[m = (1-a)^{-1}]$

For both equations (1) and (2), the starting temperature is taken as close to zero, so the temperature rise required may be considered to be effectively Ts [12].

The fact that the interparticle melted region is extremely thin results in a rapid quenching of the weld. The heat is extracted by both particles so that the weld zone is often solidified more rapidly than the starting powder, usually at a rate of well over a million degrees Celsius per second, resulting in unique, fine microstructures in the weld zone. Both calculations [17-19], (for instance, Table 2 gives the resolidification time compared to the time to melt), and measurements of dendrite arm spacings [20] have been carried out by several workers who have all confirmed these high cooling rates. One of the more interesting observations was for ceramic powders [21], which found an amorphous melt zone.

3. Mechanical Properties

The combination of a shock hardened particle surrounded by a rapidly solidified weld zone offers the potential of producing materials with unique property combinations, even when the powders compacted are conventional materials. Obviously, for these properties to be attained, the compaction conditions have to be optimized. Until the optimum conditions are achieved, the properties will vary from those of the powder to those of the a fully bonded compact [11,17].

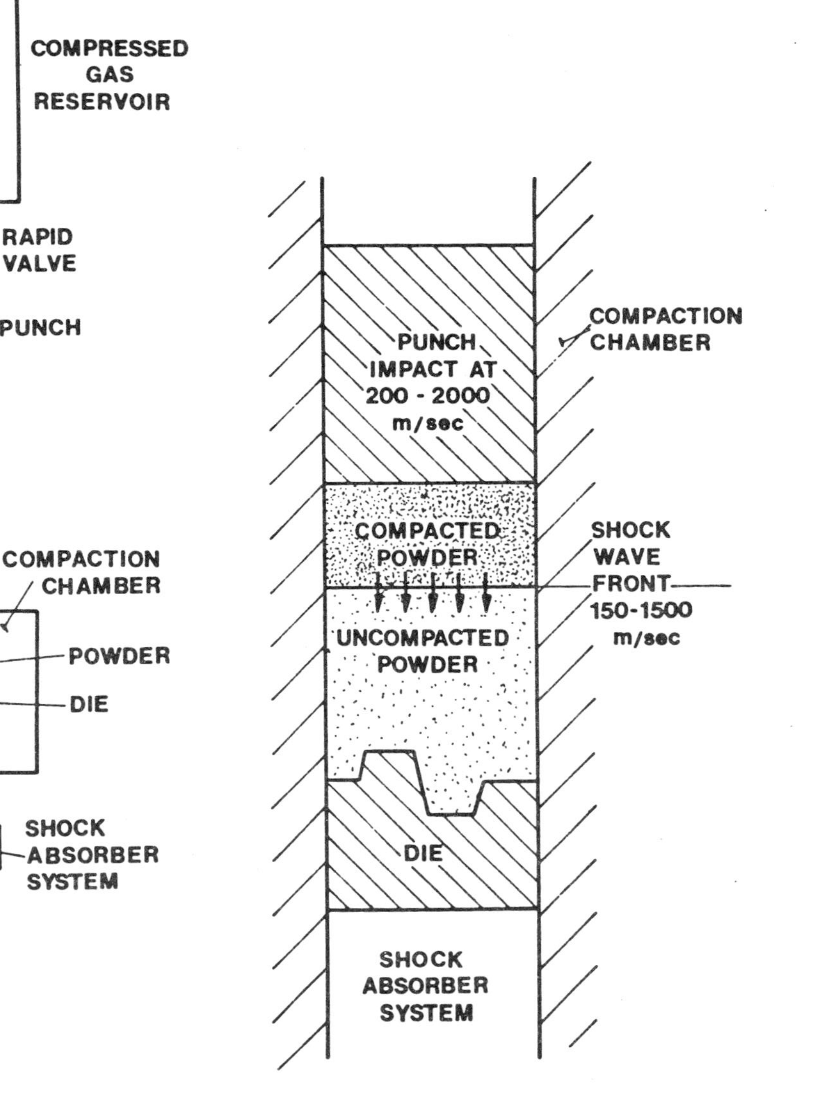

Figure 1 (a) A schematic of a gun type dynamic compactor, with a representation of the shock wave in the powder immediately after impact.

Figure 1 (b) The high speed (1,500m/s) dynamic compactor at Allied - Signal Inc. Note that the high pressure dynamic machine is easily located within a low pressure quasistatic press, which is used for compact ejection.

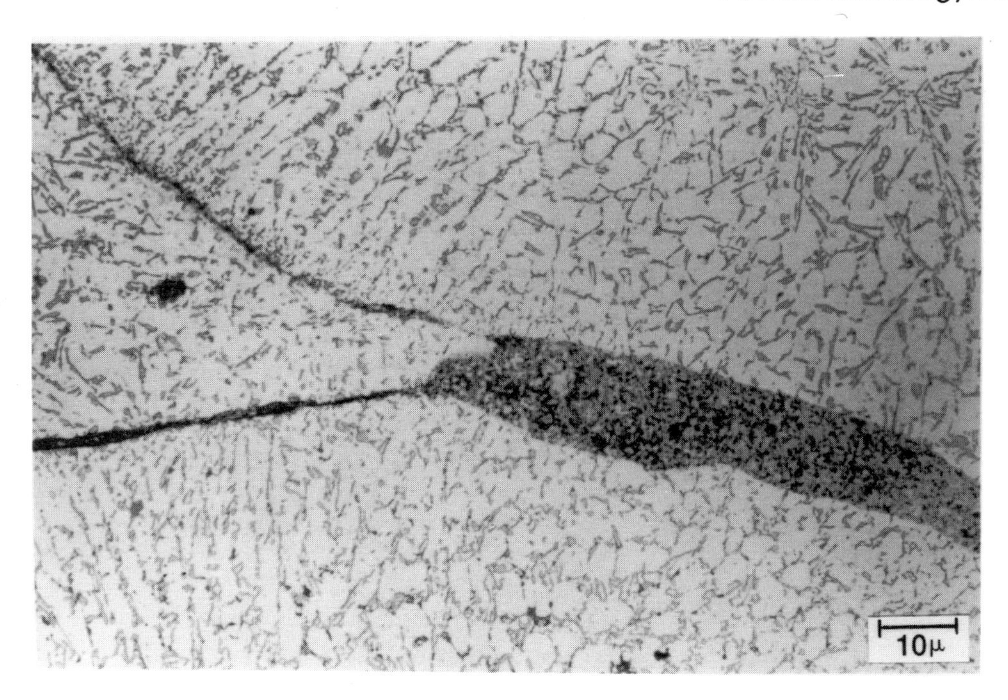

Figure 2. Interparticle melting between 3 particles of an Al 12% Si Alloy. Note the refinement of the microstructure in the welds.

Table 1 A simplified comparison of some compaction parameters for different techniques

PARAMETER	PRESS				
	HYDRAULIC	CRANK	PETROFORGE TYPE	CONTACT EXPLOSIVE	D.P.C.
IMPACTOR MASS	Large			None or Negligibl	Small
IMPACTOR VELOCITY [m/sec]	Negligi-ble	<1.7	10 to 20	300 to 1,200 ref(16)	100 to 2,000 ref.(8)
PRESSURE WAVE	None		Present	Important	
THERMAL WAVE	None			Present to Important	
NORMAL COMPACTION PRESSURE	Low to Medium			Medium to Very High	Low to High
RATE OF ENERGY DISPOSITION	Low			Very High	
VARIATION OF PROPS. IN COMPACT	Present, but Unimportant, Caused by Die Friction		Small as Die Friction is Reduced	Important Avoid by Explosive to Powder Ratio	None or Present Depends on Punch / Die Geometry

3.1 Hardness

As would be expected, the hardness of dynamic compacts is significantly above that of similar alloys produced by conventional compaction or ingot metallurgy. Surprisingly, the hardness of dynamic compacts is above that obtained by normal work or shock hardening [22].

The hardness of a superalloy, compacted by a gas gun with a punch velocity of 1200 m/s^2 increased from 400 Hv to 600 Hv. Warm explosive compaction of similar IN 718 reported a hardness of 500 Hv for compaction with no preheat, while with a 740°C preheat this decreased to 325 Hv. Recently, [15] it was shown that the hardness of pure tungsten compacts increased from the 250 Hv for the starting powder to around 550 Hv for the fully dense dynamic compact. However, in the same work it was found that powders which already have a high dislocation density, such as amorphous metal powders, undergo little or no work hardening; crystallizing these powders can decrease the starting hardness and a small increase in hardness due to shock hardening does then occur.

If the average temperature rise is sufficient, then the increase in hardness may be annealed out. This was found to be the case for copper [15], which for single shock compaction, showed no increase in hardness, but for a double shock (which results in a lower temperature rise) resulted in a doubling of the hardness of the starting powder.

3.2 Strength

For well bonded dynamic compacts, strengths equal to that obtained from wrought material have been obtained [11,17,22], (Fig.4). For instance, strengths of 1300 MN/m^2 were obtained for AISI 9310 steel [18]. Even for non-metallic powders, the strength increase according to the ratio R in equation (1), and can approach the strength of the ceramic produced by conventional techniques [12]. Further property modifications are possible by activated sintering [6,7].

3.3 Ductility

Even when good interparticle bonding occurs, the high hardness of dynamic compacts would suggest a low ductility. This does appear to be the case from the small amount of data published on ductility. For stainless steel, tensile elongations of 1 to 2% were obtained. However, if the ductility is related to the rapidly solidified weld, then, provided good interparticle bonding is obtained, the elongation could be much higher than normally associated with the high hardness. For aluminum, dynamic compacts had twice the hardness of fully work hardened wrought aluminum and an elongation of 8% compared to 3% for the "hard" wrought aluminum [5]. High strengths, U.T.S. = ~800 MN/m^2, and nominal tensile strains of 0.2 have been reported [23] for the difficult to compact Inconel 718 .

3.4 Toughness

The fracture toughness is dependent on strength and ductility. Fracture toughness tests carried out according to ASTM standard E 399-74 on stainless steel obtained values of 31 $MN/m^{3/2}$, comparable to quoted values for the wrought material, despite a low ductility. The toughness increased with the ratio R of equation (1) in the same manner as the tensile strength, (Fig.5). Interestingly, for the stainless steel compacts annealing at low temperatures, around 500 to 700°C, resulted in an increase in both strength and ductility. Presumably, this would also result in an increase in toughness, but this has not been measured.

3.5 Fatigue

Extensive fatigue testing has been carried out on dynamic compacts of aluminum. As would be anticipated from the high hardness and rapidly solidified interparticle welds high fatigue strengths were obtained [5]. At ten million cycles the fatigue strength was nearly twice that of comparable wrought aluminum. It was shown that annealing and sintering the dynamic compacts resulted in a significant decrease in the fatigue strength confirming that the high as compacted fatigue strength was due to the unique microstructure, the decrease in dislocation density with annealing temperature was followed by transmission electron microscopy.

4. Systems

4.1 Amorphous Metals

Amorphous metals are produced by rapid solidification techniques, which quench the metal so quickly that it does not have time to transform into its normal crystalline state, but, instead, retains the heavily distorted structure of a liquid metal, i.e. glass. This means that they can only be produced as thin ribbon or powders and as they transform back to the normal crystalline structure of a metal at around 450°C it is not possible to work them by conventional metalworking processes. During the early 1980s extensive work was carried out to produce 3-D bulk samples of amorphous metals. One objective was to determine the properties of bulk amorphous metals. Two areas appeared of interest. The first was the soft magnetic properties, which are the main reason for producing amorphous metals as ribbon, the other was properties related to the high hardness of amorphous metals.

The magnetic properties of bulk amorphous metals [24,25] proved to be comparable to those of existing soft magnets. In addition, it was found that fully dense compacts with good interparticle bonding were not required and could actually be detrimental to the magnetic properties. Low density compacts can be made by alternate compaction techniques [25].

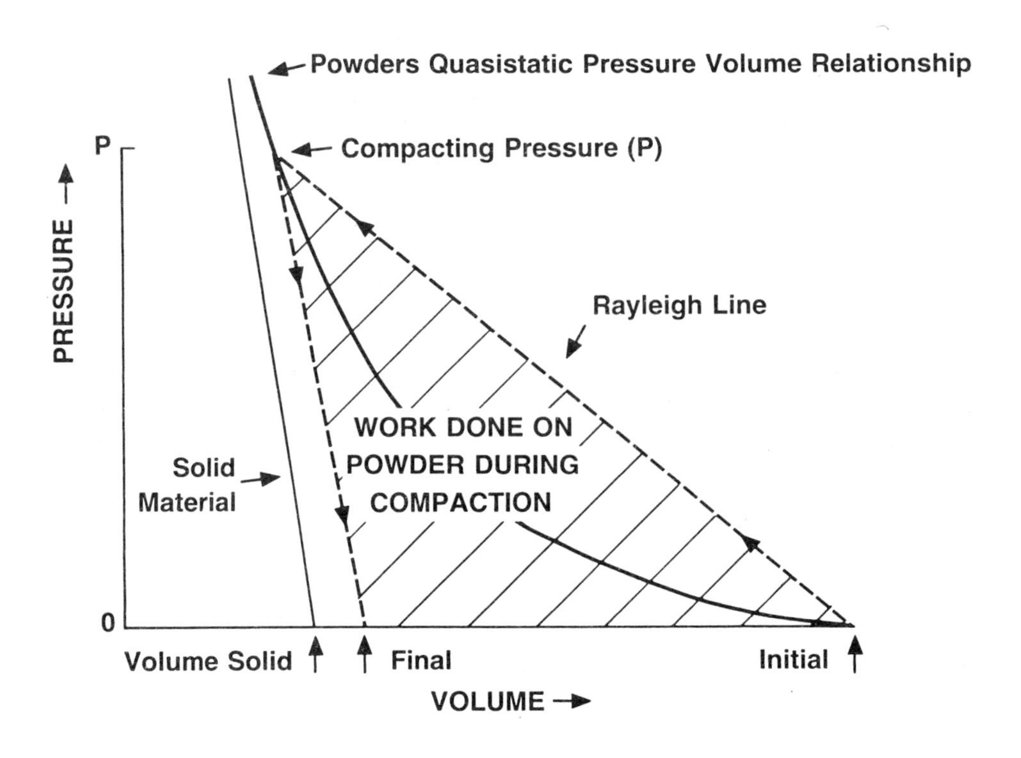

Figure 3 Pressure volume curve. During dynamic compaction the Rayleigh line is followed, while during quasistatic compaction the pressure volume curve is followed and less work is therefore done on the powder.

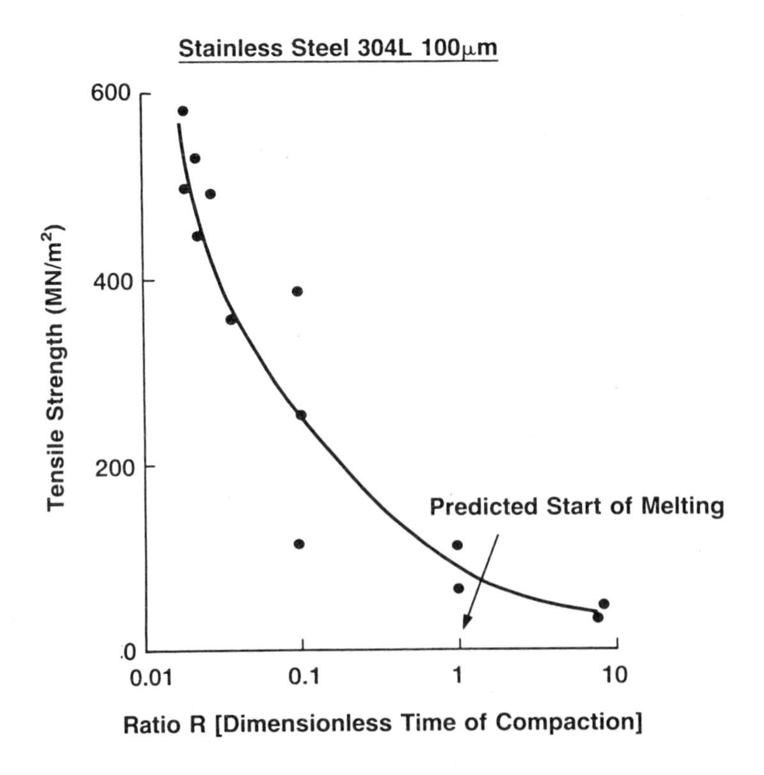

Figure 4 The compact's strength increases as a function of the ratio R of equation (1).

The strength and high hardness proved more interesting in terms of potential applications, but the competitive nature of the existing tool industry, and poor economic climate in this industry provided no incentive to explore this potential area further. One especially interesting possibility was compaction in the amorphous state and then crystallization under controlled conditions to give an extremely fine (nano crystalline) microstucture, which for a simple iron boron carbon alloy resulted in the hardness increasing from 950 Hv for the amorphous material to 1500 Hv for the nano crystalline material [26].

The attempts to consolidate amorphous powders were carried out on a wide variety of high speed presses ranging from the Petro Forge to explosive compaction. The two techniques which produced samples with measurable properties were high speed gas guns and explosives. A problem with both of these was that to achieve full densification of the hard powders, high compaction pressures had to be employed, which resulted in average temperatures after compaction close to the alloys' crystallization temperature. Extreme care was therefore necessary in obtaining a sufficiently high pressure, but not so high that the compact crystallized. To avoid this problem Russian workers [27] cooled the powder sample to liquid nitrogen temperatures prior to explosive compaction. An alternative approach, used by Raybould, involved heating the powder to a few hundred degrees Celsius prior to compaction. This takes advantage of the significant softening of amorphous metals at relatively low temperatures, i.e. below the crystallization temperature. The pressure required for compaction is significantly reduced and hence the average temperatures rise. Therefore, the compaction conditions are more easily obtained and are less critical than when high pressures are required. While room temperature compaction at 1500 m/s^2 produced a 95% T.D. and no interparticle bonding, with a 300°C preheat at 1250 m/s^2 full densification and good interparticle bonding were obtained, (Fig. 6).

4.2 Composites

Dynamic compaction has been used to produce a wide range of metal matrix composites(MMC). Usually, however, the work has been low key. In 1975, Russian workers showed [16] microstructures of glass fibres in an aluminum matrix. Similar work, (Fig.7), was carried out at Inst. CERAC, Switzerland, over a period of years on a toll basis for several clients. The ductility of compacts of aluminum plus glass fibre was good, even for high fibre contents (30 to 40% by volume). In addition, work on wear and seizure resistant alloys showed that the addition of particles of a second or third material could have little or no affect on ductility up to 20% by volume additions.

One aspect of the dynamic compaction of composites is the ability it gives of changing the powder's impedance and hence the pressure created on the impact of a projectile of a given velocity. For example, the addition of a powder denser than the matrix increases the compaction pressure, while a lower density powder decreases the compaction pressure. The impedance also depends on elastic modulus and sound speed.

If the second particles are stronger/harder than the matrix most of the deformation appears to be in the matrix, which flows plastically around the hard particles, rather than the slow rearranging, which the more isostatic conditions of quasistatic pressing produce. This is caused by impacts between the particles as the shock wave passes them, at which time the soft particle is not aware that the hard particle is unsupported, so very high impact pressures are created locally. This increases the possibility of good interparticle bonding between the matrix and the second particles.

One unique ability of dynamic compaction is to produce composites from materials which would react at the high temperatures required for sintering etc. Many such reactive mixtures have been made. Detailed studies on adding diamonds to steel and steel to aluminum have been made to determine processing parameters. Similarly, the addition of materials which are immiscible when cast in aluminum, such as lead, has been successfully demonstrated. It was found that many of these systems benefited from a low temperature heat treatment, which appeared to remove internal stress and improve the interparticle bonding, especially between the matrix and the addition. This is more beneficial than the activated sintering reported for simple alloys [7].

Many of the composites investigated required relatively low impact velocities, 600 to 1000 m/s, to achieve interesting properties.

4.3 Ceramics

A variety of ceramics have been compacted by dynamic techniques. In fact, some of the earliest work on dynamic compaction concerned the significant lowering of the sintering temperature resulting from explosively compacting or processing ceramics [6]. At present, because very high pressures are required, explosive compaction appears more promising than impact compaction using a die. Impact compaction can result in full density compacts, (Fig.8), with evidence of interparticle bonding [12], but problems may occur with ejection of the ceramic from the die, resulting in rapid die wear, unless special systems are used. In addition, the size of the part is limited, while explosive compaction has produced large plates. To avoid the cracking often associated with the sudden release of pressure with material as brittle as ceramics, Linse [28] developed an explosive technique involving preheating the powder. This resulted in the compact having sufficient ductility to withstand the unloading pressure. In addition, the pressure required for compaction is decreased so that the conditions for full densification are less extreme, and the pressure decrease on unloading is reduced.

One aspect of the activation of sintering is that after pretreatment of the ceramic powder by a shock wave, conventional compaction may be used. Sintering still occurs at a lower temperature or proceeds at a faster rate. This allows the

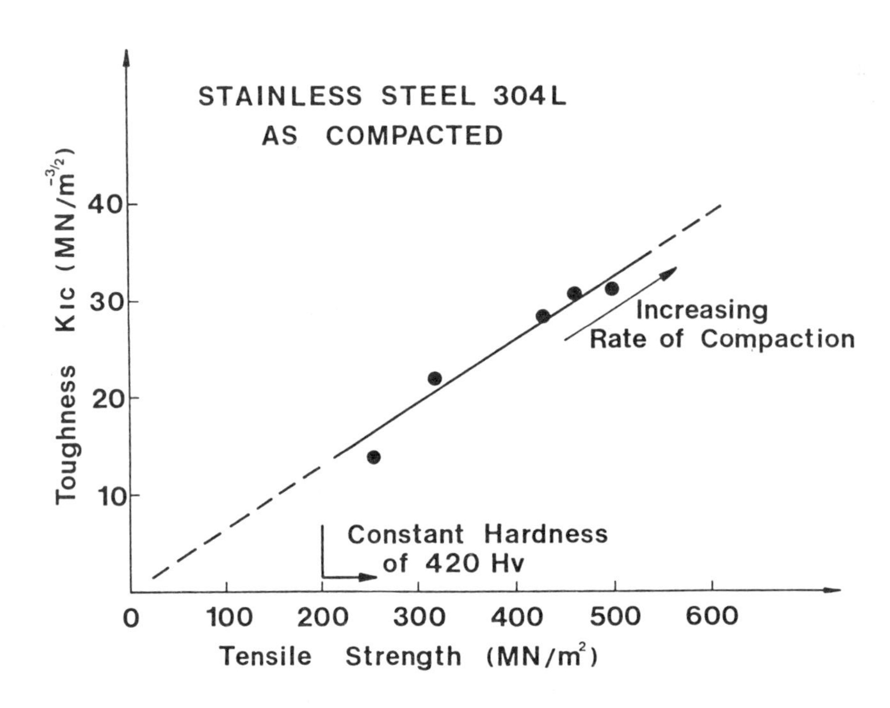

Figure 5 The fracture toughness, determined according to ASTM E 399-74 of stainless steel increases as the strength and hence with the ratio R, which indicates the degree of interparticle welding.

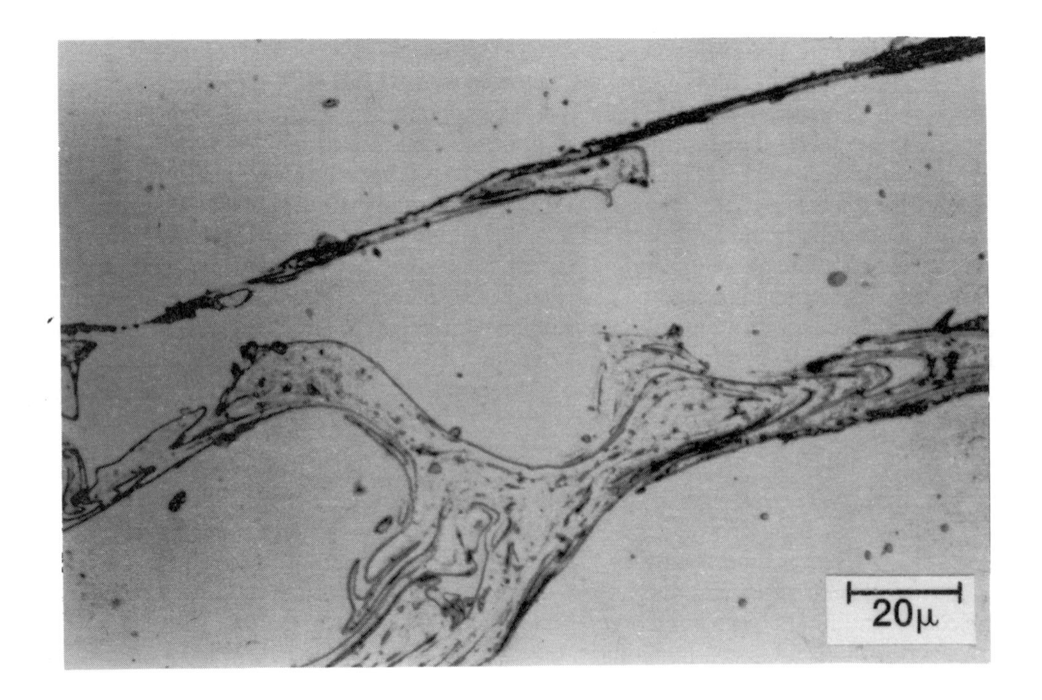

Figure 6 Extensive interparticle melting in an amorphous alloy, Devitrium 7025, an alloy designed as a tool steel replacement. Note the wave structure in the weld, as in explosive welding.

elimination of sintering additives, which can deteriorate the properties of the composite, for example; the addition of magnesia and yttria to silicon nitride to activate sintering reduces creep and oxidation resistance. The pretreatment of large quantities of powder is already carried out in the case of artificial diamonds, and provided sufficient material is treated at one time, is an economic process, despite low yields. Such an operation for ceramics would have a much higher yield and would probably require lower pressures.

4.4 Precision Parts

The production of standard PM type parts by dynamic compaction is not justified, because of the complexity of the shock wave interactions for parts with many different levels. However, due to the fact that the powder is being compacted by shock waves, special cases do arise which make shock wave compaction of complex shapes attractive. In particular, complex detail on a part of basically one level is easily achieved. It must be emphasized though that some shapes can not be made by dynamic compaction.

One advantage of shock wave compaction is the production of uniform densities, even when full densification does not occur. Conventional pressing often has a significant density variation due to friction forces which are much less important in shock wave compaction. One example of this is the rock drill produced by dynamic compaction, (Fig. 9). This was produced by compacting steel powder around cemented carbide buttons. Conventional pressing resulted in low density areas in the steel between the buttons, so that they were not held in place and, in addition, were surrounded by steel with strength and toughness values below that required for the demanding application. Dynamic compaction produced full density parts to which the carbide was well bonded.

Another advantage is the production of high densities from hard to compact powders. This can allow the production of a shape which undergoes little or no change in dimension during sintering [29]. The possibility also exists of making such parts from cheap PM dies [29].

5. Discussion

Examples of the wide range of parts fabricated by dynamic compaction have been given along with some of the basic properties that can be obtained. The technique can be seen to offer a means of tailoring properties for a given application. For example, copper, depending on the compaction conditions, can be produced with interparticle welding and a high dislocation density or fully annealed. Such tailoring of properties offers many potential applications. However, to exploit these, it is essential that the shock waves interaction in the powder are understood in terms of producing uniform densities, interparticle bonding and mechanical properties.

Dynamic compaction can be comparable economically to existing advanced metal working techniques. In the case of gun type machines this is especially so if very large parts and very high impact velocities are not required; for these, explosive techniques should probably be employed.

6. References

1. Dynamic Compaction of Metal and Ceramic Powders, NMAB-394 National Materials Advisory Board, Washington, DC, 1983.
2. Proc. 9th Int. HERF Conf. (eds.) I.V. Yakovlev & V.F. Nesterenko, Academy of Science, Novosibirsk, USSR, 1986. pub. USSR Academy of Science, Novosibirsk Branch.
3. Proc. 8th Int. HERF Conf. (eds.) I. Berman & J.W. Schroeder, ASME, New York, 1984.
4. Metallurgical Applications of Shock-Wave and High-Strain-Rate Phenomena; (eds.) L.E. Murr, K.P.Staudhammer and M.A. Meyers, Dekker, New York, 1986.
5. D. Raybould, The Fatigue Properties of Dynamically Consolidated Aluminium, J. Materials Science, 19, p.3498, 1984.
6. O.R.-Bergman and J. Barrington, J. Am. Cer. Soc., 49, p.502,1966.
7. D. Raybould, Cold Dynamic Compaction of Pre-Alloyed Titanium and Activated Sintering, in New Perspectives in Powder Metallurgy: 8, (ed.) K.M. Kulkarni, M.P.I.F., Princeton NJ, 1987.
8. B. Lemcke and D. Raybould, US Patent 4 255 374, 1980, filed in Europe in 1977.
9. S. Hagino, M. Susuki, T. Takeshita, K. Takashima and H. Tonda, Microstructures and Superconducting Properties of Y-Ba-Cu Oxide Coils Prepared by the Explosive Compaction Technique, Proc. lst Int. Conf. Superconductivity, Nageya, Japan, August 1988.
10. Shock Waves and High Strain Rate Phenomena in Metals, (eds.) M.A. Meyers and L.E. Murr, Plenum Press, New York, Appendix C, p.1045, 1981.
11. D. Raybould, The Production of Strong Parts and Non-Equilibrium Alloys by Dynamic Compaction, p.895 in Ref. 10.
12. D. Raybould and T.Z. Blazynski, Non Metallic Materials under Shock Loading in Materials at High Strain Rates, (ed.) T. Z. Blazynski, Elsevier Applied Science, London, 1987.

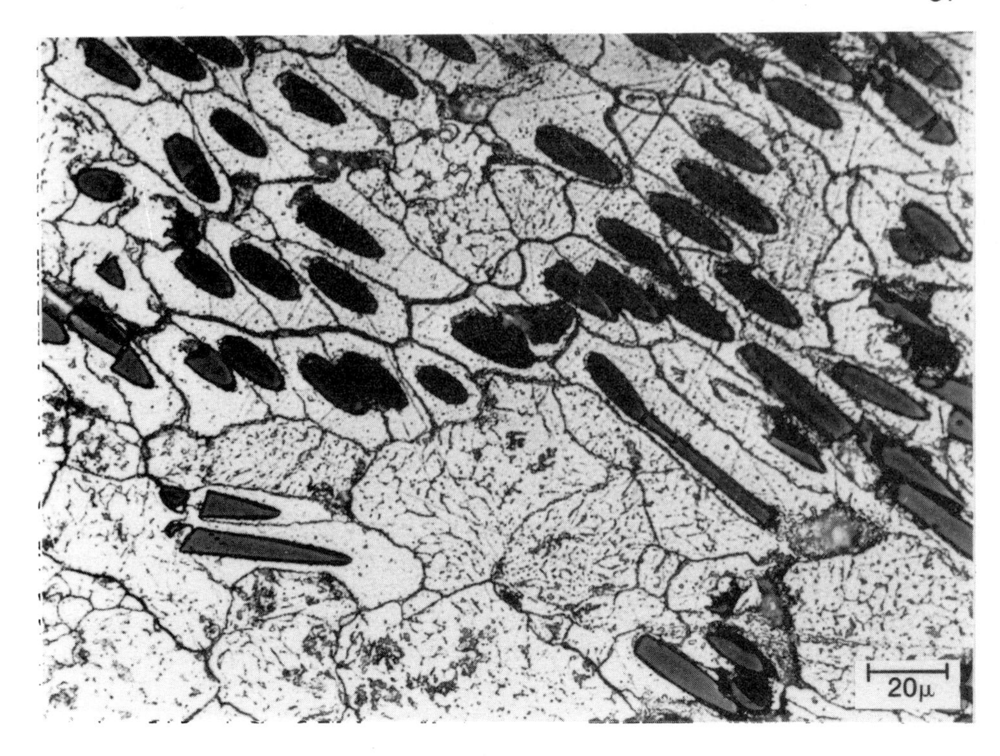

Figure 7 An aluminum glass fibre composite, one of several MMC compacted on a toll basis by Inst. CERAC.

Figure 8 A scanning electron micrograph of a dynamic compact of Alumina. The compact was 3" diameter.

Figure 9 Several precision parts produced from aluminum and steel. Note the rock drills produced by compacting steel powder around cemented carbide buttons. A dynamic compact is shown on the left and a production drill bit on the right.

13. L. Davidson, J. Appl Phys., 42, p.5503, 1971.
14. T. Sheppard, Proc. 15th Int. MTDR Conf. (eds.) S. Tobias and F. Koenigsberg, Macmillan Press, London, 1975.
15. N. Page and D. Raybould, Dynamic Powder Compaction of some Rapidly Solidified Crystalline and Amorphous Powders; Compaction Characteristics, Material Sci. & Eng. in press.
16. A.M. Straver, Physical Phenomena at the Compression of Powder Materials by Explosives, Proceedings 5th Int. H.E.R.F. Conf. Denver Research Inst. Colorado, 1975.
17. D. Raybould, The Cold Welding of Powders by Dynamic Compaction, Int. J. Powder Met . & Tech. 1, p.9, 1980.
18. R.B. Schwartz, P. Kasiraj, T. Vreeland and T.J. Ahrens, A Theory for the Shock Wave Consolidation of Powders, Acta Metal 32, p.1235, 1984. See also T. Vreeland, P. Kasiraj, A.H. Mutz and N.N. Thadhani, p.247 in Ref. 4.
19. W.H. Gourdin, Energy Deposition and Microstructural Modification in Dynamically Consolidated Alloys, J.A.P. 55, p.172, 1984.
20. D.G. Morris, Melting and Solidification during Dynamic Compaction of Tool Steels, Metal Sci., 13, 3, p.116, 1981.
21. C.S. Yust and L.A. Harris, Observation of Dislocations and Twins in Explosively Compacted Alumina, p.881 in Ref. 10.
22. D. Raybould, On the Properties of Material Fabricated by Dynamic Powder Compaction, Proc. 7th Int. HERF Conf. University of Leeds , UK , p.261, 1981.
23. S.L. Wang, M.A. Meyers and A. Szecket, Warm Consolidation of IN 718 Powder, J. Mat. Sci., p.1786, 1989.
24. D. Raybould, Consolidation of Rapidly Solidified Powders to Structural Parts and Amorphous Soft Magnets, Progress in Powder Metallurgy, M.P.I.F. Princeton, NJ, 41, p.95, 1986.
25. D. Raybould and K.S. Tan, Factors Affecting the Magnetic Properties of Consolidated Amorphous Powder Cores, J. Mat. Sci., 20, p.2776, 1985.
26. D. Raybould, US Patent 4, 594,104.
27. O.V. Roman *et al.*, Structural Properties of Amorphous Materials based on Iron, in the collection, Powder Metallurgy 6th Ed. Higher School Minsk, p.8-13, 1982.
28. V.D. Linse in Innovations in Material Processing, (eds.) G. Bruggeman & V. Weiss, Plenum, New York, 1985.
29. D. Raybould, On the Production of High Density Precision Parts in Medium Alloy Steel and Tool Steels by Dynamic Compaction, as Ref. 23, p. 249.

10

Rotary Forging in Powder Metallurgy

P. M. STANDRING AND J. R. MOON

University of Nottingham, UK

1. Introduction

Rotary forging is a metal working method which can be applied to deform solid workpieces and to make high density powder metallurgy products [1]. Active building of machines, research into applications and use in production is underway in the USA and many countries of Europe and the Far East.

2. Principles

The principles underlying the design and operation of rotary forging machines have been described many times [e.g. refs. 1,2]. The essential feature is that workpieces are deformed, or compacted, incrementally. This can be brought about in a number of ways. Most commonly, a conical tool is used, which contacts a "footprint" of small area on the surface of the workpiece. The tooling motion causes the footprint and the deformation zone beneath it to be rotated about the axis of the workpiece. The small size of the deformation zone, coupled with the mechanical advantage of the press, means that pressures of the order of several GPa can be generated from loads of the order of tens of kN.

Three sorts of rotary motions of the tooling are possible: spin, nutation and precession; these terms are defined in Fig. 1. In principle, the individual motions can be combined in any way desired, giving rise to a potential family of seven types of rotary forging machine [3], although some of these could be physically difficult to construct and control. Tool-workpiece interactions could be very complex, requiring computer simulation to help understand them [4].

The operations of two practical members of the family to compact loose powders are illustrated in Figs. 2 and 3.

In the arrangement of Fig. 2a, the die angle is fixed. The only movements are precession of the conic tool about the axis of the machine and simultaneous die closure to cause compaction of the powder [5,6,7]. The problem here is that a small gap must exist between the cylindrical die wall and the conic die. Powder particles need to be big enough not to escape through the gap or to be previously consolidated into a sintered preform. Alternatively, a die sealing pad may be used between the conic die and the powder [8] (Fig. 2b), but this spreads the load and most of the inherent advantage of rotary forging is lost.

The arrangement of Fig. 3 is the nutation-spin system [9] developed at Nottingham. This uses a cone which remains in position but spins about its own axis, whilst the workpiece also spins about its own axis. The relative spin rates can be controlled such that the kinematics of tooling motion experienced by the workpiece are identical with those of a precession-only arrangement. The significant difference is that the arrangement of Fig. 3 allows easy variation of the angle by which the conic die is tilted from the machine axis. This tilting action, or nutation, can be caused to occur at any time and at any rate during the deformation of a workpiece.

Any rotary forging machine which allows the nutation angle to be varied in a controlled way during forming can be caused at any time to operate one of four processing modules [1,9].

Module 1 - uniaxial compaction

Module 2 - nutation only

Module 3 - combination of 1 and 2

Module 4 - fixed conic die angle with axial feed (i.e. a precession or spin only operation)

Figure 3 shows a module 1 operation compacting powder unaxially and subsequent completion of module 2 or 3 to reduce the volume available for occupation by the compact. Continued rotation with axial feed would be module 4.

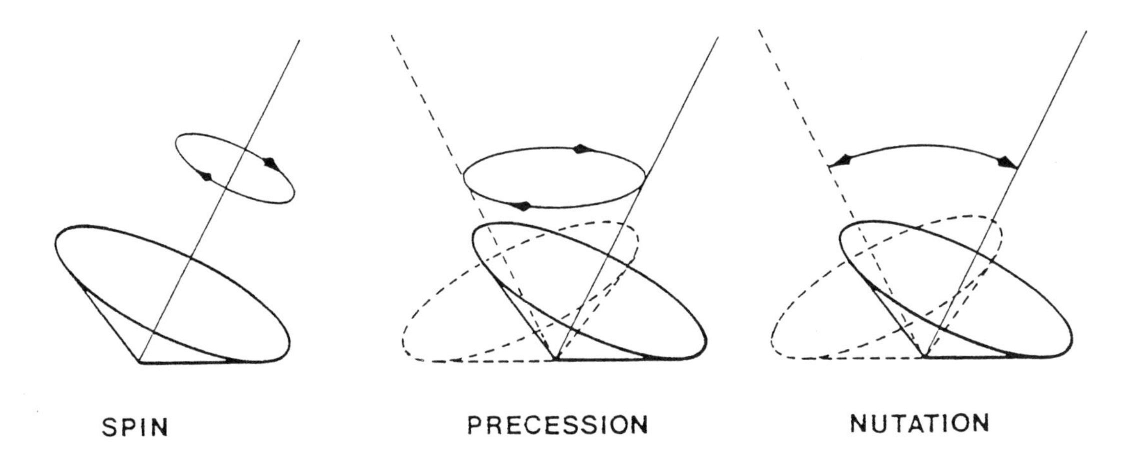

Figure 1 Definitions of terms used to describe tooling motions.

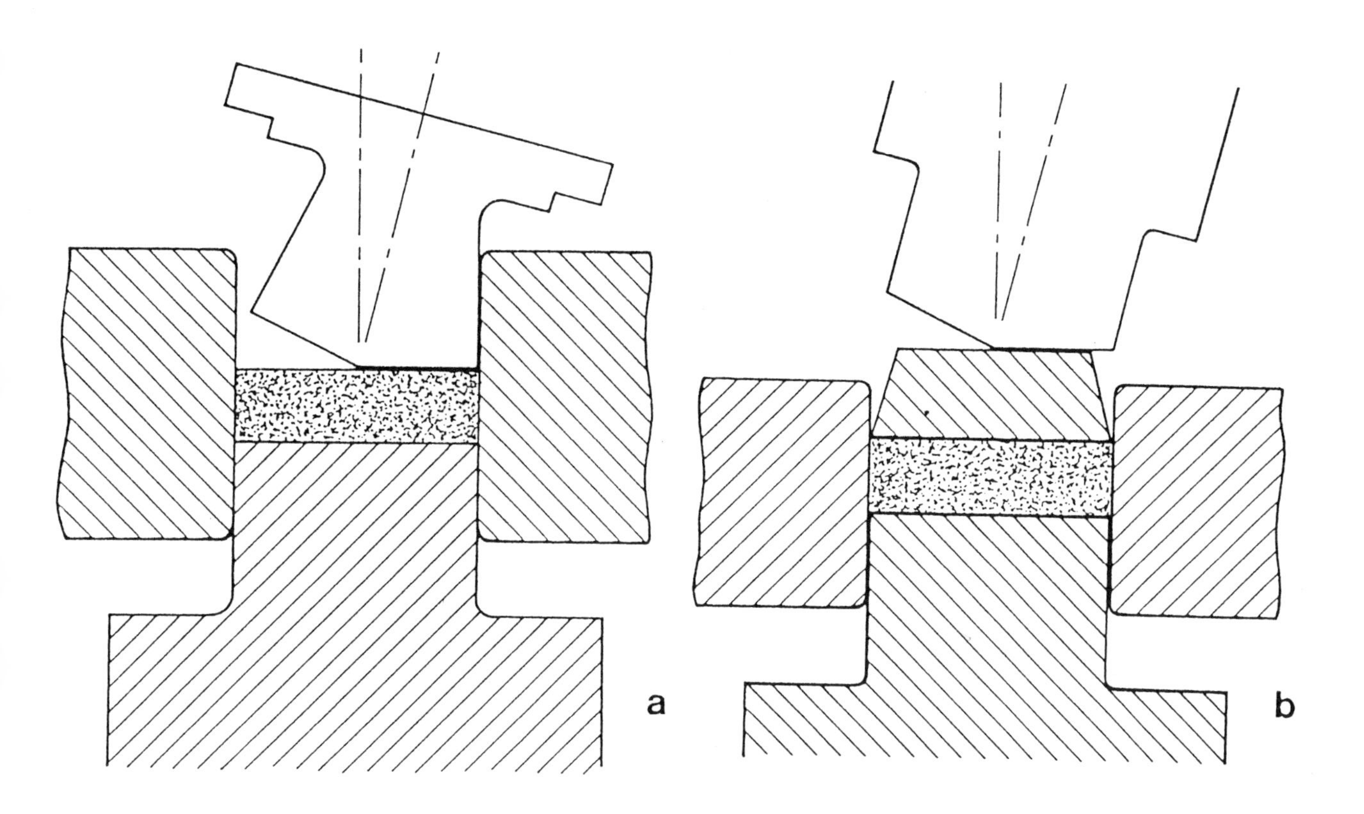

Figure 2 Tooling arrangements to compact powders using a precession only machine; (a) Refs. 5, 6; (b) Refs. 7, 8.

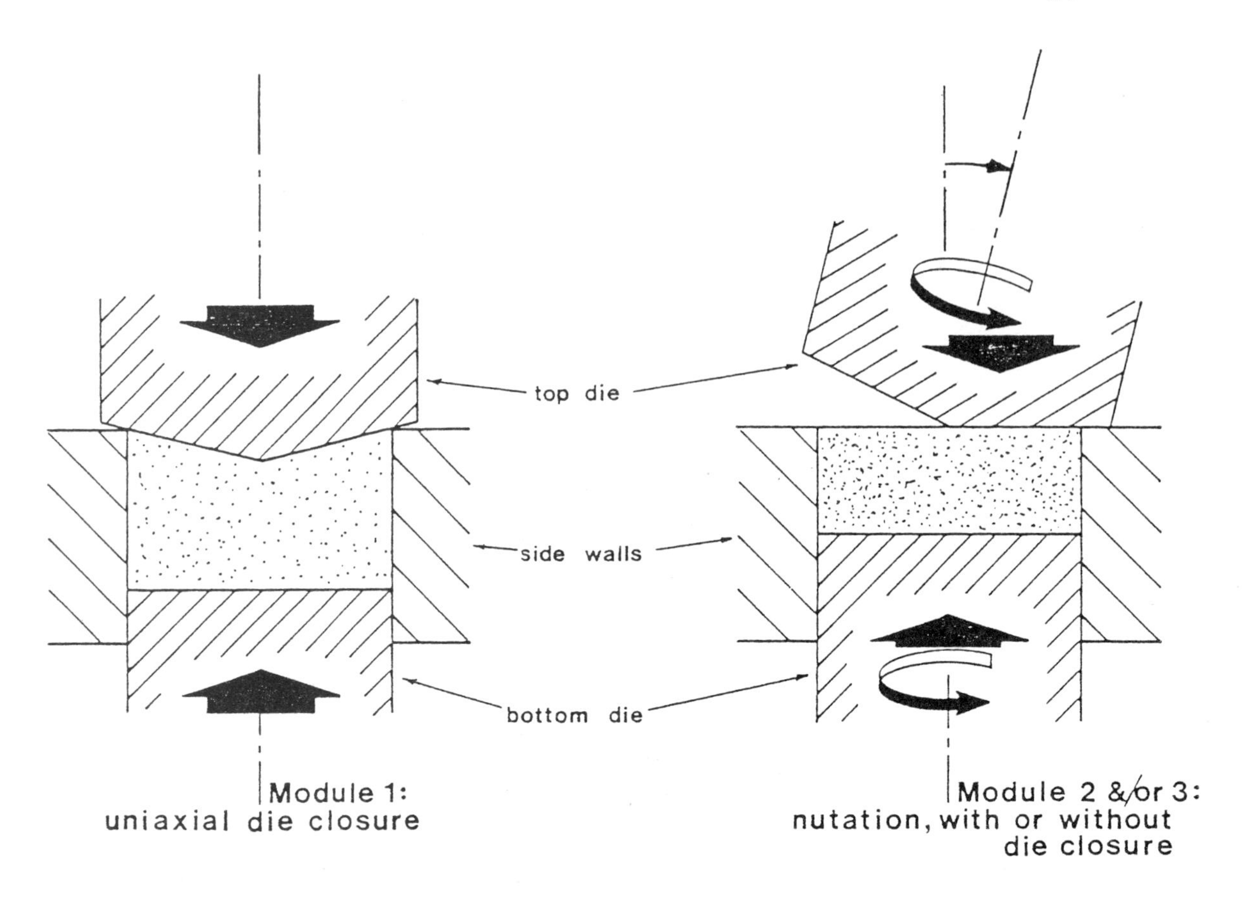

Figure 3 Typical compaction cycle using a nutation-spin machine [2,9].

Figure 5 Face gear formed by forging of sintered preform (pure iron) (x40) [Ref. 12].

3. Control

Precession only machines operate with a nutation angle pre-set by the design of the machine. Only the movement of an opposing ram is available for control.

A nutation-spin machine requires several axes to move simultaneously. Reproducible operation demands computer control of the activation systems. A teach, record and playback system to aid programming is under active development [10].

4. Densification of Sintered Preforms

The application of rotary forging techniques to powder metallurgy materials appears to have begun in Poland and Japan in the early 1970s [7,11]. Subsequent publications from these countries and the UK have been concerned largely with the use of fixed axis machines to densify powder preforms [5,6,7,8,11,12]. Sintered preforms can resist modest tensile and shear stresses. This allows preforms to be densified and shaped using suitable tooling. Floating die systems shown in Fig.4 have been developed to provide near closed die conditions [12,13].

Final relative densities approaching 100% are achieved and the formation of raised gear teeth gives microstructures with features reminiscent of ordinary forgings [12], (Fig. 5).

Examples of commercial products are rare; one report exists of the production of valve seat inserts in Japan using such a system [13].

5. Consolidation of Loose Powders

The consolidation of loose powders has been reported in the UK using fixed angle of inclination machines [15] and using variable axis machines [2,9]. It is also known that the consolidation of loose powder is being developed for commercial exploitation in the USSR.

Simple discs and rings, radial gears and face gears have been made from steels, copper and aluminium powders. Split dies can be used to solve extraction problems arising from high die-wall friction - an undesirable product of the high pressures. Their use allows production of parts with re-entrant features.

Relative densities of better than 98% can be achieved readily. Microstructures of compacts show that space has been fully occupied by deformation of particles against one another [2], (Fig. 6).

6. Prospects

Two types of application involving very different types of materials are being investigated in academic laboratories and research institutes.

One type of application is the cold consolidation of advanced materials such as rapidly solidified powders in order to exploit their advantages without disturbing their thermodynamic instability. Cold welding can, in principle, be brought about by the development of very large shear strains, but methods are required which produce such strains without also generating destructive tensile stresses in the compact.

The other area of interest is the generation of very high density compacts in common materials (e.g. steels), which will be sintered. The aim is to develop good mechanical properties and the degree of success may be judged from Table 1. This compares mechanical properties of Distaloy AB and Astaloy A, both containing 0.6%C, prepared by conventional PM processing, rotary compaction and sintering, and hot powder forging. The data is, as yet, incomplete and represents an early stage in the development of rotary compaction. For the powders concerned, the rotary forged material is nearly, but not quite, as good as powder forged material.

Commercial exploitation demands the availability of rotary forging machines and the will to use them. The current leader in sales of commercial machines is the Swiss Company, Schmid, whose machines are based on the early Polish designs.

It is known that work is being carried out using sintered preforms by industrial groups in Europe, including the USSR [15]. Experience of the use of rotary forging machines for the production of both wrought materials [16] and powder products [17] exists in China. Published information arising from USA and Japan suggests that most of the recent work in those countries is concerned with applications of the process to precision cold-forming.

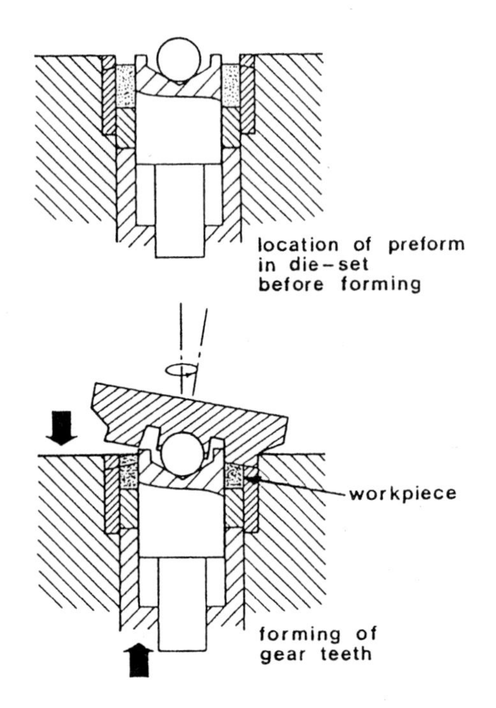

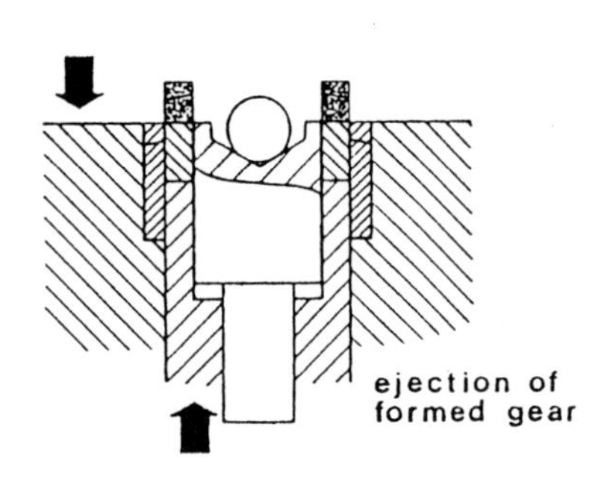

a Forming of spiral bevel gears by rotary forging.

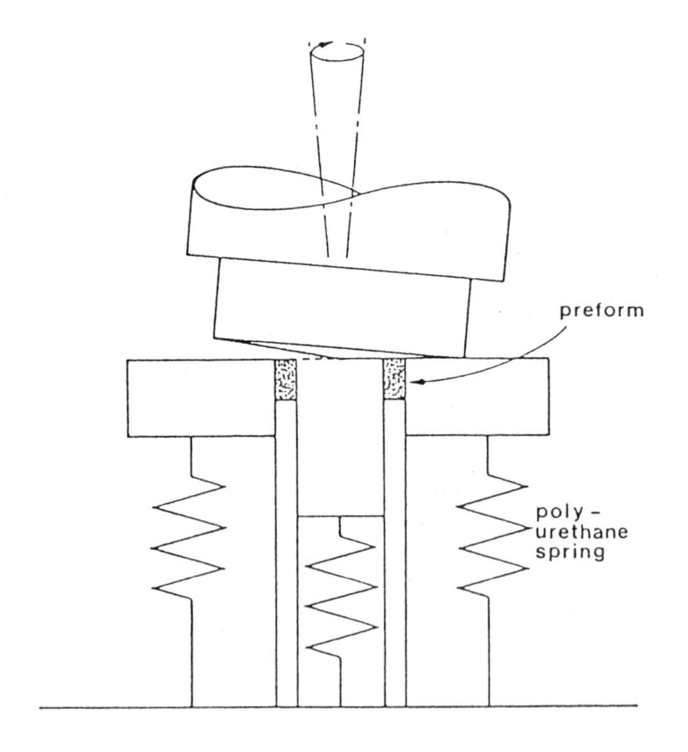

b

Figure 4 Floating die tooling arrangements used with precession only machines to densify sintered preforms; (a) Ref. 12; (b) Ref. 13.

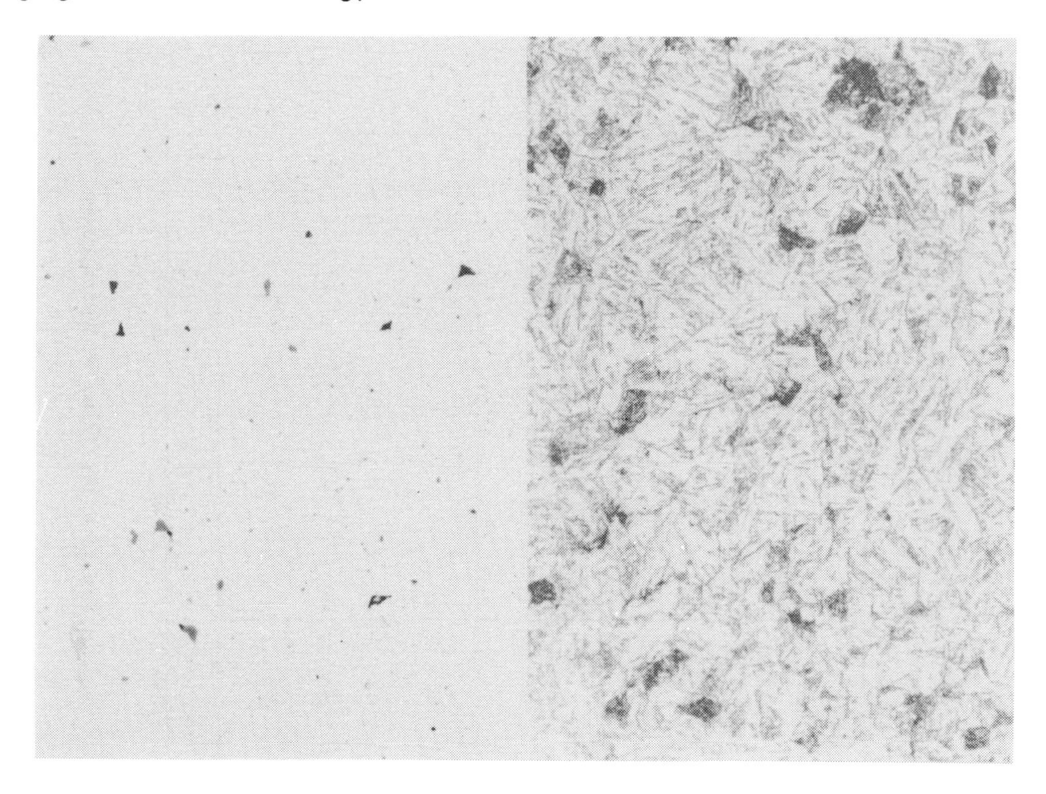

Figure 6 Microstructure of rotary compacted and sintered Astaloy A + 0.6%C, $_p$ = 0.98 (unetched & etched) (x400) (courtesy of S. Peacock).

References

1. P.M. Standring, Rotary forging - a technical review, Metallurgia, pp. 9,12,16, January 1989.
2. J.R. Moon and P.M. Standring, Rotary forging of metal powders, Powder Metallurgy, 30, pp.153-163, 1987.
3. P.M. Standring, Rotary forging - a new approach, Advanced Technology of Plasticity, 11, pp.942-947, 1984.
4. S. Mansour and P.M. Standring, The graphical simulation of tool-workpiece kinematics and deformation in rotary forging, Proceedings second international conference on production research, Edinburgh, Kogan Page, London, 1, pp.386-405, 1987.
5. K. Kubo and Y. Hirai, Sinter forging by rotary forging press, Proceedings of the twenty-sixth conference Japan Society for Technology of Plasticity, pp.393-396, 1975.
6. J. Ogrodnik and T. Karpinsky, Cold forming of sintered materials with self-aligning dies, Powder Metallurgy International, 9, pp.133-135, 1977.
7. D. Evans, J.B. Hawkyard and T.J. Davies, The rotary forging of steel powder, Powder Metallurgy International, 9, pp.174-176, 1977.
8. K. Kubo, Y. Hirai, S. Ogiso, H. Otsuka and M. Ohta, Sinter forging by rotary forging press, Proceedings Spring Conference Japan Society for Technology of Plasticity, pp.241-244, 1973.
9. J.R. Moon and P.M. Standring, Rotary forging for high density powder compacts, Metals and Materials, 2, pp.206-210, 1986.
10. M. Berg, J. Wilson and P.M. Standring, Developing a human interface for a rotary forging machine: defining the issues, Proceedings Third National Conference on Production Research, Nottingham, Kogan Page, London, pp.413-417, 1987.
11. J. Kotschy, Z. Marciniak and M. Sobczyk, Cold forming of powder metallurgy preforms by the rocking die method, International Journal of Powder Metallurgy and Powder Technology, 9, pp.135-137, 1973 and 12, pp.5-7, 1976.
12. E.R. Leheup, J.R. Moon and P.M. Standring, Production of components by rotary forging of sintered preforms, Powder Metallurgy, 26, pp.129-135, 1983.
13. F. Kiyota and K. Takemura, Densification of PM valve seat by cold rotary forging, Metal Powder Report, vol.40, pp.504-506, 1985 (also in Proceedings Third Conference on Rotary Metal Processing, Kyoto, Japan, pp.101-108, 1984.)

Table 1

MECHANICAL PROPERTIES OF POWDER METALLURGY STEELS PRODUCED BY DIFFERENT PROCESSES

HEAT-TREATMENT: as-sintered/forged

MATERIAL & PROCESS	density/ kgm^{-3}	Tensile			Toughness		Fatigue		
		yield/MPa	UTS/MPa	%elong	impact at 0°C/J	K_{IC}/MPa$\sqrt{m}$	endurance limit/MPa	ΔK_{th}/MPa$\sqrt{m}$ (R = 0.1)	Paris m (R = 0.1)
Distaloy AB-0.6C									
Conventional PM	7025	400	540	2-3		33	200	8.0	13
RF and sintered	7545	430	610	4	40			4.5	5
Powder forged	7840	510	1020	8		57		7.0	2.6
Astaloy A-0.6C									
Conventional PM	7101	370	490	3-4		32		10	10
RF and sintered	7665	530	660	8	5*	48		4.2	5
Powder forged	7860	650	970	-		62		7.5	3

*Standard Charpy v-notched test-piece: other results are for un-notched test-pieces.

Data taken from work in progress by S. Peacock, R. A. Phillips, R. T. Aylwin and J. R. Williams at Nottingham.

14. M. Ashraf and I. Thomas, An investigation of the compaction of aluminium RDR powders by rotary forging techniques. Proceedings of Powder Processing Technology Workshop (Royal Society), published by BNF Metals Technology Centre, 1985.
15. Report of OSTEM mission on High Strength Sintered Component Manufacture in Europe, Metal Powder Report, pp.90-92, February 1989.
16. Wei Jie, Lu Qi Ren, Wang Zhong Ren, Pei Xing Hua and Zao Chen, New development on rotary forging in China, Proceedings Third Conference on Rotary Metal Processing, Kyoto, Japan, (IFS Publications Ltd., Bedford). pp.61-69, 1984.
17. Liu Gui Qing, Application of rotary forging process in processing thrust bearing of steering axle, ibid, pp.93-100.

11

Sintering and Hot Isostatic Pressing Diagrams

M.F. ASHBY
Department of Engineering, Cambridge University Cambridge, UK

Abstract

Mechanisms of sintering, with and without an applied pressure, are reviewed. Equations describing the rate of densification by each mechanism are listed. The results are assembled into diagrams which show the extent of densification as a function of temperature, time and (where appropriate) pressure, and of the powder characteristics: particle size, material properties, and so forth. The diagrams are helpful in understanding and predicting the progress of densification and the factors which influence it; in planning sintering or HIPing schedules, and in summarizing a large body of information about the sintering, HIPing and sinter-HIPing of a given powder.

1. Introduction

Figure 1 shows, schematically, a powder compact which is to be consolidated. If the densification is achieved by heating alone, the process is called sintering; if it is achieved by pressure alone it is called cold isostatic pressing; if by heat and pressure together, then hot isostatic pressing (HIPing). In practice a cycle of pressure steps may be used: sintering is often preceded by cold pressing; HIPing may be preceded by sintering, and so forth. The simpler of these processes have been used for decades - some of them for centuries - as a way of making useful components from powders. They, and the more sophisticated HIP and sinter-HIP routes have lately attracted additional attention as ways of fabricating special tool-steels, advanced superalloys, novel dispersion-strengthened alloys and high performance ceramics; in short, for making useful shapes out of materials which are difficult to shape in any other way.

But there are many process-variables to get right if the product is to have adequate properties. The processing can be expensive: a single HIP-run in a large press can cost thousands of dollars, and a development programme to optimize a HIPing schedule involves many runs. The primary aim of modelling is to simplify and guide this optimization process, that is, to identify the optimal combination of process-variables: the temperature, time, and pressure, and the size, shape and condition of the powder. Ideally, the modelling should lead to a quick, accurate assessment of any proposed sintering or HIPing cycle, directing the operator to the most efficient combination of process-variables.

2. The Modelling of Packing Geometry and the Driving Force

2.1 The Stages of Densification

The process of densification is, of course, a continuous one; but the geometry changes are large, and for that reason it is helpful to think of it as occurring in the three sequential stages shown in Fig.2. The first stage, that of powder packing, is not well understood. The initial packing density depends on particle shape, on the size distribution, on the degree to which the powder has been shaken down and (particularly with fine particles) on the extent to which surface and frictional forces prevent rearrangement. An applied pressure P causes rearrangements which increase the relative packing density Δ (the fraction of space occupied by solid). It is sometimes found that Δ increases as the log of P:

$$\Delta = A\log(P) + C$$

but the origin of this behaviour is not understood.

During Stage 1 (Fig. 2) necks grow at the contact points between particles. A number of mechanisms contribute to neck growth: plastic yielding and creep dominate when the pressure is large; when it is not, densification involves various sorts of diffusion. All of these are analysed in the next section. The important characteristics of Stage 1, from the point of view of modelling, are that the necks are discrete (so that they can be treated as separate) and the porosity is connected (so that any gas in the pore space can escape).

Final densification occurs in Stage 2 (Fig. 2). By the time the relative density has reached 0.95, most of the necks have grown until they impinge and the individual pores have sealed off. From a modelling point of view, the material is now best thought of as a solid containing isolated spherical or polyhedral voids (so that the gas within them is trapped), connected by grain boundaries. Grain growth may detach the boundaries from the pores, or may drag the pores together. As in Stage 1, a number of mechanisms contribute simultaneously to the shrinkage of the pores.

In this and the next section, we develop equations for the rate of densification Δ of a powder compact, of initial relative density Δo, by each of several mechanisms. In doing so, we draw heavily on previous work [1-23] in which detailed models for various mechanisms of sintering and HIPing are developed. However, these rigourous treatments generally lead to equations which are complicated and difficult to use. Here we aim to simplify, as far as possible, the treatment of each mechanism and the resulting equations for densification-rate, while retaining adequate accuracy. We first discuss the geometry of the compact and then consider the mechanisms by which it may densify.

2.2 Stage 1 ($\Delta o < \Delta < 0.92$)

We assume that the powder particles are spherical and of a single size. Then, ignoring gross particle rearrangement (which is important only in the early stages of densification of loose powders), the continuous increase in the number of contact neighbours and the growth of the average contact area can be modelled. To do so [17-19] the arrangement of particle centres is described by a radial distribution function. The process of densification (a loss of relative volume) is modelled by imagining that the spherical particles grow in radius around fixed centres. In doing so, a given particle develops regions of overlap with its neighbours: the circles of overlap are the necks between particles. The overlapping volume is calculated and redistributed in the void space, further increasing the neck radii. One consequence of this is that the growing particles touch more neighbours, causing the number of contacts to increase.

The mathematics of all this [17-19] is complicated but the results are adequately approximated by two very simple expressions [22]. First, the number of contact neighbours per particle, Z increases, in a roughly linear way, with relative density Δ:

$$Z = 12\,\Delta \tag{1}$$

Z increases from the value 7.7 at the start of HIPing (assuming dense random packing for which $\Delta o = 0.64$) to a value of 12 when full density is reached. Second, the average area, a, of a contact is close to:

$$a = \frac{\pi}{3}\left(\frac{\Delta - \Delta o}{1 - \Delta o}\right) R^2 \tag{2}$$

(where R is the initial particle radius) giving a neck radius:

$$x = \left(\frac{a}{\pi}\right)^{1/2} = \frac{1}{\sqrt{3}}\left\{\frac{\Delta - \Delta o}{1 - \Delta o}\right\}^{1/2} R \tag{3}$$

and a total contact area per particle (normalized by the surface area $4\pi R2$) of [22]:

$$\frac{aZ}{4\pi R^2} = \frac{\Delta(\Delta - \Delta o)}{(1 - \Delta o)} \tag{4}$$

(Note that the normalized contact area correctly goes to unity as Δ approaches 1)

The curvature of the neck, ρ, is given by:

$$\rho = \frac{x^2}{2(R - x)} \tag{5}$$

This can be approximated [22] by a linear relationship between the neck curvature and the relative density which is particularly convenient for later computation

$$\rho = R\,(\Delta - \Delta o) \tag{6}$$

If an external pressure P is applied to the compact (with a current density Δ and co-ordination number Z), the average contact force, f, is easily shown [19] to be

$$f = \frac{4\pi R^2}{Z\Delta} P \tag{7}$$

Table 1

a	Average contact area at a neck (m^2)
Δ	Relative density
Δo	Initial relative density (typically 0.64)
Δi	Relative density from plastic yielding
Δc	Relative density at which pores close (around 0.95)
$\dot{\Delta}$	Densification rate (/s)
$\delta\ D_b$	Boundary diffusion coefficient times boundary thickness (m^3/s)
D_v	Volume diffusion coefficient (m^2/s)
G	Grain diameter (m)
k	Boltzmann's constant (J/K)
P	External pressure (MPa)
P_{eff}	Effective pressure on a neck (MPa)
Po	Outgassing pressure (MPa)
P_s	Equivalent pressure caused by surface tension (MPa)
P_i	Gas pressure inside a closed pore (MPa)
R	Particle radius (m)
R'	Radius of unit sphere containing one particle (m)
T	Absolute temperature (K)
T_m	Absolute melting temperature (K)
t	Time (s)
V	Volume (m^3)
X	Neck radius (m)
Z	Number of contacts per particle
γ	Surface free energy of particle (J/m^2)
εo, σo, n	Creep parameters (s^{-1}, MPa)
Ω	Atomic or molecular volume (m^3)
σ_y	Yield strength (MPa)

The contact force produces a contact pressure, P_{eff}, on each particle contact of:

$$P_{eff} = \frac{f}{a} = \frac{4\pi R^2}{aZ\Delta} P \qquad (8)$$

Using equations (1) and (2) we obtain:

$$P_{eff} = \frac{P(1-\Delta o)}{\Delta^2(\Delta-\Delta o)} \qquad (9)$$

(which correctly goes to P as Δ goes to 1).

The surface energy γ of the particles provides an additional driving force. When the applied pressure P is large (the HIPing regime), the contribution of surface tension can be neglected; but when P is small or absent (the sintering regime), it takes over as the main driving force for densification. It is most conveniently introduced by replacing it by an equivalent external pressure, P_s [23]. This is calculated by equating the work done by, $P_s dV$, in an increment of densification, $d\Delta$, to the change in surface free energy γdA where dA is the change in surface area of the particles in the increment of densification. Thus:

$$P_s \left(\frac{dV}{d\Delta}\right) d\Delta = \gamma \left(\frac{dA}{d\Delta}\right) d\Delta \qquad (10)$$

Now, for one particle,

$$V = \frac{4\pi}{3\Delta} R^3 \qquad (11)$$

from which $dV/d\Delta$ can be found; and the surface area A removed by contacts is simply aZ (equation (4)). Using this we obtain:

$$P_s = \frac{3\gamma}{R} \Delta^2 \left(\frac{2\Delta-\Delta o}{1-\Delta o}\right) \qquad (12)$$

This equivalent sintering pressure simply adds to any applied pressure P in the densification equations which follow. Strictly, the pressure P should be replaced by the external pressure minus any internal pressure, P_o.

It is convenient to define a dimensionless driving force for densification, in stage 1, as:

$$\tilde{F}_1 = \left((P - P_o) + 3\Delta^2 \left(\frac{2\Delta-\Delta o}{1-\Delta o}\right) \frac{\gamma}{R}\right) \frac{\Omega}{KT} \qquad (13)$$

where Ω is the atomic volume, k is Boltzmann's constant and T is the temperature.

2.3 Stage 2 (0.92 < Δ < 1)

At relative densities above about 0.92 the pores seal off. The particle shape is now best approximated by a polyhedron. We assume that the porosity is closed and uniformly distributed, such that the pores (spherical and of equal size) are situated at the corners of a tetrakaidecahedron [following refs. 19 and 20]. The pore radius is [20]:

$$r = R\left(\frac{1-\Delta}{6\Delta}\right)^{1/3} \qquad (14)$$

In this stage, the effective pressure is the external pressure, P, plus an equivalent-surface-energy pressure P_s, minus the internal pressure in the pores, P_i. The equivalent surface-energy-pressure is found from equation (10) to be:

$$P_s = \frac{2\gamma}{r} \qquad (15)$$

where r is the pore radius (equation (14)). The internal pressure in the pores, P_i, increases with the density, and can, eventually, stop densification completely. This internal pressure is given by [14,16,22]:

$$P_i = P_o \frac{(1-\Delta_c)\Delta}{(1-\Delta)\Delta_c} \tag{16}$$

where P_o is the outgassing pressure and Δ^c is the density at which the pores close.

As before, it is convenient to define a dimensionless driving force for densification, which, in Stage 2, has the form (assembling the above results):

$$\widetilde{F}_2 = \left[(P-P_i) + 2\left(\frac{6\Delta}{1-\Delta}\right)^{1/3}\frac{\gamma}{R}\right]\frac{\Omega}{kT} \tag{17}$$

The sintering literature contains references to the "efficiency factor" in sintering [2,8,9,10,21,23]. It is the ratio of the volume shrinkage of a powder compact to the volume of material removed from the necks where particles touch; it is high at the start of densification because, when particles touch only at points, the removal of a tiny volume of material from these points allows the particles to move towards each other, giving a relatively larger volume change. There is an analogous effect in HIPing: the external pressure P is transmitted through the small contact points between powder particles where the local pressure, P_{eff} (equation (9)), is much larger than P. The treatment given here properly includes this effect, both in sintering and HIPing.

4. The Densification Rate Equations

4.1 Plastic Yielding

When pressure is applied to a powder compact it will first deform by plastic yielding. This causes the average contact area, a, to grow and the effective pressure (equation (9)) to fall until the yield stress, σ_y, of the material is no longer exceeded [1,13-15,17]. Yielding will occur during the initial stage provided:

$$P_{eff} \geq 3\sigma_y \tag{18}$$

The external pressure which will just cause yielding is thus (using equation (9)):

$$P_{lim} = 3\Delta^2 \frac{(\Delta - \Delta o)}{(1-\Delta o)}\sigma_y \tag{19}$$

$$\approx 1.3\frac{(\Delta^3 - \Delta o^3)}{(1-\Delta o)}\sigma_y \tag{20}$$

Densification by yielding is instantaneous, so that densification by time-dependent mechanisms begins from the density resulting from instantaneous plastic yielding. This starting density is given by inverting equation [13,14,22] to give:

$$\Delta_{yield} = \left(\frac{(1-\Delta o)P}{1.3\sigma_y} + \Delta o^3\right)^{1/3} \tag{21}$$

The compact enters final stage sintering during plastic yielding only if the pressure is high enough to cause yielding of the spherical shell surrounding each pore. In this case the limiting pressure for yielding is given by [1,13,14]:

$$P_{lim} = \frac{2\sigma_y}{3}\ln\left(\frac{1}{1-\Delta}\right) \tag{22}$$

which gives a starting density for the time dependent mechanisms of:

$$\Delta_{yield} = 1 - \exp\left(\frac{-3P}{2\sigma_y}\right) \tag{23}$$

4.2 Diffusion from Interparticle Boundaries

Densification can occur by diffusion of material from the contact areas between powder particles, such that the particles move closer together and the pores fill up [5-23]. Both grain-boundary and lattice diffusion contribute to the process.

4.2.1 Stage 1

The densification rate by boundary diffusion during Stage 1 is obtained by equating the rate of volume deposition of material on one particle by boundary diffusion to the rate of removal of the overlapping volume from the contact zones of one particle. There have been numerous attempts to solve this problem, both for sintering [5,7-12,20,23] and HIPing [7-16,22], and to generalize them to include lattice diffusion. They are in broad agreement. The results are best described by the densification-rate equation:

$$\dot{\Delta} = 43\left(\frac{1-\Delta o}{\Delta-\Delta o}\right)\frac{(\delta D_b + 3\rho D_v/4)}{R^3}\,\widetilde{F}_1 \qquad (24)$$

Here δD_b is the grain boundary diffusion coefficient times the boundary thickness, D_v is the lattice diffusion coefficient, Ω is the volume of the diffusing atom or molecule, k is Boltzmann's constant and T is the absolute temperature. Inserting equation 6 for ρ gives the two diffusion-controlled densification rate equations for Stage 1 listed in Table 2.

During Stage 2, matter diffuses from the necks to fill the spherical porosity at grain corners. The numerous attempts to treat this problem, [5,20,22] too, broadly agree. They are best summarized in the densification-rate equation:

$$\dot{\Delta} = 4\,\frac{(\delta D_b + 3rD_v/4)}{R^3}\,\widetilde{F}_2 \qquad (25)$$

Inserting equation (14) for r gives the two diffusion controlled densification-rate equations for Stage 2 listed in Table 2.

4.3 Power-law Creep

Densification under pressure can occur by the creep-deformation of the particle contacts [7-16,22]. The constitutive equation for power-law creep is conventionally written as:

$$\dot{\epsilon} = A\,S^n \exp-\frac{Q_c}{RT} \qquad (26)$$

where $\dot{\epsilon}$ is the strain rate, S the stress, Q_{CRP} is the activation energy for creep, R the gas constant, and A and n are creep constants for the material. The rate constant A is particularly difficult either to estimate or to check, so we recast the equation in the following way. Define S_{ref} as the stress which will cause a tensile strain-rate of 10^{-6}/sec at a temperature of exactly one half of the absolute melting point (T = Tm/2):

$$10^{-6} = A\,S_{ref}^{\,n} \exp-\frac{2Q_c}{RTm}$$

Dividing the first equation by the second gives:

$$\dot{\epsilon} = 10^{-6}\left(\frac{S}{S_{ref}}\right)^n \exp-\left(\frac{Q_c}{RTm}\left(\frac{Tm}{T}-2\right)\right) \qquad (27)$$

which is condensed to:

$$\dot{\epsilon} = D_c\left(\frac{S}{S_{ref}}\right)^n \qquad (28)$$

with:

$$D_c = 10^{-6}\exp-\left(\frac{Q_c}{RTm}\left(\frac{Tm}{T}-2\right)\right)/s \qquad (29)$$

This is the form of the creep equation used below. Its great advantage [24] is that it minimizes extrapolation errors (because HIPing, typically, is carried out at temperatures around Tm/2 and involves strain rates near 10^{-6}/s) and it makes checking simpler (because S_{ref} is usually of order Sy/2).

In this terminology, the rate of densification by power-law creep in Stage 1 [13-15,22] is given by:

$$\dot{\Delta} = 3.1\,\Delta\left(\frac{\Delta-\Delta o}{1-\Delta o}\right)^{1/2} D_c\left(\left(\frac{1-\Delta o}{\Delta-\Delta o}\right)\frac{P-Po}{3\Delta^2 S_{ref}}\right)^n \qquad (30)$$

Table 2

STAGE 1 | **STAGE 2**

INITIAL COMPACTION

Stage 1:

$$\Delta_i = \left[\frac{(1 - \Delta o)}{1.3\ S_y} P + \Delta_o^3\right]^{1/3}$$

Stage 2:

$$\Delta_i = 1 - \exp - \left[\frac{3}{2}\frac{P}{S_y}\right]$$

VOLUME DIFFUSION

Stage 1:

$$\dot{\Delta} = 32\ (1 - \Delta o)\ \frac{D_v}{R_2} \tilde{F}_1\ S_1$$

Stage 2:

$$\dot{\Delta} = 3 \left[\frac{1 - \Delta}{6\Delta}\right]^{1/3} \frac{D_v}{R^2} \tilde{F}_2\ S_2$$

BOUNDARY DIFFUSION

Stage 1:

$$\dot{\Delta} = 43\ \frac{(1 - \Delta o)}{(\Delta - \Delta o)}\ \frac{\delta D_b}{R^3}\ \tilde{F}_1\ S_1$$

Stage 2:

$$\dot{\Delta} = 4\ \frac{\delta D_b}{R^3} \tilde{F}_2\ S_2$$

POWER-LAW CREEP

Stage 1:

$$\dot{\Delta} = 3.1\Delta \left[\frac{\Delta - \Delta o}{1 - \Delta o}\right]^{1/2} D_c \left[\left[\frac{1 - \Delta o}{\Delta - \Delta o}\right] \frac{(P - P_o)}{3\Delta^2\ S_{ref}}\right]^n S_1$$

Stage 2:

$$\dot{\Delta} = 1.5\ \Delta(1 - \Delta)\ D_c \left[\frac{1.5}{n}\ \frac{(P - P_i)}{S_{ref}\ (1 - (1 - \Delta)^{1/n})}\right]^n S_2$$

NABARRO-HERRING or DIFFUSIONAL CREEP

Stage 1:

$$\dot{\Delta} = \frac{14.4}{\Delta} \left[\frac{1 - \Delta o}{\Delta - \Delta o}\right]^{1/2} \left[\frac{D_v}{G^2} + \frac{\pi \delta D_b}{G^3}\right] \tilde{F}_1\ S_1$$

Stage 2:

$$\dot{\Delta} = 32\ (1 - \Delta) \left[\frac{D_v}{G^2} + \frac{\pi \delta D_b}{G^3}\right] \tilde{F}_2\ S_2$$

STAGE 1

$$\tilde{F}_1 = \frac{(P - P_o)\Omega}{kT} + \frac{3\Delta^2(2\Delta - \Delta o)}{(1 - \Delta o)}\ \frac{\gamma}{R}\ \frac{\Omega}{kT}$$

STAGE 2

$$\tilde{F}_2 = \frac{(P - P_1)\Omega}{kT} + \left[2 \left[\frac{6\Delta}{1 - \Delta}\right]^{1/3} \frac{\gamma}{R}\right] \frac{\Omega}{kT}$$

$$D_v = D_{ov} \exp - \frac{Q_v}{RT}$$

$$\delta\ D_b = \delta D_{ob} \exp - \frac{Q_b}{RT}$$

$$D_c = 10^{-6} \exp - \frac{Q_c}{RT_m} \left[\frac{T_m}{T} - 2\right]$$

$$P_i = \left[\frac{1 - \Delta_c}{1 - \Delta}\right] \frac{\Delta}{\Delta_c} P_o$$

and in Stage 2 [13-15,22] by:

$$\dot{\Delta} = 1.5\,\Delta\,(1-\Delta)\,D_c\left[\frac{1.5}{n}\cdot\frac{(P-P_i)}{S_{ref}\,(1-(1-\Delta)^{1/n})}\right]^n \qquad (31)$$

Surface forces are too small to cause significant power-law creep.

4.4 Nabarro-Herring Creep or Diffusional Flow

We have observed that the grain size of copper powder is often significantly smaller than the particle size. Then a new deformation mechanism - diffusion flow (or Nabarro-Herring/Coble creep) contributes to densification [22].

Allowing for both volume (Nabarro-Herring creep) and grain boundary (Coble creep) diffusion the rate-equation for diffusion flow is [25]:

$$\dot{\epsilon} = \frac{14\sigma\Omega}{kT}\left(\frac{D_v}{G^2} + \frac{\pi\delta D_b}{G^3}\right) \qquad (32)$$

where G is the mean grain size. Then if G << 2R, this deformation adds to power-law creep as a way of deforming the contact zone. The same constitutive equation applies as for power-law creep (equation 28) but in this case n = 1 and S_{ref} is given by equating equations (28) and (32). Using equation (30) for the initial stage, we obtain:

$$\dot{\Delta} = \frac{14.3}{\Delta}\left(\frac{1-\Delta_o}{\Delta-\Delta_o}\right)^{1/2}\left(\frac{D_v}{G^2} + \frac{\pi\delta D_b}{G^3}\right)\frac{(P-P_o)\,\Omega}{kT} \qquad (33)$$

and using equation (31) for the final stage,

$$\dot{\Delta} = 32\,(1-\Delta)\left(\frac{D_v}{G^2} + \frac{\pi\delta D_b}{G^3}\right)\frac{(P-P_i)\,\Omega}{kT} \qquad (34)$$

When computing the diagrams shown in Section 5 this mechanism was considered to operate only when the grain size, G, was smaller than the neck size, 2x.

4.5 Pore Separation and Grain Growth

During HIPing the compact is usually held for a considerable time (an hour or so) at a high temperature (0.5-0.7 T_m). This can result in grain growth and pore separation from the grain boundaries, isolating the pores in mid-grain [22,26,27]. When the pores no longer lie on grain boundaries, densification by diffusive mechanism is suppressed, though power-law creep and plastic flow are not, of course, affected.

Three factors inhibit boundary motion and pore separation: the intrinsic drag on grain boundaries, the drag caused by solute, and the pinning force exerted by the pores themselves. As the pores shrink in size the third contribution decreases, and boundaries start to move. Brook [26] and Yan *et al.* [27] have analysed the conditions for pore separation and have shown that a minimum intrinsic and solute drag force is required to prevent it. If the drag force lies below the minimum (as it does in most pure one-component systems), pore separation will occur when a critical density has been exceeded. When pore separation occurs, some pores are detached from the boundaries, but not all. Let the fraction of pores on boundaries be f_p. Then only a fraction f_p of all pores continue to shrink by diffusional densifying processes after exceeding the critical density, and the densification rate by these "mechanisms" is reduced to $f_p\dot{\Delta}$.

It is possible to estimate the fraction f_p for single-crystal powders for which grain growth can be assumed to be controlled by Zener pinning. The stable grain size [28] is:

$$\bar{G} = \frac{4r}{3f} \qquad (35)$$

where r is the radius of a void during final stage of sintering, and f = 1 - Δ is the volume fraction of voids. The fraction of pores on grain boundaries is obviously proportional to the area of grain boundary per unit volume ($1/\bar{G}$). We know that all pores are on grain boundaries when $\bar{G} = 2R$. So (using also equation (10) and (29)):

$$f_p = \frac{2R}{\bar{G}} \approx 2.7\,(1-\Delta)^{2/3} \qquad (36)$$

provided that $f_p \leq 1$; otherwise, $f_p = 1$.

Now in computing HIP diagrams, we can allow for pore separation by writing, for the appropriate diffusional densifying mechanisms:

$$\dot{\Delta} = f_p \, \dot{\Delta}o \tag{37}$$

Here, $\dot{\Delta}o$ is the densification rate when all pores are on boundaries. In computing the diagrams of Section 6 we have considered pore separation in the case of single crystal powders. It should be noted, however, that in polycrystalline powders, grain growth cannot be controlled by a Zener pinning mechanism until the grains have attained the size of the particles.

The equations for densification rate are summarized in Table 2.

5. Densification Maps

All this modelling is most readily accessed through densification maps. The density of a given aggregate of powder, after a time t, is related to the temperature T and the pressure P of consolidation:

$$\Delta = f\,(T, P, t, \text{Material Properties})$$

It is the end result of the joint action of the mechanisms described earlier, some acting simultaneously, some sequentially. This behaviour can be displayed in two ways. The first, of which Fig. 3 is an example, is best suited for the analysis of sintering, though HIPing can be shown too. The axes are density Δ and temperature T; the contours show the time t; and it is constructed for a fixed value of the pressure P. The second, illustrated by Fig. 5, is best suited for the analysis of HIPing, though sintering can be shown too. The axes are density Δ and pressure P; the contours, again are time; and it is constructed for a fixed value of the temperature T. The two maps can be thought of as orthogonal sections through Δ - P - T space, with surfaces of constant time

The maps are constructed by evaluating the contributions to the densification rate from each of the mechanisms described in Section 4 (equations (19), (21), (24), (26) and (27)), and summing them (since they are independent) to give the total densification rate; pore separation is included in the way described in Section 4.5. The method of construction is outlined by Helle *et al.* [22] and, in more detail, by Ashby [24].

Examples of the diagrams are shown in Figs. 3 to 11. They show, first, the field of dominance of each mechanism (that is, the range of P, T and D in which a given mechanism contributes more to the densification rate than another). The field boundaries are the lines along which two mechanisms contribute equally to the total; the boundary of the "YIELD" field is the density reached by yielding alone (equations (15) and (17)). Superimposed on the fields are contours of constant time; they show the density reached in a given time (we have used 1/4, 1/2, 1, 2 and 4 h).

The material data used to construct maps for tool steel, alumina and copper are listed in Table 3. The material properties which enter the equations are, of course, those at the temperature of HIPing, so, as far as possible, a temperature coefficient is used to correct properties like the modulus. Data for the temperature-dependence of the high-temperature yield strength proved hard to find, and we have, instead, used a fixed, average, high-temperature strength. Small errors here do not affect the results much since underestimating the starting density (by yield) causes more rapid densification by the other mechanisms, largely compensating for the error.

How accurate are the diagrams? The biggest uncertainty lies in the material properties: diffusion coefficients (particularly for grain boundary diffusion) and power-law creep constants are, often, very uncertain. We have exercised the maximum critical judgement in selecting the data, extracting the material parameters by the methods used by Frost and Ashby [25]. We then find that the maps give a tolerably good fit to the HIPing data. Further refinement of the material parameters to improve the fit is possible (by using the HIPing data itself to correct diffusion and power-law creep parameters), but we have not attempted it here.

6. Applications: The Analysis of Sintering and HIPing Schedules

Models are only useful if they have predictive value. In a process as complicated as sintering or HIPing, involving many mechanisms each with its own difficulties of modelling, it is unrealistic to imagine that the models accurately reproduce all the details of the real process. But it is a reasonable expectation that they contain the right physical quantities (pressure, temperature, particle size, density and so forth) combined in the correct way; it is the constants of proportionality which are less certain. This means that the models may need to be *calibrated*, using a subset of data. If valid, the models will then describe the remaining data accurately and can then be expected to predict the outcome of further, hypothetical, or proposed, runs.

There is a further difficulty: it is the imprecision of values for material properties (diffusion coefficients, creep constants and so forth). Even the best bulk data can be inaccurate. And powders never have exactly the same properties as the bulk solid from which they were made because of the unusual, often savage, processes used to make them. For this

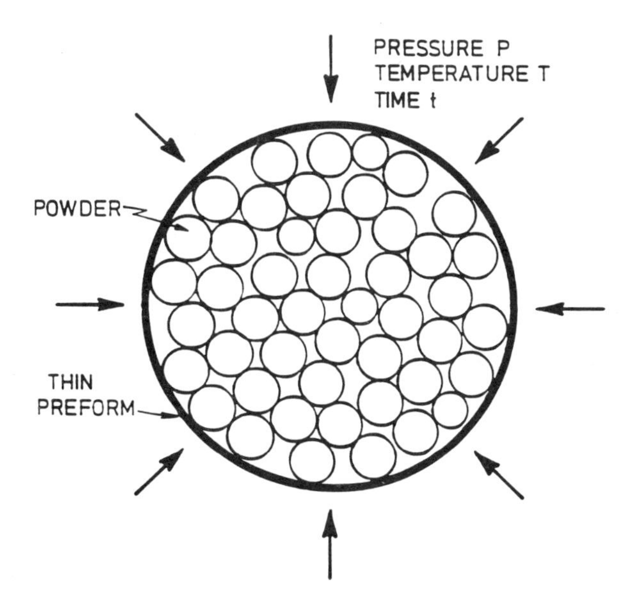

Figure 1 The HIPing process. Powder is contained in a thin preform and subjected to a pressure P at a temperature T for a time t.

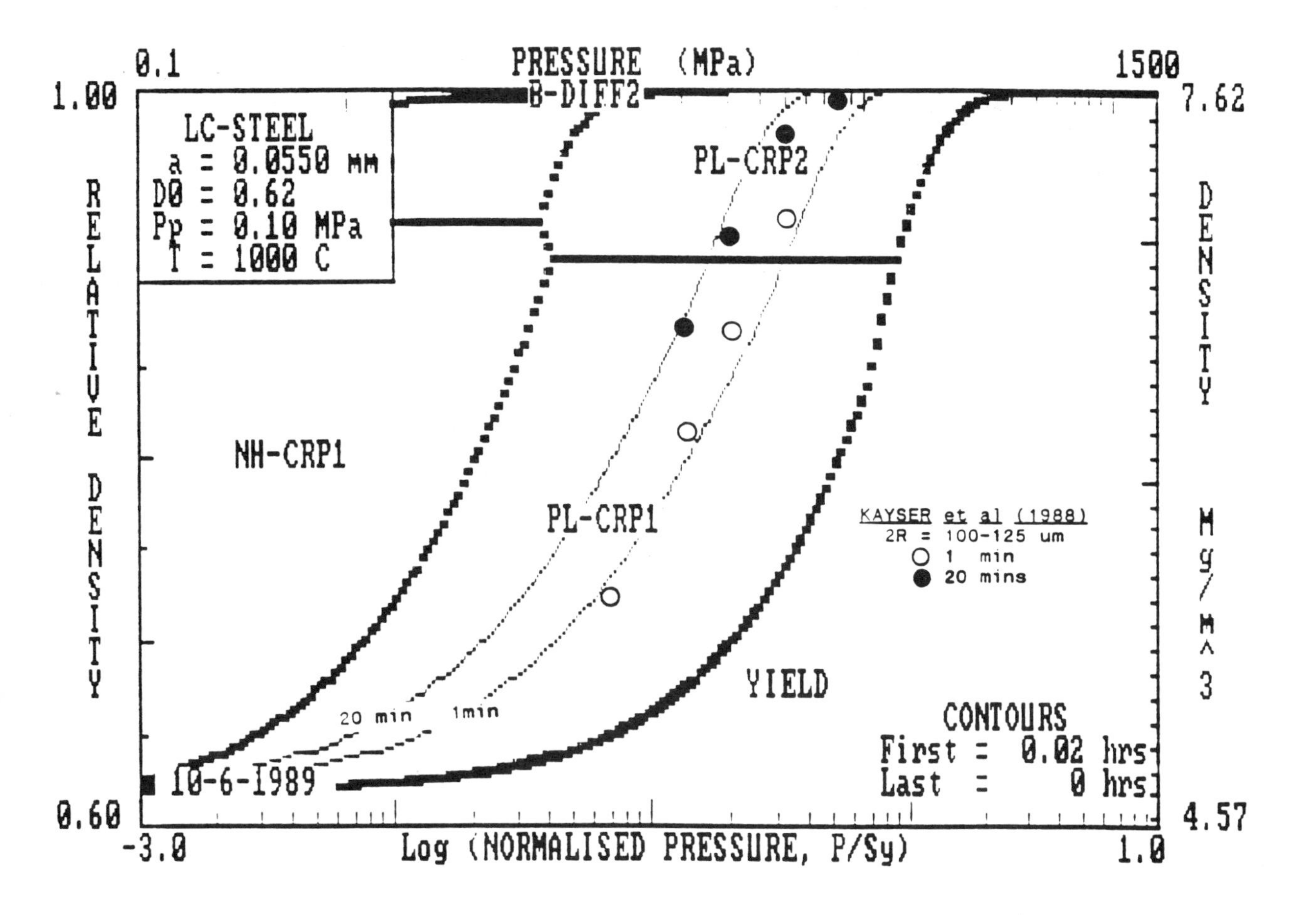

Figure 3 A density/pressure map for low-carbon steel powder with a 100 - 125 μm particle size. Data from Kaysser et al. [29].

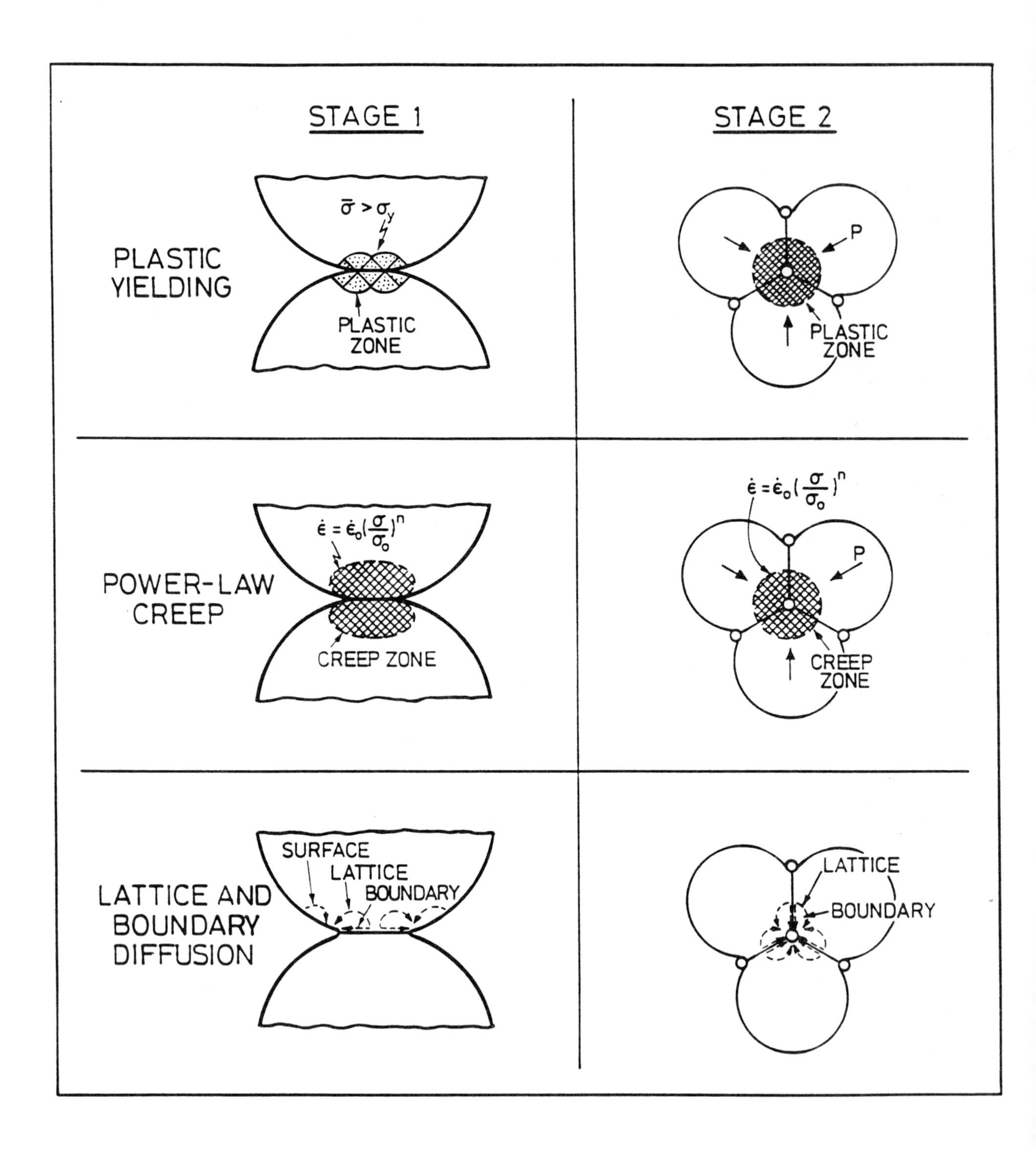

Figure 2 Densification is a continuous process, but it is convenient to divide it into three stages. The first (Stage 0) describes the density reached by the packing of the loose powder. The second (Stage 1) describes the early phase of densification, while the porosity is still connected. The final stage (Stage 2) describes the final densification, when the residual porosity is in the form of small holes.

reason, the material property values may have to be "tuned", using information from a small number of sintering or HIPing experiments. Once tuned, the data set can be used with the calibrated models to predict the outcome of any proposed sintering or HIP run.

With all these problems, it might be asked whether modelling is worth the effort. But the potential gains in understanding and in the ability to control powder consolidation to give a reliable, high quality product, are enormous. And - as shown below - the models, even in their present imperfect form - already give a remarkably good first description of the densification process.

Now, three examples in which densification maps are used to analyse data drawn from the open literature and industrial reports. All the maps were constructed using the program HIP 487, available from the author [24]. The values of material properties are listed in Table 3. The meaning of the short names for the mechanisms which appear on the maps is given in Table 4.

6.1 The HIPing of a Low-Carbon Steel

Kaysser *et al.* [29] HIPed a C1018 steel powder with a particle size range of 100-125 μm and a grain size of 30 μm at 1000°C. They used two pressing times: 1 minute and 20 minutes. Their measurements of relative density are listed in Table 5, and plotted on the appropriate map in Fig. 3: open circles mark the data for 1 minute; full circles, those for 20 minutes. They should be compared with the two contours, which are for these two times. Agreement is excellent. Densification is by power-law creep, with the consequence that it is independent of particle size but strongly dependent on pressure (as P^n). The close agreement between model predictions and experimental measurement gives confidence that the models for power-law creep, at least, are realistic. No attempt has been made to tune the material properties in this case.

Agreement is seldom this good. Partly, this is because continuous read-out of densification during HIPing is not yet available (though coming), so the measurements are from separate, discrete, runs, giving much scatter. And partly it is because the material properties of powders are so variable. The next example illustrates this.

6.2 The Sintering and HIPing of Copper

Copper, more than any other metal, has been used as a model material in sintering and HIPing studies. Data from three sources are listed in Tables 6, 7 and 8.

Moon and Choi [30] cold-isostatically pressed copper powders at pressures between 0 and 600 MPa, at room temperature. Their measurements of density are plotted on the appropriate map in Fig. 4. The map describes the data well - but to achieve this fit by the material property, the yield strength of the particles, has been tuned. Commercially pure copper, in bulk, has a yield strength of 45 MPa when annealed, rising to about 250 MPa in the "half hard" (15% reduction) condition. In computing the diagram, a single fixed, yield strength of 180 MPa has been used. The effect of work hardening can be seen in the deviation of the data from the line at high densities.

Sintering is shown in Fig. 5. Rhines *et al.* [31] measured the densification of 70-88 μm copper powders in the temperature interval 915 to 1040°C. Their data are shown on Fig. 5: they lie a little above the predicted 100 hour line, but - without calibration or property-tuning - this agreement is not bad. It is not clear at present whether the sintering equations or the material data for copper are at fault here (experiments are in progress to find out). Calibration and tuning will not change the main conclusion: densification is dominated by volume diffusion above 1000°C, and by boundary diffusion below 900°C. This powder cannot be sintered to full density without a pressure (it would take about 1000 hours).

Finally, the HIPing data. Figures 6 and 7 show the measurements of Kayser *et al.* [29] who HIPed 150-200 μm copper powders at 750°C (Fig. 6) and 850°C (Fig. 7). The lowest of the three constants, that for 20 minutes, corresponds to the conditions of the tests. At low densities, agreement is good, but at high it is not; the reasons for the discrepancies are not yet clear. But conclusions can still be drawn: the dominant mechanism is power-law creep, so that particle size is not an important variable.

6.3 Densification Maps for Alumina, an Oxide Ceramic

Powder methods come into their own with the processing of engineering ceramics. Among these, the aluminas are the cheapest and most widely available; they are, as it were, the "mild steels" of the ceramic business; and they are fabricated, almost always, from powders. Here we analyse the densification of two different high-purity alumina powders, the words "high purity" meaning that no additives, designed to give a liquid phase, were present.

Uematsu *et al.* [32] HIPed 1.25 μm high purity alumina powder at pressures between 5 and 200 MPa, and temperatures from 1100 to 1400°C. Their observations are summarized in Table 9. Maps showing their data appear as Figs. 8 and 9.

The conclusions are these. Diffusional mechanisms dominate in the sintering and HIPing of fine-particle high-purity aluminas. The densification times then depend strongly on particle size; and linearly on pressure. Grain growth as full density is approached is a potential problem: it slows the diffusional mechanisms. Here HIPing helps: by speeding up the densification rates, it leaves less time for grain growth and helps preserve a fine-grained micro structure. The property file and models describe the densification of pure alumina adequately. This means that they can be used to predict the outcome of any other proposed sintering or HIPing cycle, or, properly implemented, a sinter-HIP sequence.

Figures 8 and 9 show appropriate densification maps, computed from the data in Table 9. At the low end of the pressure range (5 MPa), surface-tension driven sintering is important; at the high end, the pressure far outweighs surface tension

Table 3

HIP 587 MATERIAL PROPERTIES for			LC-STEEL		COPPER		ALUMINA
GENERAL PROPERTIES							
1 Structure type (Manual Tbl.1)		=	10	=	10	=	40
2 Solid density	kg/m^3	=	7620.00	=	8960.00	=	3975.00
3 Melting point	K	=	1800.00	=	1356.00	=	2320.00
4 Surface energy	J/m^2	=	2.10	=	1.72	=	0.90
MECHANICAL PROPERTIES							
5 Youngs modulus at R.T.	GPa	=	200.00	=	145.00	=	395.00
6 Yield stress at R.T.	MPa	=	150.00	=	180.00	=	2100.00
7 T-dependence of yield		=	0.81	=	0.54	=	0.35
DIFFUSION PROPERTIES							
8 Atomic volume	m^3	=	1.18E-029	=	1.18E-029	=	4.25E-029
9 Pre-exp. volume diffusion	m^2/s	=	1.80E-005	=	6.20E-005	=	2.80E-003
10 Activ. energy, vol. diff.	kJ/mol	=	270.00	=	207.00	=	477.00
11 Pre-exp. bdry diffusion	m^3/s	=	7.50E-014	=	5.12E-015	=	8.60E-010
12 Activ. energy, bdry. diff.	kJ/mol	=	159.00	=	104.00	=	419.00
CREEP PROPERTIES							
13 Power-Law Creep exponent		=	4.50	=	4.80	=	3.00
14 Reference stress, P-L creep	MPa	=	60.00	=	120.00	=	1250.00
15 Activ. energy for P-L creep	kJ/mol	=	270.00	=	197.00	=	477.00
PARTICLE CHARACTERISTICS							
16 Grain diameter in particle	m	=	3.00E-005	=	5.00E-005	=	1.00E-003
17 Particle radius	m	=	5.50E-005	=	8.70E-005	=	1.00E-006

LOW-CARBON STEEL (C1018) (File name: LC-STEEL)
The data are from Frost H.J. and Ashby M.F., "Deformation Mechanism Maps", Pergamon Press, Oxford (1982); and from Kayser W., Aslan M. and Petzow G., Proc. Antwerp Conf on Hot Isostatic Pressing of Materials, Royal Flemish Soc. of Engineers, Antwerp, Belgium (1988).*

COMMERCIAL PURITY COPPER (File name: COPPER)
Data are from Frost H.J. and Ashby M.F. "Deformation Mechanism Maps", Pergamon Press, Oxford (1982); and from Swinkels F.B. and Ashby M.F. Acta Metall. 29 (1981) 259; and Helle A.S., Easterling K.E. and Ashby M.F. Acta Metall. 33 (1983) 2163.*

ALUMINIUM OXIDE, AL2O3 (File: ALUMINA).
Most data are from Frost H.J. and Ashby, M.F. "Deformation Mechanism Maps", Pergamon Press, Oxford (1982) p. 99. See also Helle A.S., Easterling K.E. and Ashby M.F. Acta Met vol 33 (1985) p. 2163, Table 1.*

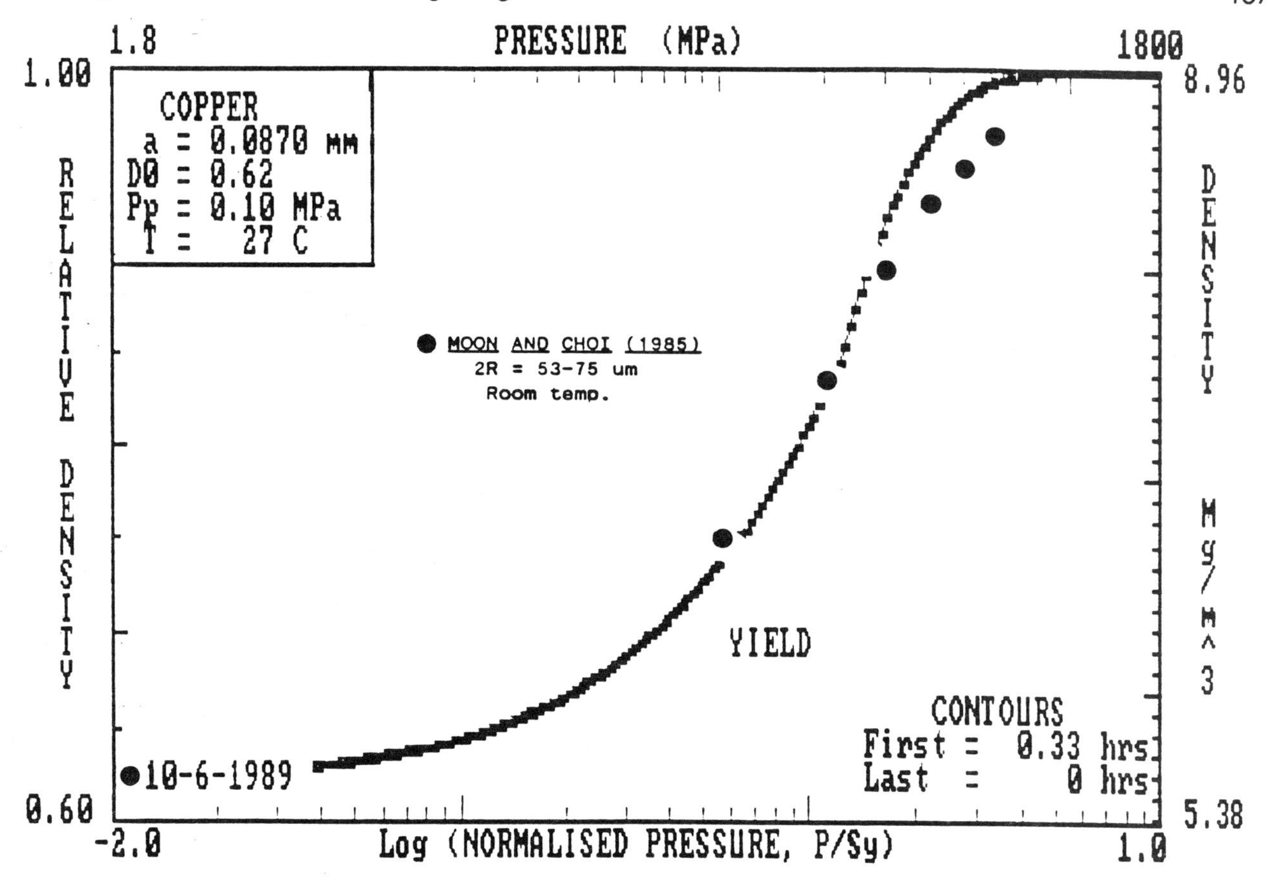

Figure 4 A density/pressure map for the cold isostatic pressing of copper powder. Data from Moon and Choi [30].

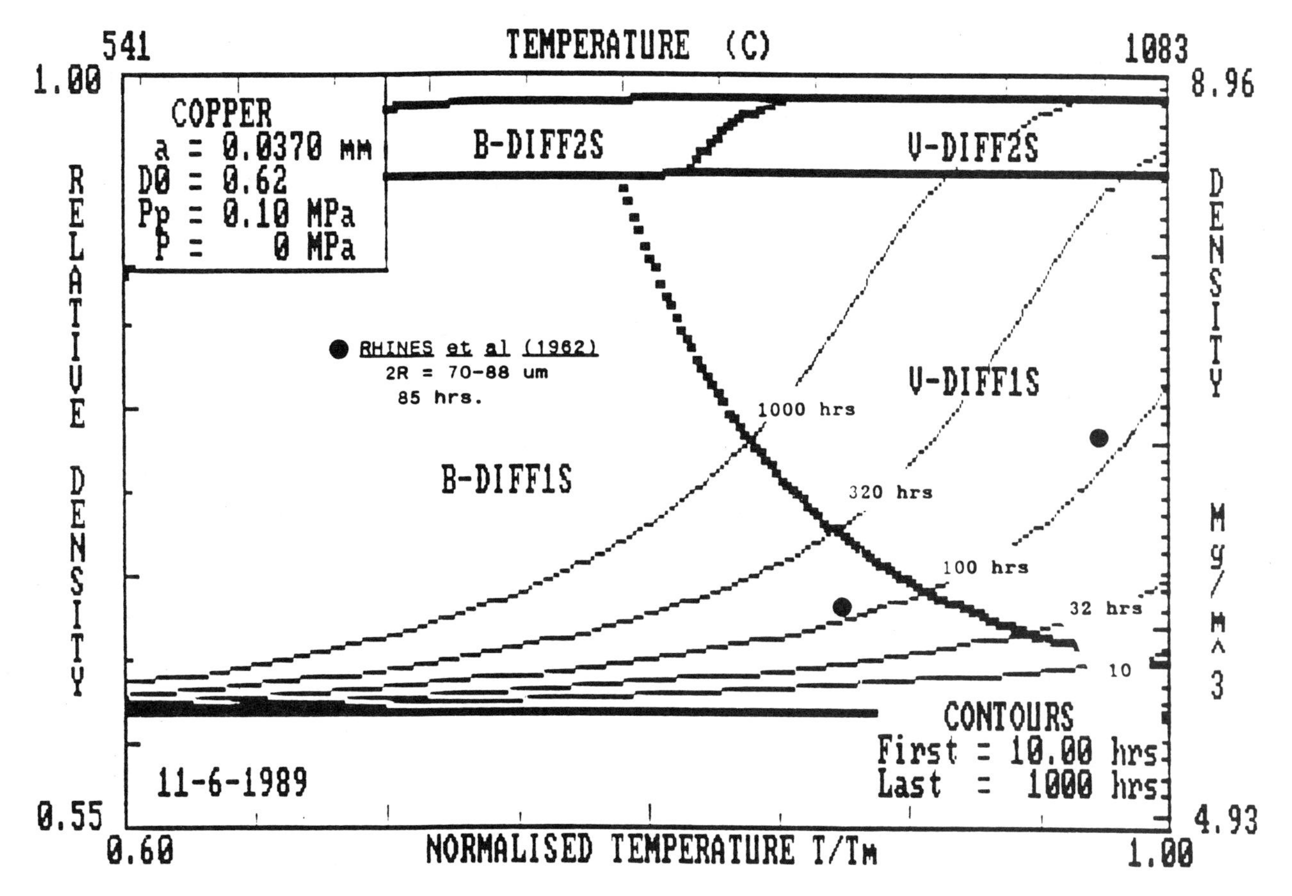

Figure 5 A density/temperature map for copper powder with a 70 -80 μm particle size. The data are from Rhines et al. [31].

Table 4

NAME	DESCRIPTION	EQUATION
YIELD	Plastic yielding	(21) and (23)
PL-CRP 1	Power-law creep, Stage 1	(30)
PL-CRP 2	Power-law creep, Stage 2	(31)
B-DIFF 1	Boundary-diffusion, Stage 1, pressure driven	(24)
B-DIFF 2	Boundary-diffusion, Stage 2, pressure driven	(25)
B-DIFF 1S	Boundary-diffusion, Stage 1, surface tension driven	(24)
B-DIFF 2S	Boundary-diffusion, Stage 2, Surface tension driven	(25)
V-DIFF 1	Volume-diffusion, Stage 1, pressure driven	(24)
V-DIFF 2	Volume-diffusion, Stage 2, pressure driven	(25)
V-DIFF 1S	Volume-diffusion, Stage 1, surface-tension driven	(24)
V-DIFF 2S	Volume diffusion, Stage 2, surface-tension driven	(25)
NH-CRP 1	Narbarro-Herring creep, Stage 1	(33)
NH-CRP 2	Narbarro-Herring creep, Stage 2	(34)

Table 5

Pressure P (MPa)	Temperature T (°C)	Time t (hrs)	Rel Density Δ
10	1000	0.02	0.72
20			0.82
30			0.87
50			0.93
20	1000	0.33	0.87
30			0.92
50			0.97
80			1.0

Table 6

Pressure P (MPa)	Temperature T (C)	Time t (hrs)	Rel Density Δ
0	300	0.2	0.63
100			0.75
200			0.83
300			0.88
400			0.92
500			0.94
600			0.95

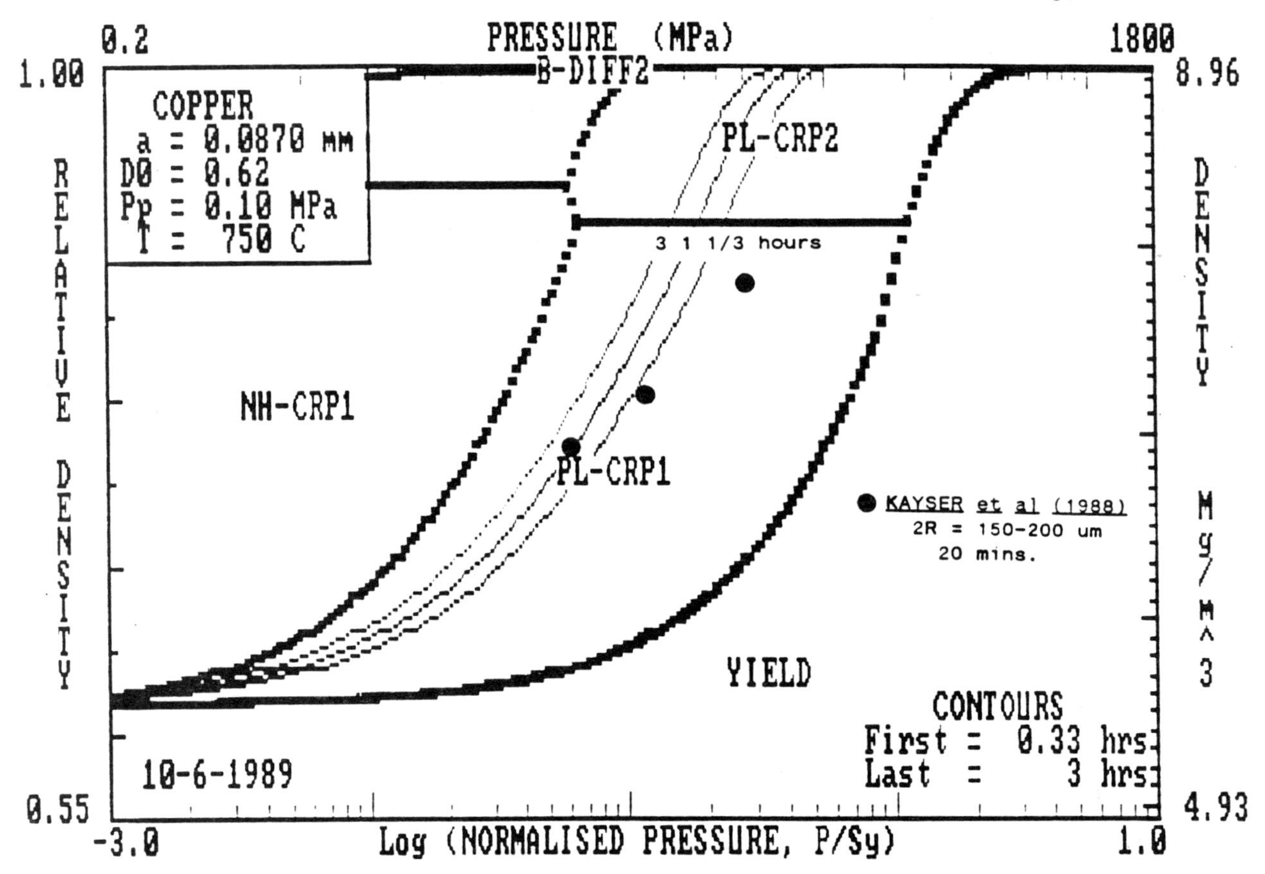

Figure 6 A density/pressure map for copper powder with a particle size of 150 - 200 μm, at 750°C. The data are from Kaysser et al. [29].

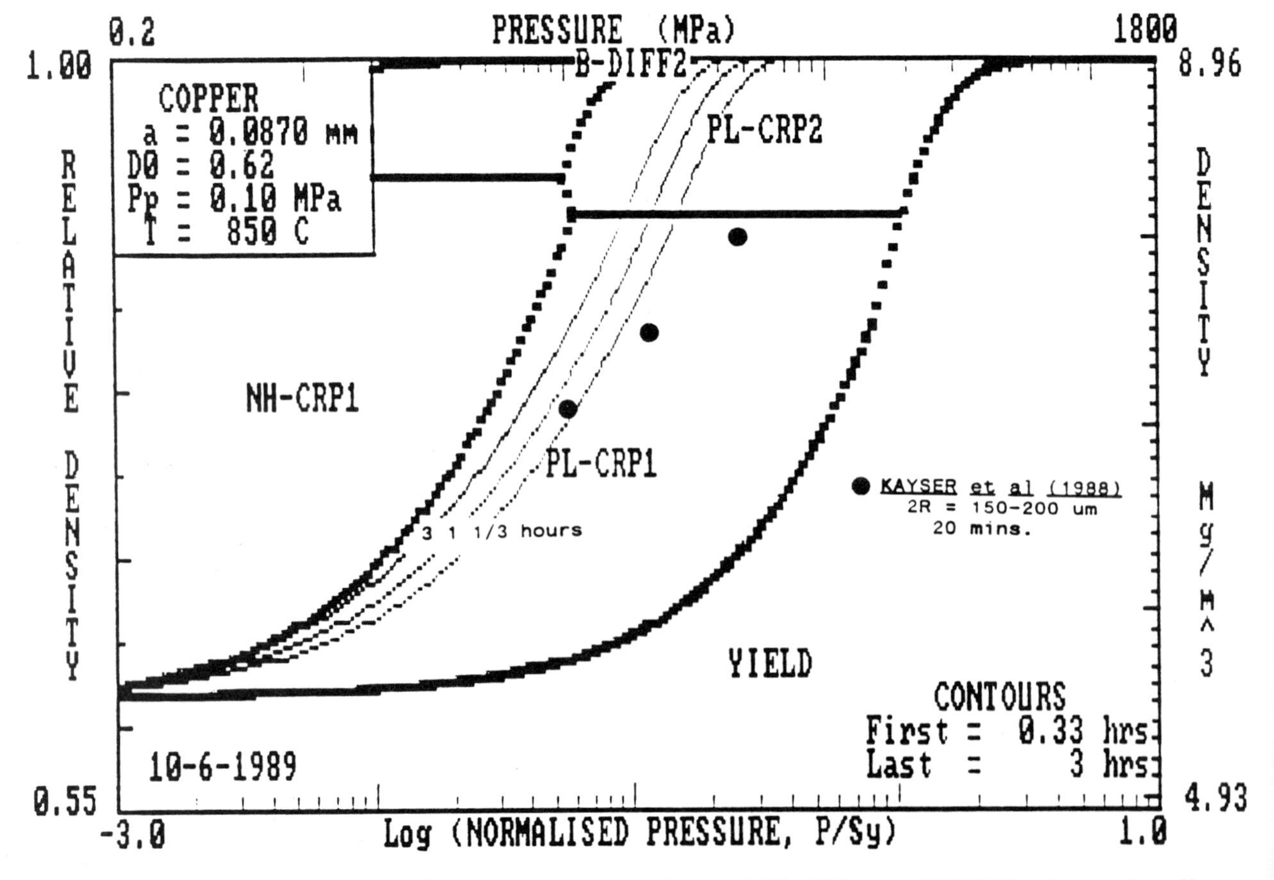

Figure 7 A density/pressure map for copper powder with a particle size of 150 - 200 μm, at 850°C. The data are from Kaysser et al.[29].

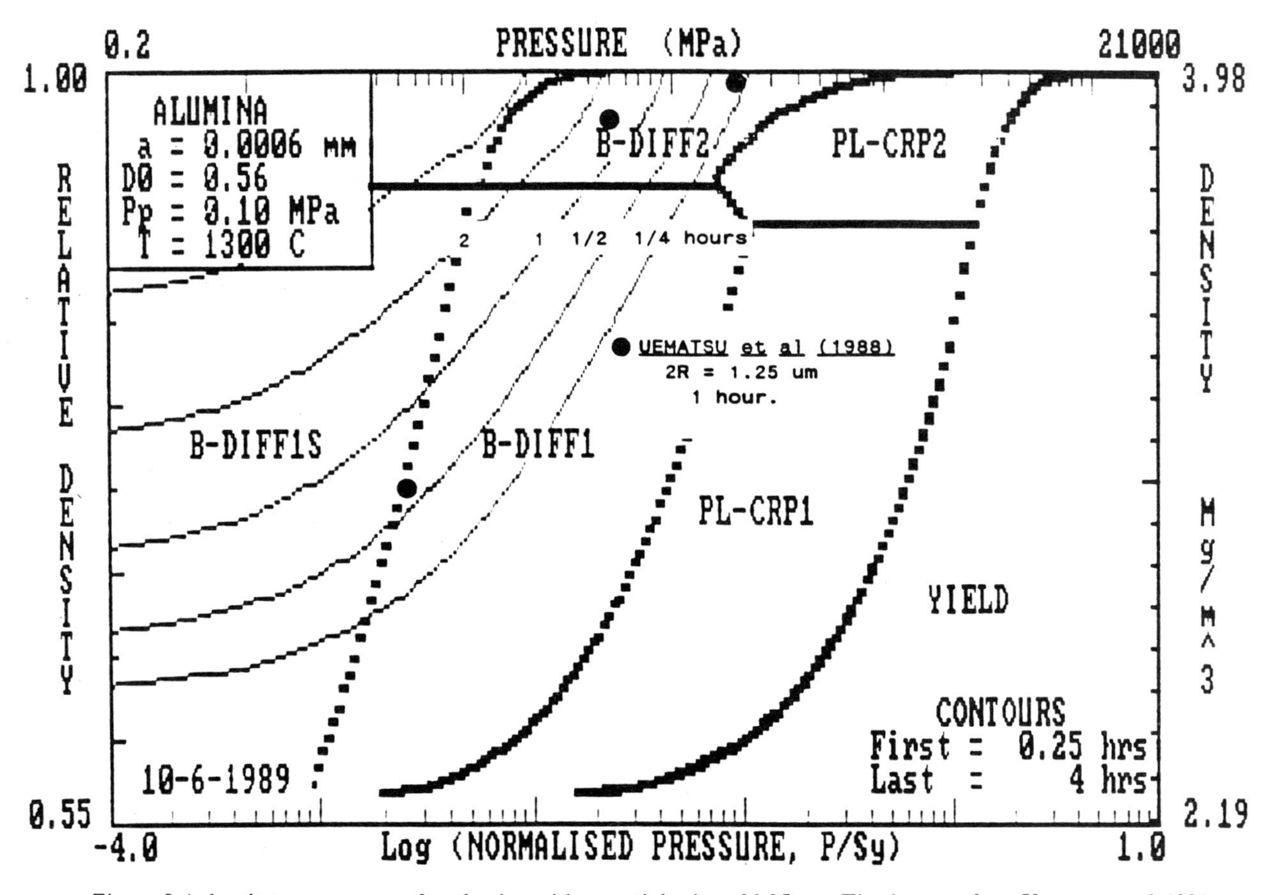

Figure 8 A density/pressure map for alumina with a particle size of 1.25 µm. The data are from Uematsu et al. [32].

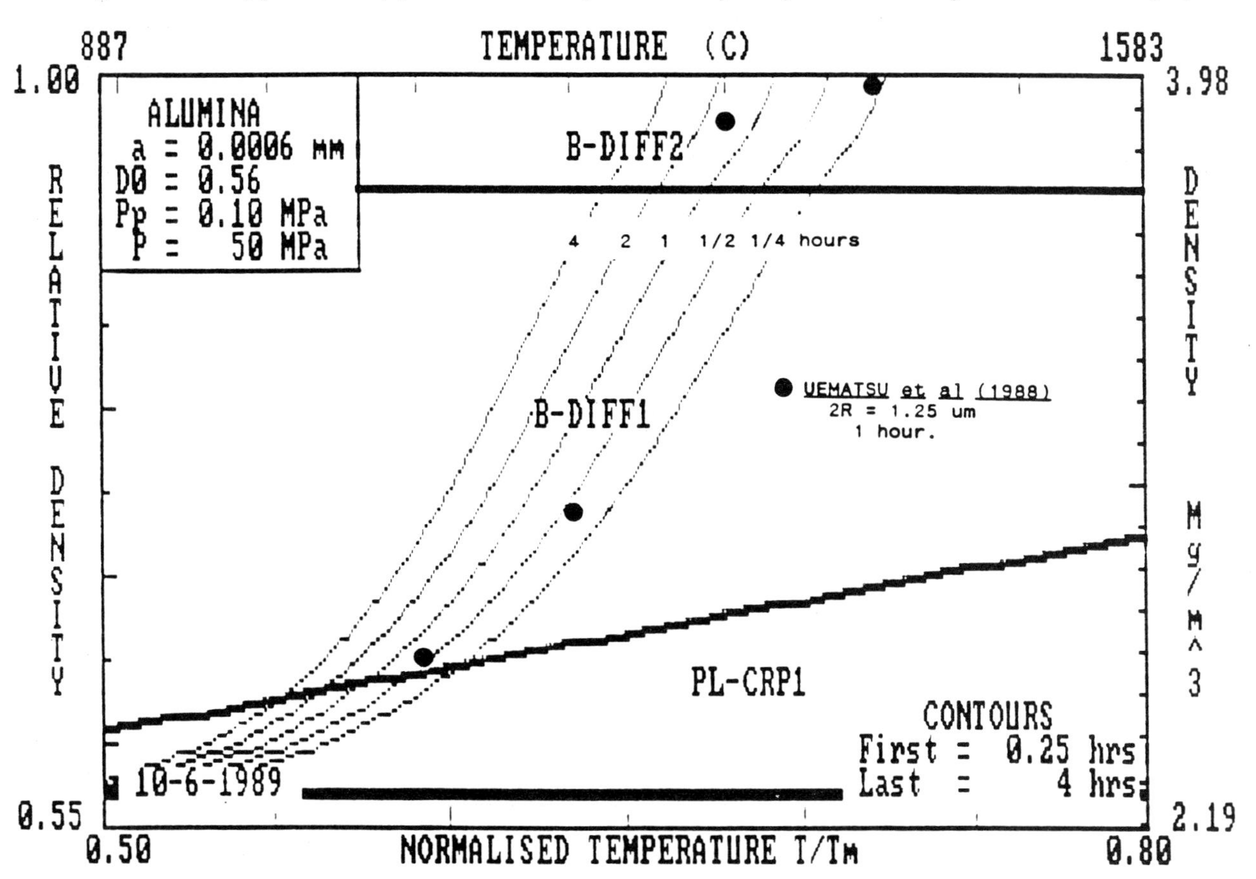

Figure 9 A density/temperature map for the same alumina as Fig. 8.

Table 7

Pressure P (MPa)	Temperature T (°C)	Time t (hrs)	Rel Density Δ
0	915	85	0.68
0	1040	85	0.78

Table 8

Pressure P (MPa)	Temperature T (°C)	Time t (hrs)	Rel Density Δ
10	750	0.33	0.765
20		0.33	0.80
50		0.33	0.865
10	850	0.33	0.788
20		0.33	0.83
50		0.33	0.984

Table 9

Pressure P (MPa)	Temperature T (°C)	Time t (hrs)	Rel Density Δ
50	1100	1	0.65
	1200	1	0.74
	1300	1/2	0.92
	1300	1	0.98
	1300	4	0.995
	1400	1	0.995
5	1300	1	0.75
50	1300	1	0.98
200	1300	1	1.0

Table 10

Pressure P (MPa)	Temperature T (°C)	Time t (hrs)	Rel Density Δ
100	1200	2	0.975
	1200	4	0.981
	1300	2 and 4	0.996
	1400	2 and 4	0.997
200	1200	2 and 4	0.996
	1300	2	0.995
	1400	2	0.998

as the driving force. With small particles such as these, diffusional mechanisms dominate, in this case, it is that in involving boundary diffusion. This means that particle size plays an important role: the densification rate, by this mechanism, varies as 1/R. Uematsu *et al.* could not measure density continuously (the data points were derived from separate HIP runs) and for this reason there is considerable scatter. Allowing for this, agreement is good. The maps summarize well the way density varies with both pressure and temperature.

A second HIPing study - that of Larker [33] - used a different batch of high-purity alumina with a larger particle size (0.75 - 2.5 μm, mean 1.25 μm). The data, listed in Table 10, are all for conditions which lead to almost full density (0.975 is the lowest), so Figs. 10 and 11 are expanded to show only the later stages of densification, using the same material property values as before.

7. Conclusions

Much progress has been made in the modelling of sintering and hot-isostatic pressing. The mechanisms which contribute to each stage of densification have been identified and studied, and most are now well characterised. Transient effects leading to non-uniform densification and shape-change have been studied, and the preferred process conditions (including guide-lines for the rates of heating and of application of the pressure) can be formulated [34]. Experiments are currently being conducted to check and refine the calculations. Once this has been done, the remaining problem is that of technology transfer: how can these complex results be presented in a way which helps the materials engineer design HIP cycles and select materials for the HIPing process. The best way of doing this appears to be through the careful preparation of computer software which, for a selected material, will evaluate the consequences of any proposed HIP cycle, showing the operator the final density, other possible changes of structure, and suggesting to him the way in which the process variables can be traded off against each other to optimize the process [24]; it must, however, be recognised that material properties are so variable that a tuning process, in which the equations are calibrated to the results of real HIP runs, will always be necessary.

Software of this sort is under development. It appears possible that the procedure could be integrated with information derived from sensors in the press itself, so that some degree of real-time control, based on physical modelling, might be possible. This route of "smart" processing is an attractive one in processes as complex as this, and provides a route whereby the fundamental work on a process such as HIPing can contribute to the industrial practice.

8. Acknowledgements

I wish to acknowledge the support of the National Institute of Standards, the Office of Naval Research and the Defense Advanced Research Projects Agency through the University Research Initiative Program centred at the University of California at Santa Barbara.

9. References

1. C. Torre, U. Berg and H.H. Huttenman, Hochschule Leoben, 93, p.62, 1948.
2. J. MacKenzie and R. Shuttleworth, Proc. Phys. Soc., Lond., 62, p.833, 1949.
3. A.K. Kakar and A.C.D. Challander, Trans. Am. Inst. Min. Engrs 242, p.1117, 1968.
4. M.S. Koval'chenko and G.V. Samsanov, Povoshkhov. Metall., 1, p.3, 1961.
5. R.L. Coble, J. Appl. Phys., 41, 4798, 1970.
6. R.L. Hewitt, W. Wallace and M.C. de Malherbe, Powder Metall. 16, p.4798, 1973.
7. D.L. Johnson, J. Appl. Phys, 40, p.192, 1969.
8. R.T. Hoff, in Sintering and Heterogeneous Catalysis, (eds) G.C. Kuczynski, A.E. Miller and G.A. Sargent, Gordon and Breach, NY, p.423, 1967.
9. W. Beere, Acta Metall., 23, p.139, 1975.
10. W. Beere, Acta Metall., 23, p.139, 1975.
11. R.L. Eadie and G.C. Weatherly, Scripta. Met., 9, p.285, 1975.
12. R.L. Eadie, D.S. Wilkinson and G.C. Weatherly, Acta Metall., 22, p.1185, 1974.
13. D.S. Wilkinson and M.F. Ashby, Acta Metall., 23, p.1277, 1975.
14. D.S. Wilkinson,Ph.D. thesis, Univ. of Cambridge, 1977.
15. D.S. Wilkinson and M.F. Ashby, Proc. 4th Int. Conf. on Sintering and Related Phenomena, p.473. Plenum, New York, 1975; Science of Sintering, 4th Int. Round Table on Sintering, 10, pp.67-76, 1978.
16. E. Arzt, M.F. Ashby and K.E. Easterling, Metall. Trans. A, 14A, p.211, 1983.
17. E. Arzt, Acta Metall., 30, p.1883, 1982.
18. H.F. Fischmeister, E. Arzt and L.R. Olsson, Powder Metall., 21, p.179, 1978.
19. H.F. Fischmeister and E. Arzt, Powder Metall., 26, p.82, 1983.
20. F.B. Swinkels and M.F. Ashby, Acta Metall., 28, p.259, 1981.

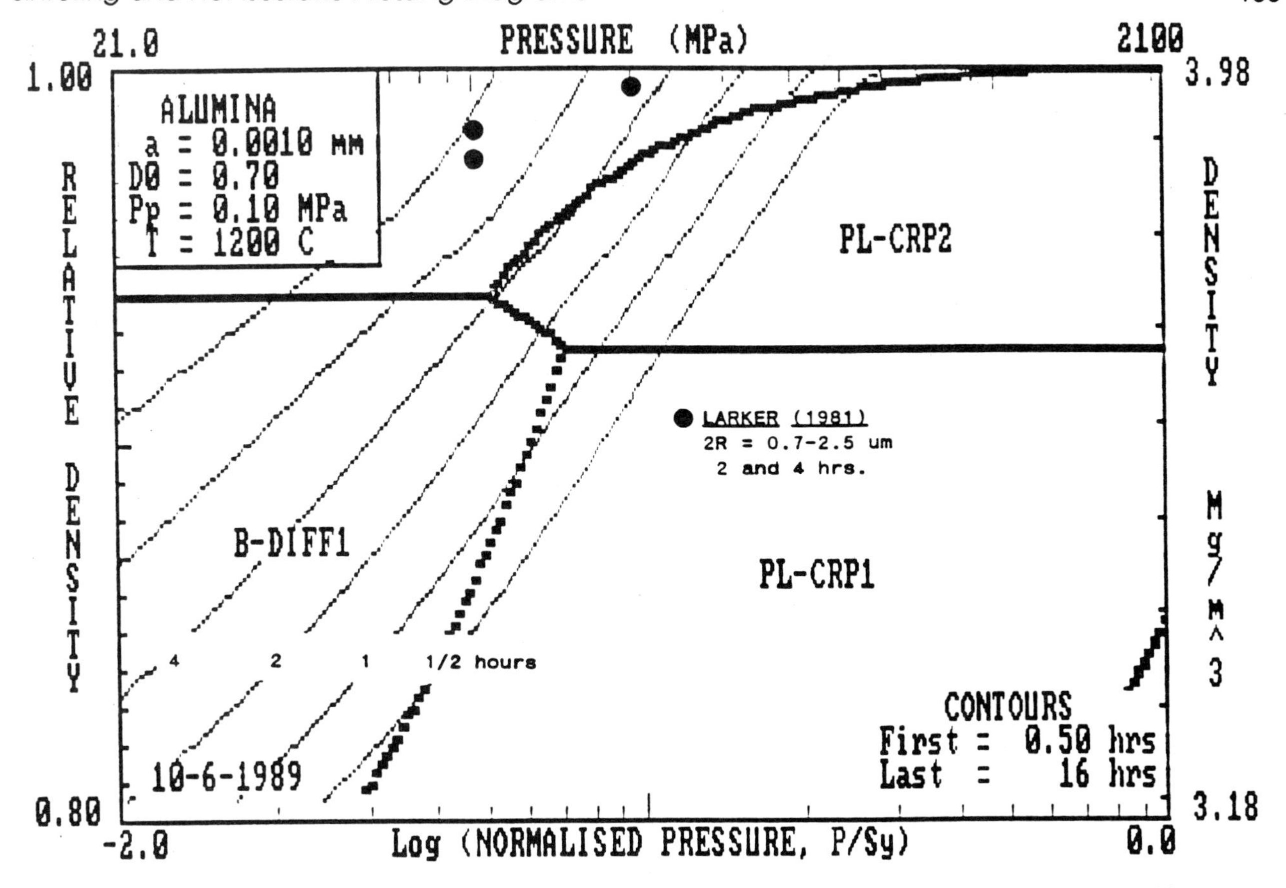

Figure 10 A density/pressure map for alumina powder with a particle size of 2 μm. The data are from Larker [33].

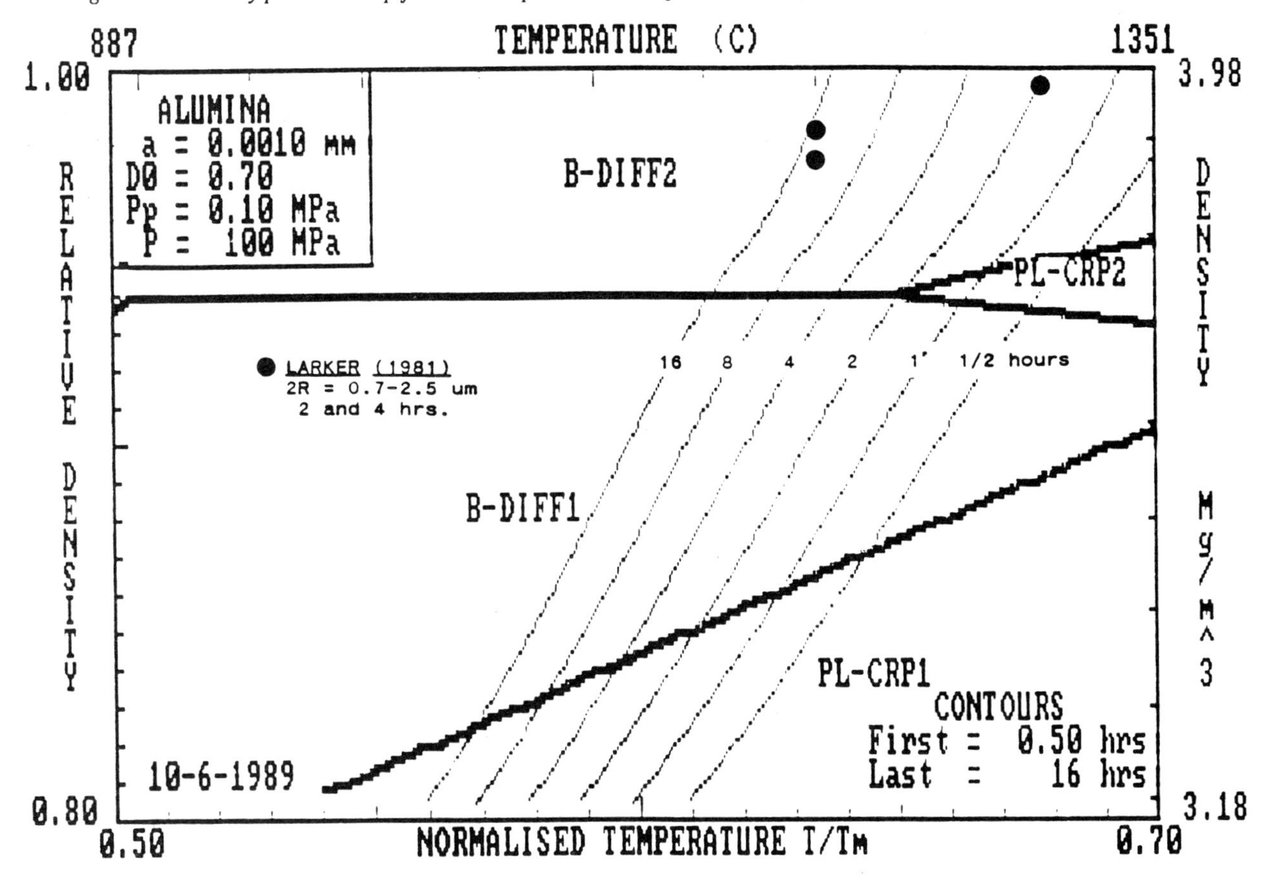

Figure 11 A density/temperature map for the same alumina as Fig. 10.

21. J.M. Viera and R.J. Brook, J. Am. Ceram. Soc. 67, p.245, 1984.
22. A.S. Helle, K.E. Easterling and M.F. Ashby, Acta Metall., 26, p.2163, 1985.
23. U. Eisele, Ph.D. Thesis, Ceramics Dept., Univ. of Leeds, Leeds, UK 1989.
24. M.F. Ashby, HIP-MAPS, Software and Operating Manual, Cambridge University Engineering Department, 1987.
25. H.J. Frost and M.F. Ashby, Deformation-Mechanism Maps, Pergamon Press, Oxford, 1982.
26. R.J. Brook, J. Am. Ceram. Soc., 52, p.56, 1969.
27. M.F. Yan, R.M. Cannon, Jr, H.K. Bowen and V. Chowdhry, Mater. Sci. Engng., 60, p.275, 1983.
28. C. Zener, quoted by C.S. Smith, Trans. Am. Inst. Min. Engrs., 175, p.15, 1948.
29. W.A. Kaysser. Private Communication, 1985.
30. I.H. Moon and J.S. Choi, Powder Met., 28, p.21, 1985.
31. F.N. Rhines, R.T. DeHoff and R.A. Rummel, in Agglomeration, (ed.) W.A. Knepper, Interscience N.Y, 1962.
32. K. Uematsu, K. Hakara, N. Uchida, K. Saito, A. Miyamota and T. Miyashita, Proc Sintering '87, Tokyo, Elsevier, 1988.
33. H. Larker, ASEA, Robertsfors, Sweden, private communication, 1981.
34. W.B. Li, M.F. Ashby and K.E. Easterling, Acta Metall., 35, p.2831.

Section 4

Sintering

12

Solid State Sintering

W. A. KAYSSER

Max-Planck-Institut für Metallforschung, Institut für Werkstoffwissenschaft, Stuttgart, Germany

Abstract

With respect to the particular importance of dimensional changes, microstructural coarsening, and homogenization during solid state sintering of crystalline materials, sintered products are classified into parts with inherent functional porosity, structural parts and high performance materials. Basic mechanisms leading to shrinkage and grain coarsening are described for single phase systems. Their effect on the microstructural design by different heating schedules is discussed. Solid state sintering in the presence of additives is outlined and the influence of a second solid phase is described.

1. Introduction

Present day industrial powder metallurgy is characterized by pressureless or pressure-assisted sintering of multicomponent parts and materials. Preforms of ceramics or metallic powders are heated to high temperatures where shrinkage, grain growth, annealing recovery, compositional homogenization, and chemical reactions to form new phases or compounds may occur. During cooling, changes such as precipitation hardening or other phase transformations may take place in specific systems. The efficient use of the technology depends on the combination of the scientific understanding of complex physical and chemical problems of powder production, handling, forming and consolidation with highly developed complex engineering capabilities. The basic steps of conventional powder metallurgy are powder preparation, powder handling, shaping and consolidation. Powders may be produced by chemical reactions from precursors or by disintegrating solid or liquid in bulk. The powders are shaped in dies under a uniaxial load or filled in thin walled capsules and isostatically pressed. The shaped compacts contain pores between the individual particles which amount to porosities of between 60 and 10 vol.-%. It is the purpose of subsequent sintering - a heat treatment below the melting temperature of the major constituents - to initiate and enlarge solid contacts between the particles (driven by a reduction of the sum of the surface energies of solid/vapour and solid/solid interfaces). It may be a second goal of the sintering procedure to eliminate the pores, driven by a reduction in solid/vapour interface area. Porosity reduction requires an increasing fraction of solid material in each unit volume by rearrangement or by shape change of the particles as illustrated in Fig. 1.

Figure 2 gives a rudimentary outline of the major production routes. Different objectives require different initial powders and different dimensional changes during the heat treatment. In other words the objective of sintering is dependent upon the particular type of product required.

The powder metallurgical production route can be used to produce parts with a functional inherent porosity. As starting material, powder particles of various shapes between spherical and largely elongated fibre are used. Sintering is interrupted at a stage when a certain mechanical strength of the contacts between adjacent particles is reached, but open porosity still exists. In the case of open porosity, pore channels connect the surface of a part with pores in the interior. An example of such a product are electrodes for pace makers which are sintered from spherical Pt-alloy powder. The porosity is required to ensure a stable positioning of the sensing and transmitting electrode by growth of human tissue into the pore space. Secondly, the high surface area of the electrode improves its sensing and transmitting capability. A further well-known example is the self-lubricating bearing. Figure 2 illustrates the initial powder size and dimensional change during sintering for different types of products.

The powder metallurgical production route can be used to produce structural parts of an accurate shape and size (labelled "cheaper" in Table 1). The material of these parts is of at least slightly inferior quality to the material of parts machined from cast and wrought products. The main emphasis in the control of the dimensional changes during sintering in order to avoid machining. The dimensions of structural parts, e.g. of gears are well controlled, if either shrinkage or swelling during pressureless sintering is negligibly small [1]. Therefore, porosity after cold compaction and after sintering should be approximately equal. The simultaneous requirement to keep dimensional changes at a minimum and to have

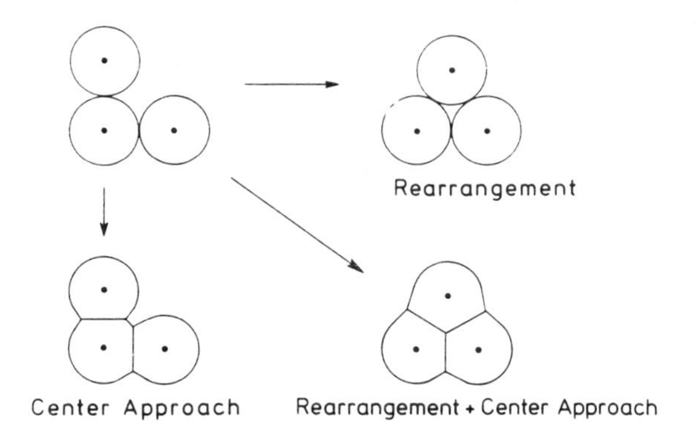

Figure 1 Shrinkage by rearrangement and center approach (shape change) during sintering.

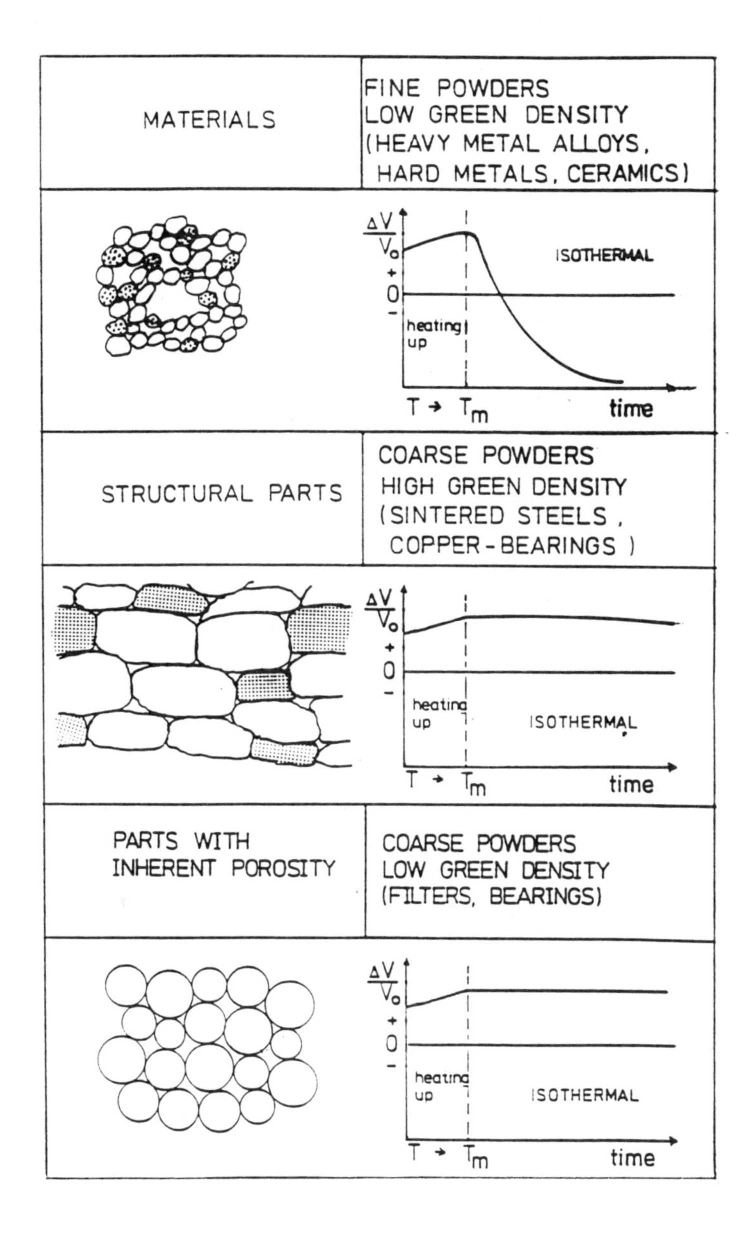

Figure 2 Intital powder size and dimensional change during sintering for different types of products.

a high final density to achieve good mechanical properties leads to the use of unalloyed, coarse powders (larger than 100 μm) which are cold compacted to high density into the final shape of the part.

To obtain alloys, usually mixtures of different unalloyed powders are used. During sintering chemical bonds form between the particles. Subsequent homogenization of the components is a necessary precondition for the development of reasonable mechanical properties by solid solution hardening, precipitation hardening or the formation of martensite. Dimensional control and the required hardenability of sintered structural parts require knowledge of the diffusion processes leading to dimensional changes and homogenization during solid state sintering [2].

Table 1 Powder types and dimensional changes during sintering of PM products

Powder	Dimensional changes	Cheaper	Superior
Coarse	Small	Dimensional control (Gears)	Dimensional control (Filter, bearings)
Fine	Coarse	Densification Dimensional control (Injection moulding)	Densification Grain growth (Hard metal)

The major objective in the use of the powder metallurgical route can also be the production of parts or semi-finished material with superior properties to those of materials produced by casting [3] (labelled "superior" in Table 1). Fine grained segregation-free material can be obtained, particularly for materials and phase combinations which cannot be prepared by melting, casting, and solidification. Examples are refractory metals (W, Mo), hardmetal (WC-Co), heavy metal alloys and many hi-tech ceramics. These high performance materials must be sintered to full density, and a small grain size must be maintained. Of major interest for the production of these PM materials are, therefore, the mechanisms and processes that control densification and grain coarsening during sintering [2].

2. Phenomenology of Solid State Sintering

The main essentials of the sintering process can be deduced from phenomenological observations. Figure 3a shows the densification of compacts (150 MPa) from Cu powder (average particle size 45 μm) during isothermal sintering at temperatures between 700 and 1000°C. The shrinkage rate reduces continuously during isothermal sintering, indicating a reducing rate of material transport leading to changes in the arrangement or shape of the powder particles. The reduced transport rates may result either from a loss of the driving force, from longer transport paths or from the increasing volume of material to be transported to produce equal macroscopic density changes. The shrinkage rate at equal density increases with increasing temperature indicating the transport process to be thermally activated. Figure 3b shows the linear shrinkage of W powder compacts after sintering at 1800°C for 5h. Smaller powders yield an increased density, indicating that either transport rates are enhanced or the volumes of material to be transported to result in equal macroscopic density changes are reduced with decreasing particle size.

Metallographic sections of the Cu compacts at densities of 70, 87 and 96% of the theoretical density - that means with 30, 13 and 4% porosity respectively - schematically shown in Fig. 4., characterize three subsequent stages of different pore arrangement during the sintering of the materials.

In Stage 1, solid material bridges form in the necks between the individual particles. The reduction in porosity is modest until the necks are approximately one third of the particle diameter in size [4]. Grain growth to sizes above the size of the initial powder is usually suppressed. If a loose arrangement of spheres is considered, with an initial density between 58 and 68% [5], the transition to Stage 2 occurs at densities of approximately 75% of the theoretical. The individual initial particles can no longer be distinguished. Pore channels are situated along triple junctions of adjacent grains. Stage 2 is characterized by rapid shrinkage and slow grain growth [6,7]. At densities between 91 and 95% Stage 3 starts, when the interconnected pore channels become unstable forming individual isolated pores which lie at the grain boundaries or in the interior of the grains [8,9]. The open porosity thus transforms into a closed porosity. During the final stage of sintering both slow densification (pore shrinkage) and fast coarsening (pore coarsening and grain coarsening occur). During the production both of parts with inherent porosity and also of structural parts, sintering is terminated during Stage 1 or the early Stage 2. Due to the high initial green density of structural parts, the transition from Stage 1 to Stage 2 occurs at densities between 86 and 92%. As previously mentioned materials are subject to all stages, which will be discussed in more detail later.

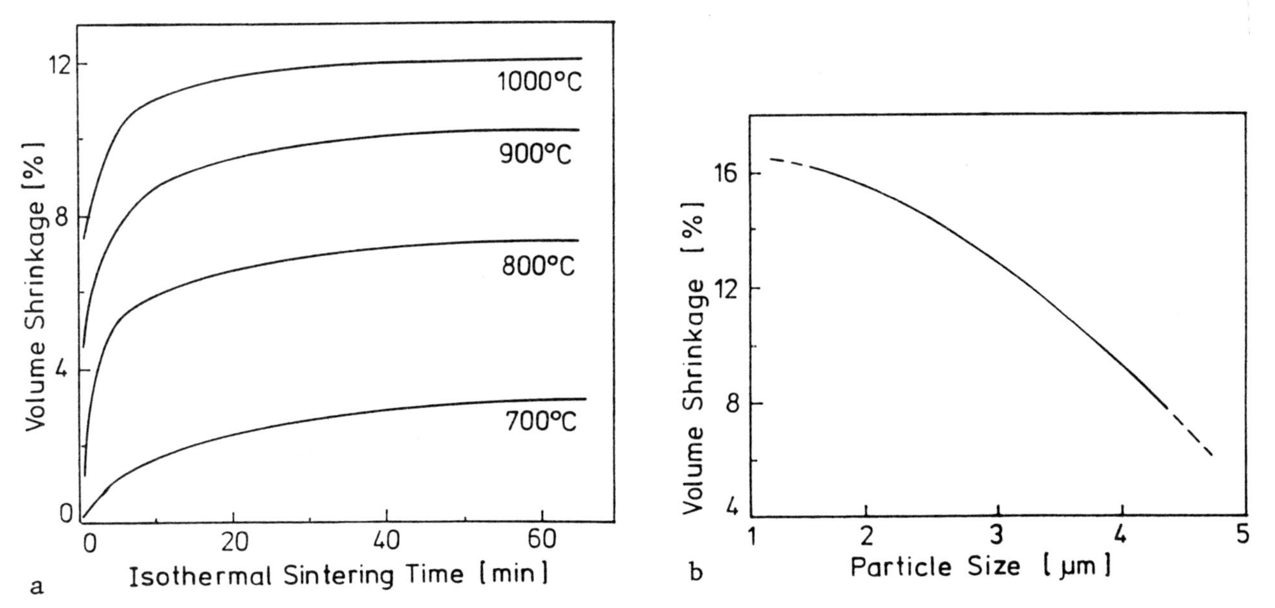

Figure 3a and b Densification of green compacts during sintering. a) Cu powder compacts at different temperatures b) W compacts of different initial particle size.

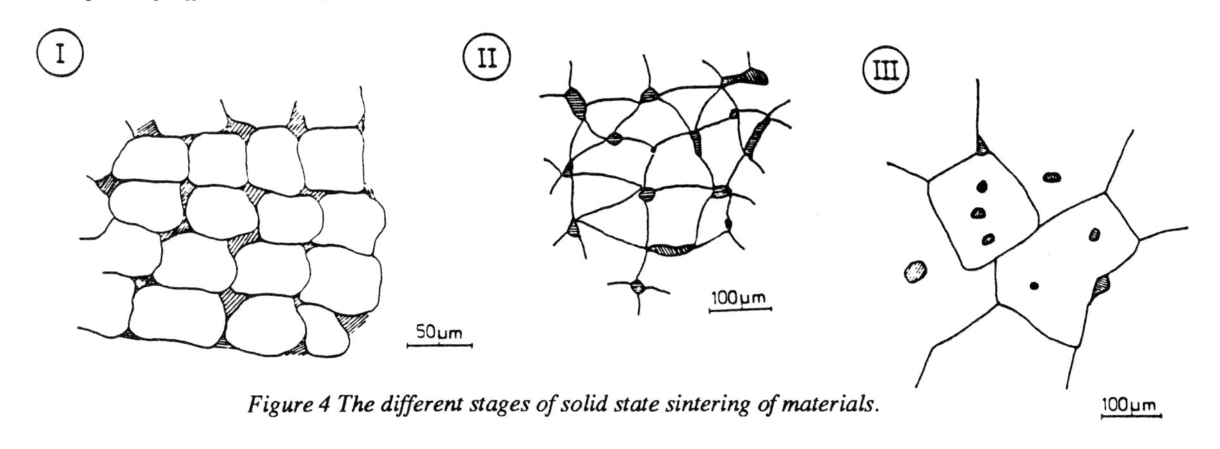

Figure 4 The different stages of solid state sintering of materials.

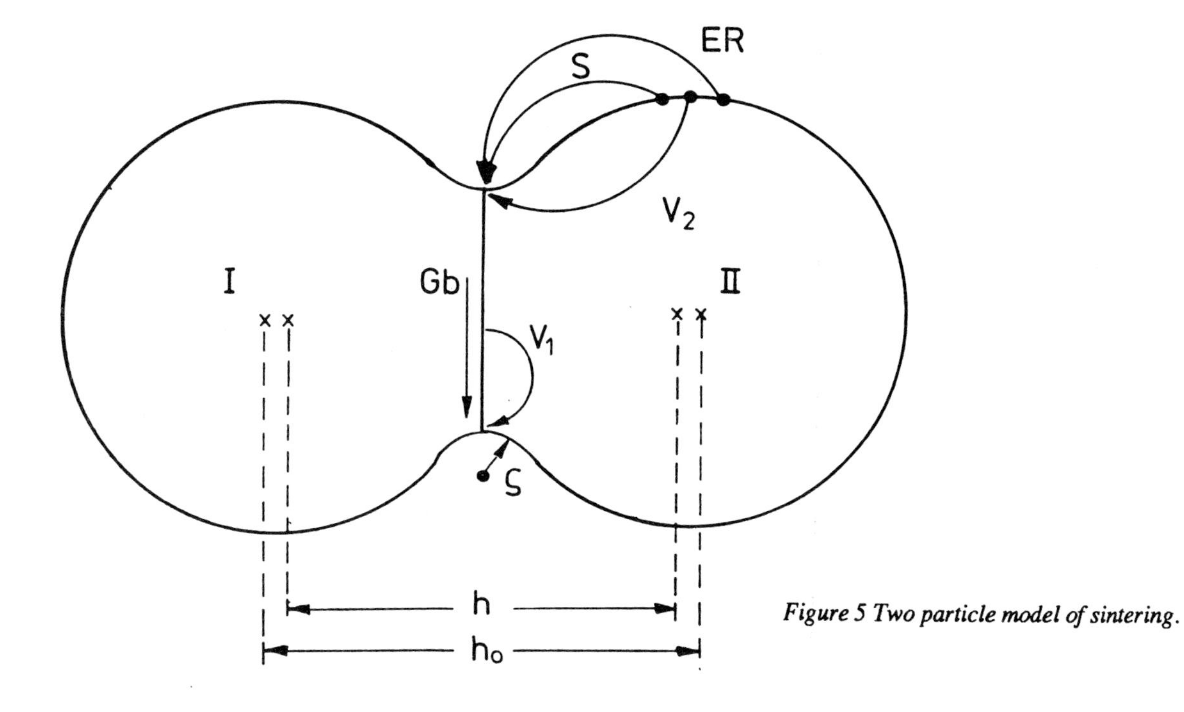

Figure 5 Two particle model of sintering.

3. Theory of Solid State Sintering: Stage 1

It is the objective of the sintering theory to provide a helpful general picture of the material transport processes active during sintering. The influence of different parameters, such as particle size, initial density (green density), temperature and time is described with a view to indicating how dimensional changes and microstructural development may be controlled. Sintering is a multi-particle problem and hence theory will neither explain all the details, especially in more complex systems, nor provide quantitative predictions bearing a high degree of accuracy.

Figure 1 shows the two shrinkage phenomena: rearrangement and centre approach of the particles. In solid state sintering the centre approach of the particles is the dominating phenomenon. It requires the flow of matter out of the contact region of adjacent particles. The yield strength is much too high to obtain plastic flow and the bulk creep processes such as power-law creep are much too slow under the stress fields originating in the neck regions to contribute measurably to material transport in the neck region [10]. Centre approach is achieved when vacancies which emanate from the surfaces in the neck regions, are eliminated at the grain boundaries in the particle necks [11]. The five transport paths for atoms (as a counterflux to the flux of vacancies) shown in Fig. 5: Gb) grain boundary diffusion; V_1) bulk diffusion from the neck to the grain boundary; V2) bulk diffusion from distant surfaces to the neck surface; S) surface diffusion and ER) evaporation and recondensation, all lead to neck growth. Only the first two, Gb and V_1 however, lead to a centre approach of the particles and thus to the shrinkage of a powder compact during sintering.

The curvature of the surfaces in the neck region causes stress gradients which initiate the flux of vacancies and the equivalent counterflux of atoms. Generally a curved interface between a condensed and a gaseous matter results in a Laplacian stress in the condensed part of the system.

$$\Delta p = \gamma_{sv} \cdot (1/r_1 + 1/r_2) \tag{1}$$

where:
r_1 and r_2 - the radii of the principle curvatures of the interface at a point P_u.
γ_{sv} - the specific interface energy solid/vapour.

A convex-shaped interface of the solid results in compression of the condensed matter and a concave-curved interface results in dilation. Vacancies will diffuse along stress gradients from regions of low relative pressure to regions of high relative pressure. In the model of two sintering particles (Fig. 5), the neck region shows two curvatures namely $1/r_1 = 1_p$ yielding dilational strain and $1/r_2 = -1/_x$ yielding a compressive strain. During the first stage of sintering p is always much smaller than x, yielding tensile stresses at the neck surface and compressive stresses in the central areas of the neck. The gradient of the chemical potential of the vacancies $V\mu_v$ is related to the stress gradient $V\sigma$ by $V\mu_v = V\sigma \cdot \Omega$.

The flux of vacancies is defined by

$$J_v \cdot \Omega = dV/(Adt) = D \cdot c_{vo} \cdot \Omega \cdot V\mu_v/RT \tag{2}$$

where Ω is the molar volume of the vacancies. With a number of simplifications equations describing the neck size, x/a, and the relative shrinkage, $\Delta h/a$, (see Fig. 5) of the following general forms [4]

$$(x/a)^n = C \cdot t/a^m \tag{3}$$

and

$$(\Delta h/a)^{n/2} = (C \cdot t)/(2^n \cdot a^m) \tag{4}$$

were obtained for the Stage 1 sintering of spherical particles. The parameters for the different transport processes for equations (3) and (4) are given in Table 2 [12]. η is the viscosity, γ_{sv} is the specific surface energy, D_g, D_s and D_v are respectively the grain boundary, surface and bulk diffusion coefficients ϵ is a reaction rate constant, δ is the effective thickness of the grain boundary, Ω is the molar volume, R is the gas constant, k is the Boltzman constant, T is the absolute temperature. Z is $\gamma_{sv}\Omega/RT$.

4. Theory of Solid State Sintering: Stages 2 and 3

Coble developed a model for the second sintering stage [16]. He assumed the pores to be channels of constant diameter situated along the junction lines of an arrangement of space-filling Kelvin bodies, in particular tetrakaidekahedrons. The porosity reduction rate dP/dt was obtained with similar assumptions on geometrical changes and diffusion kinetics as in the first sintering stage yielding:

Table 2 Parameters for initial stage sintering (Stage 1) of spherical particles for eqs. (3) and (4)

Transport	Measured		Plausible		C	Ref
	n	m	n	m		
Viscous flow	2	1	2	1	$3\gamma_{sv}/2\eta$	13
Evaporation Condensation	3-7	2-4	3	2	$2.4\,\epsilon Z$	14
Grain boundary diffusion at grain boundary	6	4	6	4	$48D_g\delta Z$	15
Volume diffusion from neck	4-5	3	5	3	$16D_vZ$	15
Volume diffusion from surface	4	3	4	3	$20D_vZ$	15
Surface diffusion	3-7	2-4	7	4	$23D_s(\Omega k/R)^{1/3}Z$	15

$$dP/dt = K_1 \cdot D_v/(T \cdot G^3) \quad (5)$$

for bulk diffusion from pores to the grain boundaries, where G is the average grain diameter. A similar approach was made for Stage 3, when monosized, isolated pores were assumed to be present at all corners of the space filling polyhedrons. The approach for Stage 3 does not take into account, however, the important feature of grain boundary/pore separation and is therefore rather unrealistic. During the final stage of sintering, both densification (pore shrinkage) and coarsening (pore coarsening and grain coarsening) occur simultaneously. Shrinkage occurs by migration of vacancies from the pores mainly along the grain boundaries to vacancy sinks (grain boundaries and free surfaces). Coarsening either occurs by mass transfer between isolated pores and/or by pore coalescence during grain growth. Figure 6 shows a scheme initially proposed by Brook [17] for pores but which is extended to apply to second phase inclusions [18]. It shows fourteen mechanisms which may control the migration of a boundary. In all cases a decreasing pore size at a constant total porosity results in shorter efficient pathways and an increased mobility of the pores. Mechanisms 6 and 7 show two pathways which lead to coarsening of the pores by grain boundary or lattice diffusion. It is obvious that small amounts of additives can influence all these mechanisms including the intrinsic boundary movement {1} and the impurity drag {2}. A second set of coarsening mechanisms is based on coalescence of pores by collision when the boundaries move [18]. Similar arguments are valid, if instead of pores, second phase inclusions are present at the grain boundaries.

During grain growth, pores either remain at the boundaries or become separated and "trapped" inside the grains. Pores that are trapped, usually shrink at much lower rates than pores which are at grain boundaries, since bulk diffusivity is much smaller than grain boundary diffusivity. Pore drag and pore/grain boundary separation in W-0.15wt.%Ni after annealing at 1400°C for 3h are shown in Fig. 7a. There are two regimes where pores can remain attached to the grain boundaries (Fig. 7b). Pores which are large enough are immobile but retard the motion of the boundary (pore control in Fig. 7b) [8,9,19]. Pores which are small enough move along with the grain boundaries without stopping their motion (boundary control in Fig. 7b). The mechanisms of pore migration are shown in Fig. 6. The intersection of the two regimes of pore-boundary interaction leads to a pore-boundary separation region with G* as the smallest grain size where grain boundary/pore separation occurs (Fig. 7a). The trajectory of the grain size/pore size development has to be kept off the grain boundary/pore separation field. Effects which decrease the grain boundary mobility and increase the pore mobility shift the border of the separation field to larger pores, that means to an increased value of G*. Effects which suppress pore coalescence and which support pore size reduction by vacancy elimination lead to a passage of the trajectory at lower temperatures.

5. Sintering Diagrams

Ashby has developed a graphical representation of the sintering behaviour of powders using sintering diagrams [20,21]. The diagrams plot the relative neck radius (x/a) or the relative density (D/D_{th}), as a function of temperature for various times at various temperatures. Figure 8 shows isochrones of the neck size for sintering of Cu. Areas are identified in which a single transport mechanism dominates the neck growth rate. The thick lines indicate the temperature and neck size parameter at which the adjacent mechanisms provide equal contributions. The sintering diagrams are valuable tools for the engineer to obtain a first idea of the conditions required for sintering of a particular material of known composition and grain size. The neck growth and shrinkage rate values are calculated with similar equations as those shown in Table 2. The sintering diagrams are therefore by no means more accurate than the models from which the equations were obtained and the diffusion and interface energy data used for the calculations.

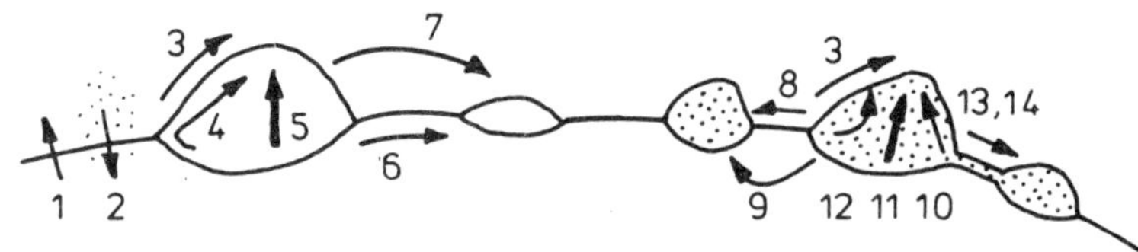

Figure 6 Mechanisms of grain boundary migration control involving pores, or second phase inclusions. 1 Intrinsic boundary movement, 2 impurity drag, 3 pore movement by lattice diffusion in the host material, 4 pore movement by surface diffusion, 5 pore movement by vapour transport, 6 pore coarsening by grain boundary diffusion, 7 pore coarsening by lattice diffusion, 8 inclusion coarsening by grain boundary diffusion, 9 pore coarsening by lattice diffusion, 10 inclusion movement by interface diffusion, 11 inclusion movement by grain boundary diffusion in the inclusion, 12 inclusion coarsening by lattice diffusion, 13 inclusion coarsening by interface diffusion, 14 inclusion coarsening by bulk diffusion in the second phase.

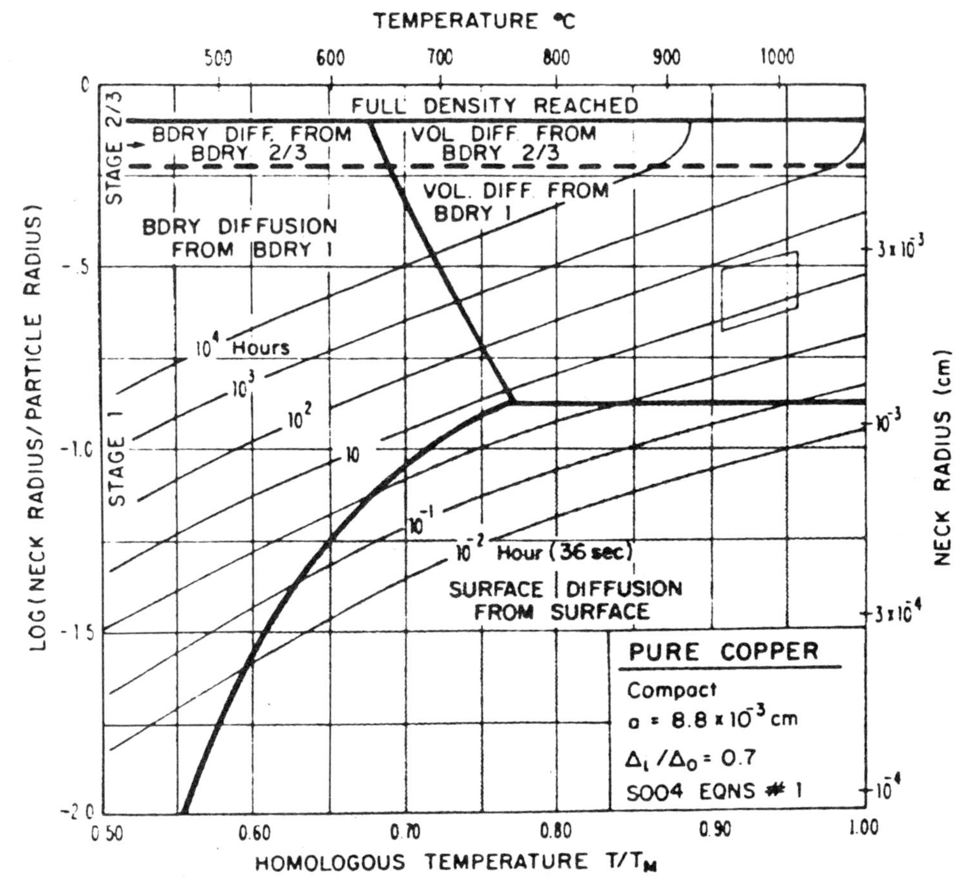

Figure 8 Sintering diagram of Cu (after Ashby [20]).

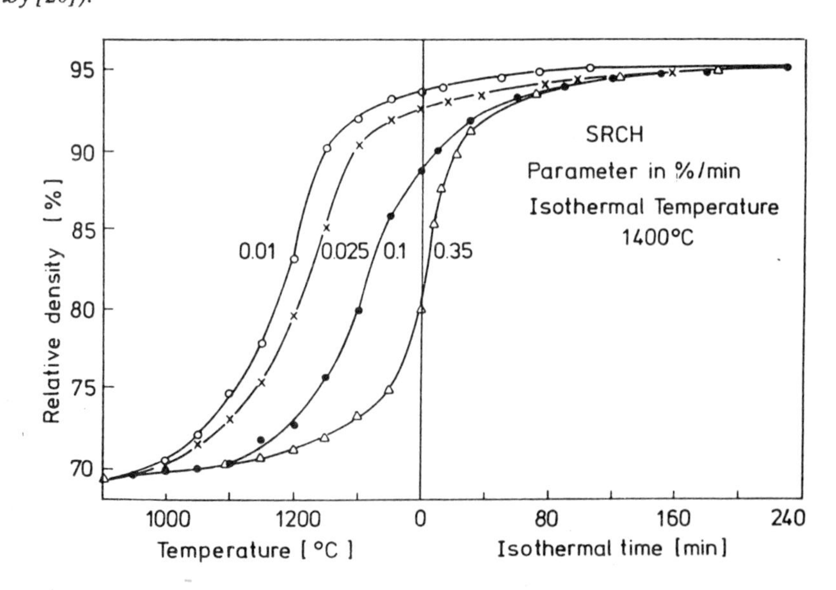

Figure 9 Shrinkage rate controlled heating of W-0.15wt.%Ni to 1400°C and subsequent isothermal annealing.

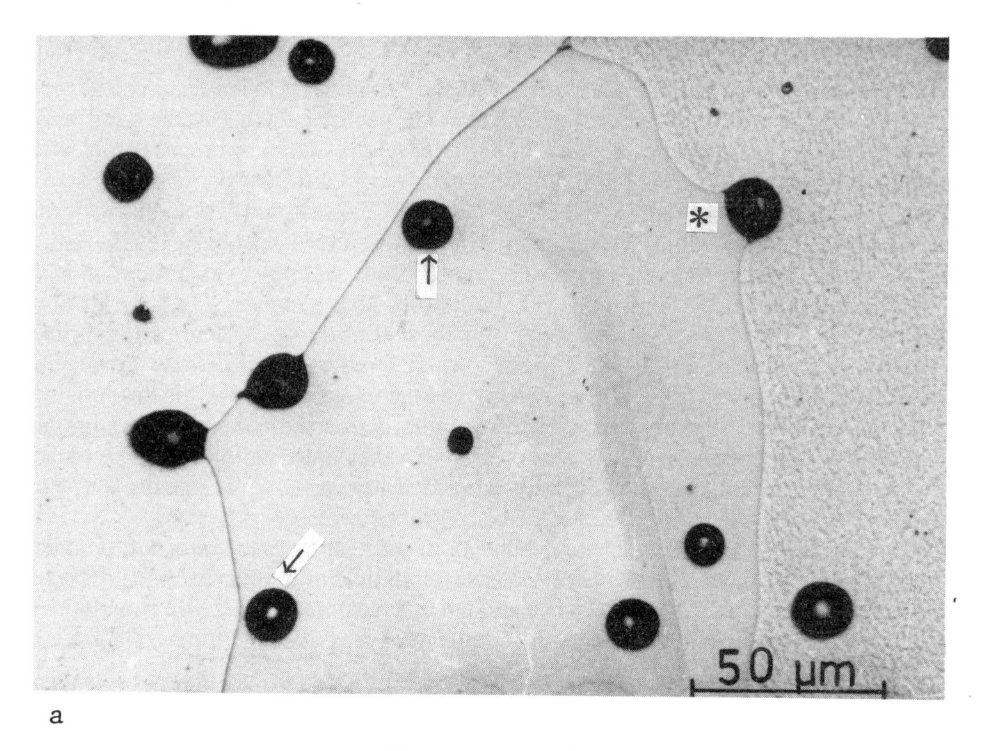

a

Figure 7a Microstructure showing pore drag (asterisk) and pore/grain boundary separation (arrow) after final stage sintering of W-0.15 wt.%Ni at 1400°C for 3h.

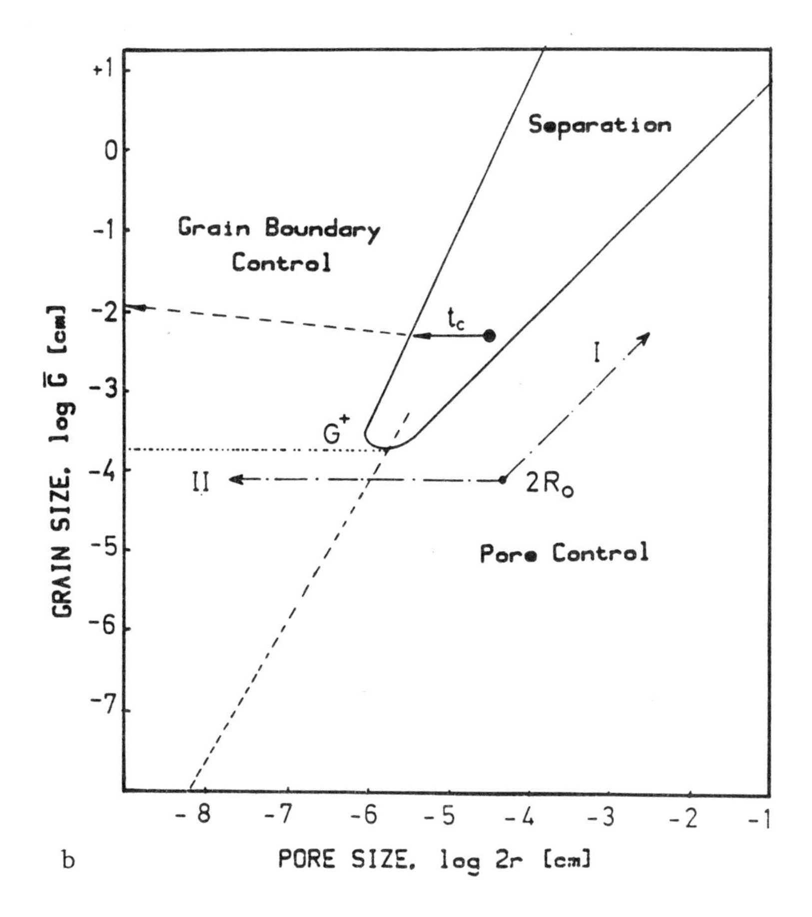

b

Figure 7b Pore/grain boundary seperation diagram. $2R_o$ is the intial grain size.

6. Influence of Heating Rate on Sintering

It can be expected that the heating rate will influence the shrinkage behaviour. Slower heating leads to higher densities at equal temperatures (Fig. 9) due to the longer sintering time intervals per temperature. At the same density, the sintering rate will increase with increasing temperature. Beyond these trivial rules a number of more complex cases exist where changes in the heating schedule lead to parts of different final density and different density/grain size ratios. A special case is the shrinkage rate controlled sintering (SRCS) [22]. During SRCS a constant shrinkage rate is maintained during heating up to temperature by a feed back controlled variation of the heating rate. It was the basic assumption of its inventors that a feed back SRCS would result in sintered materials with properties superior to materials sintered with other heat treatment schedules. The success of SRCS is, however, limited to those systems where the diffusion mechanisms controlling grain growth have a different activation energy from those controlling the dominating shrinkage mechanism.

Figure 10a shows the grain size of W-0.15wt.%Ni samples sintered to equal density by very different heating schedules. The direct dependence of the grain size on density and the unimportance of the heat treatment which led to the densification is also shown in Fig. 10b where the average maximum diameter of the grains of a large number of samples is plotted versus the density [23]. In this figure it is also clearly seen that sintering in Stage 3 is characterized by rapid grain coarsening. A similar strict relation between porosity, pore size and grain size was found when MgO-doped Al_2O_3-Anorthite(glass) composites were sintered by shrinkage rate controlled heating.

A strict relation between density and grain size, independent of the heating procedure, was predicted by Coble and Kuczynski [6,7]. Coble found that the rate of porosity decrease, dP/dt, and of grain growth, da/dt, in powder compacts have the same temperature dependence. Therefore, he concluded that the pores in the grain boundaries must inhibit and thus control the grain growth. Consequently a Zener-like relation [24] was assumed for a grain growth equation:

$$a_a = K \cdot r_{pa}/(P)^s \tag{6}$$

where:
rpa - average pore size,
aa - average grain size,
K - constant,
s is 0.5 when all pores are located in the grain boundaries and s is 1 when they are randomly distributed.

In contrast to the heating of MgO-doped Al_2O_3-Anorthite (glass) composites, glass-free samples showed smaller grain sizes at equal densities after faster heating to temperature. The extreme use of increased heating rates during heating up of MgO-doped Al_2O_3 to temperatures, fast firing (firing for a matter of minutes at temperatures higher than those conventionally used), received attention several years ago because of the microstructural benefits or because of the high production rates that can be achieved. Harmer, Roberts and Brook [25] showed that Al_2O_3 can be fired to theoretical density at 1850°C in some 15 min. Harmer and Brook [26] developed a simple explanation of the effects of fast firing and the influence of different heating rates based on the assumption of different activation enthalpy changes of processes controlling densification and grain growth. The grain growth rate is given as:

$$da/dt = (da_o/dt) \cdot \exp(-Hg/RT) \tag{7}$$

where da_o/dt is the growth rate for T going to infinity. For grain growth in the pore free state Hillert assumed da/dt = M • F, where F is the force on the grain boundary and M is the mobility of the boundary [27]. As shown by Cahn [28] and Lücke [29] a window of migration rates exists where:

$$M = M_o \cdot \exp(-G_a/RT) \tag{8}$$

holds with G_a as the free activation enthalpy of the atomistic step controlling boundary mobility. Except for very pure materials, the grain boundary movement is controlled by the intrinsic diffusivity in the grain boundary and the diffusivity of segregated impurity atoms in the adjacent bulk. In MgO-doped dense Al_2O_3 it is now well accepted that Mg decreases the migration rate at least of certain grain boundaries [30,31]. This yields an activation energy for the atomic movements controlling the grain boundary migration which is related to the atomic mobility in the grain boundary and which is presumably close to the activation energy H_g for grain boundary diffusion. Similar to the separation of the grain growth equation into a temperature dependent and into a temperature independent term, a formal separation of the shrinkage rate, -dP/dt, into:

$$dP/dt = -(dP_o/dt) \cdot \exp(-H_d/RT) \tag{9}$$

was proposed, where $-dP_o/dt$ is the shrinkage rate for T approaching infinity and H_d is an activation enthalpy for the rate controlling step. As discussed earlier by Harmer and Brook [26] the controlling mechanism for densification has been

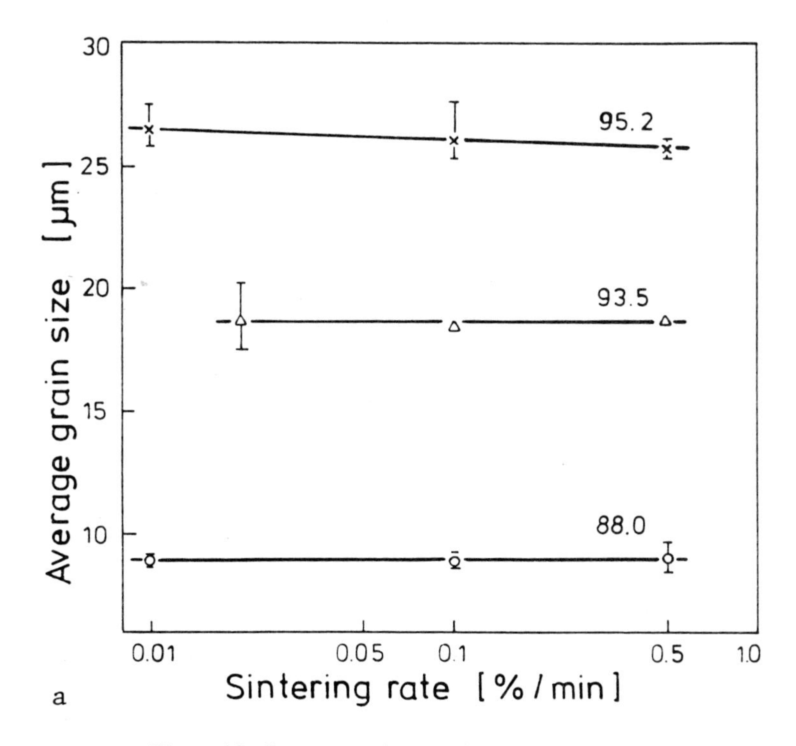

Figure 10a Average grain size of samples of equal density.

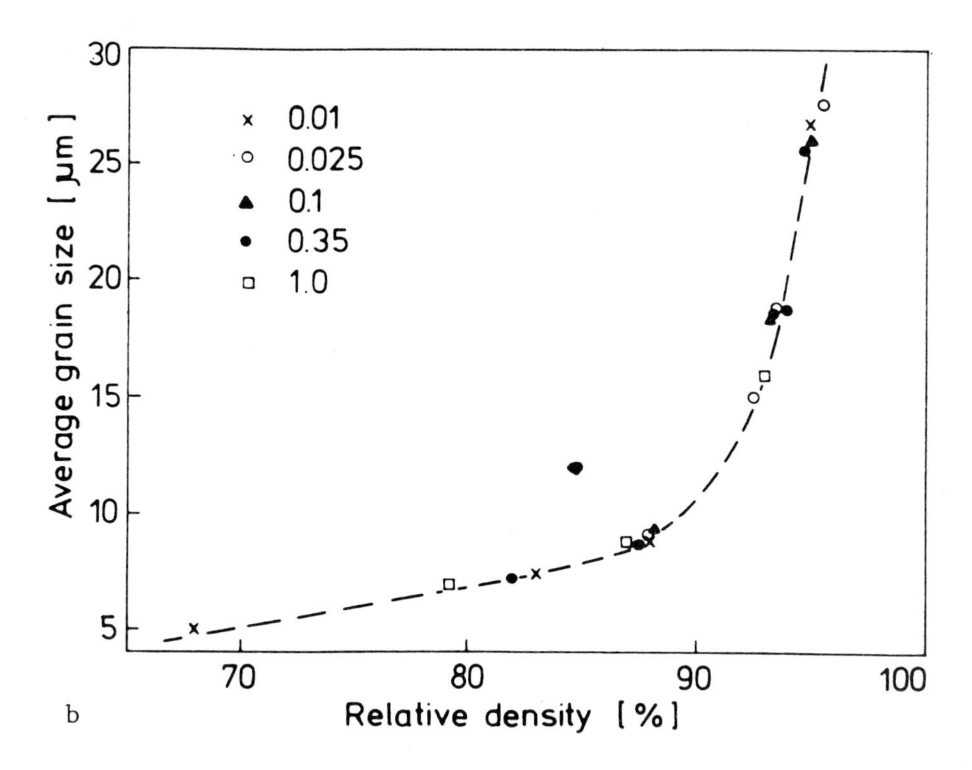

Figure 10b Average grain size of grains versus density.

Figure 10a and b Influence of intial heating to 1400°C on the microstructure of W-0.15wt.% which was heated by shrinkage rate contolled heating with different shrinkage rates to 1400°C and subsequntly isothermally sintered.

found to be lattice diffusion of the aluminium ions which yields an expected activation enthalpy equal to the activation enthalpy of that lattice diffusion. The latter is thought to be much higher than the activation enthalpy for surface or grain boundary diffusion. As a consequence, the relative rates of grain growth to densification are higher at lower temperatures and decrease with increasing temperature. Material of the same density will have smaller grains if heated more rapidly through the lower temperature range, in accordance with the experimental observations made. In the presence of the anorthite as an intergranular second phase, grain growth and densification by the shape change of the Al_2O_3 grains are both controlled by the transport of Al_2O_3 through the anorthite [31]. The activation energy for densification and grain growth is therefore the same, hence the relation between density and grain size is independent of the heating schedule.

6. Solid State Sintering with a Small Amount of Additive

Shrinkage and grain growth of materials as well as dimensional changes and homogenization of metallic structural parts are often manipulated by additives. They are intentionally added to the sintering system in contrast to impurities which are unintentionally present. The usual impurity level is several thousand ppm; impurity levels below 300 ppm must be considered as low [3]. The amount of additive may vary from 0.1 to 10 wt.%. Some additives are present as solid phases only, as solid solutions with the host metals or as segregates at the interfaces of the host metals. Other additives form liquid phases which often penetrate as thin liquid films between the particles and grains of the host powder [32]. Often additives are used to accelerate shrinkage, e.g. in Ni-doped W.

It was deduced by various investigators that the transport of W or Mo in their grain boundaries or in thin second phase layers at particle contacts must be fast if Ni or other so called "activating" elements are to be effective in enhancing the shrinkage of these materials [33,34,35,36]. Two independent measurements confirm that the presence of Ni at W or Mo grain boundaries increases the grain boundary self diffusion coefficient of these materials [33,38]. Our experiments and the earlier experiments of Schintelmeister and Richter [38] on grain boundary diffusion of Ni-free and Ni-doped W grain boundaries show that the grain boundary self-diffusion of W at 1300°C is increased by a factor of 500 to 5000 if Ni is present. Both investigations established grain boundary self diffusion as the essential mechanism which causes shrinkage in activated sintering.

Unfortunately, the transport of vacancies to the grain boundary sinks by bulk diffusion or by grain boundary diffusion does not provide complete densification in metal powders of sizes between 1 and 100 μm. In the final stage of sintering, when the porosity is below 6%, the continuous pore channels transform into isolated pores which lie at grain boundaries. If the pores are too small to hold the grain boundaries, but too large (that means too immobile) to move along with the grain boundaries, grain boundaries and pores separate during regular grain growth (Fig. 7). Pores that are trapped inside the grains shrink at much lower rates than pores which are at grain boundaries and are therefore deleterious for densification. Powder arrangements can be sintered to full density if the diffusion mechanism controlling the vacancy transport from the pores to the grain boundaries is enhanced and if the grain boundary mobility is decreased by additives. These basic principles were successfully applied to obtain translucent Al_2O_3 ceramic materials for Na vapour pressure lamps. The dense Alumina (Lucalox) can be produced by pressureless sintering when the bulk diffusivity is enhanced [38] and the grain boundary mobility is reduced [39] by doping the powder with less than 1000 ppm MgO.

The bulk diffusivity of metals, which controls one major transport path for vacancies to move from the pore to the grain boundary via diffusion through the bulk, can only be moderately increased by doping with small amounts of an alloying metal additive. The enhancement of the grain boundary self diffusion, which controls the second major transport path for vacancies, as in W doped with Ni [40], however, also increases the grain growth rate leading to extended pore/grain boundary separation.

The metallic materials produced by pressureless sintering, such as hard metals, heavy metal alloys, and permanent magnets are all sintered in the presence of a liquid phase, which demonstrates the problems of sintering to a pore free material in solid state, but also indicates the ability to obtain dense materials by liquid phase sintering.

7. Solid State Sintering with a Second Phase

Densification, dimensional control and homogenization can also be influenced by additives which are present as a second solid phase. The additives may act as a fast diffusion path for the host atoms. The additive may also form inclusions on obstacles which retard densification and grain growth. The broad variety of possible chemical interactions between the solid phases yields a complicated influence on densification and dimensional control.

An example of a simple interaction is found in W-Cu [41]. At temperatures below 1100°C, W powder does not sinter and the mutual solubility of W and Cu is negligible. Despite the insolubility, W-Cu compacts densify by solid state sintering during heating up to temperature in the solid state (Fig. 11). The densification rate increases with increasing volume content of Cu. In the vicinity of the Cu particles Cu spreads onto W particles. The comparison of shrinkage and microstructural development supports a two stage model where, initially rapidly enlarging zones of low Cu content yield an initial rearrangement and shrinkage. A second massive spreading of Cu from the original Cu particles leads to the local formation of denser Cu-W particle aggregates and to a continuous macroscopic shrinkage after prolonged sintering when most aggregates form an interconnected network (Fig. 12a and b).

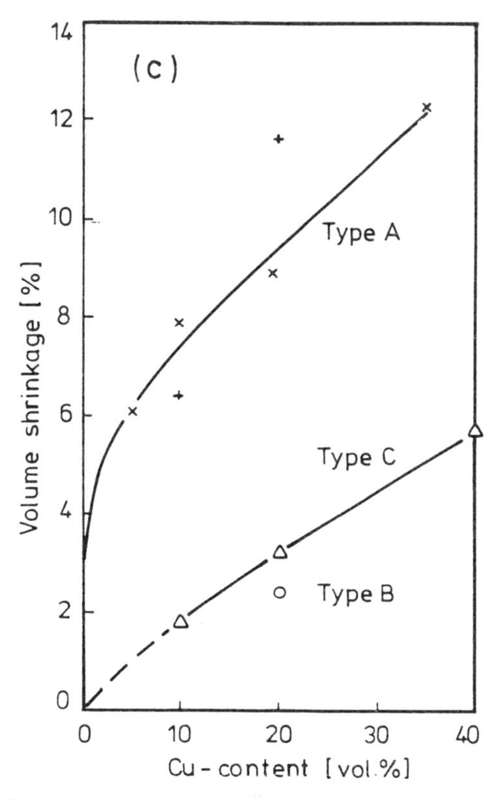

Figure 11 Volume shrinkage of W-Cu compacts after sintering at 1050°C for 1h. A, B and C are W powders with an average particle size of 0.8, 5.0 and 2.4μm.

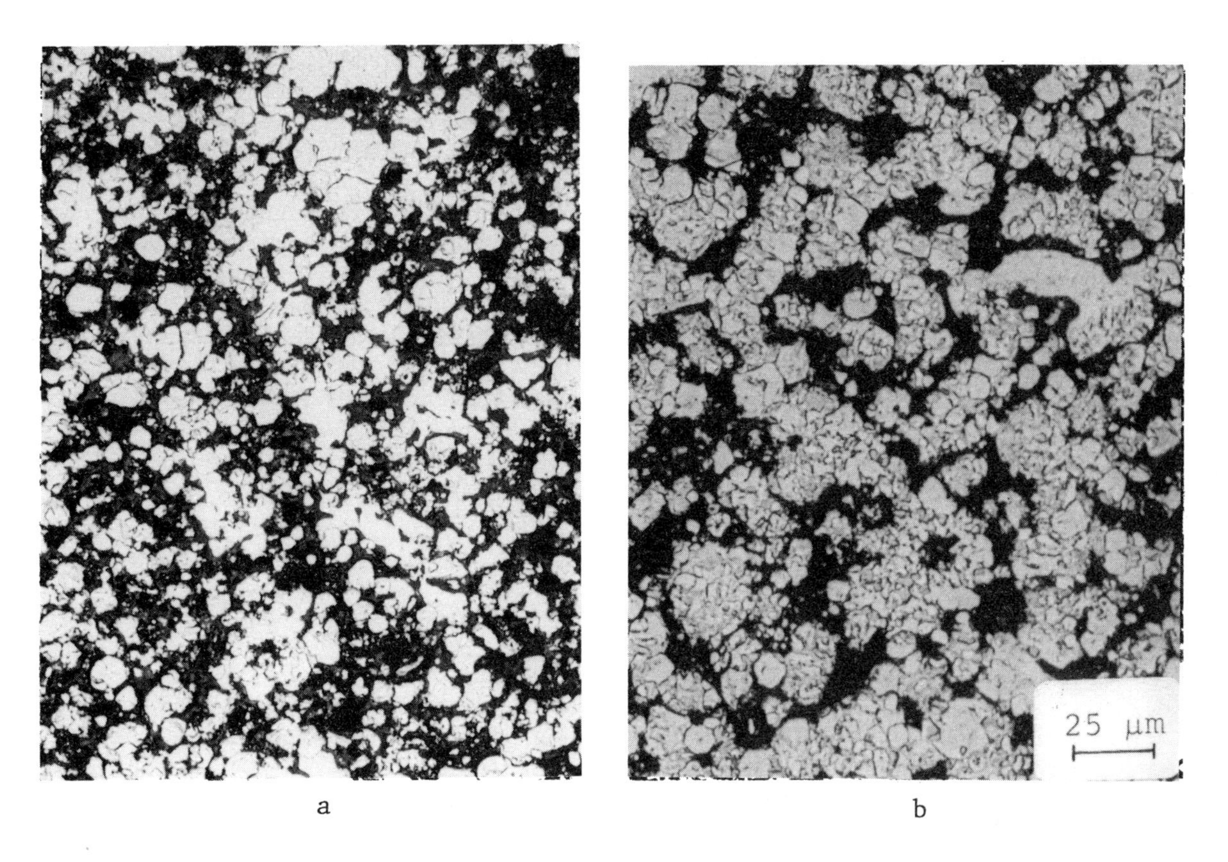

Figure 12

During the final stage of sintering the migration of grain boundaries may be retarded by second phase inclusions. If the bulk diffusion of the host material in the second phase inclusions is rapid the second phase inclusions will migrate with the grain boundaries and coalesce similarly to pores. Figure 13a shows W-1wt.%Ni after sintering at 1400°C for 122 h. The "bell shape" of the inclusions indicates the drag of the moving boundaries on the second phase. As a consequence of the drag forces of the inclusions on the migrating boundary the grain growth rate is reduced if the Ni concentration exceeds the solubility limit of Ni in W at 1400°C (Fig.13b) [41].

References

1. W.B. James, in Powder Metallurgy 1986 - State of the Art, (eds) W.J. Huppmann, W.A. Kaysser and G. Petzow, Verlag Schmid, Freiburg, p.71, 1986.
2. G. Petzow and W.A. Kaysser, in Sintered Metal - Ceramic Composites, (ed.) G.S. Upadhyaya, Elsevier Science Publishers, Amsterdam, 51, 1984.
3. E.Klar, in Metals Handbook, Ninth Edition, 7, Powder Metallurgy, coordinator E.Klar, American Society for Metals, Metals Park, Ohio, 24, 1984.
4. G.C. Kuczynski, Trans AIME, 185, p.169, 1949.
5. M.Mitkov, M. Aslan and W.A. Kaysser, Powder Met. Int, 21, p.7, 1989.
6. G.C. Kuczynski, Z. Metallkde., 67, p.606, 1976.
7. R.L. Coble, in Proc. Second Conference on Sintering and Related Phenomena, Notre Dame 1965, Gordon & Breach, New York, p.423, 1967.
8. R.J. Brook, J.Am.Ceram.Soc., 52, p.56, 1969.
9. C.H. Hsueh, A.G. Evans and R.L. Coble, Acta Metall., 30, p.1269, 1982.
10. D.S. Wilkensen and M.F. Ashby, Acta Metall., 23, p.1277, 1975.
11. B.H. Alexander and R.W. Balluffi, Acta Metall., 5, p.666, 1957.
12. H.E. Exner, Grundlagen von Sintervorgängen, in Materialkundliche Stuttgarter Technische Reihe, (Hrsg.) G. Petzow, Gebr. Borntraeger, Berlin, 4, 1978.
13. J.Frenkel, J. Phys. USSR, 9, p.385, 1945.
14. W.D. Kingery and M. Berg, J. Appl. Phys., 26, p.1205, 1955.
15. G.R. Rockland, Z. Metallkde., 58, p.476, 1969.
16. R.L. Coble, J. Appl. Phys., 32, p.787, 1961.
17. R. Brook, S.P. Howlett and Su Xing Wu, in Sintering - Theory and Practice, (eds) D. Kolar, S. Pejovnik and M.M. Ristic, Material Science Monographs, Elsevier Scientific Publishing Company, Amsterdam, p.135, 1982.
18. W.A. Kaysser, G. Petzow and M. Mitkov, in Ceramics and Metals; Proceedings of the VIII Yugoslav-German Meeting on Materials Science and Development 1987, Brdo pri Kranju, Yugoslavia, (eds) D. Kolar, M.Kosec and J.Krawczynski, KFA-Julich, ISBN 3-89336-001-8, p.47-61, 1988.
19. K. Uematsu, R.M. Cannon, R.D. Bagley, M.F. Yan, U. Chowdry, and H.K. Bowen, in Int. Symposium of Factors in Densification of Oxide and Non-Oxide Ceramics, (eds) H. Saito and S. Somiya, Hakone, Japan, 190, 1978.
20. M.F. Ashby, Acta Metall., 22, p.275, 1974.
21. V. Smolej, Z.Metallkde., 74, p.689, 1983.
22. H. Palmour III and T.M. Hare, in Sintering 85, (eds) G.C. Kuczynski, D.P. Uskokovic, H. Palmour III, and M.M. Ristic, Plenum Press, New York, 12, 1987.
23. W.A. Kaysser and I.S. Ahn, in Modern Developments in Powder Metallurgy, (eds) P.U. Gummeson and D.A. Gustavson, 18, p.235, 1988.
24. C. Zener, Private communication to C.S. Smith, Trans.AIME, 175, p.15, 1949.
25. M.P. Harmer, E.W. Roberts, and R.J. Brook, Trans. J. Brit. Ceram. Soc., 78, p.22, 1979.
26. M.P. Harmer and R.J. Brook J. Brit. Ceram. Soc., 80, p.147, 1981.
27. M. Hillert, Acta Metall., 13, p.227, 1965.
28. J.W. Cahn, Acta Metall., 10, p.789, 1962.
29. K. Lücke and H.P.Stüwe, in Recovery and Recrystallization of Metals, (ed.) L. Himmel, John Wiley and Sons, New York, p.171, 1963.
30. S.J. Bennison and M.P. Harmer, J. Am. Ceram. Soc., 68, C22, 1983.
31. W.A. Kaysser, M. Sprissler, C.A. Handwerker, and J.E. Blendell, J. Am. Cer. Soc., 70, 339, 1987.
32. W.A. Kaysser, in Ceramic Powder Science, (eds) G.L. Messing, E.R. Fuller and H. Hausner, Ceramic Transactions 2, Am. Ceram. Soc, Westerville, OH, p.955, 1988.
33. W.A. Kaysser, M. Hofmann-Amtenbrink and G. Petzow, Powder Metallurgy, 8, p.199, 1985.
34. J.H. Brophy, L.A. Shepard, and J. Wulff, in Powder Metallurgy, (ed.) W. Leszynski, AIME-MPI, Interscience, NY., p.113, 1961.
35. I.J. Toth and N.A. Lockington, J. Less-Com. Met., 12, p.353, 1967.

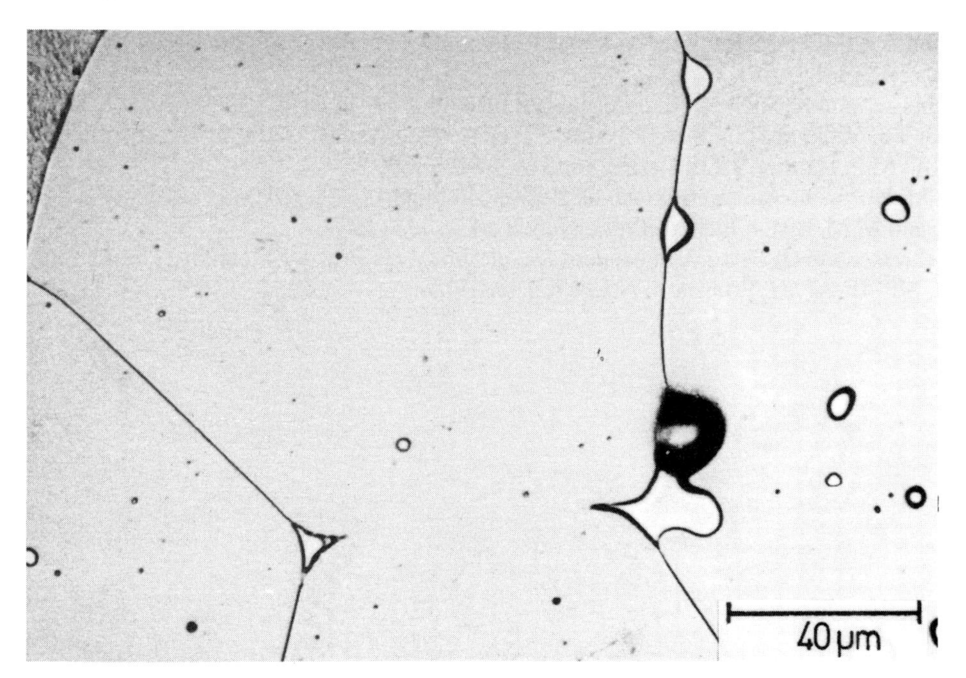

Figure 13a Second phase drag and separation of second phase inclusions from the boundary in W-1wt.%Ni after sintering at 1400°C for 122 h.

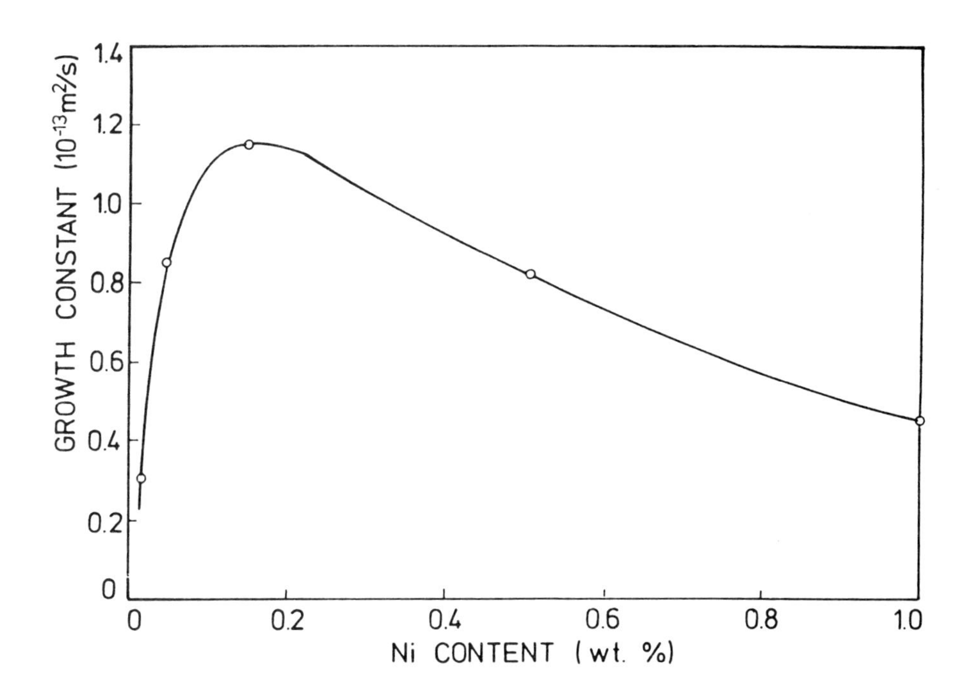

Figure 13b Grain growth constants of W-Ni compacts during sintering at 1250°C.

36. R.M. German and Z.A. Munir, in Reviews on Powder Metallurgy and Physical Ceramics, Publishing House Freund, Saint-Saphorin, 2, p.9, 1982.
37. V.W. Schintelmeister and K. Richter, Planseeberichte für Pulvermetallurgie, 13, p.3, 1970.
38. F.A. Kroger and V.J. Vink, in Solid State Physics, 3 , (eds) F.Seitz and D. Turnbull, Academic Press, p.307, 1966.
39. S.J. Bennison and M.P. Harmer, J.Am.Ceram.Soc., 66, C90, 1983.
40. W.A. Kaysser, M. Hofmann-Amtenbrink and G. Petzow, in Sintering 85, (eds) G.C. Kuczynski, D.P. Uskokovic, H. Palmour III, and M.M. Ristic, Plenum Press, New York, p.121, 1987.
41. J.S. Lee,W.A. Kaysser and G. Petzow, in Modern Developments in Powder Metallurgy, (eds) E.N. Aqua and C.M. Whitman, MPFIF-APMI, Princeton, N.J., 17, p.99, 1985.

13

Liquid Phase Sintering

W.A. KAYSSER

Max-Planck-Institut für Metallforschung, Institut fur Werkstoffwissenschaft, Stuttgart, Germany

Abstract

Pressureless sintering of metals, ceramics or composites often requires the presence of a liquid phase to eliminate all porosity or to achieve a desired degree of chemical homogenization. The shrinkage and the microstructural development during liquid phase sintering of almost all materials are governed by a number of basic principles, which will be outlined and illustrated by experimental results from sintering metal, ceramic, metal-ceramic and ceramic-glass systems. Shrinkage will be related to primary and secondary rearrangement, with major emphasis on estimating the effects of coarsening versus contact flattening, and the effects of melt penetration and liquid film migration. Swelling and chemical homogenization of structural parts are explained by basic principles of melt penetration leading to contact and particle dilatation. Melt penetration along grain boundaries also leads to grain boundary alloying and liquid film migration, which can improve the mechanical properties of sintered materials considerably.

1. Introduction

Similar to PM products obtained by solid state sintering, products obtained by liquid phase sintering may be classified into parts with functional inherent porosity, materials and structural parts. Whereas liquid phase sintering of parts with functional porosity is an exception, almost all materials and the majority of the structural parts are sintered in the presence of a liquid phase.

Powder metallurgy provides compositional and microstructural control in forming multiphase or composite materials which is superior to the casting method. A classic example of a PM composite material is the hard metals with a very hard carbide phase embedded in a softer metal matrix [1]. More recent examples are aluminium alloys reinforced by SiC fibres or Si_3N_4 reinforced with SiC whiskers [2,3]. Powders of the different phases are mixed and consolidated by pressureless sintering or by pressure consolidation (hot pressing, warm extrusion, hot isostatic pressing) to full density [4]. In the more economic pressureless sintering often the presence of a liquid phase is required to eliminate all porosity [5]. The microstructural development and the shrinkage of PM materials during sintering in the presence of a liquid phase are governed by a number of basic principles, which apply to liquid phase sintering of almost all materials [6,7]. These principles will be outlined and illustrated by experimental results from sintering metal, ceramic, metal-ceramic and ceramic-glass systems.

Sintered structural parts are usually in economic competition with similar parts manufactured by casting and machining. To avoid machining, the size and shape of structural PM parts after sintering must fit the required tolerances after a simple coining operation. The dimensions of structural parts are well controlled if shrinkage or swelling during sintering are negligibly small. To obtain alloys of adequate strength, green compacts are pressed from mixtures of different unalloyed powders. Homogenization during sintering is a necessary precondition to obtain reasonable mechanical properties by solid solution hardening, precipitation hardening or the formation of martensite. Again, dimensional control and the required homogenization are often achieved by the presence of a liquid phase. The melt accelerates the material transport during sintering and compensates part of the shrinkage by swelling induced by melt penetration between particles and along grain boundaries [6].

2. Shrinkage Phenomena

2.1 Basic Principles

Densification during liquid phase sintering is based on re-arrangement and shape change of the solid constituent particles (Fig. 1). [8] The driving force for both phenomena results from a decrease in energy resulting from the changes in the areas of the liquid/vapour, liquid/solid and solid/vapour interfaces during re-arrangement and/or shape change. Melt may originate from low melting point particles or areas which have formed low melting point compositions by diffusion during heating. With the condition of good wetting, the liquid phase is pulled by capillary forces into particle necks and small pores. As a reaction to the capillary forces, particles are re-arranged if their mobility allows it. This re-arrangement of the solid particles is a necessary pre-requisite for effective densification throughout the sintering process even during later sintering stages, when, for example the rates of particle shape change also limit the re-arrangement rate. To distinguish clearly between re-arrangement, which is essentially rate controlled by the mechanical movement produced by the capillary forces, and re-arrangement, which is rate controlled by dissolution reprecipitation processes, these particle movements are described as primary and secondary re-arrangements [6]. Shrinkage due to the shape change of the solid constituents always involves solution-reprecipitation processes and is therefore connected to secondary rearrangement in most cases.

2.2 Driving Forces

A number of calculations show that high capillary forces are present when the amount of melt phase is small [9,10]. Pores which are much larger than the grains may exert no or small capillary forces if the wetting angle is >0 deg [11,12]. As a rule of thumb, the capillary pressure P in the melt resulting from a pore is estimated by $P = 2*\Gamma_{lv}/r_p$. Γ_{lv} is the specific energy of the liquid/vapour interface and r_p is the radius of the pore. If the melt is interconnected and at rest, then the capillary pressure is equal in the whole system [12].

2.3 Primary Re-arrangement

In systems with homogeneously distributed liquid phase contents of less than 10 vol.% primary re-arrangement leads essentially to a local densification of a limited number of particles by movement of individual particles in directions of higher liquid bridge coordination [13]. The local densification provides macroscopic densification of a much lower degree, hence areas of increased particle density form together with adjacent areas of reduced particle density. Areas of higher particle density soak in additional liquid from areas of lower particle density which results in further enhanced densification of areas which are already denser than their vicinity.

Figure 2a and b show the formation of denser and less dense areas as well as the formation of a crack, due to re-arrangement during liquid phase sintering of a monolayer of Cu-coated W spheres (120 μm) [14]. Primary re-arrangement can be observed particularly well during solid state sintering of systems consisting of particles with a high melting point such as W and a second phase of high mobility (diffusivity) such as Cu. Despite the mutual insolubility, W-Cu compacts densify by solid state sintering. A massive spreading of Cu from initial Cu particles leads to a local formation of larger pores and denser Cu-W aggregates resulting in continuous macroscopic shrinkage after prolonged sintering when most aggregates form an interconnected network (Fig. 2c) [15]. Figures 2d and e show the inhomogeneous density distribution due to liquid flow and primary rearrangement in liquid phase sintered W-20vol.%Cu and Al_2O_3-5vol.% anorthite.

3. Secondary Re-arrangement: Coarsening Effects Versus Contact Flattening

3.1 Coarsening

Grain growth during liquid phase sintering of PM materials determines their densification and their properties after processing. In many systems grain coarsening follows the relation:

$$G_m{}^3 = (G_{mo}{}^3 + kt)^{1/3} \tag{1}$$

which indicates diffusion-controlled Ostwald ripening as the rate controlling step [16,17]. G_{mo} and G_m are the initial and the actual average particle size. Experiments with Fe-20wt.%Cu and Fe-30wt.%Cu samples also showed that during liquid phase sintering the particle size distribution, D(S), developed a steady state form which could be described by:

$$D(S) = 2.136\, S^2 \exp(-0.712\, S^3) \tag{2}$$

where 2S is the reduced particle size, G/G_m [17].

When Price, Smithells and Williams published their famous micrographs of a W-Cu-Ni heavy metal alloy in 1937 they already recognized that grain growth and shape change of the solid constituents had occurred simultaneously [18]. From recent experiments with mixtures of small tungsten powder (10 μm) and large single crystal W seeds (220 μm) that were liquid phase sintered in the presence of a Ni-rich melt, this idea was rejuvenated [19]. It was emphasized from the microstructural development of these samples that shrinkage is directly related to grain growth and shape accommodation. During liquid phase sintering the initial spherical shape of the single crystal seeds changed toward polyhedral shapes (Fig. 3). Special etching showed the shape change of the spheres to be due only to the precipitated material [19], that was supplied by dissolving smaller particles. Calculations of microstructural development showed similar general features as found in the liquid phase sintering experiments [20]. The calculations were conducted with the assumption that the large particles were taking part in the usual Ostwald ripening of the arrangement. No pressure in the flat contact areas was assumed beyond a value which kept the thickness of the thin liquid films in the contact areas between the particles at a constant value. As seen already from the experiments, the calculation showed that the deposition of material in the contact areas is extremely small in this system of large and small particles (Fig. 3).

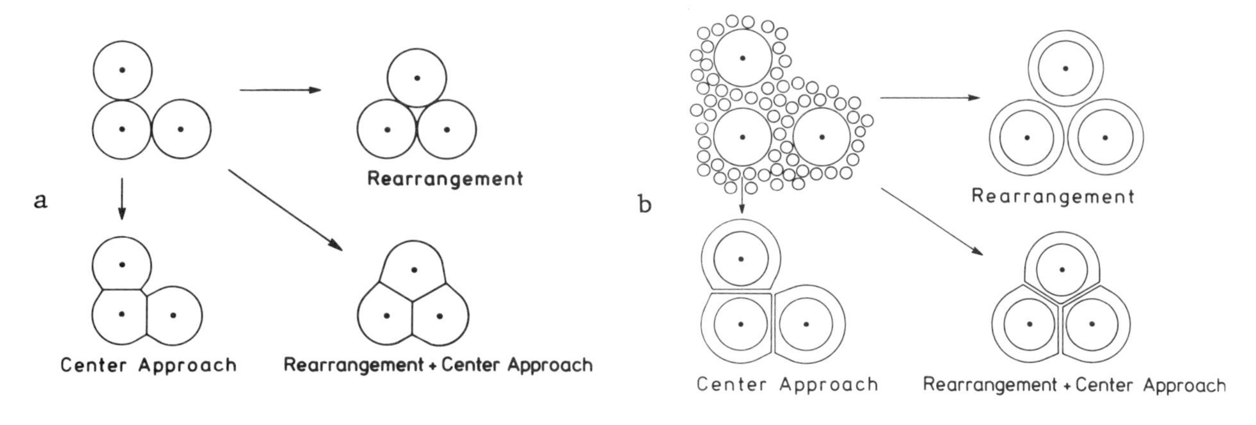

Figure 1 Shrinkage by rearrangement and centre approach (shape change) during liquid phase sintering. (a) Monosized grains without coarsening. (b) Grains with broad size distribution and coarsening.

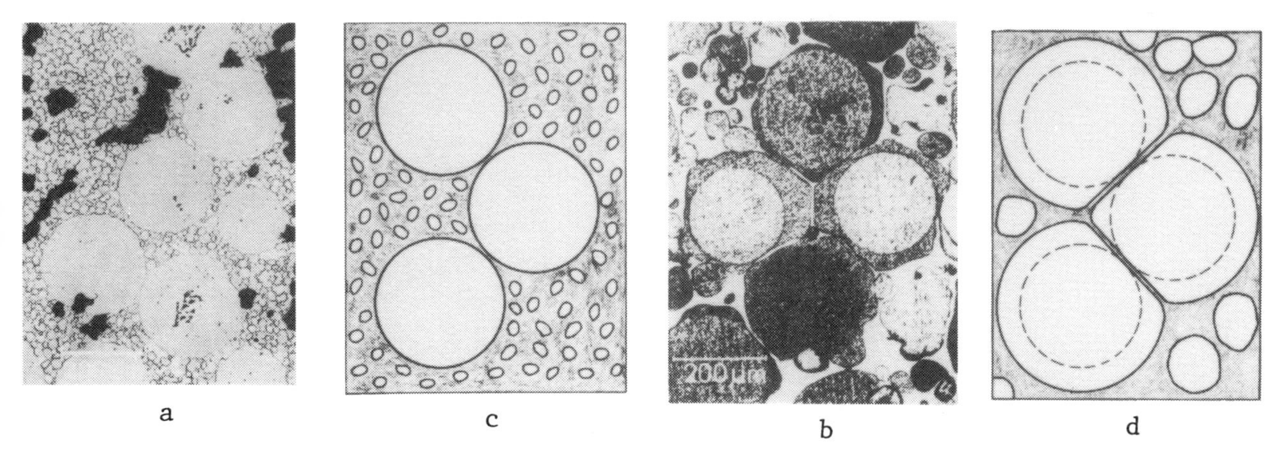

Figure 3 Grain growth and shape accomodation during liquid phase sintering of W-Ni at 1670°C. (a),(c) Microstructure after 3 and 120 min. (b),(d) Calculated microstructure after 1 and 120 min.

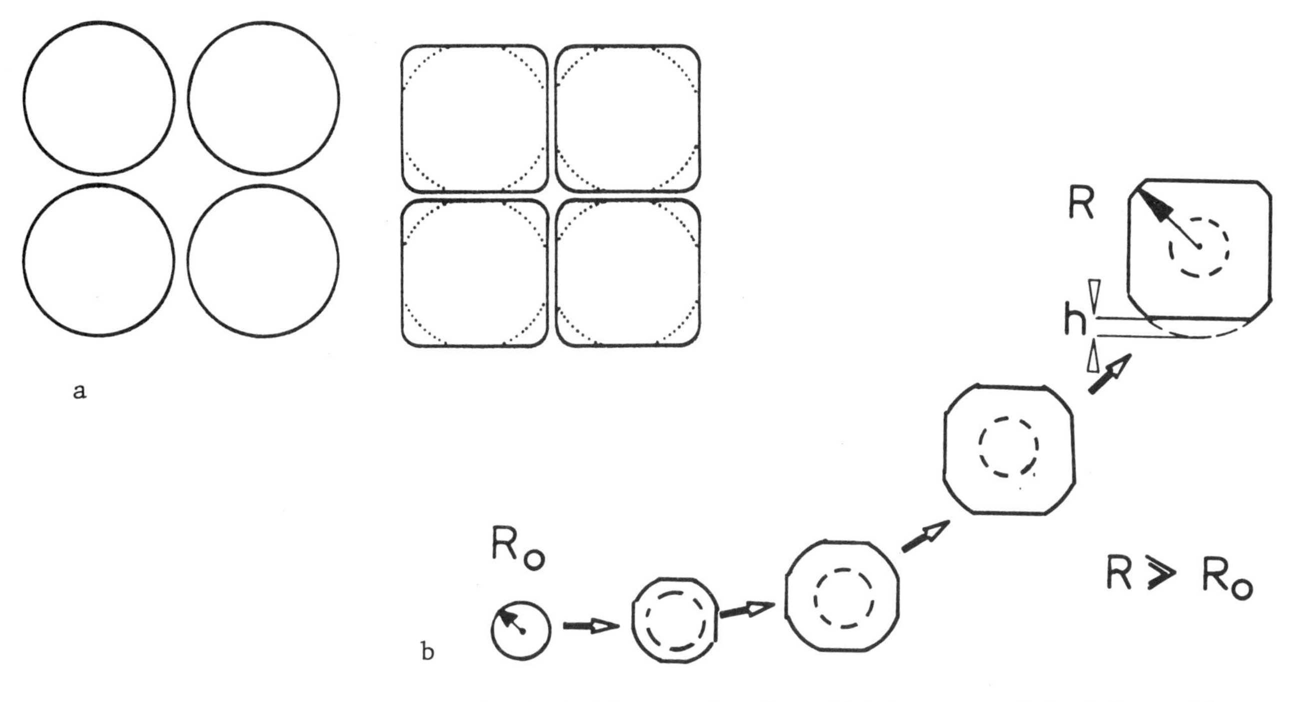

Figure 5 Shape change during liquid phase sintering by (a) contact flattening and (b) shape accomodation during particle growth.

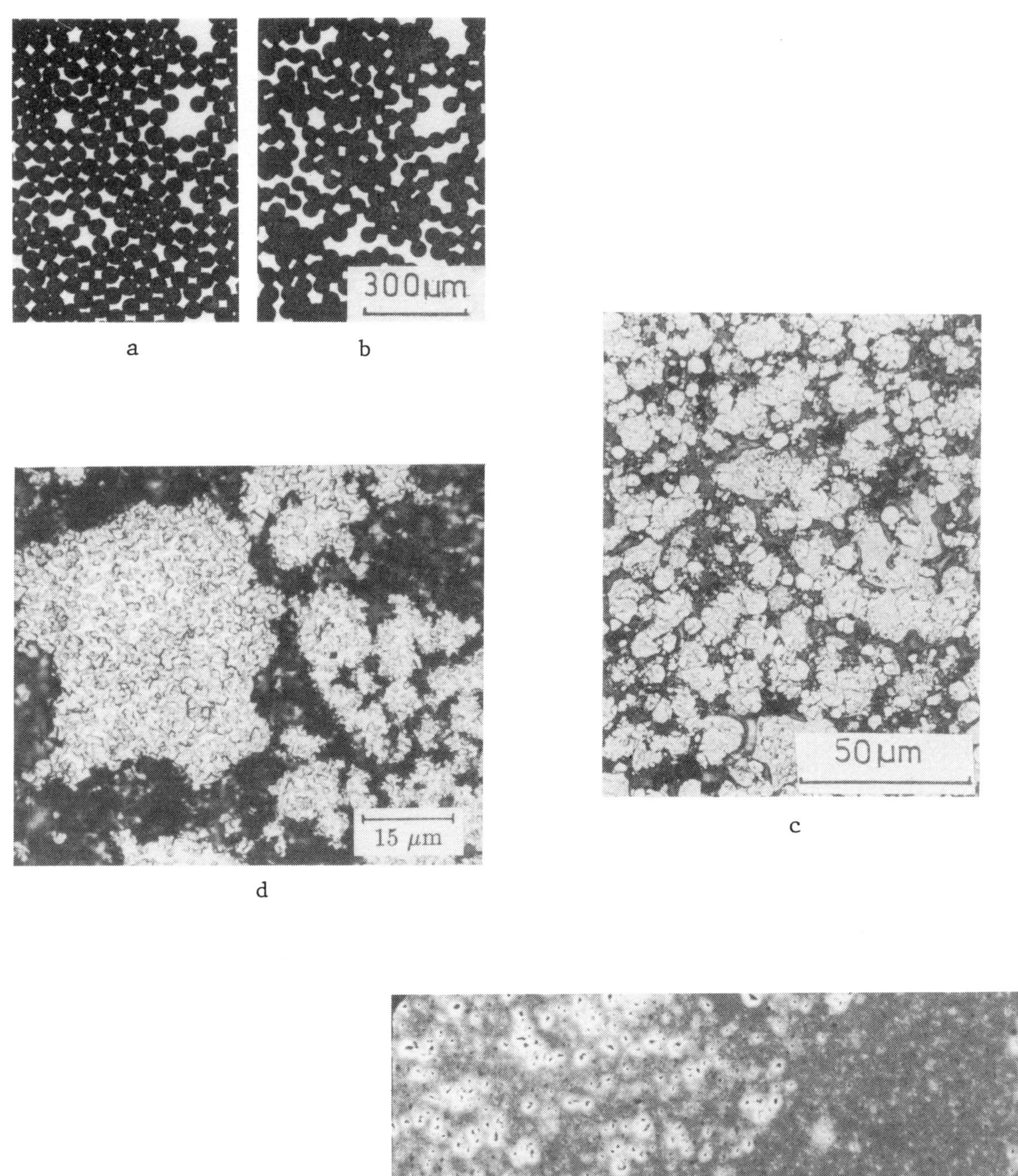

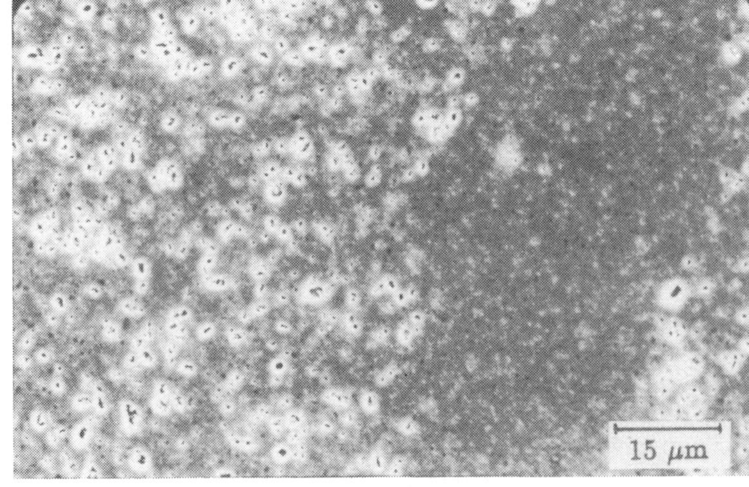

Figure 2 Inhomogeneous densification and liquid phase distribution after primary rearrangement. (a) Initial state of a monolayer of Cu-coated W spheres (120 µm). (b) Monolayer after liquid phase sintering at 1100°C for 4 min.[13]. (c) Aggregate formation during solid state sintering of W-20vol.%Cu at 1000°C for 5h [14]. (d) Microstructure of W-20vol.%Cu after liquid phase sintering at 1150°C for 10 min. [14]. (e) Microstructure of Al_2O_3-5vol.% anorthite after sintering at 1500°C in air for 6 min.[15].

During liquid phase sintering of a system with a steady state particle size distribution, similar shape accommodation was observed. During coarsening many grains rapidly accommodated their shape to a changing vicinity (Fig. 4a) [21]. Even inert obstacles did not hinder further grain growth and were entrapped into growing grains [22]. This extreme of shape accommodation to an adjacent particle is shown in Fig.4b in Mo-Ni-Al_2O_3. The shape development of the Mo grains during annealing is indicated by etching boundaries. During liquid phase sintering with 2wt.%Ni at 1460°C the Mo grains trapped small Al_2O_3 spheres (black). No evidence of contact flattening was found.

3.2 Contact Flattening

Contact flattening is a classical approach to explain shape change and shrinkage during liquid phase sintering [23]. The contact flattening assumes thin liquid films in contact areas separating the particles (Fig. 5a). All particles transmit forces caused by the presence of pores. Close to the contact areas and through the thin liquid layers, the forces between the particles are thought to be transmitted whereby the stress is intensified in the contact areas compared to other areas of the particles. Thermodynamically, the increased pressure usually results in an increased solubility of the solid phase in the liquid. The concentration gradient due to the stress gradient in the liquid film and the adjacent solid areas yields a flow from the contact areas toward other lower stressed areas of the melt or the solid. In most cases the flow through the melt is assumed to be diffusion controlled. The continuous flow of dissolved material leaving the contact areas leads to a shape change of the particles and to the centre approach of the particles.

Although many researchers have tried to find clear microstructural evidence for contact flattening as a major mechanism during liquid phase sintering, no unambiguous results have been published up to now. One of the major criticisms is that grain growth is completely neglected in the classical approach; although grain growth is much faster than the predicted contact flattening rates. A detailed calculation which involved all assumptions of the classical contact flattening theory but included in addition rapid grain growth showed that the contact flattening approach is incorrect [17]. The results of those calculations which included grain growth were in sharp contradiction to the variety of experimental results for those systems where shrinkage and grain growth during sintering had been measured.

3.3 Coarsening and Shrinkage

It must be concluded that shape accommodation during sintering occurs during grain coarsening and is due to a small re-precipitation rate at particle contacts (Fig. 5b). Shrinkage may be vaguely described as secondary re-arrangement which occurs when small particles are dissolved giving way for the movement of larger particles (Fig. 1b). The model of contact flattening cannot describe the experimental observations of shape accommodation. If calculations based on the contact flattening concept, which include the observed rapid grain growth, were performed, then the results are in sharp contradiction with the experimentally measured shrinkage behaviour.

4. Secondary Re-arrangement: Melt Penetration Effects

Grain boundary penetration by the liquid leads to skeleton disintegration, but also to contact dilatation, particle dilatation and subsequently to particle disintegration. Important consequences of grain boundary penetration are boundary alloying and liquid film migration.

4.1 Skeleton Disintegration

An example of skeleton disintegration is shown in Fig. 6. In many systems a rigid skeleton of small particles forms as a result of *solid state* sintering during heating to the temperature of liquid formation. When, in a Fe-30wt.%Cu material the Cu melts, it is instantaneously drawn into the small pores between the 2 μm Fe-particles which form a rigid skeleton. No shrinkage or change of particle morphology is observed for times up to 2 min. This period ends with the collapse of particle chains as the melt penetrates along the grain boundaries, leading to rapid shrinkage of the sample by secondary re-arrangement within the specimen [6].

4.2 Grain Detachment and Particle Disintegration

Early grain boundary attack does not necessarily lead to densification. But if grain boundary attack is completed along all grain boundaries of a polycrystalline particle, particle disintegration and densification by secondary re-arrangement may occur. This is illustrated in Fig. 7a to c. In Al_2O_3-glass mixtures the shrinkage after particle disintegration due to re-arrangement could be separated from shrinkage due to other primary re-arrangement or solution-re-precipitation processes [6]. Up to 1400°C, the shape and size of the polycrystalline Al_2O_3 spheres was maintained and only primary re-arrangement occurred (Fig. 7b). Above this temperature grain boundary attack and particle disintegration occurred (Fig. 7c) and secondary re-arrangement led to considerable additional shrinkage (Fig. 7a) [7,24].

5. Swelling by Melt Penetration

The attack of grain boundaries by a melt during sintering has been most extensively studied in the Fe-Cu system in which compacts of Fe spheres with 10vol.%Cu spheres (Fig. 8a), both of 100 μm diameter were sintered at 1165°C for times up to 60 min. [25]. After annealing for 3 min. above the melting point of Cu, all contact regions between the Fe particles

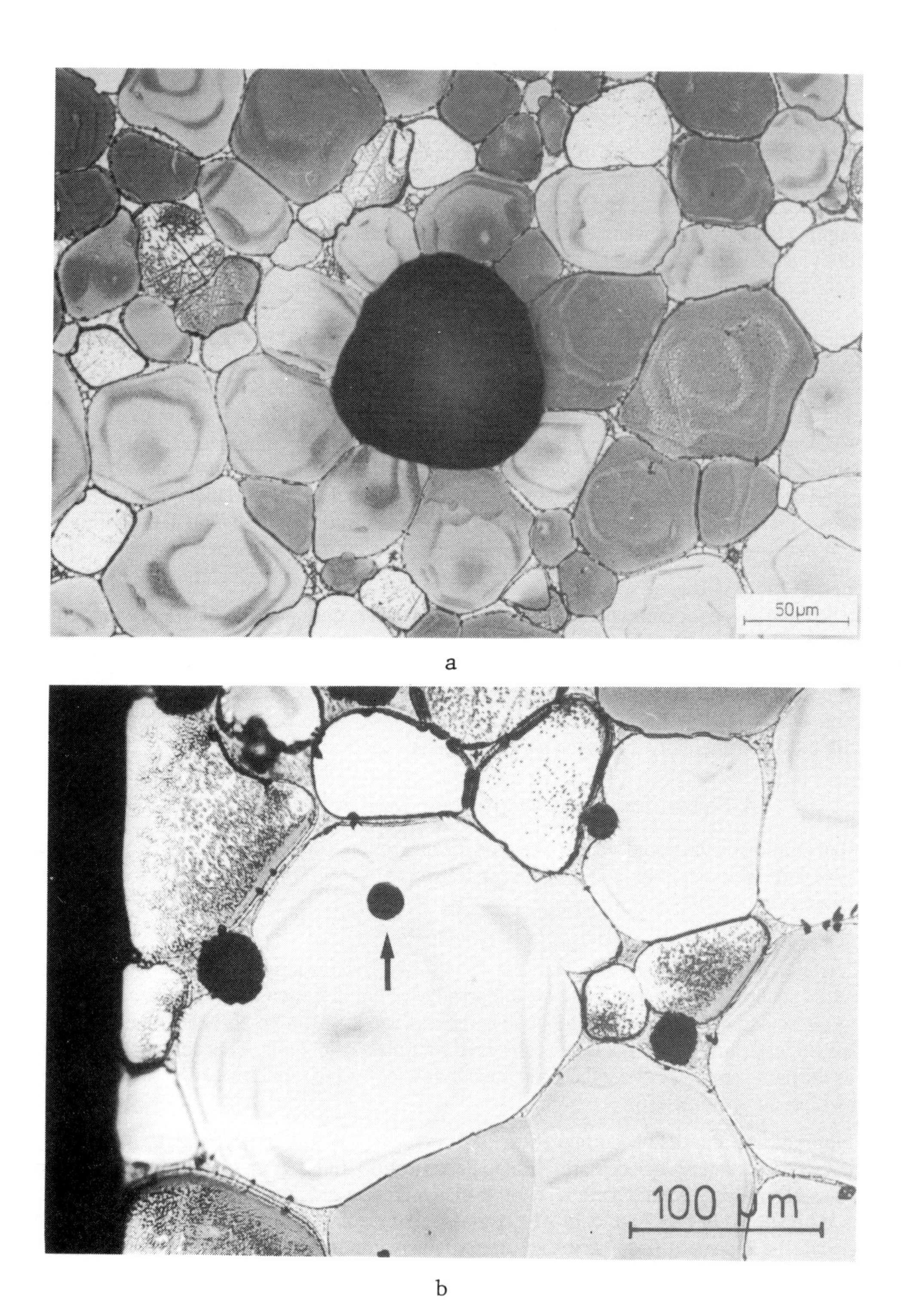

Figure 4 Microstructure of Mo-4wt.%Ni after cyclic liquid phase sintering between 1300 and 1460°C. Held at 1460°C for 30 min in each cycle. Etched with Murakami's solution. (a) Microstructure around a pore [21]. (b) Microstructure around Al_2O_3 spheres deliberately added to the Mo-Ni powder mixture before sintering [22].

were filled with liquid (Fig. 8b). After 8 min., grain boundary attack had led to severe changes in the contact area geometry; instead of flat interparticle contact areas, wavy layers of liquid were present and bumps consisting of Fe(Cu) solid-solution formed opposite each grain boundary. The additional melt in prior contact areas and at prior grain boundaries in the interior of the Fe particles caused contact and particle dilatation leading to macroscopic swelling [25]. A comparison between the microstructural and the macroscopic dimensional changes allowed the quantitative prediction of the swelling behaviour of Fe-10wt.-%Cu alloys during liquid phase sintering (Fig. 9).

6. Homogenization Effects of Melt Penetration

6.1 Boundary Alloying

Melt penetration along grain boundaries results in preferential alloying of areas close to the prior grain boundaries. If the annealing time is short and bulk diffusion is slow, then unexpected phases may be present at the boundaries. These phases can determine the mechanical properties of liquid phase sintered composite materials as shown in the subsequent example. An Fe-10Cu-0.1C composition was mixed from spherical, polycrystalline Fe-0.011C (100 μm) and from spherical Cu (100 μm) powders. The compact (400 MPa) was sintered in the presence of a liquid phase at 1110°C for 30s and cooled in the furnace. The white cementite at the boundaries in the interior of the Fe particle and the black Cu areas are clearly visible (Fig. 10). During the short liquid phase sintering time Fe-Cu austenite has only formed in the intimate vicinity of the penetrated thin melt layers, whereas the regions in the interior of the grains remained Cu-free. Due to the fact that the formation of ferrite and pearlite in Fe-Cu austenite occurs at lower temperatures than in Cu-free austenite [26], formation of ferrite starts in the interior of the grains. The low solubility of carbon in ferrite results in an enrichment of carbon in the austenitic area of each grain which remains during growth of the ferrite phase. The last remaining austenite areas are the Fe-Cu solid solution layers in the vicinity of the boundaries. The enrichment of carbon in these layers results in a degenerated pearlite formation in the boundary areas. It should be noted that the sample contains only 0.1wt.% carbon.

The penetration of melt along grain boundaries may also initiate liquid film migration. Liquid film migration is a more general phenomenon, however, which also occurs under much different conditions. Therefore liquid film migration will be described in a separate section.

6.2 Liquid Film Migration

In recent years boundary migration in metals and ceramics, which is associated with a noticeable concentration change in the wake of the moving boundaries, has been investigated intensively. It is known as diffusion induced grain boundary migration (DIGM) [27] and liquid film migration (LFM) [28,29]. The migrations are of considerable importance to solid state and liquid phase sintering of many composites [30].

When a well recrystallized Mo foil is annealed at 1380°C essentially no grain growth can be detected, but if a thin (unreacted) Ni film is applied to the surface (or if the sample is exposed to Ni vapour), the grain boundaries develop bulges often looking like worms or snakes [31] (Fig. 11a). The length of the grain boundary and hence the grain boundary area is increased due to the formation of the bulges. Microprobe measurements show that the Ni concentration is increased in the boundaries, indicating a thin liquid film. Behind the migrating boundaries, a saturated solid solution is formed (Fig. 11b). The Mo crystals in front of the migrating boundaries are completely free of Ni. From those measurements it became obvious that the formation of a solid solution may provide the driving force which allows the increase of the boundary area. Other measurements showed this type of boundary migration occurs when the solid solution left in the wake of the boundary has a lower free enthalpy than the material in front of the boundary [29]. Thus, the concentration in the wake of the migrating boundary may be higher or lower in solute concentration than in the front of the boundary. A second prerequisite is that the boundary provides a sink or source for the required changes of the solute.

The free enthalpy change is necessary but does not yield itself the atomistic kinetics which cause boundary migration. The present atomistic model is based on coherency strains due to changes in the lattice parameters in the solid in front of the migrating boundaries [27 - 31].

The solid in front of the migrating boundaries is not in equilibrium with the solute concentration of the boundary and solute will either deplete from or diffuse into this solid. When the boundary migrates at a constant velocity a steady state concentration profile will develop. The concentration gradients in the solid close to the liquid/solid interface will be very steep. If the lattice parameter changes with the concentration of the solute and coherency of the peripheral layers of the solid is maintained, coherency strains will arise. These coherency strains raise the free enthalpy of the solid which is in contact with the boundary. The solid in the wake of the migrating boundaries is thought to lose the coherency with the initial grain rapidly due to the incorporation of misfit dislocations.

This model was first proposed by Hillert [32] and recently refined by Handwerker *et al.* [29]. It explains nicely the effects of liquid film migration. Some modifications are required to explain the more complex stress states at grain boundaries during diffusion-induced grain boundary migration.

The penetration of melt into contact areas and grain boundaries yields perfect conditions for liquid film migration [6]. Figure 9 shows an area in an Fe particle after liquid phase sintering in the presence of Cu for 2 min. The thin liquid films

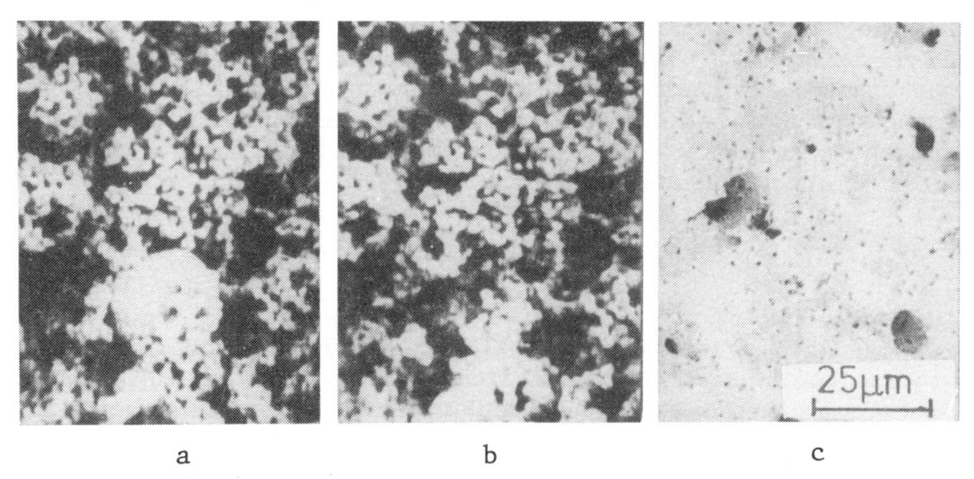

Figure 6 Direct observation of skeleton disintegration during early stage of liquid phase sintering of carbonyl Fe (< 5µm) plus 30wt.% Cu (10 µm) at 1150 °C in a SEM with hot stage device [6].

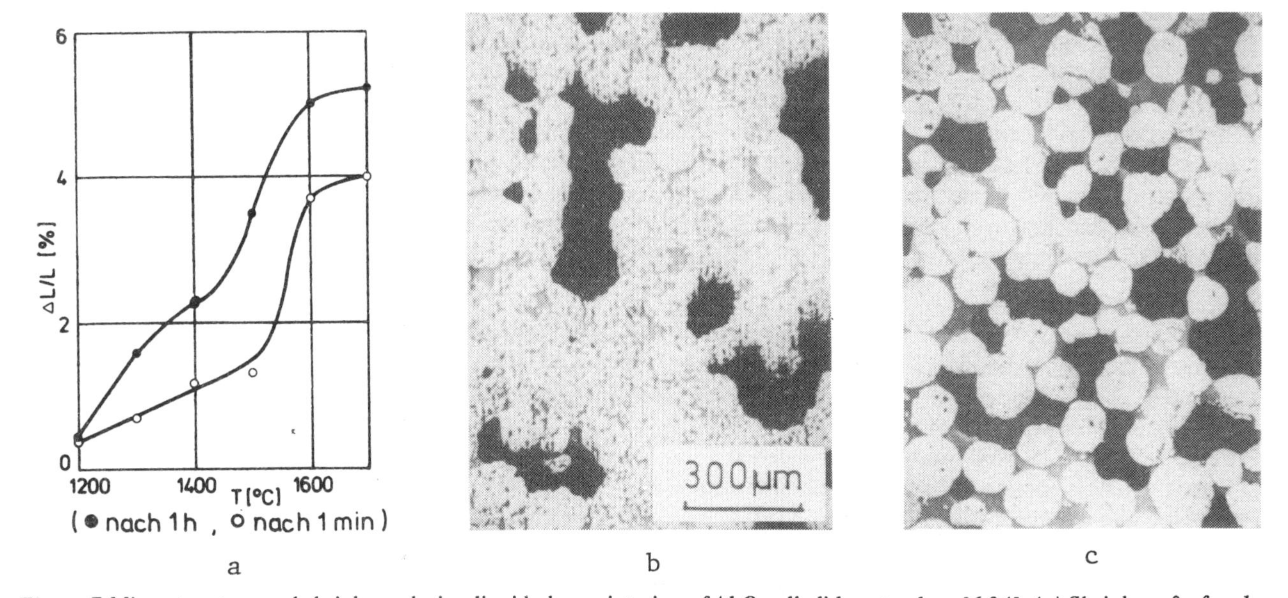

Figure 7 Microstructure and shrinkage during liquid phase sintering of Al_2O_3-alkali borate glass [6,24]. (a) Shrinkage ° after 1 min, ° after 120 min. (b) Microstructure after 1h at 1400°C. (c) Microstructure after 1h at 1700°C.

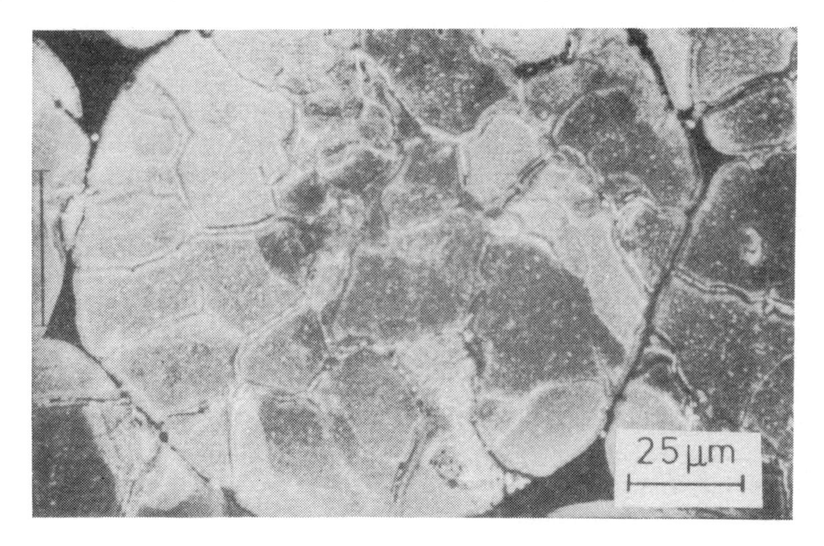

Figure 10 Microstructure of Fe-10wt.%Cu-0.1wt.%C after liquid phase sintering at 1110°C for 30s [24]. The bright phase is cementite, the dark phase is Cu (etched with Klemm3 solution [25]).

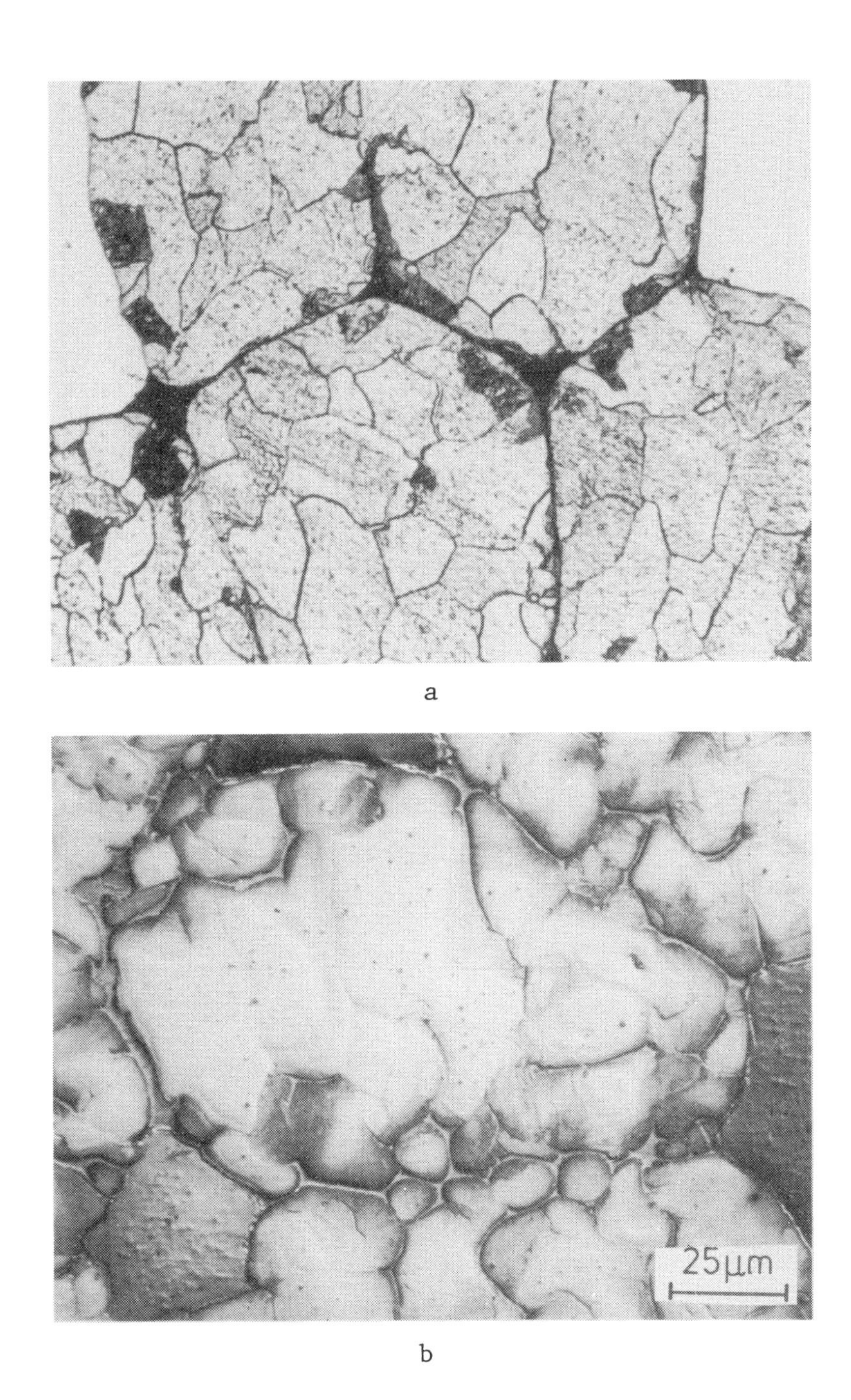

Figure 8 Grain boundary penetration during liquid phase sintering of a Fe-10wt.%Cu compact at 1165°C [25]. (a) Before sintering. (b) After liquid phase sintering for 8 min.

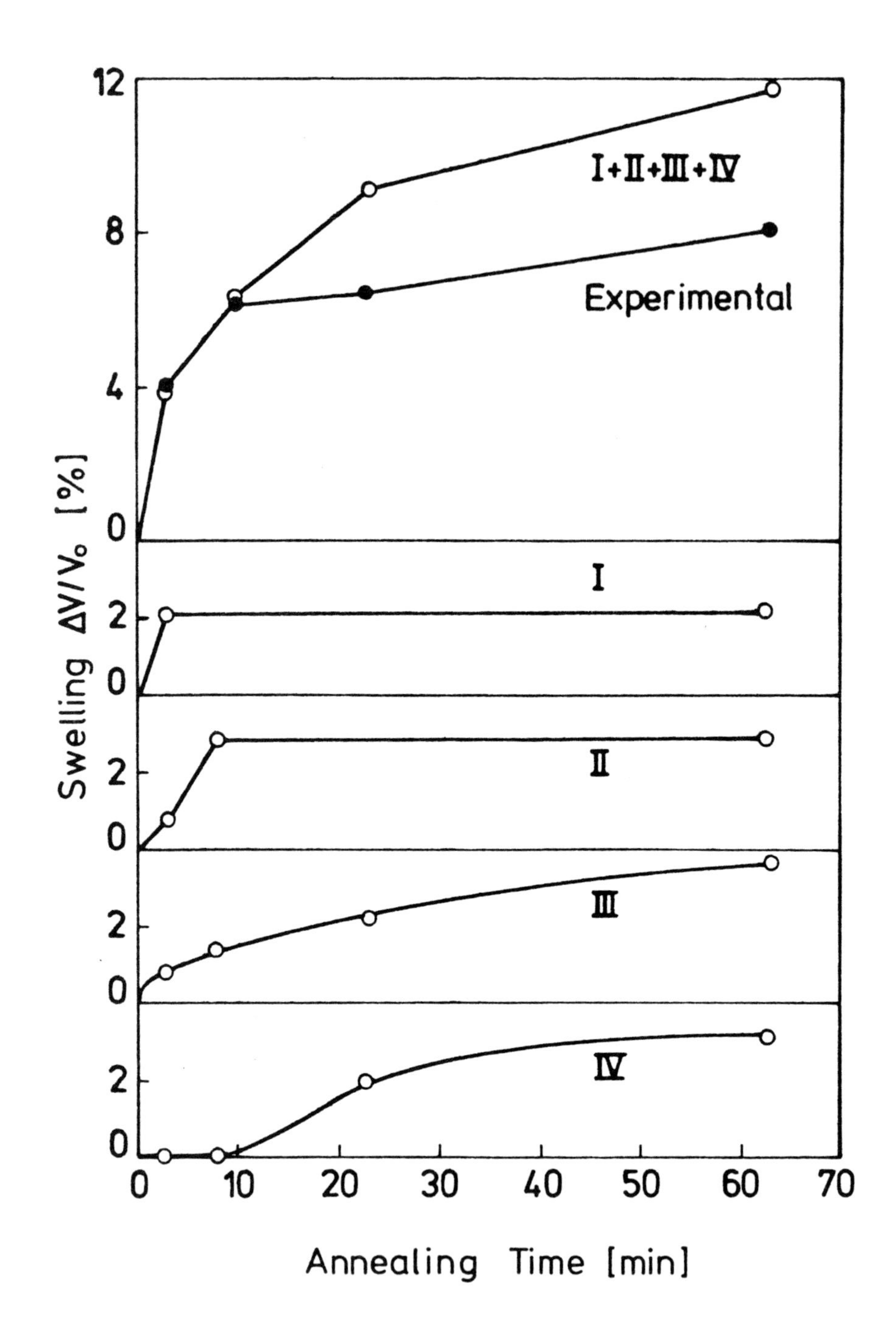

Figure 9 Dimensional changes during liquid phase sintering of Fe-10wt.-%Cu at 1165°C (compacted with 785 MPa) [25]. Calculated contributions to swelling: I Penetration of melt into particle contact areas. II Penetration of melt along particle contact areas. III Diffusion of Cu into Fe starting from contact areas. IV Diffusion of Cu into disintegrated Fe grains.

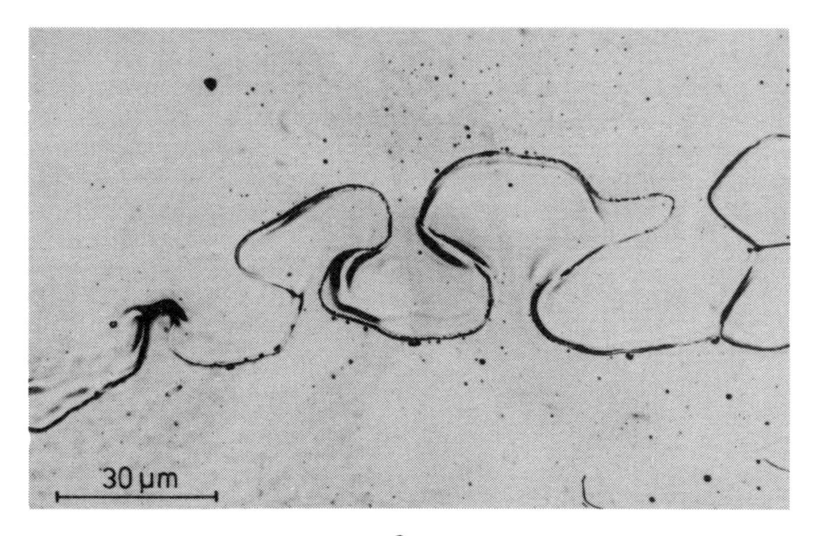

a

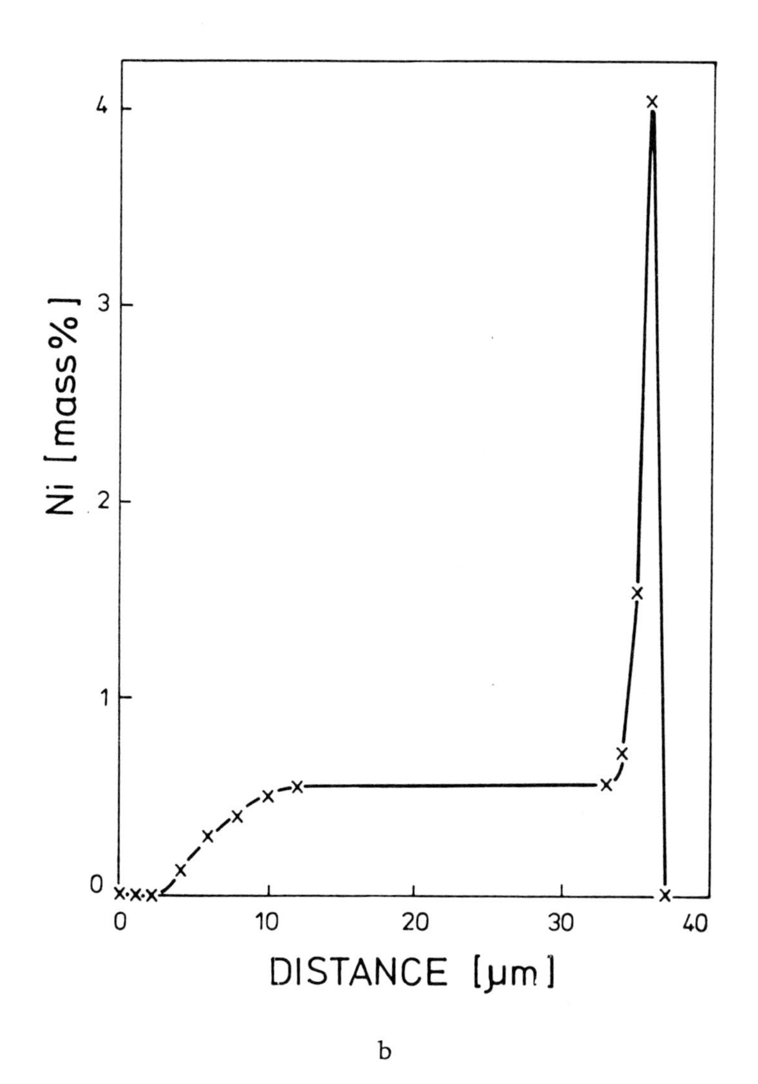

b

Figure 11 Mo foil with a thin initially unreacted coating Ni layer after sintering at 1420°C for 10 min. [28]. (a) Surface morphology. (b) Concentration profile across the migrated boundary shown in (a).

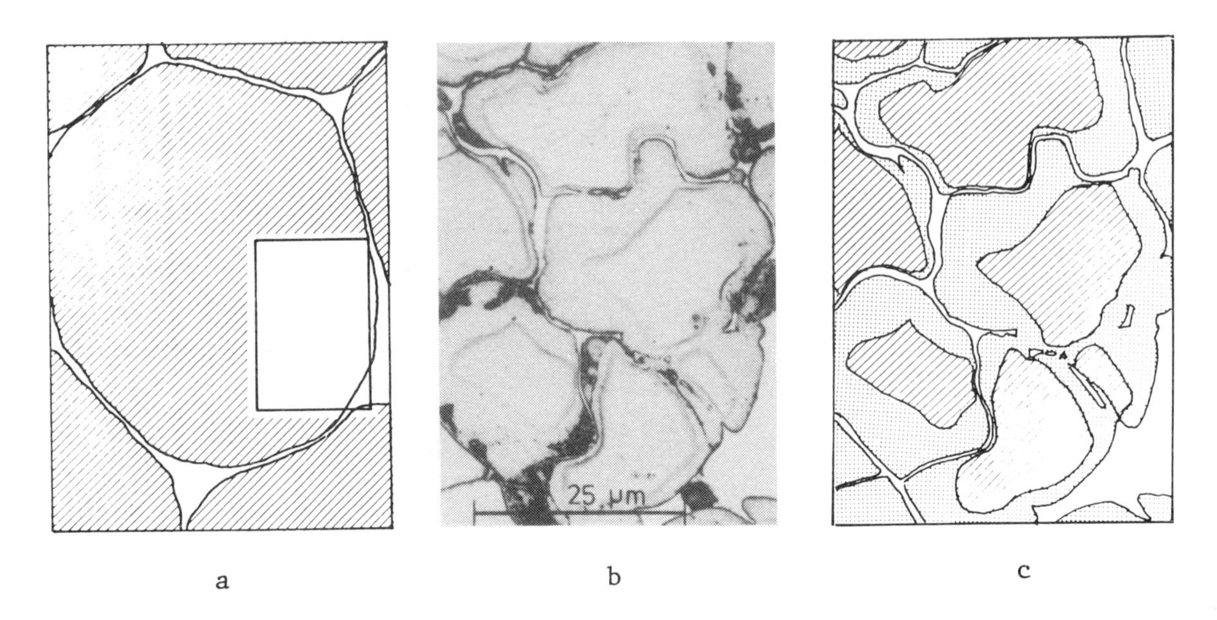

Figure 12 Liquid film migration in Fe-10wt.%Cu during liquid phase sintering at 1120°C [7,30]. (a) Overview, schematic. (b) Microstructure. (c) Formation of solid solution, schematic; white: melt, dotted: solid solution Fe(Cu), lined: unalloyed Fe.

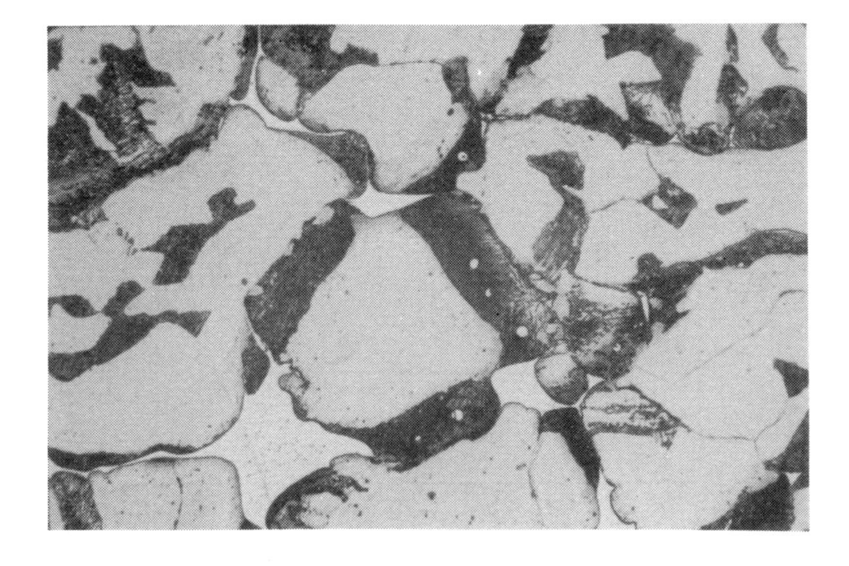

Figure 13 Liquid film migration in Fe-10Cu-0.45C sintered at 1130°C for 20 min.[6].

have formed the bulges characteristic of liquid film migration. Solid solutions of Fe-Cu have formed in the wake of the migrating boundaries [6].

It must be emphasized that the bulk diffusivity of Cu in Fe is extremely small; thus, no solid solution formation is expected by bulk diffusion during the sintering times and temperatures applied usually in practice. The rapid homogenization is obtained by liquid film migration which seems to be of great importance for this system.

In Fe-Cu the penetration of melt along grain boundaries is suppressed if carbon is present in the alloy. However, penetration of contact areas is possible. Figure 13 shows the microstructure of Fe-10Cu-0.4C after liquid phase sintering for 20 min. at 1130°C. The thin liquid layers in the contact areas are shifted by liquid film migration. In the wake of the moving boundaries a saturated solid solution of Fe-Cu is left, which mainly contains the pearlite which preferentially formed in these areas during cooling due to the fact that the eutectoid of Fe-C forms at a higher temperature than the eutectoid of Fe-Cu-C [6].

Liquid film migration influences the mechanical properties of a number of systems. It is well known that W and Mo embrittle if sintered with activators like Ni. The Ni segregates at the grain boundaries. If an appropriate heat treatment is done, then liquid film migration produces bulged grain boundaries (Fig. 14a) which mechanically interlock the grains. This tailored microstructure improves the mechanical properties considerably. Figure 14b shows the bend strength σ_B of Mo sintered as usual with Ni, and Mo sintered with Ni and heat treated to obtain liquid film migration. Similar improvements might be possible in other systems susceptible to liquid film migration such as liquid phase sintered superalloys.

7. Acknowledgement

The continuous interest of Prof. G. Petzow is gratefully acknowledged. Parts of the work were supported by the Deutsche Forschungsgemeinschaft and the Internationales Buero of KFA Juelich.

8. References

1. F. Benesovsky, Carbide, Ullmanns Enzyklopaedie der technischen Chemie, Weinheim, FRG, Verlag Chemie, 13, pp.13-30, 1970.
2. J.J. Lewandowski, C. Liu and W.H. Hunt,Jr., Microstructural Effects on the Fracture Micromechanics in 7XXX Al PM'-SiC Particulate Metal Matrix Composites, in Processing and Properties for Powder Metallurgy Composites, (eds) P. Kumar, K. Vedula and A. Ritter, The Metallurgical Society, Warrendale, PA, USA, pp.117-137, 1987.
3. G.C. Wei and P.F. Becher, Development of SiC Whisker Reinforced Ceramics, Am. Ceram. Soc. Bull., 64, pp.298-304, 1985.
4. W.B. James, New Shaping Methods for PM Components, Powder Metallurgy 1986 - State of the Art, (eds) W.J. Huppmann, W.A. Kaysser and G. Petzow, Freiburg, FRG, Verlag Schmid, pp.71-100, 1986.
5. R.M. German, Liquid Phase Sintering, New York, Plenum Press, 1986.
6. W.A. Kaysser and G. Petzow, Present State of Liquid Phase Sintering, Powder Metallurgy, 28, pp.145-150, 1985.
7. G. Petzow and W.A. Kaysser, Basic Mechanisms of Liquid Phase Sintering, Sintered Metal-Ceramic Composites, (ed.) G.S. Upadhyaya, Amsterdam, Elsevier Science Publishers, pp.51-70, 1984.
8. G. Petzow and W.A. Kaysser, Liquid Phase Sintering, Science of Ceramics 10, (ed.) H. Hausner, Weiden, FRG, Deutsche Keramische Gesellschaft, pp.269-280, 1980.
9. W.J. Huppmann and H. Riegger, Modelling of Rearrangement Processes in Liquid Phase Sintering, Acta Metall., 23, pp.965-971, 1975.
10. K.S. Hwang, R.M. German and F.V. Lenel, Capillary Forces Between Spheres During Agglomeration and Liquid Phase Sintering, Met. Trans., 18A, pp.11-17, 1984.
11. H.H. Park, O.J. Kwon and D.N. Yoon, The Critical Grain Size for Liquid Flow into Pores During Liquid Phase Sintering, Met.Trans., 17A, pp.1915-1919, 1986.
12. W.A. Kaysser and G. Petzow, Geometry Models for the Elimination of Pores during Liquid Phase Sintering in Systems with Incomplete Wetting, Science of Sintering, 16, pp.167-175, 1984.
13. V. Smolej, S. Pejovnik, and W.A. Kaysser, Rearrangement During Liquid Phase Sintering of Large Particles, Powder Met. Int., 14, pp.126-128, 1981.
14. W.J. Huppmann and H. Riegger, Progress in Liquid Phase Sintering, Acta Metall., 23, p.965, 1975.
15. J.S. Lee, W.A. Kaysser, and G. Petzow, Microstructural Changes in W-Cu and W-Cu-Ni Compacts During Heating up for Liquid Phase Sintering, in Modern Developments in Powder Metallurgy, (eds) E.N. Aqua and C.M. Whitman, MPIF-APMI, Princeton, N.J., 15, pp.489-506, 1985.
16. W.A. Kaysser, S. Takajo and G. Petzow, Particle Growth by Coalescence During Liquid Phase Sintering of Fe-Cu, Acta Metall., 32, pp.115-122, 1984.
17. W.A. Kaysser and G. Petzow, Ostwald Ripening and Shrinkage During Liquid Phase Sintering, Z. Metallkde., 76, pp.687-692, 1985.

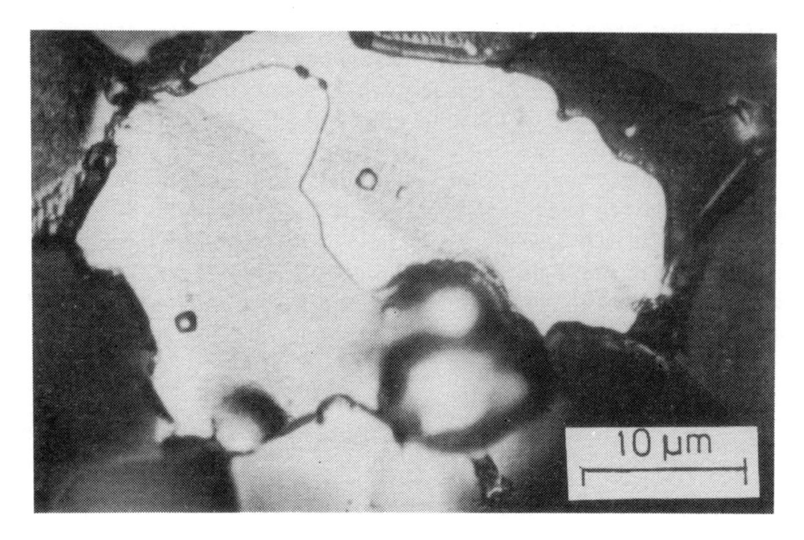

a

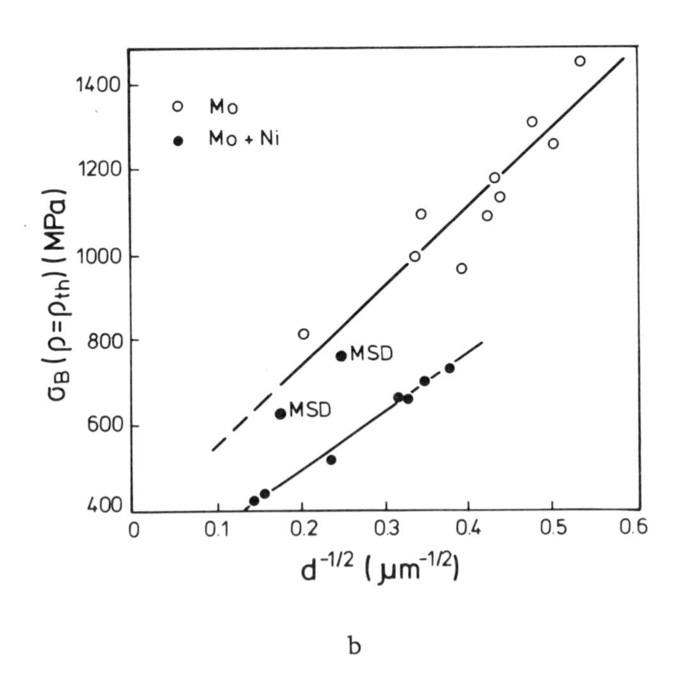

b

Figure 14 Microstructure and bend strength of Mo-Ni after slow cooling. (a) Microstructure. (b) Bend strength, σ_B; d is the average grain size, MSD cooling procedure resulted in bulged interlocking grain boundaries.

18. G.H.S. Price, C.J. Smithells and S.V. Williams, Copper-Nickel - Tungsten Alloy Sintered with a Liquid Phase Present, J. Inst. Met., 62, pp.239-245, 1938.
19. D.N. Yoon and W.J. Huppmann, Grain Growth and Densification During Liquid Phase Sintering of W-Ni, Acta Metall., 27, pp.693-698, 1979.
20. W.A. Kaysser *et al.*, Shape Accomodation during Liquid Phase Sintering, Contemporary Inorganic Materials: Progress in Ceramics. Metals and Composites, (eds) G. Ondracek and O. Voehringer, Juelich, Bilateral Seminars of the International Bureau KFA-Juelich CIM, pp.217-222, 1986.
21. S.-J.L. Kang, W.A. Kaysser, G. Petzow and D.N. Yoon, Liquid Phase Sintering of Mo-Ni Alloys for Elimination of Isolated Pores, Modern Developments in Powder Metalluray, (eds.) E.N. Aqua and C.M. Whitman, Princeton, NJ, MPIF-APMI, 15, pp.477-488, 1985.
22. S.-J.L. Kang, W.A. Kaysser, G. Petzow and D.N. Yoon, Growth of Mo Grains Around Al_2O_3 Particles During Liquid Phase Sintering, Acta Metall., 33, pp.1919-1926, 1985.
23. W.D. Kingery, Densification During Sintering in the Presence of a Liquid Phase I, J. Appl. Phys., 30, pp.301-309, 1959.
24. S. Pejovnik, D. Kolar, W.J. Huppmann and G. Petzow, Sintering of Al_2O_3 in Presence of Liquid Phase, Sintering - Theory and Practice (eds) D. Kolar, S. Pejovnik and M.M. Ristic, Amsterdam, Elsevier Scientific Publishers, pp.361-365, 1982.
25. W.A. Kaysser, W.J. Huppmann and G. Petzow, Analysis of Dimensional Changes During Sintering of Fe-Cu, Powder Metallurgy, 23, pp.86-91, 1980.
26. Y.A. Chang *et al.*, Phase Diagrams and Thermodynamic Properties of Ternary Copper-Metal Systems, NY, New York, International Copper Research Organization, pp.367-386, 1979.
27. M. Hillert and G. Purdy, Chemically Induced Grain Boundary Migration, Acta Metall., 26, pp.333-340, 1978.
28. D. N. Yoon, J.W. Cahn, C.A. Handwerker, J.E. Blendell and Y.J. Baik, Coherency Strain Induced Migration of Liquid Films through Solids, Interface Migration and Control of Microstructure, Washington, NBS, 1985.
29. C.A. Handwerker, J.W. Cahn, D.N. Yoon and J.E. Blendell, The Effect of Coherency Strains on Alloy Formation: Migration of Liquid Films, Atomic Transport in Alloys: Recent Developments, (eds) G.E. Murch, M.A. Dayananda, TMS/AIME Publications, Dayton, OH, 1985.
30. W.A. Kaysser, Boundary Alloying by Melt Penetration, to be published in Z . Metallkde.
31. M. Hofmann-Amtenbrink, W.A. Kaysser and G. Petzow, Grain Boundary Migration in Recrystallized Mo Foils in the Presence of Ni, J. Physique, C4, p.545, 1985.
32. M. Hillert, On the Driving Force for Diffusion Induced Grain Boundary Migration, Scripta Metall., 17, pp.237-240, 1985.

14

Sintering Aids in Powder Metallurgy

The Role of the Platinum Metals in the Activated Sintering of Refractory Metals

C. W. CORTI
Johnson Matthey Technology Centre, Sonning Common, Nr. Reading, Berks, U.K.

Abstract-p807

Abstract

When a metallic powder is subjected to a sufficiently high pressure a certain amount of adhesion takes place between individual particles. If this compact is then sintered the bond is improved by diffusion and intergranular grain growth. The earliest known platinum objects were fabricated by such a powder metallurgical process, and when European scientists first addressed the problem of manufacturing platinum bars they also used powder metallurgy to overcome their inability to melt the metal. Now powder metallurgical methods are widely used for fabricating a variety of materials, and this paper reviews studies made of the sintering of refractory metals when this process is promoted by the addition of a minor amount of a platinum group metal activator.

1. Introduction

The manufacture of engineering metals and alloys in fabricated forms generally commences with the melting and casting of ingot material for subsequent shaping by mechanical techniques, such as forging, rolling and extrusion, although in many instances molten metal can be cast directly to a final shape. However, in the case of the refractory elements, such as tungsten, molybdenum and rhenium, their very high melting points (in excess of 2000°C), as well as their resistance to deformation, generally precludes the melting approach as a practical route to material and component manufacture. This has led to the development of processes in which consolidation of powder materials is achieved by sintering at temperatures below their melting points.

To promote and assist the sintering process, two techniques have been developed which involve the use of metallic sintering additives. These are known as Liquid Phase Sintering and Activated Sintering. In Liquid Phase Sintering the refractory metal powders are sintered in the presence of one or more metals - generally transition metals such as copper or iron - at temperatures above the melting point of the additive, so that sintering occurs in a molten binder phase which may be present in substantial amounts, for example up to 40 wt-%.

In contrast, Activated Sintering is performed in the presence of small amounts of metal additives, again often transition metals, but in the solid state at temperatures *below* the melting point of the additive. Thus, as can be seen from Table 1, Activated Sintering can be accomplished at lower temperatures than Liquid Phase Sintering, although not necessarily so, depending on the particular metal additive used. In both cases, however, the temperatures employed are substantially lower than would otherwise be required if the refractory powders were sintered without additives. Kurtz, for example, showed in 1946 that 99 per cent dense tungsten parts could be achieved by sintering below 1400°C with less than 1wt-% addition of nickel [1], whereas temperatures above 2800°C are required to achieve a comparable density in untreated tungsten powder.

2. Activated Sintering

Since Vacek reported the enhancement of sintering by additions of small quantities of transition metals to tungsten in 1959 [2], making it possible to lower the sintering temperature substantially, a great deal of work has been carried out into the activated sintering of tungsten and other refractory metals, particularly with additions of Group VIII transition metals. Much of this work has involved the use of platinum group metals which have been shown to be very effective as sintering activators. This paper reviews the published work on the effect of the platinum group metals on the activated sintering of tungsten and other refractory metals, in particular on the kinetics and mechanisms of sintering. The properties and microstructure of the sintered materials are also examined. Finally, the scientific basis for the beneficial effect of the platinum group metals in the activated sintering of the refractory metals is examined in terms of current theories and phenomenological models.

2.1 Tungsten

Much of the early work on the activated sintering of tungsten was carried out by Brophy, Hayden and co-workers [3-7]. Their initial work focused on the sintering of tungsten powder coated with nickel. They found that, on sintering at 1100°C, the tungsten underwent rapid densification to more than 90 % theoretical density. Moreover, they found that the amount of nickel required to promote this accelerated sintering was roughly equivalent to a nickel coating thickness of about 1 atom monolayer [3]. Nickel coatings thicker than this did not produce any further enhancement; indeed, there was a tendency for the sintering rate to decrease from the optimum. They also found that densification occurred in two stages, the second stage coinciding with the onset of grain growth in the tungsten [4].

In an attempt to clarify the mechanism of activated sintering, which they had earlier attributed to the activating metal acting as a carrier phase for the diffusion of tungsten to the interparticle "necks", Hayden and Brophy examined the influence of ruthenium, rhodium, platinum and palladium additions on the kinetics of sintering in the temperature range 850 to 1100°C [7].

As in their previous work, the platinum group metal sintering additives were added to the tungsten powder in the form of aqueous solutions of salts (chlorides and nitrates) in the requisite amount; this was dried at 150°C and pre-reduced in hydrogen at 800°C to form a metallic coating on the tungsten powder. Sintering was carried out under hydrogen.

For all the platinum group metals examined, Hayden and Brophy found that a minimum level of platinum group metal was required to promote full activation, (Fig. 1), as had been observed in the case of nickel, and that larger amounts did not produce any further enhancement. Interestingly, palladium was the most effective element; this is clearly illustrated in Fig. 2 which shows the temperature dependence of shrinkage for each platinum group metal additive after sintering for 1 hour. Ruthenium was the least effective element. Significantly, the authors found that palladium was better than nickel in promoting densification. For example, the densities of samples sintered at 1100°C for 30 min. and 16 h. were 93.5 and 99.5 %, respectively, in the case of palladium in tungsten compared to 92 and 98 %, respectively, for nickel in tungsten. Untreated tungsten would only be pre-sintered at this temperature.

Analysis of the sintering kinetics in terms of the process controlling mechanism in their carrier phase model of activated sintering - which applies also to liquid phase sintering - showed that for all the platinum group metals examined, the sintering rate was not dependent upon composition, but was proportional to the cube root of time, except for rhodium in a low shrinkage regime. This time dependence was interpreted in terms of the diffusion controlled transport of tungsten in the interface between the tungsten and the platinum group metal coating layer. In the case of rhodium there is a transition in the rate controlling process, from the dissolution of tungsten in the rhodium layer at low shrinkages to interface diffusion at large shrinkages. Table 2 summarizes these results, the slope, S, being the time dependence of the sintering curves.

Also shown in Table 2 are the calculated activation energies for each platinum group metal additive. These lie in the range 86 to 114kcal/mol, which the authors believed to be comparable to the activation energy for tungsten grain boundary self-diffusion.

The effectiveness of the platinum metals and nickel in promoting enhanced sintering of tungsten were found to be in the order:

Pd > Ni > Rh > Pt > Ru

The reason for the platinum group metals being such effective activators for the sintering of tungsten was not established in this work, although it was suggested that it may be linked to their relatively high (10 to 20 %) solubility for tungsten and their low solubility in tungsten.

Subsequently Hayden and Brophy extended their work on platinum group metal activators to the sintering of tungsten with iridium additions [8]. In contrast to the other platinum metals, they found that iridium actually decreased the rate of densification of tungsten, the effect reaching a minimum value at about 2wt-% iridium, larger additions having no further effect. The measured activation energy of 133kcal/mol was close to that for volume self-diffusion of tungsten (135kcal/mol).

Further work on the activated sintering of tungsten by palladium and nickel additions was carried out by Toth and Lockington, who also found that there were optimum concentrations of both palladium and nickel for maximum densification during sintering at 1000°C [9]. Calculations showed these optimum concentrations to correspond approximately to a monolayer of the activating element on the tungsten surface, as also found earlier by Brophy, Shepherd and Wulff [3]. Once again, palladium was found to be more effective than nickel, especially at and below a temperature of 950°C. Toth and Lockington found the time dependence of the densification to be 0.5 for both palladium and nickel, in contrast to the value of 0.33 for palladium found by Brophy [7]. The apparent activation energies were lower, 62.5kcal/mol compared to 86kcal/mol for palladium and 50.6kcal/mol as against 68kcal/mol for nickel. Microprobe analysis of the fracture surfaces of sintered specimens showed segregation of the activating elements on grain boundary surfaces. The authors concluded that Brophy's model for activated sintering was not applicable; rather, they favoured a mechanism

Table I

Typical Sintering Temperatures for Activated Sintering and Liquid Phase Sintering of Refractory Metals

Refractory element	Melting point, °C	Activated sintering: Additive	Activated sintering: Sintering temperature, °C	Liquid phase sintering: Additive	Liquid phase sintering: Sintering temperature, °C
Tungsten	3410	Nickel	1100	Nickel	1550
		Palladium	1100	Copper	1100
Molybdenum	2610	Nickel	1200	Nickel	1460
		Palladium	1200		
Tungsten carbide	—	—	—	Cobalt*	>1350

*The tungsten carbide-cobalt eutectic temperature is 1320°C

Table II

Summary of the Effect of Platinum Group Metal Additives on the Sintering of Tungsten (Data from References 7 and 3)

Additive	Slope 'S'	Control of process	Activation energy, kcal/mol
Palladium	0.33	Diffusion	86
Ruthenium	0.39	Diffusion	114
Platinum	0.33	Diffusion	92
Rhodium [a]	0.5	Solution	85
Rhodium [b]	0.33	Diffusion	98
Nickel	0.5	Solution	68

(a) Rhodium : low shrinkage regime
(b) Rhodium : high shrinkage regime

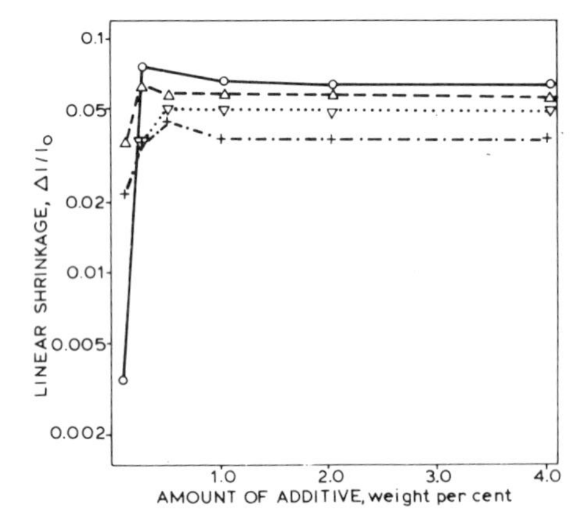

Fig. 1 During the activated sintering of tungsten the linear shrinkage is dependent upon the activator, and the amount added; a minimum level being required to promote full activation. Data from Hayden and Brophy (7).

Sintered for 1 hour
O Palladium, 950°C
+ Ruthenium, 1100°C
▽ Platinum, 1100°C
Δ Rhodium, 1100°C

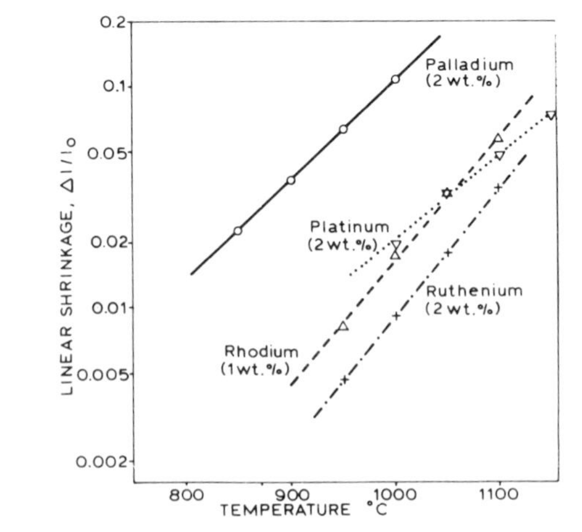

Fig. 2 The dependence of linear shrinkage upon temperature during the activated sintering of tungsten. This shows that after sintering for 1 hour, palladium is the most effective activator (7)

in which the surface diffusion of tungsten on the activator surface is the controlling step; both are shown schematically in Fig. 3.

The influence of a wide range of transition metal additions, including all the platinum group metals, on the sintering of tungsten at temperatures between 1000 and 2000°C was studied by Samsonov and Jakowlev [10]. They found, in agreement with earlier findings, that additions of Group VIII elements - including the platinum group metals - promoted densification of tungsten, with the exception of osmium which was neutral. Iridium had a small beneficial effect at the highest temperature studied, 2000°C, which is not inconsistent with the earlier work of Hayden and Brophy [8], since extrapolation of their Arrhenius plots predicts a transition from a detrimental to a beneficial effect at temperatures above about 1400°C. The effectiveness of the platinum group metals in enhancing sintering was found to be in the order:

Ru < Rh < Pd
and Os < Ir < Pt

with the upper row of elements being superior to the lower row. This is shown in Table 3, which also gives the measured values of compressive strength, hardness and grain size. On this basis, nickel appears to be slightly more effective as an activator than palladium, in contrast to the earlier work, but this is based on results obtained at higher sintering temperatures than those of the earlier studies.

These results show that the stronger activators also enhance the associated grain growth in the final stage of sintering. The higher densities (lower porosity) achieved are also reflected in higher values of compressive strength and hardness.

Samsonov and Jakowlev summarized their findings in terms of the position of the activating element in the Periodic Table, (Fig. 4). The arrows indicate an increasing degree of activation. They interpreted these results in terms of the electron structure of the activators and tungsten; an increase of the stable d-bonds in the system lowers the free energy, activating the sintering process in which diffusion is accelerated by the activators for which tungsten acts as an electron donor.

More recently, German and his co-workers have investigated the activated sintering of tungsten in more detail [11-13]. German and Ham [11] confirmed that palladium is the best metallic activator for the sintering of tungsten in the range 1100 to 1400°C, as shown in Fig. 5. This shows that, for both palladium and nickel, enhancement of sintering starts at approximately 1 monolayer thickness of additive and peaks at a thickness of 4 monolayers. Sintering in a moist hydrogen atmosphere was found to be detrimental to palladium activation for one tungsten powder, but beneficial for a second. The apparent activation energy is lowered on sintering in a moist atmosphere.

German and Munir [12] extended this work to other Group VIII elements including platinum, and confirmed that enhanced sintering commenced at about 1 monolayer thickness and peaked at 4 monolayers. They found the effectiveness of the activator to be in the order:

Pd > Ni > Pt = Co > Fe > Cu

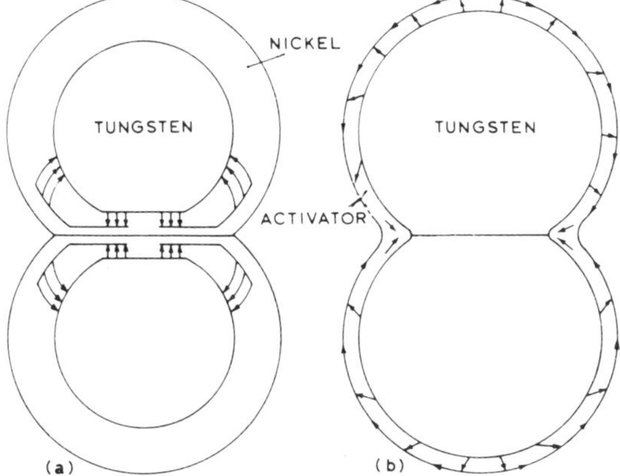

Fig. 3 These two models show different representations of the sintering of tungsten particles which have been coated with a metallic activator.
(a) The model of Brophy, Hayden and Wulff (3) has tungsten diffusing through the carrier (activator) phase, away from the line joining the centres of adjacent particles, to be redeposited elsewhere on the particles as indicated by the arrows.
(b) The model of Toth and Lockington (9), where dissolution of tungsten at the activator-tungsten interface is followed by volume diffusion outwards through the activator layer and subsequent surface diffusion, this being the rate controlling step. Diffusion through the activator layer to the contact point between adjacent particles results in the formation of sintering "necks"

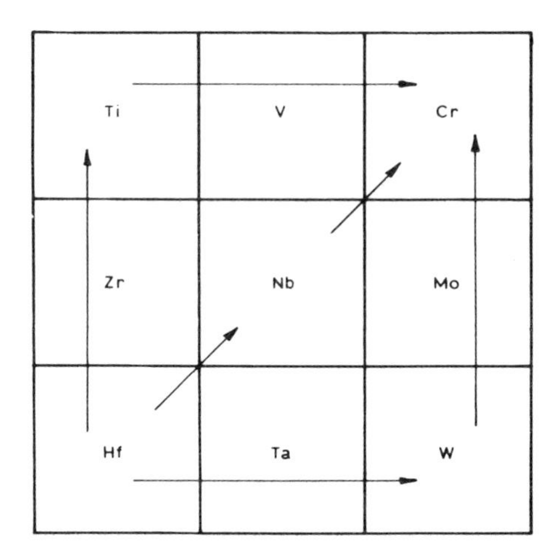

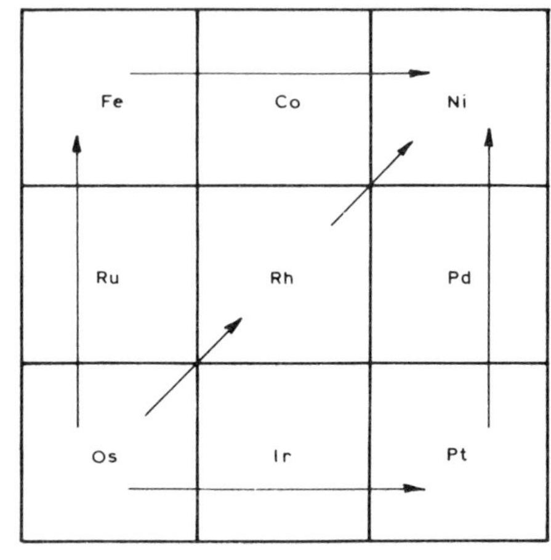

Fig. 4 Trends in the activated sintering of tungsten are related to the position of the activating element in the Periodic Table, as proposed by Samsonov and Jakowlev (10). The arrows indicate increasing degrees of activation

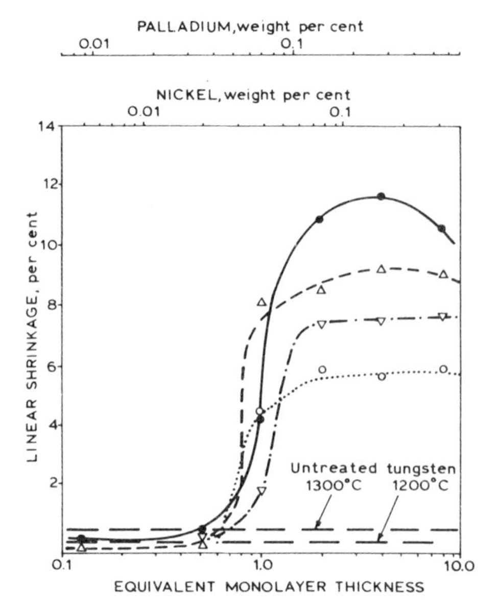

Fig. 5 The effect of the thickness of the palladium or nickel activators upon the linear shrinkage of tungsten powder sintered at 1200 and 1300°C in dry hydrogen is shown, after German and Ham (11)

• Tungsten + palladium, 1300°C
△ Tungsten + nickel, 1300°C
▽ Tungsten + palladium, 1200°C
○ Tungsten + nickel, 1200°C

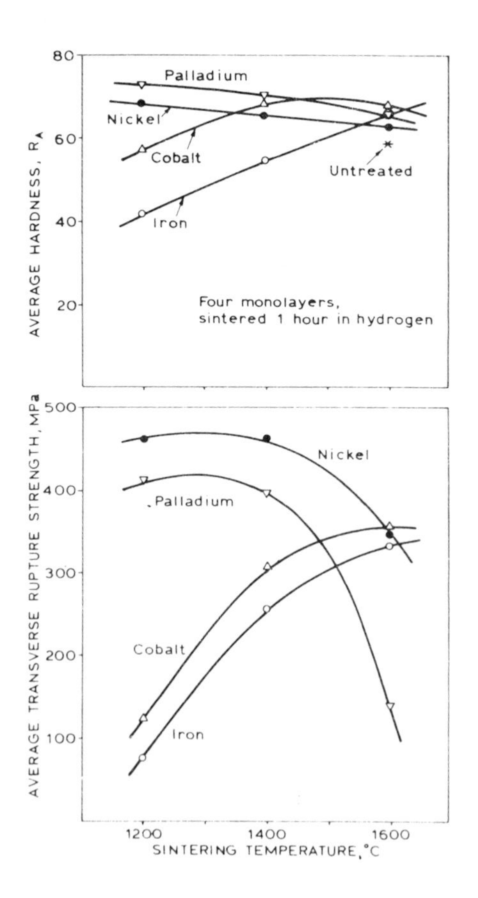

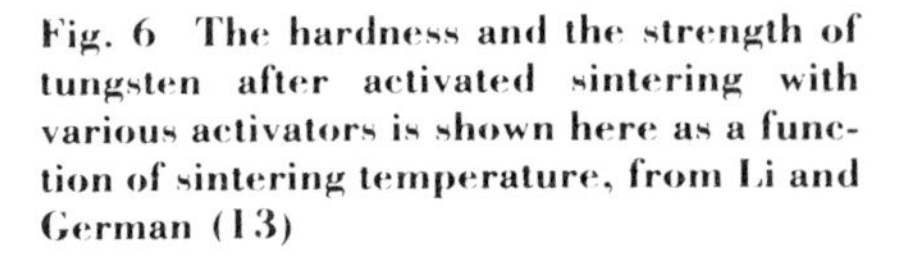
Fig. 6 The hardness and the strength of tungsten after activated sintering with various activators is shown here as a function of sintering temperature, from Li and German (13)

Table III

Dependence of the Properties of Tungsten on the Activating Element at the Optimum Concentration and Sintering Temperature (10)

Activator, weight per cent	Sintering temperature, °C	Density, g/cm³	Compressive strength, kg/mm²	Hardness, kg/mm²	Tungsten grain size, μm
W	2000	16.1	80	181	5–7
Fe (0.5–1.0)	1600	17.4–17.95	92–96	310–390	12–15
Co (0.3–0.4)	1600	17.4–17.95	83–87	277–282	20–25
Ni (0.2–0.4)	1400–1600	18.1–18.4	91–95	280–306	100–120
Ru (1.0)	1600–1800	17.4	77	309	15–20
Rh (0.5)	1600	17.8	71	290	10–15
Pd (0.3–0.4)	1400–1600	18.1–18.35	83–90	290–300	20–25
Os (1.0)	2000	16.1	80	181	5–7
Ir (1.0)	2000	16.7	98	238	5.5–7
Pt (1.0)	1800–2000	17.6–17.9	88–93	305–330	20–30

Below 1300°C, iron was more effective than platinum and cobalt. In the case of palladium and nickel, where sintering progressed to the second stage, extensive tungsten grain growth was observed. The onset of grain growth was associated with a decline in the shrinkage rate. The authors found a time dependence of the shrinkage for all activators, similar to that found by Toth and Lockington, favouring volume diffusion of tungsten through the activator layer as the rate controlling mechanism. The addition of 0.4 wt-% palladium was found to increase the apparent grain boundary diffusion rate by about 6 orders of magnitude, corresponding to a decreased activation energy. The measured activation energies decreased in the order of increasing activator effectiveness, palladium having the lowest value. This was related to the electron structure modifications as postulated by Samsonov and Jakowlev [10], that is the transition metals with unfilled d-shells are the optimal activators for tungsten. Based on this concept, the authors suggest that palladium and nickel are optimum activators for all refractory metal powders.

Later work by Li and German examined the properties of palladium- and nickel-activated tungsten sintered with optimum activator content in the temperature range 1200 to 1600°C [13]. Hardness levels were in the order Pd>Ni>Co>Fe at lower temperatures, as can be seen in Fig. 6, but were closer together at the higher temperatures. In the case of transverse rupture strength, nickel-activated tungsten was stronger than palladium-activated tungsten, the strength decreasing with increasing sintering temperature above 1400°C due to rapid grain coarsening. For the 0.43 wt-% palladium-activated material, the grain size increased from 4.5μm at 1200°C to 18.0μm at 1400°C and to 28.5μm at 1600°C.

Recent work on activated secondary recrystallization of heavily-drawn doped tungsten wire has provided additional evidence for the influence of the activators during sintering [14]. In this work the tungsten wire was coated with palladium, platinum or nickel prior to annealing and the rate of secondary recrystallization measured. The highest rate of recrystallization was found in the presence of palladium, followed by nickel and then platinum, grain growth being induced at temperatures several hundred degrees lower than uncoated tungsten wire. The process was controlled by the penetration of the activating elements into the wire. The diffusivities of these were found to be much higher than in pre-recrystallized tungsten, which is attributed to the high diffusivity paths through an intergranular phase formed by the activator which segregates to the grain boundaries. Auger electron spectroscopy revealed this layer to be about 2nm thick for both palladium and nickel. The measured diffusivities of the activators were in the order:

Pd > Ni > Pt > Co

Studies by Gessinger and Buxbaum on electron emission from thoriated tungsten cathodes has shown that platinum can also activate enhanced diffusion of thorium to the surface along grain boundaries, enabling the temperature limit for electron emission to be extended from 1950 to 2150K and the maximum emission current to be increased from 3 to 7.5 A/cm^2 [15]. This work demonstrates that platinum group metals not only enhance the diffusion of tungsten, but can also enhance the diffusion of other elements in the tungsten grain boundaries.

2.2 Molybdenum and Other Refractory Metals

As with tungsten, several investigators have shown that both palladium and nickel can enhance the sintering of molybdenum, [for example see Refs. 16-18]. Further, more detailed work on the activated sintering of molybdenum by platinum group metal additions has been carried out by German and his co-workers in the USA [19,20]. Their work on the heterodiffusion modelling of tungsten was extended to molybdenum where the effect of 13 transition metal additions including rhodium, palladium, iridium and platinum was examined in the temperature range 1000 to 1350°C [19]. Again, they found that activation of sintering commenced at activator concentrations equivalent to 1 monolayer thickness and reached the maximum effect at about 10 monolayers' thickness, although this plateau shifted to greater thicknesses with increasing sintering temperature. They confirmed that palladium was the best activator for molybdenum, with the degree of effectiveness being in the order:

Pd > Ni > Ph > Co > Pt > Au > Fe

As in the case of tungsten, iridium was detrimental to the sintering of molybdenum. The activation energy for sintering decreased with increasing effectiveness of the activator, that for palladium-activation being 280kJ/mol (66.9kcal/mol) compared to 405kJ/mol (96.8kcal/mol) for untreated molybdenum. This decrease in activation energy for palladium, nickel and platinum was observed to be concentration dependent, a rapid decrease occurring at about 1 monolayer thickness and reaching a minimum value at the optimum concentration of about 10 monolayers. The authors concluded that the sintering process was in accord with the grain boundary heterodiffusion model developed earlier for tungsten [21].

Later studies by German and Labombard [20] on palladium, nickel and platinum additions to two different molybdenum powders of the same particle size sintered at low temperatures (1050 to 1150°C) confirmed the earlier findings, namely that palladium is the best activator, followed by nickel and platinum, and that sintering behaviour conformed to the heterodiffusion model.

German and Munir also studied the activated sintering of hafnium [22] and tantalum [23] with transition metals as part of their broader investigation into the mechanisms, particularly the d-electron exchange model proposed by Samsonov [10]. In the case of hafnium, the activators were added to a thickness equivalent of 4 monolayers. Isothermal sintering experiments in the range 1050 to 1450°C showed densification after 1 hour in the following order of enhancement:

Ni > Pd > Co > Pt

Cobalt and platinum were only beneficial at temperatures above about 1300°C. Unusually, a non-Arrhenius temperature dependence was found for the activated sintering, and this was confirmed by constant heating rate experiments which showed sharp peaks for some activators at varying temperatures. These peaks occurred at about 1375°C for palladium and about 1240°C for nickel, for example, and are indicative of an optimum sintering temperature for maximum densification enhancement. It was concluded that the activated sintering of hafnium does correlate with the electron structure model, although the activators only impart a limited benefit.

In their study of the activated sintering of tantalum in the range 1250 to 1700°C, German and Munir found only slight enhancement with platinum, palladium and rhodium [23]. The poor enhancement of palladium was particularly surprising, but examination of fractured surfaces suggested that palladium enhanced only surface diffusion, not bulk diffusion.

The use of palladium and nickel as activators in the sintering of chromium has been studied at the Tokohu University in Japan [24]. It was found that palladium enhanced sintering considerably, the degree of enhancement reaching a plateau at about 0.8 wt-% palladium over the range 1050 to 1200°C. Above 1200°C, on sintering for 1 hour, the extent of enhancement became suppressed; this was mainly an effect of the higher density levels achieved at the higher temperatures and a reflection of the retardation due to grain growth in the second stage. In contrast, nickel had little activating effect. The relative behaviour of palladium and nickel was interpreted by the authors in terms of the mutual solubility criterion suggested earlier by Hayden and Brophy [7]. Palladium-chromium fulfils this requirement whereas nickel-chromium does not.

The activated sintering of rhenium and tungsten-rhenium mixtures has been studied by several investigators [25-28]. Dushina and Nevskaya found that on sintering rhenium for 2 hours in the range 1300 to 2000°C, substantial enhancement of sintering occurred with palladium contents of 0.1 to 0.5 wt-% [25]. Maximum enhancement was found in the range 0.2 to 0.4 wt-% palladium. This enhancement was observed to be accompanied by substantial grain growth, grain sizes of 10 to 15μm being observed compared to 1 to 2μm for untreated rhenium. Sintering at 1800°C produced densities of 92 % in 0.2 wt-% palladium-activated rhenium and only 81 % for untreated rhenium.

In their study on the sintering of rhenium, German and Munir found that at 1000°C only platinum enhanced sintering, while elements such as palladium, nickel, iron and cobalt inhibited densification [26]. The enhancement effect of platinum commenced at a thickness of about 1 monolayer, reaching a peak at about 2 monolayers. At 1400°C, both platinum and palladium enhanced sintering, platinum becoming effective at about 1 monolayer, rising to a plateau of maximum effectiveness at about 4 monolayers. Palladium acts less rapidly, reaching that of platinum at about 10 monolayers thickness, which suggests that palladium would be better than platinum at higher concentrations. The results of this study were considered to correlate with the electron structure model reasonably well.

The study of palladium additions to co-reduced tungsten-rhenium powders by Shnaiderman and Skorokhod again illustrates the beneficial effect of palladium in activated sintering of refractory metals, although in this instance palladium-rich alloy interlayers are formed at the grain boundaries [28].

3. Models of Activated Sintering

As we have seen, the activation of sintering refractory metal powders by transition metal elements has been interpreted in terms of several models, which are generally qualitative in nature. The results of the many studies on several refractory metals have shown a reasonably consistent pattern in that the most beneficial activators are palladium, nickel and platinum, generally in that order. Since these three elements sit in the same column of the Periodic Table, it is reasonable to assume that their role is related to their electronic structure and its ability to promote rapid diffusion of the refractory element. The time, temperature and activator concentration dependencies are also similar for all the refractory metals studied, which suggests that there is a common basis for a generalized model [29].

The initial model postulated by Hayden and Brophy was based on a solution-precipitation approach in which the relative solubilities of the activator in the refractory element and the refractory metal in the activating element should be low and high, respectively [7]. In this model, illustrated schematically in Fig. 3a, the refractory element diffuses away from the interparticle boundary and is redeposited elsewhere on the particle surface - as indicated by the arrows in the Figure. Experimentally the rate controlling step is found to be either refractory metal diffusion at the interface with the activator layer, or refractory metal solution in the activator layer.

Later, Samsonov and Jakowlev proposed that activated sintering was a consequence of the electronic structure stabilization of the refractory metal caused by the additive metal [10]. They based this approach on the argument that a metallic system containing partially filled d-subshells becomes more stable as the number of d^5 and d^{10} electron configurations increase. The refractory element acts as an electron donor, and this ease of electron transfer gives rise to the high solubility in the activating element.

A more recent proposal by German and Munir takes this model further [21] and applies the Engel-Brewer theory [30] to the prediction of the activation energies for the diffusion of refractory metals through the activator layer. In this quantitative model the activator has a role in providing enhanced grain boundary diffusion of the refractory metal. This is shown schematically in Fig. 7, with the activator layer wetting the interparticle grain boundary. This is taken from Ref. 29, where a more detailed description of these models is given. The relative solubility criterion is a prerequisite for enhanced diffusion of the refractory metal. Enhanced mass transport, and hence densification, results from the lowering of the activation energy for the refractory metal in the activator.

German and Munir have shown [29] that their calculated values of activation energy for diffusion of molybdenum agree well with experimentally determined values [19]. These calculations indicate that palladium, nickel and platinum are the best activators, as shown experimentally for several refractory metals. The predicted shrinkages for molybdenum also agree well with experiment, as shown in Table 4 [29]. Figure 8 shows the good agreement between their predictions and the experimental consensus for tungsten in terms of the position of the activator in the Periodic Table. This clearly demonstrates the superiority of palladium as an activator, with rhodium and platinum also in significant positions.

More recently, Miodownik has proposed a quantitative figure of merit for assessing the potential of additive elements as activators [31]. This parameter, ϕ, has been derived by combining the relevant heats of solution, surface energies and the energy of vacancy formation in the activator, and is based on the underlying thermodynamic parameters that are responsible for the phase equilibria; the solubility criterion of the earlier models is an aspect of the latter:

$$\phi = \Delta H_1 + \Delta H_2 + \Delta H_3$$

where ΔH_1, ΔH_2 and ΔH_3 are thermodynamic functions related to solubility, segregation and diffusion, respectively. Using calculated values of ϕ for the sintering of tungsten, Miodownik's predictions are correct for 12 out of 14 activator elements shown in Fig. 8, the only discrepancies being manganese and gold. Once again palladium is predicted to be the most effective activator.

4. Properties of Sintered Materials

While there has been a considerable number of studies into the phenomenon of activated sintering, relatively few studies have measured the properties of the sintered materials. Fracture is generally intergranular in nature, suggesting that the activator-rich grain boundaries are paths of easy fracture. Strength and hardness are very density dependent and the effectiveness of the activator on densification clearly plays a major role. Thus, both type and concentration of the activator influence sintered strength, as shown in Table 3 and Fig. 6. Strengths as high as 1050MPa have been shown for palladium-activated tungsten [10].

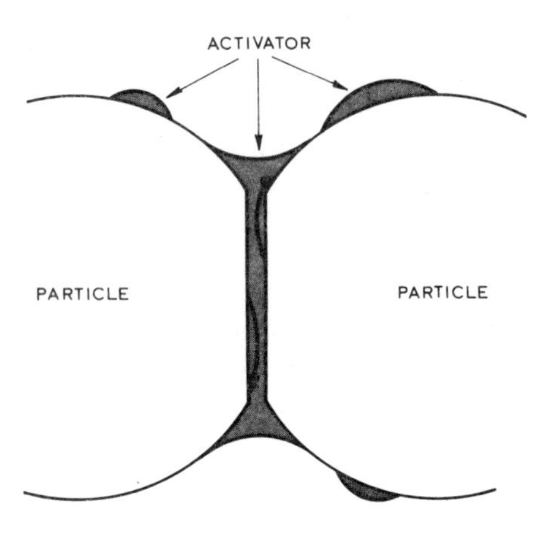

Fig. 7 This geometric model of the heterodiffusion controlled activated sintering process shows the activator layer wetting the interparticle grain boundary, after German and Munir (29)

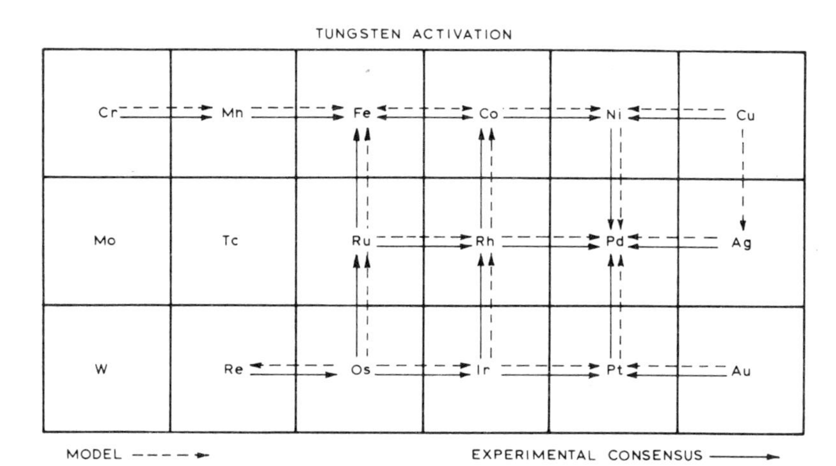

Fig. 8 The position of the activator in the Periodic Table affects the densification of the tungsten. There is good agreement between the predictions of the heterodiffusion model (29) and experimental consensus

Table IV

Comparison of Predicted and Measured Shrinkage for the Activated Sintering of Molybdenum Powder (29)

2.2μm size: sintered 1 hour at 1250°C in hydrogen

Activator	Shrinkage, per cent	
	Predicted	Measured
Untreated Mo	1.6	2.1
Pd	8.9	8.2
Ni	6.9	7.2
Pt	4.7	3.6
Co	4.6	4.3
Fe	3.5	2.2
Cr	1.1	1.1

No data have been presented for high temperature creep properties, but the enhancement of grain boundary diffusion in the presence of activators would be expected to promote diffusional creep of the Coble type.

As stated above, the use of platinum group activators can promote enhanced electrical properties: Gessinger and Buxbaum have utilized the increased grain boundary diffusivity in platinum-activated thoriated-tungsten emitters to improve electron emission [15].

5. Summary

This paper has reviewed the numerous studies of the activated sintering of refractory metals by transition metal additions and has shown that the platinum group metals, and particularly palladium and platinum, even at amounts of less than 1 wt.%, are very effective in promoting densification at temperatures several hundred degrees lower than would otherwise be required. The models developed to describe the phenomenon have been examined and those of German and Munir [29] and Miodownik [31] have been shown to predict the order of effectiveness remarkably well. Significantly, palladium is predicted to be the best activator element for several refractory metals including tungsten and molybdenum, in accord with experimental findings. The enhanced densification that results from the use of platinum group metal activators leads to improved strength properties.

References

1. J. Kurtz, Proc. Second Annu. Spring Meeting, Metal Powder Assoc., 40, 1946.
2. J. Vacek, Planseeber. Pulvermetall., 7, 6, 1959.
3. J.H. Brophy, L.A. Shepherd and J. Wulff, Powder Metallurgy, (ed.) W. Leszinski, AIME-MPI, Interscience, New York, p.113, 1961.
4. J.H. Brophy, H.W. Hayden and J. Wulff, Trans. AIME, 224, 797, 1962.
5. H.W. Hayden, S.B. Thesis, Massachussetts Institute of Technology, Metallurgy Dept., 1960.
6. J.H. Brophy, H.W. Hayden and J. Wulff, Trans. AIME, 221, 1225, 1961.
7. H.W. Hayden and J.H. Brophy, J. Electrochem. Soc., 110, 7, 805, 1963.
8. H.W. Hayden and J.H. Brophy, J. Less-Common Met., 6, 214, 1964.
9. I.J. Toth and N.A. Lockington, J. Less-Common Met., 12, 353, 1967.
10. G.W. Samsonov and W.I. Jakowlev, Z. Metallkd., 62, 8, 621, 1971.
11. R.M. German and V. Ham, Int. J. Powder Metall. Powder Technol., 12, 2, 115, 1976.
12. R.M. German and Z.A. Munir, Metall. Trans. A, 7A, 1873, 1976.
13. C. Li and R.M. German, Metall. Trans. A, 14A, 2031, 1983.
14. L. Kozma, R. Warren and E. Th. Henig, presented at the 7th Int. Risφ Symposium, Roskilde, Denmark, Sept. 1986.
15. G.H. Gessinger and Ch. Buxbaum, Sintering and Catalysis, ed. G. C. Kuczynski, Plenum, New York, p.295, 1975.
16. V.V. Panichkina, V.V. Skorokhod and A.F. Khrienko, Soviet Powder Metall. Ceram., 6, 558, 1967.
17. O. Neshich, V.V. Panichkina and V.V. Skorokhod, International Team for Studying Sintering, ITS 27, 1972.
18. V.V. Skorokhod, S.M. Solonin, L.I. Chernyshev, L.L. Kolomiets and L.I. Schnaiderman, Soviet Powder Metall. Ceram., 15, 435, 1976.
19. R.M. German and Z.A. Munir, J. Less-Common Met., 58, 61, 1978.
20. R.M. German and C.A. Labombard, Int. J. Powder Metall. Powder Technol., 18, 2,147, 1982.
21. R.M. German, Sintering-New Developments, 4th Int. Conf. on Sintering, ed. M.M. Rustic, Dubrovnik, Sept.1977, Elsevier, p.257, 1979.
22. R.M. German and Z.A. Munir, J. Less-Common Met., 46, 333, 1976.
23. R.M. German and Z.A. Munir, Powder Metall., 20, 3, 145, 1977.
24. R. Watanabe, K. Toguchi and Y. Masuda, Sci. Sintering, 15, 2, 73, 1983.
25. O.V. Dushina and L.V. Nevskaya, Soviet Powder Metall. Ceram., 8, 642, 1969.
26. R.M. German and Z.A. Munir, J. Less-Common Met., 53, 141, 1977.
27. V.V. Panichkina, L.I. Shnaiderman and V.V. Skorokhod, Dokl. Akad. Nauk. Ukr. SSR, A, 5, 469, 1975.
28. L.I. Shnaiderman and V.V. Skorokhod, Soviet Powder Metall. Ceram., 19, 1, 27, 1980.
29. R.M. German and Z.A. Munir, Rev. Powder Metall. Phys. Ceram., 2, 1, 9, 1982.
30. L. Brewer, High Strength Materials, (ed.) V. F. Zackay, J. Wiley, New York, p.12, 1965.
31. A.P. Miodownik, Sintering: Theory and Practice, Harrogate, October 1984; Powder Metall., 28, 3, 152, 1985.

15

Selection of Atmospheres for Sintering

P. R. WILYMAN AND M. VANDERMEIREN

Air Products PLC, Brunel Science Park, Uxbridge, Middlesex, UK

1. Introduction

Powder metallurgy is defined as the manufacture of metal powders and their subsequent processing to form parts that have significant mechanical strength.

Sintering is the process of bonding together the particles in a powder compact by heating to a temperature somewhat below the melting point of the major constituent of the powder. Because materials are generally very reactive at these temperatures and because of the high surface area exposed by the powder to the atmosphere the choice of atmosphere under which sintering is to take place is vital in order to obtain the required final product. This choice has been made more critical in recent years because the advances made in metal powder quality have required similar advances in the quality of the atmosphere used.

Depending on the material being sintered the sintering atmosphere must perform some or all of the following functions:

- Generally have no reaction with the material at the sintering temperature though sometimes the atmosphere is used to provide a small refinement to the chemical composition of the powder being processed;
- Exclude air from the furnace or compensate for the effect of its ingress;
- Reduce any surface oxides on the powder to improve metal to metal contact;
- Physically and or chemically to remove volatilized lubricants and binders from the compressed compact and from the furnace;
- Provide a convective means of heat transfer in the furnace.

The choice of atmosphere is largely determined by the material to be processed. While a prime requirement is, of course, the ability of the atmosphere to fulfil the requirements set out above, other factors such as economics, safety and availability must also be taken into account. The value of the sintered product is also a major determinant in the choice of atmosphere. A high value product will be more likely to justify an expensive atmosphere than a lower value product which may be sintered in an atmosphere which, although not perfect in terms of chemical reactivity in the furnace, may be a satisfactory compromise in terms of commercial and practical considerations.

2. Selection of the Atmosphere

The bulk of all sintered products are structural or mechanical parts which are based predominantly on iron, but also on copper and aluminium, although there are other equally important materials which are produced in smaller volumes. These other materials can be classified into groups such as refractory metals (e.g. molybdenum, tungsten), composite materials, consisting of two materials which are immiscible even in the liquid state (e.g. hard metals, friction materials), high duty alloys which are high strength alloys based on iron or nickel, and porous materials.

For all these materials, the atmospheres available range from the expensive hydrogen and vacuum to the much cheaper generated atmospheres such as exothermic and endothermic gases. In general the atmosphere chosen is the cheapest atmosphere that is chemically suitable. Figure 1 summarizes the atmospheres commonly used for a variety of sintered products.

2.1 Batch Processes

Apart from structural powder metallurgy parts most powder metallurgy components are produced in such small numbers that batch rather than continuous sintering is preferred. Because the batch furnace can be sealed better than a continuous furnace, the atmosphere requirement in terms of volume is much less than for continuous sintering and as the value of each component may be much greater than continuously sintered products the cost of the atmosphere as a proportion of the total cost of manufacture is not large. Therefore in batch processes the tendency is to opt for the most chemically "ideal" atmosphere, with less regard to its cost, and often the atmosphere chosen is hydrogen or vacuum (which can be regarded as a special case of atmosphere), either of which becomes more hazardous or difficult to use in a continuous process (Fig. 1).

2.2 Continuous Sintering Process

Neither hydrogen nor vacuum are widespread atmospheres for the bulk of structural or mechanical sintered parts. For these products sintering is usually a continuous process and large volumes of atmosphere are required. Except in exceptional circumstances something cheaper must be used.

To determine the choice of atmosphere for continuous furnaces it is first necessary to consider what is required of the atmosphere in the sintering furnace. Figure 2 is a diagrammatic representation of a typical sintering furnace.

(i) The first part of the furnace operates at relatively low temperature (typically 500°C - 600°C) and it is in this zone that the binders and lubricants in the compressed powder compact are volatilized and removed from the furnace. Because the binders are invariably organic compounds a slightly oxidizing atmosphere is advantageous in this zone to prevent the deposition of soot on the parts and on the furnace surfaces. It is also necessary to have a significant flow of atmosphere from this area to the furnace entrance to encourage the volatilized binder to be physically removed from the furnace.

(ii) The second zone or sintering zone of the furnace operates at higher temperatures (> 1100°C). The atmosphere in this zone must fulfil the requirements described earlier i.e. it must reduce any surface oxides or ingressing air and it must be largely neutral to the product. Because of the high reactivity at these temperatures it is the chemical reactions in this zone that principally determine the atmosphere to be used. This will be discussed at length in a later section.

(iii) After sintering the product has to be cooled to room temperature in an atmosphere that will not cause further chemical change. In practice this means that it must not oxidize the sintered parts. Thus a reducing atmosphere is required, ideally with a high thermal conductivity to avoid needlessly long furnaces.

Although all furnaces have these three major zones, some furnaces may have, for example a pre-sinter zone for oxide reduction or carbon diffusion and some may have a post-sinter zone for carbon restoration. The chemical reactions in each part of the furnace ideally require a different atmosphere.

Choice of Atmosphere

Although, a sintering atmosphere has to be largely neutral, nevertheless chemical reactions must occur in the furnace. The principal reactions taking place are:

1. $M_xO + H_2 \Leftrightarrow XM + H_2O$ (reduction of metal oxides)

2. $O_2 + 2H_2 \Leftrightarrow 2H_2O$ (ingress of air), and if carbon is present:

3. $CO \Leftrightarrow CO_2 + C$

4. $H_2O + C \Leftrightarrow H_2 + CO$

For materials where the carbon level is either zero or not critical, an atmosphere comprised of nitrogen and hydrogen is suitable for sintering applications. The percentage of hydrogen necessary is that required to have a reducing potential sufficient to reduce any metal oxides and oxygen present and to prevent any oxidation of the components of the furnace. The oxidation potential of the atmosphere is given from equation (2) and is the ratio of water vapour content to hydrogen content, there being usually insignificant amounts of other oxidants. Figure 3 gives the oxidation potential required for various metals [1] and Table 1 gives the water content as a function of the dew point of the atmosphere. In general, because of the relative cost of hydrogen versus nitrogen and for safety considerations the hydrogen content of the atmosphere is as low as possible. Typical atmospheres would be 2% hydrogen in nitrogen for copper alloys through 10% hydrogen for nickel alloys to 30% hydrogen for stainless steel, though if nitride formation is a problem stainless steels may have to be sintered in a pure hydrogen atmosphere. Although these atmospheres offer acceptable results, increasing the hydrogen level can offer some advantages in terms of, for example, tensile strength because of the increased reduction of oxides [2] (Fig. 4). Argon as an alternative to nitrogen, although it is less reactive, is usually excluded on the grounds of cost.

When carbon control of the sintered product is required, equations (3) and (4) above have to be considered. Although hydrogen itself is neutral to carbon it may be that the oxidation potential of the atmosphere is sufficiently high to oxidize carbon. As a nitrogen-hydrogen atmosphere has no carburizing ability to compensate, a carbon potential has to be created by the addition of a carbon bearing species. The best addition to the atmosphere is carbon monoxide because of its reversible reaction with carbon dioxide and carbon.[3] Measurement of the carbon dioxide and monoxide levels can give

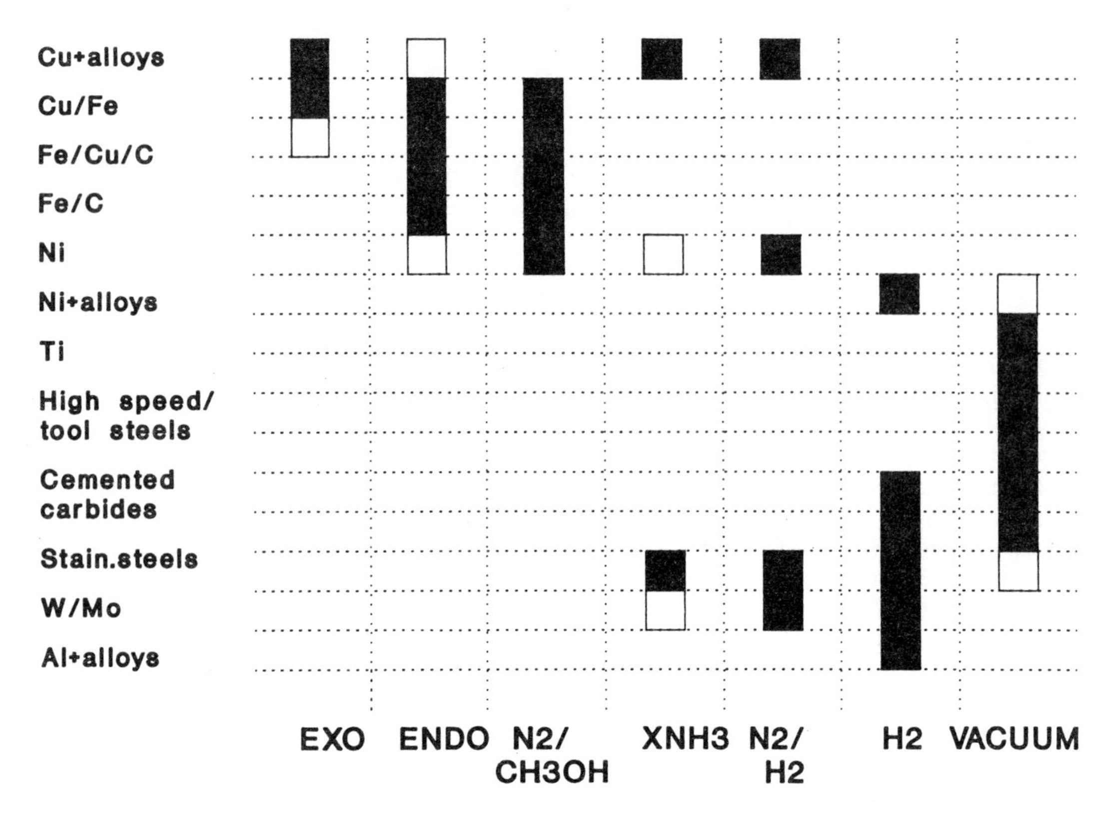

Figure 1 Summary of atmospheres for powder metal sintering (cost of atmosphere increases from left to right).

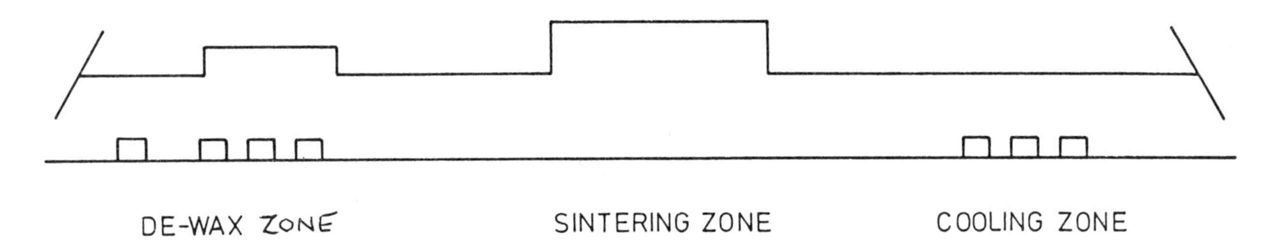

Figure 2 Typical furnace schematic for sintering steel.

Table 1

Dew point		Water
°C	°F	Vol %
- 80	- 112	0,00005
- 60	- 76	0,00106
- 50	- 58	0,00388
- 40	- 40	0,01270
- 30	- 22	0,03760
- 20	- 4	0,1020
- 10	+ 14	0,2570
0	+ 32	0,6020
+ 10	+ 50	1,2120
+ 20	+ 68	2,3080

the carbon potential [4] (Fig. 5) and if necessary the carbon potential can be increased by the further addition of a hydrocarbon.

$$\text{e.g. } CH_4 = C + 2H_2 \quad (5)$$

Alternatively control of the carbon potential of the atmosphere can be obtained solely by the addition of hydrocarbon (usually methane or propane) to a nitrogen-hydrogen mixture but because reaction 4 is essentially an irreversible reaction, precise carbon control is slightly harder to obtain [5].

3. Obtaining the Atmosphere

Whatever atmosphere is used it is obtained in one of two ways.

Firstly and perhaps traditionally it can be generated on the site of the use point. Alternatively each constituent of the atmosphere can be delivered to storage tanks from which they are separately piped to be mixed close to the furnace to produce what might be termed a synthetic atmosphere.

3.1 Generated Atmospheres

Although it is possible to generate hydrogen or any hydrogen/nitrogen mixture on site, with one major exception, these types of atmospheres for sintering are not generally obtained by purely on-site generation. The common generated atmospheres are:

1. Dissociated ammonia

The one exception is the atmosphere of 75% Hydrogen 25% Nitrogen produced by the dissociation of ammonia which is widely used for sintering purposes. If ammonia is heated in the presence of a catalyst it dissociates according to the reaction:

$$2NH_3 \rightarrow 3H_2 + N_2 \quad (6)$$

With a correctly operated dissociator the resulting mixture is dry (dew point -50°C) and contains insignificant quantities of residual ammonia. The major limitation is that the composition of the mixture is fixed, and the presence of nitrogen may lead to a reduction in mechanical properties of some materials particularly stainless steel.

2. Endothermic atmospheres

Endothermic atmospheres are used for the bulk of sintered carbon-steel. The atmosphere is formed in an endothermic generator by passing a hydrocarbon (usually natural gas or propane) over an externally heated catalyst in the presence of sufficient air to give incomplete combustion. Usually the air to hydrocarbon ratio is adjusted to give the minimum possible CO_2 and water vapour levels without soot formation. If natural gas is used as the generator feedstock the resulting atmosphere is typically:

40% H_2, 20% CO, 0.5% CO_2, Dewpoint +5°C, balance N_2

If propane is the feedstock the mixture produced is:

31% H_2, 23% CO, 0.5% CO_2, Dewpoint +5°C, balance N_2

The equilibrium carbon potential of endothermic atmospheres at normal sintering temperatures is in the region of 0.2%C. This is not necessarily significantly decarburizing to higher carbon steels because this composition gives a higher carbon potential at lower temperatures as cooling takes place (Fig. 6). If however a higher carbon potential is required small additions of natural gas or other hydrocarbon can be made.

3. Exothermic atmospheres

Exothermic atmospheres are produced by incomplete combustion of a hydrocarbon in a similar manner to endothermic atmospheres but in this case sufficient air is used such that the reaction is exothermic and no catalyst is required. Typical exogas compositions are between:

2-15% H_2, 2-12% CO, 6-12% CO_2, balance N_2 & water vapour

In practice the dew point is reduced to typically 5°C by cooling the gas but because of the high levels of CO_2, exothermic gas is unacceptable for sintering steels. It can be used for copper and its alloys that do not require a highly reducing atmosphere.

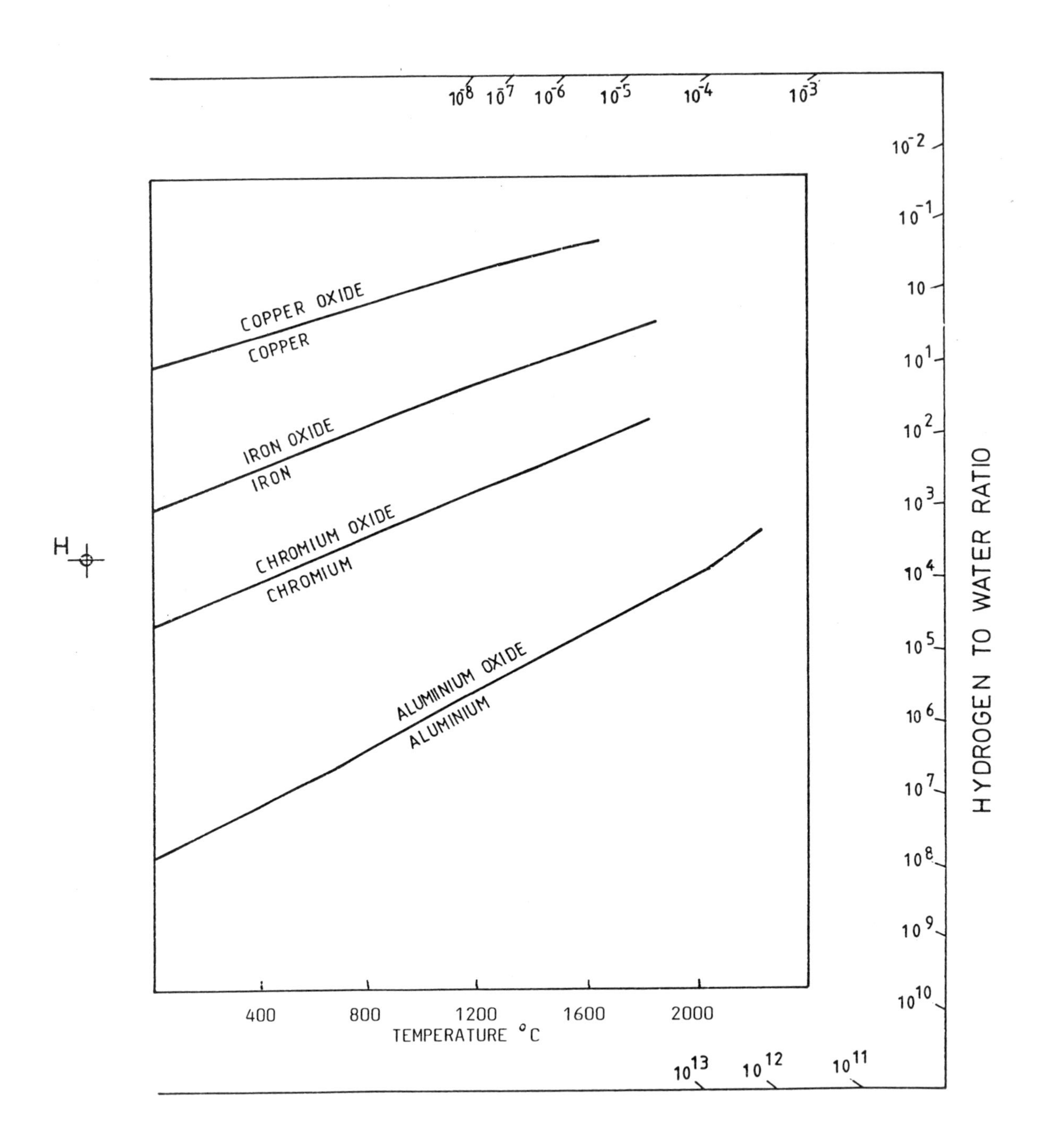

Figure 3 Oxidation potential for various metals, in terms of hydrogen to water vapour ratio.

3.2 Synthetic atmospheres

A much wider range of atmospheres, more precisely tailored to the user's needs can be obtained when the constituent gases in the atmosphere are piped from individual storage tanks at the user's site. Although volume for volume the cost of these "synthetic" atmospheres may be more than generated gases, this disadvantage is usually outweighed by the flexibility and simplicity that they can offer. For instance, nitrogen-hydrogen atmospheres can be supplied to the furnace in any proportion required simply by adjustment of a simple mixing panel. Similarly nitrogen-hydrogen-hydrocarbon mixtures can be supplied in the same way though if methane were required, the supply would normally be from mains natural gas. Carbon monoxide is not readily available in commercial quantities so if a nitrogen/hydrogen/carbon monoxide atmosphere is required it can be obtained by dissociation of methanol in the furnace according to the equation:

$$CH_3OH \rightarrow CO + 2H_2 \quad (7)$$

Although the relative proportions of hydrogen and carbon monoxide cannot be altered they will invariably be diluted with nitrogen to give whatever reducing or carburizing conditions are required. As with endogas, hydrocarbon additions can be added if necessary. A methanol based atmosphere is easier to control than a generated atmosphere and a comparative study has shown that the composition and properties of products sintered in a methanol atmosphere shows much less variation than products sintered in an endothermic atmosphere. [6] (Fig. 7 and Table 2)

A major advantage of these synthetic atmospheres is that they can offer the facility of putting different gas mixtures into different parts of the furnace. For example it can be seen that the slightly oxidizing atmosphere required in the de-wax zone of the furnace is incompatible with the reducing atmosphere required in the sintering zone. With generated atmospheres it is necessary to rely on the physical movement of the atmosphere to remove lubricants. However, although this also applies to synthetic atmospheres, the composition can be modified in the de-wax zone to assist in the removal of lubricants by chemical means. This can be done by the addition of humidified nitrogen to this zone. Similarly, although a low hydrogen atmosphere might be adequate for sintering purposes, a higher hydrogen atmosphere might be required in the cooling zone to give more rapid cooling and/or because the oxidation potential increases as temperature reduces. The concept of supplying different mixtures to different parts of the furnace is too involved to be discussed fully here as to a certain extent each furnace and each product has to be individually considered but used correctly this concept can give the best possible sintering conditions while being economic in the use of atmosphere. Another advantage is that as sintering generally requires an atmosphere which is largely neutral to the product being sintered, synthetic atmospheres, which usually contain more nitrogen than generated atmospheres, are less reactive because of the reduced availability of the active constituents. [7]

To a certain extent the advantages of both generated and synthetic atmospheres can be gained by simply using the generated atmosphere as one component of the synthetic system. In practice this means adding nitrogen to a furnace using the generated gas. In this way low levels of hydrogen can be obtained from dissociated ammonia, for example, or nitrogen can be added to a furnace using endogas to encourage the flow of atmosphere out of the furnace mouth to assist in binder removal.

Care should be taken, however, when diluting atmospheres with nitrogen as, for example, Table 3 shows. As the percentage of endothermic gas in the blend reduces, sooting increases only to decrease again at below 30% of endogas.[8] It has been suggested that the longitudinal flow of the atmosphere within the furnace, which varies according to the flammability and density of the atmosphere, may result in the incomplete removal of some of the lubricants and binders present in the unsintered compact.[9]

However, because of the increased flexibility offered by synthetic systems nowadays,nearly all commercial sinterers use, if not a fully synthetic system, some trucked-in gas, usually nitrogen, to increase the flexibility of their generated atmosphere. Nethertheless, because of the differences between even apparently identical furnaces, the exact atmosphere and flow-rate to be used can never be determined without evaluation on the furnace concerned.

4. Safety

It is important to consider the safety of furnace atmospheres especially as virtually every component of an atmosphere is either flammable, toxic or an asphyxiant. In practice these hazards are overcome by ensuring that the gases do not leak into the environment or if they do they are burned in a controlled manner as they escape. In this way the flammable and toxic hazards are reduced, leaving only the hazard of asphyxiation. In practice, sintering areas are sufficiently well ventilated so that no real asphyxiation hazard usually exists. However special precautions must be taken especially when large proportions of nitrogen are used because although this may well reduce explosion risks it may reduce the flammability of the gases such that they do not burn at all and toxic gases may leak into the working area. If this is likely, gas monitoring equipment should be used and training should be given in appropriate first aid.

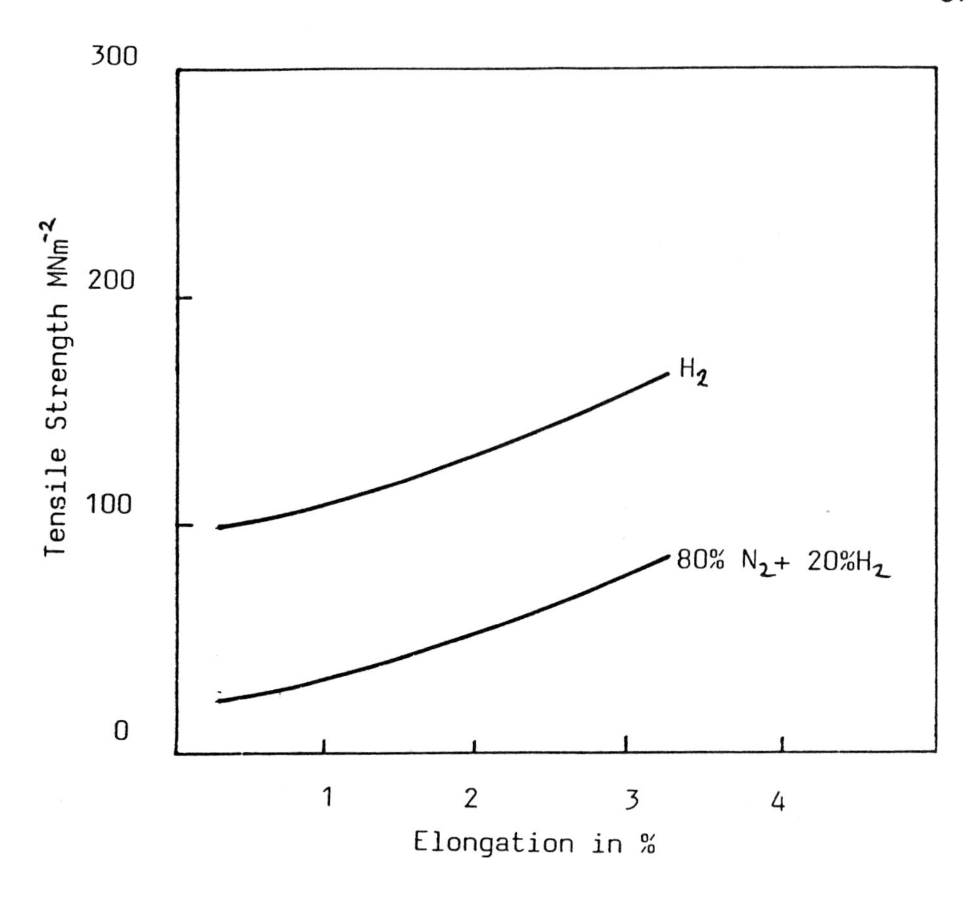

Figure 4 Influence of atmosphere composition on tensile strength (90%Cu, 10%Sn).

Table 2

Nominal diameter of component (mm)	Endogas Actual diameter (mm)		N_2-CH_3OH Actual diameter (mm)	
	MEAN	ST. DEV.	MEAN	ST. DEV.
12.1	12.185	0.0079	12.188	0.0025
36	36.126	0.0165	36.12	0.0037
39.5	39.746	0.0167	39.741	0.0099

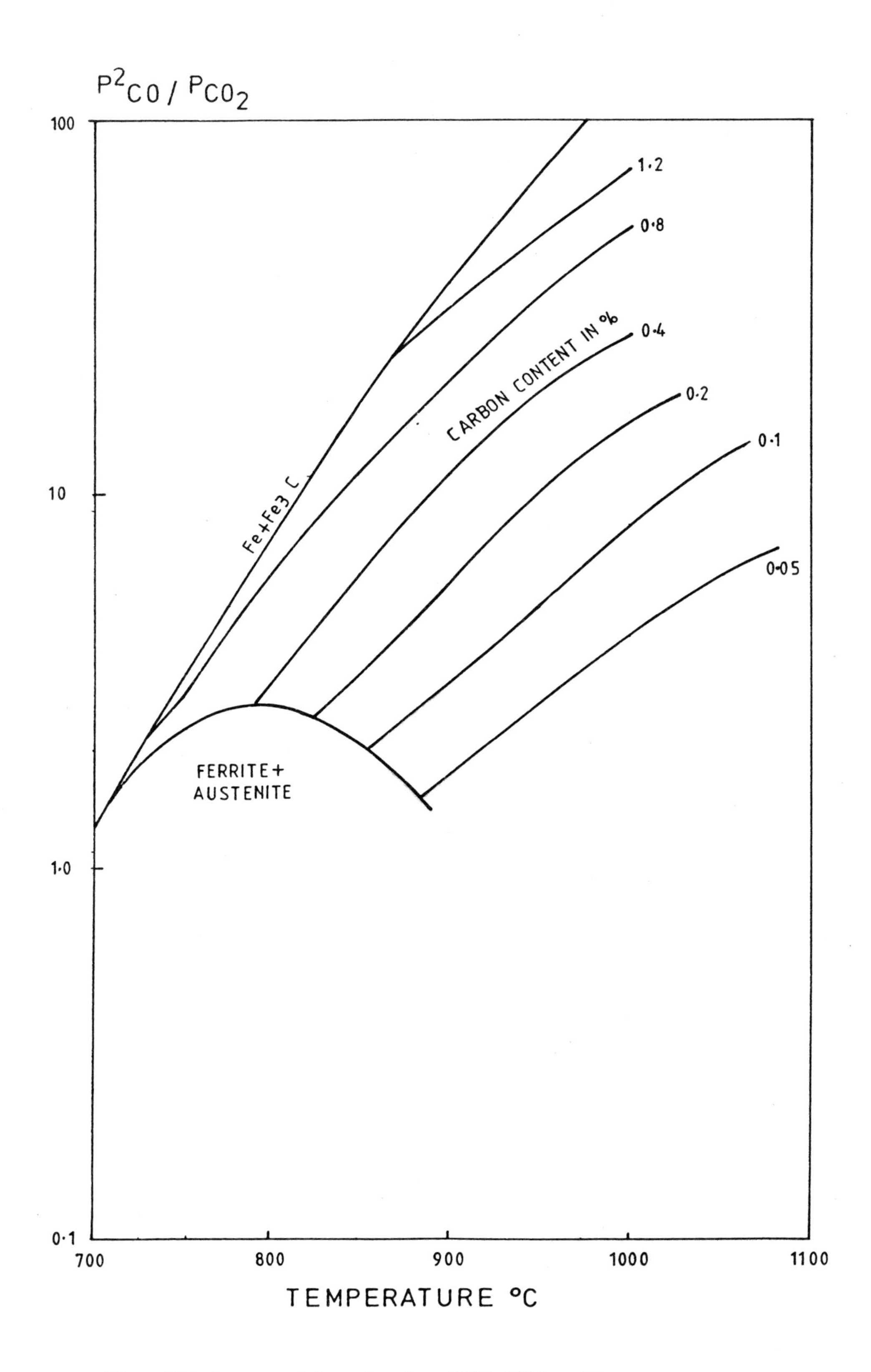

Figure 5 Carbon potential as a function of PCO_2/PCO_2 at different temperatures.

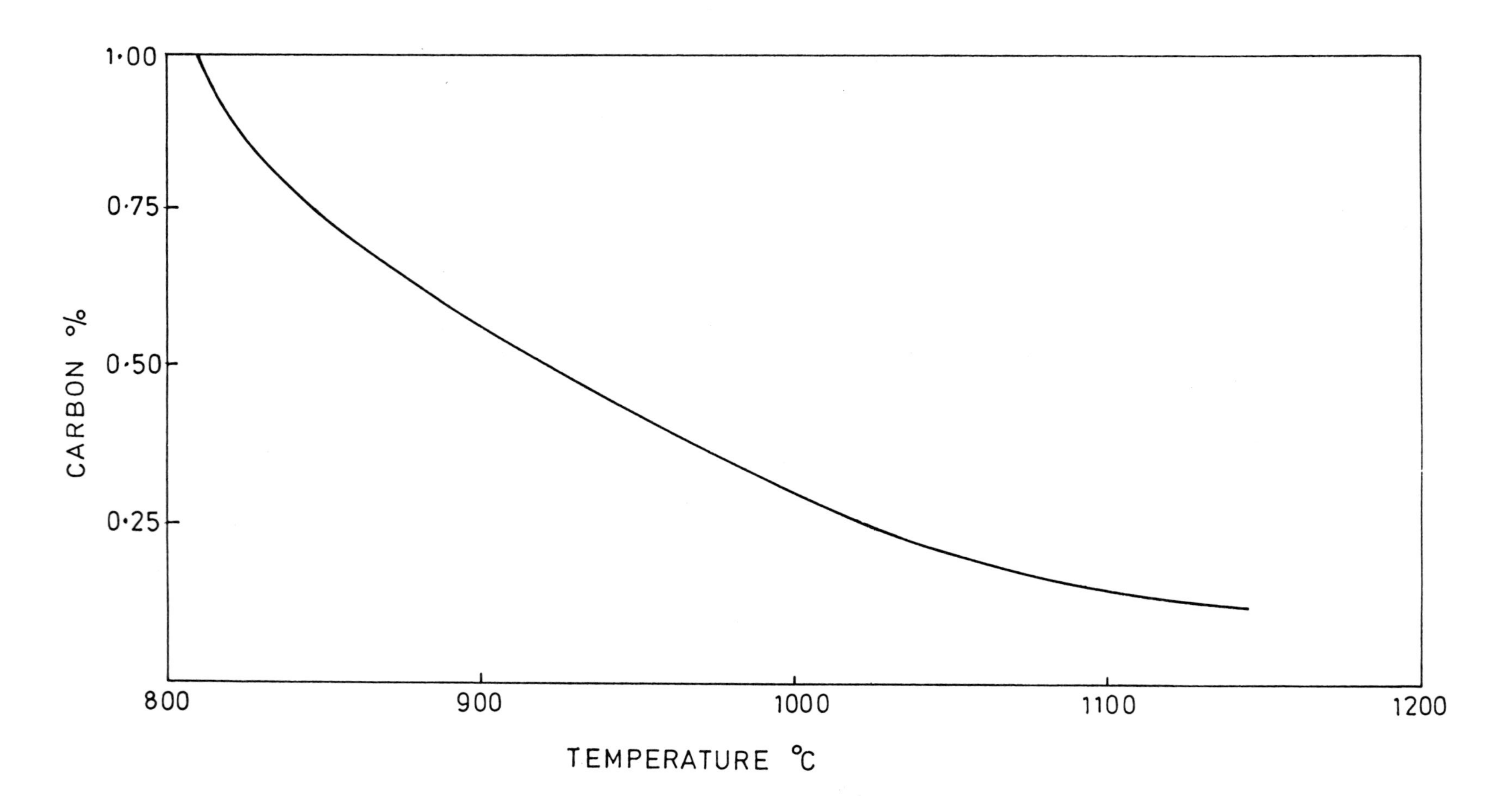

Figure 6 Carbon potential of endogas at various temperatures (40%H_2,20%CO, dew point ~5°C).

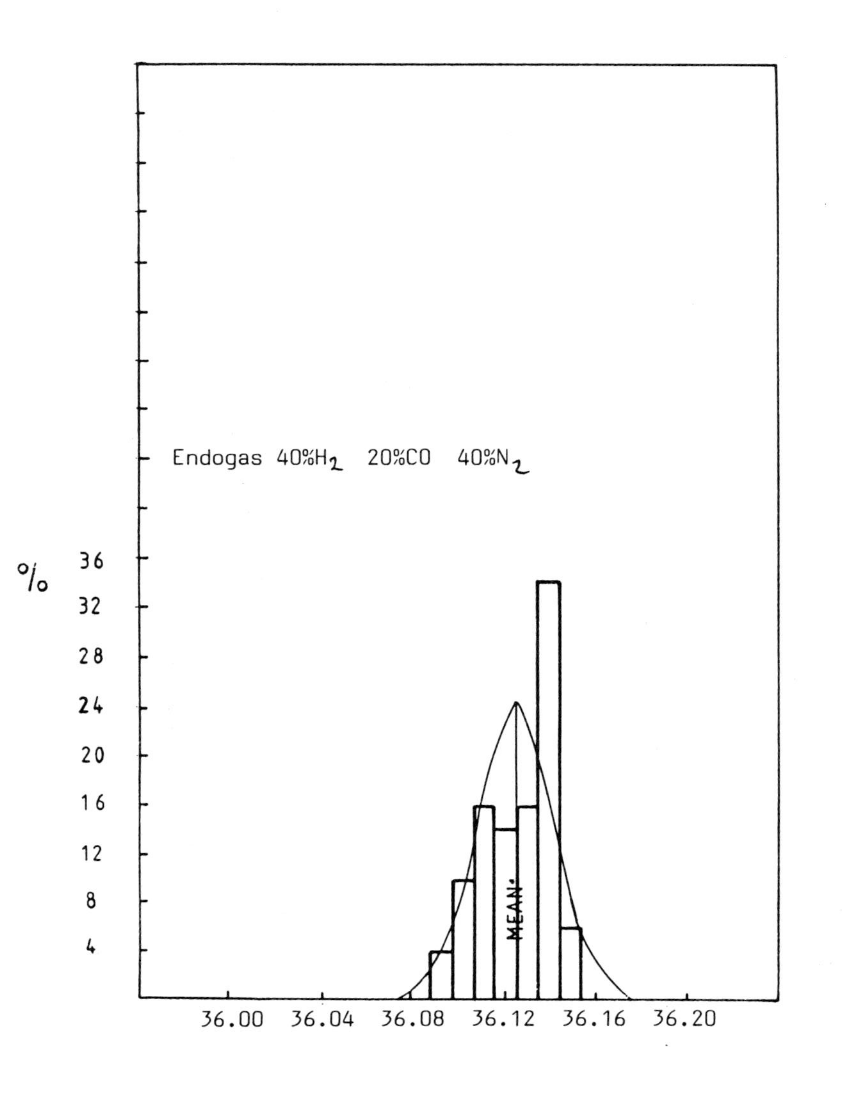

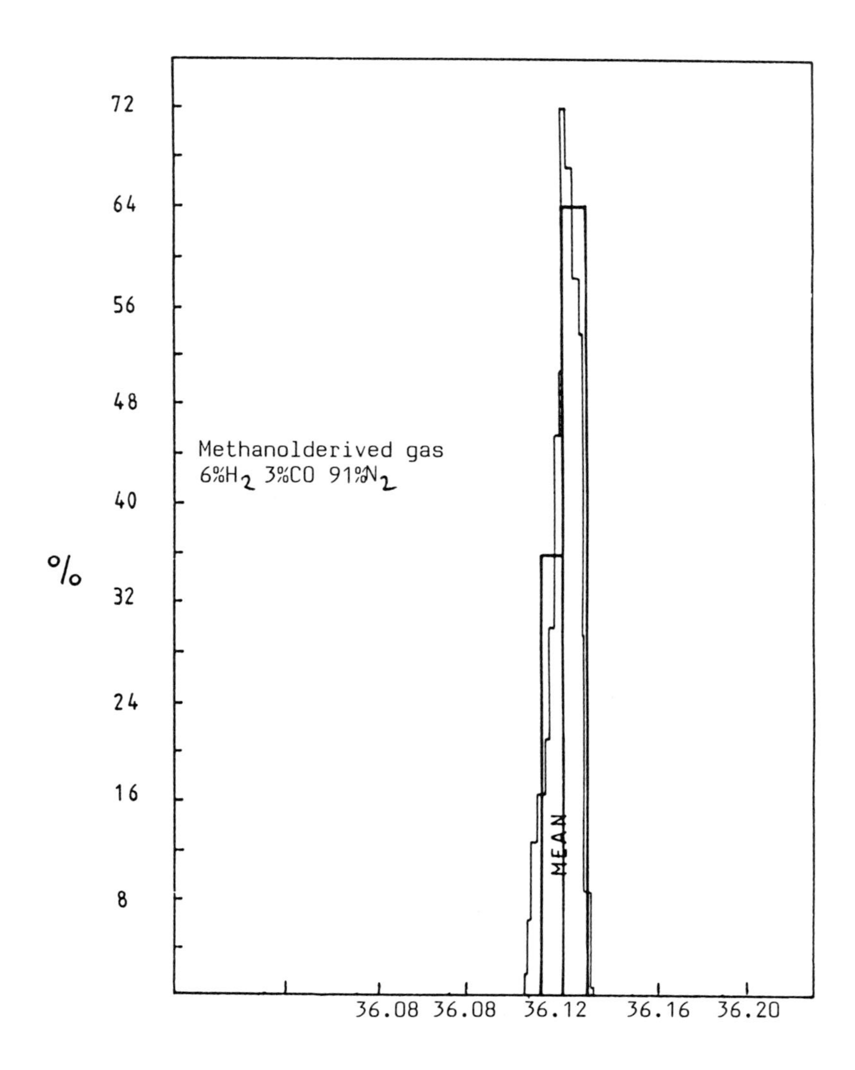

Figure 7 Variation in dimensions of product after sintering in endothermic gas and methanol derived atmospheres.

References

1. F.D. Richardson and J.H.E. Jeffes, The Thermodynamics of substances of interest in iron and steelmaking from 0°C to 2400°C, Journal of the Iron and Steel Institute, 160, p.261, 1948.
2. Norddeutsche Affinerie Aktiengesellschaft, publication Alsterterrasse 2, 2000 Hamburg 36, p.105 June 1986.
3. J.H. Kaspersma and R.H. Shay, Carburisation of iron by CO-based mixtures in nitrogen at 925°C, Met Trans, 12B, 1, pp.77-83, March 1981.
4. J.B. Austin, M.J. Day, Controlled Atmospheres, Amer. Soc. Met, pp.20-49, 1941.
5. A.J.F. Paterson and C.G. Smith, An Appraisal of Gaseous Carriers for Carburising, Heat Treatment of Metals, 2, pp.27-32, 1981.
6. M. Vandermeiren, unpublished data.
7. S.M. Adams, H.S. Nayar, Sintering Carbon Steels under Atmospheres with lowered Carbon Monoxide, Progress in Powder Metallurgy, 33, pp.815-827, 1984.
8. J.S. Becker, Choosing the Proper Atmosphere for Sintering, Precision Metal, 37, pp.74-79, Nov 1979.
9. P.R. Wilyman, Sintering with Nitrogen Based Atmospheres, Powder Metallurgy, 28, pp.85-89, 1985.

Table 3

	Percent endo in blend			
	100	60	30	20
H_2, %	35.5	24.5	13.8	10.8
CO, %	18.3	10.7	6.7	5.5
O_2, %	0.47	0.28	0.15	0.12
Dew Point, °C	-8	-25	-37	-40
R_b	68-75	70-80	70-78	73-80
Soot, index	1.0	1.2	2.0	1.0
Carbon, %	0.6	0.55-0.6	0.5-0.55	0.6

Data is for total flow of 1500 SCFH, 18-in. furnace with constant CH_4 addition of 12 SCFH.

Section 5

Applications

16

Current Status of PM Technology of Light Metals

F.H. FROES, Y.-W. KIM*, S. KRISHNAMURTHY*
AND R. SUNDARESAN

Institute for Materials and Advanced Processes, University of Idaho, Idaho, U. S. A.
*Metcut-Materials Research Group, Wright-Patterson Air Force Base, USA**

Abstract

This paper reviews the advances which have occurred in the past few years in the development of new and improved alloys of the lightweight metals, aluminium, titanium, and magnesium using the rapid solidification (RS) and mechanical alloying (MA) approaches. It also includes sections on metal matrix composites (MMC) fabricated using the powder metallurgy (PM) route. This review is an update of a number of similar attempts to define progress in these areas. Because of the differences in degree of maturity of alloys based on the three elements, there is a trend from technology-oriented to research-oriented in going from aluminium to titanium and finally magnesium-based systems. The paper has been separated into sections on RS of the three base systems, followed by a section which deals with MA of alloys based on all of the systems. It is concluded that the increased chemistry/processing window offered by these techniques should allow the metallurgist to produce materials to compete more effectively with the polymer matrix composites.

1. Introduction

At the present time the competition between various classes of materials, specifically metals, polymers and ceramics, has reached a critical stage [1-6]. To stem the inroads being made into traditional metals areas by materials based on polymers and ceramics, including composite concepts, the metallurgist must reach out with innovative chemistry/processing concepts to enhance the behaviour of metallic systems. Three such concepts are the techniques of rapid solidification (RS) and mechanical alloying (MA), including metal matrix composites (MMC) (in the present paper based on the powder metallurgy [PM] approach), which separately and together widen the chemistry/processing window beyond that available using conventional ingot metallurgy (IM) methods. The purpose of the present paper is to review the state-of-the-art of the RS and MA of the light metals aluminium, titanium, and magnesium; building on previous publications in the same area [7-9, 10-12].

1.1 Rapid Solidification (RS)

The advantages which can be achieved by RS can most conveniently be divided into constitutional and microstructural effects [13]. RS allows large departures from equilibrium constitution (the identity and compositional ranges of phase formed) resulting in large extensions of solid solubility, formation of non-equilibrium or metastable crystalline phases and metallic glasses, and retention of disordered crystalline structures in normally ordered materials such as intermetallic compounds. Additionally it leads to changes in the morphology and dimensions of the microstructure (the size and location of the phases present) in the direction of a more uniform and finer microstructure with a large reduction in solute segregation effects. The microstructural features which are refined include the grain size, dendrite spacing, constituent particles, and precipitates and dispersoids.

Other advantages of RS are increased tolerance to tramp elements, a big gain in days of recycling [7]. Since RS is also a powder metallurgy (PM) technique [7], attributes also include circumvention of working problems by production of net or near-net shapes, and elimination of highly textured products especially in hexagonal close packed materials such as titanium. RS production methods have been summarized in a number of recent publications [13,14].

1.2 Mechanical Alloying (MA)

Mechanical alloying (MA) can also lead to similar constitutional and microstructural effects as RS, with the magnitude of these effects often being more extensive than in the RS case [4-6,11,12]. Extension of solid solubility limits, microstructural refinement, and formation of non-equilibrium amorphous phases can be achieved by MA entirely in the solid state. The process of MA has been reviewed elsewhere [11,15] and so will not be covered here. At present the main effect of MA that is used in alloy development is the capability of the process for grain refinement and for introducing extremely fine dispersions. MA aluminium alloys, incorporating fine grain structure and fine dispersions, are already being commercially produced [16]. Work is currently going on to extend the application of MA to further light metals systems [12]. Studies are also in progress to take full advantage of the potential of the process including development of alloys not possible by other techniques, and to exploit combining MA with RS to even further enhance mechanical property behaviour.

2. Aluminium Rapid Solidification

Aluminium has a stable FCC structure with no element completely miscible in the solid state. Only five elements have a maximum terminal solid solubility (TSS) greater than 5 at% and at room temperature (RT) few TSS's exceed 0.5 at%, limiting aluminium alloys as precipitation strengthened alloys. The limited number of soluble elements is reflected in the wide variety of intermetallic phases [17,18] whose stability and compositional range appears to be determined by size and valency factors. Aluminium binary systems can be conveniently grouped into three types; eutectic, peritectic, and monotectic. Most of the elements form eutectic-type systems and all the elements having the maximum TSS higher than 1 at% belong to this category. Some low melting temperature elements form simple eutectic systems with aluminium, while complex eutectic systems, with intermediate intermetallic phase(s), are formed with many elements including inner-transition metals, and the actinide and lanthanide elements. Outer-transition metals with high melting points form peritectic systems with aluminium, with all TSS's occurring below 0.6 at%. Other elements are partly miscible with aluminium in the liquid state and solidify monotectically, with TSS's less than 0.02 at%.

Investigation of the RS of aluminium-base alloys has been more extensive than for the other light metals in both fundamental and practical aspects for various reasons including high liquid solubilities and low solid solubilities of many metallic and non-metallic elements [9]. Another important reason is substantial solid solubility extensions upon RS as shown in Table 1 for cooling rates of $\sim 10^6$ K/s [9,19,20]. Alloying elements having solid solubilities greater than 1 at% exhibit solubility extensions between 10 at% and 45 at% often falling around the eutectic composition. Those with solubilities less than 1 at% show extensions to the level of 1-10 at% generally exceeding the eutectic or peritectic composition substantially. It is expected that higher solubility extensions can be obtained if a greater undercooling is combined with higher rates of both solidification and solid state cooling [9]. The high solubility extensions are generally associated with planar solidification resulting in microscopically featureless microstructure [21-23]. Another advantage of RS in aluminium alloys is its drastic effect on morphological changes and microstructural refinement such as in grains, dendrites, and second phases [13,24-29]

New or improved aluminium alloys that have been developed during the last ten years using the benefits of RS can be divided into five classes:

(i) High strength/corrosion resistant alloys. These alloys are of hypoeutectic compositions containing one or more alloying elements with solid solubilities greater than 2 at% and are generally prepared by atomization [30-33] although Osprey techniques [34] have also been evaluated.

(ii) Low density/high stiffness alloys. These alloys are basically hypoeutectic compositions containing 2.8-4.0 wt%Li and other elements (Cu, Mg, Zr, etc.) and alloy powders are produced by atomization [35-37] or jet flow casting [36,38].

(iii) High temperature alloys. These alloys are of hypereutectic or hyperperitectic compositions consisting of two or more alloying elements with solid solubilities less than 2 at% and are produced by atomization [39-43] or planar flow casting [27,36,44].

(iv) Intermetallic compound alloys. These alloys are based on Al_3Ti [45-47] although other compounds such as $Al_{10}V$ [48,49] have also been investigated; atomization has been used to produce alloy powder [46-48].

(v) Metal matrix composites (MMC).

The extensive developments for the above RS PM aluminium alloys have been possible due, in addition to its ability to accommodate the RS benefits, to the relatively low chemical reactivity of molten aluminium with ceramics or high melting point metallic materials. The major concern, however, has been surface hydroxides that make process/property optimization complicated [31,32,50-54]. Because of this concern, RS processing has been conducted in vacuum or inert environment for most alloys, with some exceptions for the high strength PM alloys.

These MMC alloys are based on aluminium alloys, either age hardenable or non-heat treatable, with 10-40 vol% of ceramic particulates or short fibres that are introduced by either IM route or PM route [2,55].

2.1 High Strength/Corrosion Resistant Aluminium Alloys

Some of the conventional composition alloys from the 2000-series and 5000-series have shown improved properties with RS [56,57]. Although some work on 2000-series is still underway, most of the efforts recently have been towards 7000-series (containing Zn, Cu, and Mg) with small additions (Co, Ni, Zr or Mn), such as alloys 7090, 7091, 7064, and CW67 [17,30,33,35,50,58,59]. These alloys are produced by atomization and subsequent PM process. Atomized alloy powder has a wide range of particle size distribution and is generally screened to -100 mesh. These powder particles have an average particle diameter (APD) between 15 μm and 20 μm and exhibit dendritic microstructure with 1-2 μm dendrite arm spacing as shown in Fig. 1a for 7091 alloy powder after cold compacting. The standard PM process consists of, sequentially, cold compacting of powder particles, vacuum degassing, vacuum hot pressing (VHP), and hot working such as extrusion and/or forging [31-33,53,54]. Critical conditions during the PM process are cold compact density, degassing temperature, and hot work parameters such as extrusion ratio [53,54]. A cold compact density of 75% or lower is a necessity to ensure a continuous path for easy outgassing. Degassing temperature should be higher than 500°C (930°F) in order to decompose surface hydroxides completely. Such high temperature degassing is quite acceptable for these age hardenable alloys. The degassing not only minimizes residual hydrogen content (<1 ppm) but also makes the surface oxide film easily breakable during subsequent vacuum hot pressing. Redistribution of the surface oxides along powder particle boundaries after VHP reduces the detrimental effects of localized oxide distribution [60]. This can be done by applying adequate deformation during hot working such as high extrusion ratio and/or high aspect ratio [53]. Figure 1b shows fine and uniform microstructure of 7091 alloy powder after extrusion at 20:1 extrusion ratio and 5:1 aspect ratio.

Table 2 lists room temperature mechanical properties of these alloys obtained after process optimization. All the PM alloys have greater strength, at least 25% improvement in yield strength, over that of their IM counterpart 7075 with comparable elongation. Fine microstructure (Fig. 1) and higher solute content (Table 2), that is, RS benefits are responsible for the improved strength. Except for 7090, fracture toughness is also greatly improved. Improved corrosion resistance and stress corrosion cracking resistance is another important achievement in these alloys. The improved corrosion resistance makes the high strength PM alloys particularly attractive for many of the potential applications where the combination of high strength, fracture toughness, and corrosion resistance are all pertinent design characteristics. For example, PM7090 forgings will be used for a landing gear part of the Boeing 757 airliner [59] to replace IM7175 forgings. PM7091 extrusions are being evaluated to be used in transport airplanes and for leading edge flaps on the A-7 aircraft [59]. CW67 forgings have been successfully evaluated to be used as aircraft structural components such as T-38 outboard engine mount support beams, Fig. 2.

Unfortunately, the high cost of the PM alloys, typically 1.3 times or more than IM 7000-series alloys, reduces cost-effectiveness based on weight savings using the current PM alloys. Another factor impacting the attractiveness of 7090 and 7091 is the potential of low density aluminium-lithium alloys for saving weight. To be cost-effective, the high strength PM alloys have to be approximately 30% stronger, much tougher, and more resistant to fatigue than the IM7075 alloy.

PM CW67 appears to meet the cost-effectiveness requirements since in the extruded condition with a T7 temper, the alloy exhibits up to a 30% improvement in strength and as much as a 45% advantage in toughness compared to the IM7075 alloy. The CW67-T7 extrusions also show excellent fatigue crack growth resistance in a tension spectrum as well as in a tension-compression spectrum loading condition [59,61]. The attractive combination of properties exhibited by CW67 and related PM/RS alloys has resulted in considerable interest in substituting these alloys for existing airframe components which are experiencing costly maintenance problems. Flight tests are already underway on an advanced leading edge flap design (Vought), extruded forward longerons (Warner Robbins Air Logistic Center), a main landing gear door actuator (Boeing), and a variety of other trials are in the planning process [2].

2.2 Low Density/High Stiffness Al-Li Alloys

The aluminium-lithium alloys that are currently under active development and evaluation are predominantly ingot metallurgy alloys that contain 1.5 wt% to 2.7 wt% lithium as a primary alloying element [62]. These additions of lithium lower the density by as much as 9% compared to conventional IM alloys such as 7075 and increase elastic modulus. These IM Al-Li alloys (2090, 2091, and 8090) were developed to directly replace such IM alloys as 2024-T3, 7075-T6, and 7075-T73, and they have already demonstrated their excellent mechanical properties in longitudinal direction [62]. Table 3 compares the properties of IM2090 with those of IM7075.

A major competitor for low density PM/RS Al-Li alloys is low density IM Al-Li alloys [35]. The current IM Al-Li alloy development and commercialization effort is active and shows promise of achieving near term commercial success. This high level of activity in the IM area has had the effect of dampening the development efforts in the PM arena. The basic reason for this is that PM Al-Li alloys must demonstrate improved properties in order to justify their expected higher cost, and these improved properties, in general, have not been forthcoming.

Unfortunately, to date, no IM Al-Li alloys exhibit good short transverse properties in forged products. There is a substantial industry need for this product form and the IM Al-Li alloys have had difficulty in achieving the required corrosion resistance properties together with the required toughness and fatigue properties. The poor ductility and fracture toughness of Al-Li alloys are mainly due to shearing of δ' (Al_3Li) precipitates during deformation. This planar slip can be reduced by effectively introducing second phases resistant to dislocation shear. A homogeneous distribution of the

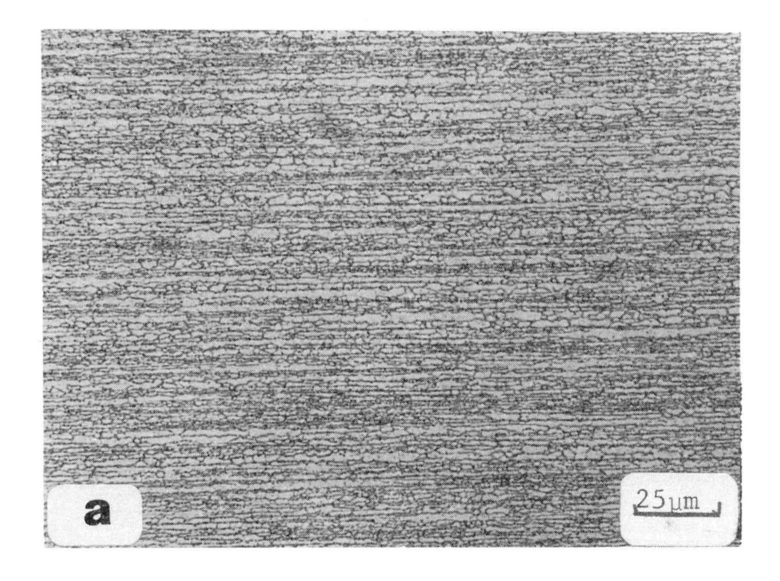

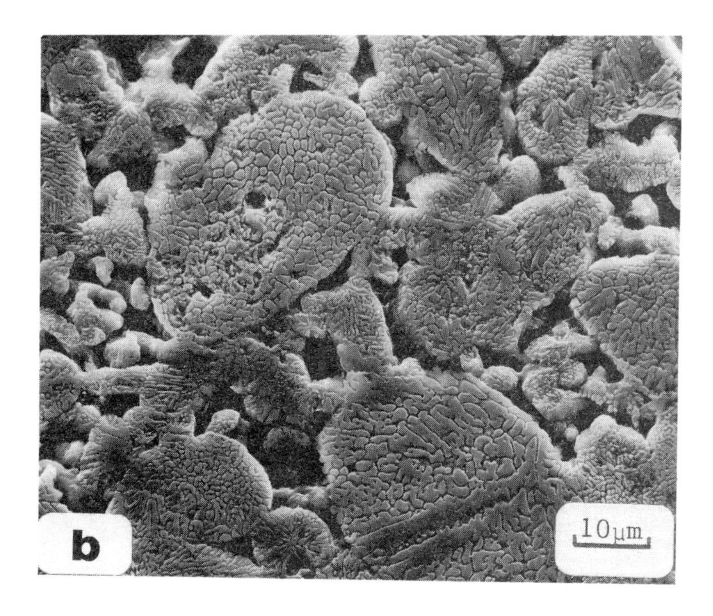

Figure 1 7091 alloy powder after (a) cold compacting and (b) vacuum hot pressing followed by extrusion at 20:1 extrusion ratio and 5:1 aspect ratio.

Table 1 Maxiumum solubility limits (at.%) of solutes in binary aluminum alloys under equilibium and rapid solidification condition [19.20]

Solute	Equilibrium Solubility	Equilibrium Ce or Cp[a]	Extended Solubility
Eutectic			
Zn	66.4	88.5	38-43.5
Ag	23.8	37.0	25-40
Mg	16.3	36.4	36.8-40
Cu	2.5	17.5	17-18
Si	1.6	12.0	10-16
Mn	0.9	0.95	6-10
Fe	~0.02	0.9	4-6
Co	<0.01	0.45	0.5-5
Ni	~0.02	2.8	1.2-7.7
Ce	0.01	2.6	1.9
Peritectic			
Ti	0.6	0.06	0.2-2
Cr	0.4	0.19	5-7
V	0.25	0.05	1.4-2
Zr	0.09	0.03	1.2-1.5
Mo	0.07	0.03	1.0-1.5
W	0.02	0.01	0.9-1.9

[a]Ce = eutectic composition; Cp = peritectic composition.

Table 2 Room temperature longitudinal mechanical properties of high strength/corrosion resistant RS/PM aluminum alloys [17, 59]

Alloy	Density, g/cm^3	Temper	UTS, MPa	YS, MPa	El., %	K_{Ic}, MPa$\sqrt{m}$	SCC[a], MPa
IM7075[b]	2.80	T73 Forg	505	435	13	32	290
IM2090[b,c]	2.60	T841 Forg	552	517	8	38	---
PM7090[b]	2.85	T7 Ext	627	586	10	26	310
PM7091[b]	2.82	T7 Ext	593	545	12	~44	310
PMCW67[b]	2.88	T7 Ext	614	580	12	47	310
PM7064[b]	---	T7 Forg	600	552	6	---	---

[a]Stress level for no stress corrosion cracking failure in short transverse direction.

[b]IM7075 (Al-5.6Zn-2.5Mg-1.6Cu-0.23Cr)
IM2090 (Al-2.7Cu-1.8Li-0.25Mg-0.12Zr-0.15Ti-0.6 Others)
PM7090 (Al-8.0Zn-2.5Mg-1.0Cu-1.5Co)
PM7091 (Al-6.5Zn-2.5Mg-1.5Cu-0.4Co)
PMCW67 (Al-9.0Zn-2.5Mg-1.5Cu-0.14Zr-0.1Ni)
PM7064 (Al-7.4Zn-2.4Mg-2.1Cu-0.3Co-0.2Zr)

[c]From Alcoa Alloy 2090, Alcoa Green Letter No. 226, 1988.

non-shearable S' (Al_2Cu Mg) precipitates by stretching the alloy is quite effective in dispensing dislocation movement. In forged parts the stretching is difficult or not possible. Another way of improving ductility is to add zirconium to Al-Li alloys so as to form metastable Al_3Zr, which is isostructural with the major Al-Li strengthening phase, Al_3Li. During aging treatments, Al_3Li precipitates around coherent Al_3Zr, forming non-shearable $Al_3(Li,Zr)$ precipitates, which induce Orowan by-passing of dislocations, resulting in an improvement in ductility [10]. A maximum of 0.25 wt%Zr can be added by the IM method limiting the number density of $Al_3(Li,Zr)$ precipitates. For further reduction of planar slip, accordingly, the stretching process is still required to distribute the S' ppts homogeneously.

Thus the two major opportunities for PM/RS Al-Li alloys are for forging applications and maximum density reduction. The first case is related to improve ductility and toughness without stretching forging parts. This relies on the ability of RS in solid solubility extension, microstructural refinement, and chemistry homogenization. RS can extend the solubility of Zr up to 0.51 wt% without detrimental effect [44]. For maximum density reduction, it has been demonstrated that RS can dissolve lithium in aluminium as much as 4.0% (maximum solubility = 4.2%) without segregation [37]. Atomization [35-37] and jet casting [36,38] techniques have been used to produce rapidly solidified Al-Li base alloy powders or ribbon. Both atomized alloy powder and ribbon after chopping are consolidated by a PM process similar to that used for the HS/CR alloy powders. Room temperature mechanical properties of selective RS/PM Al-Li base alloys after hot work and aging are listed in Table 4.

2.3 Elevated Temperature Aluminium Alloys

These alloys are of hypereutectic or hyperperitectic compositions consisting of two or more of the insoluble elements from transition metals (Fe, Ni, Ti, Zr, Cr, V, Mo, etc.) and lanthanide series (Ce, Gd, etc.), occasionally with nonmetallic additions such as Si. Production of these alloys is only possible by rapid solidification such as atomization and planar flow casting because of low solid solubilities of alloying elements. Because of the low solid solubility/diffusivity combined with quite extensive solubility extensions of the alloying elements, they offer good elevated temperature behaviour.

Presently these RS alloy are the most attractive commercially since it shows a clear density advantage over titanium alloys (the high strength/corrosion resistant alloys to not show a clear advantage over competing IM aluminium alloys, while in the low density arena IM Al-Li alloys are receiving the vast majority of attention because of cost considerations and compatibility with existing equipment).

Most elevated temperature alloys developed and/or characterized so far are of Al-(7-12)wt%Fe base ternary or quaternary compositions, with a few exceptions such as Al-Cr base alloys. The Al-Fe base alloys include:

(i) Al-Fe-X type with X being an eutectic forming element from either rare earths such as Ce [39,40,63] and Gd [29] or transition elements [39,51] such as Ni.

(ii) Al-Fe-Z type with Z being a peritectic forming element like Mo [64], V [27], Zr [27], and Ti [65].

(iii) Al-Fe-Z-Z type with Z-Z being two of the peritectic forming elements such as Mo-V [40] and V-Zr [27].

(iv) Al-Fe-Si-Z type [66] with Z being a peritectic forming element (V, Cr or Mo). These alloys typically contain total solute contents (5-9 at%) around the maximum extended solubility (Table 1). These solutes yield fine intermetallic compounds during RS and/or post-RS processing to form about 20-40 vol% strengthening dispersoids, which tend to coarsen, however, exclusively above about 370°C (700°F) [29,51,66]. The peritectic Al-(4-5)wt%Cr base alloys contain a small amount of Zr and additionally Mn [41,43]. These alloys contain relatively small solute contents (~3.5 at%) within the maximum extended solubility, which, however, transforms into adequate amounts of dispersoids due to the formation of aluminium ultra-rich compounds such as $Al_{13}Cr_2$.

Among the number of alloys tested in this class the most promising engineering alloys to date include Al-Fe-Ce, Al-Fe-V-Si, and Al-Cr-Zr alloys. These alloys exhibit attractive tensile properties at elevated temperatures up to 350°C (660°F). Poor fracture toughness has been a concern [40,53,66]; however, significant improvements have been achieved by chemistry control and process optimization including degassing and microstructure uniformization.

2.3.1 Al-Fe-Ce Alloys

The aluminium-rich Al-Fe-Ce powder alloys developed for elevated temperature applications have hypereutectic compositions with a total solute (Fe and Ce) content between 5 at% and 7 at%, all of which is used to form strengthening intermetallic compounds [40,63]. These alloy powders are produced by a flue gas atomization and are usually screened to -200 mesh size or 74 μm to 2 μm (Fig. 3). This results in nonuniform microstructure widely varying powder to powder [23,53,67-71], which causes wide hardness variations [67,69]. The nonuniform microstructure is developed due to three different types of solidification in the powder, i.e., planar solidification (PS), cellular solidification (CS), and dendritic solidification (DS). The PS microstructure contains extremely fine (<10 nm) dispersoids while the DS microstructure contains coarser (10-100 nm) dispersoids in interdendritic regions [23,53,63]. Fine powder particles (<6 μm) consist of an entirely PS microstructure, intermediate sized (6-10 μm) powders contain both PS and CS microstructures, and larger

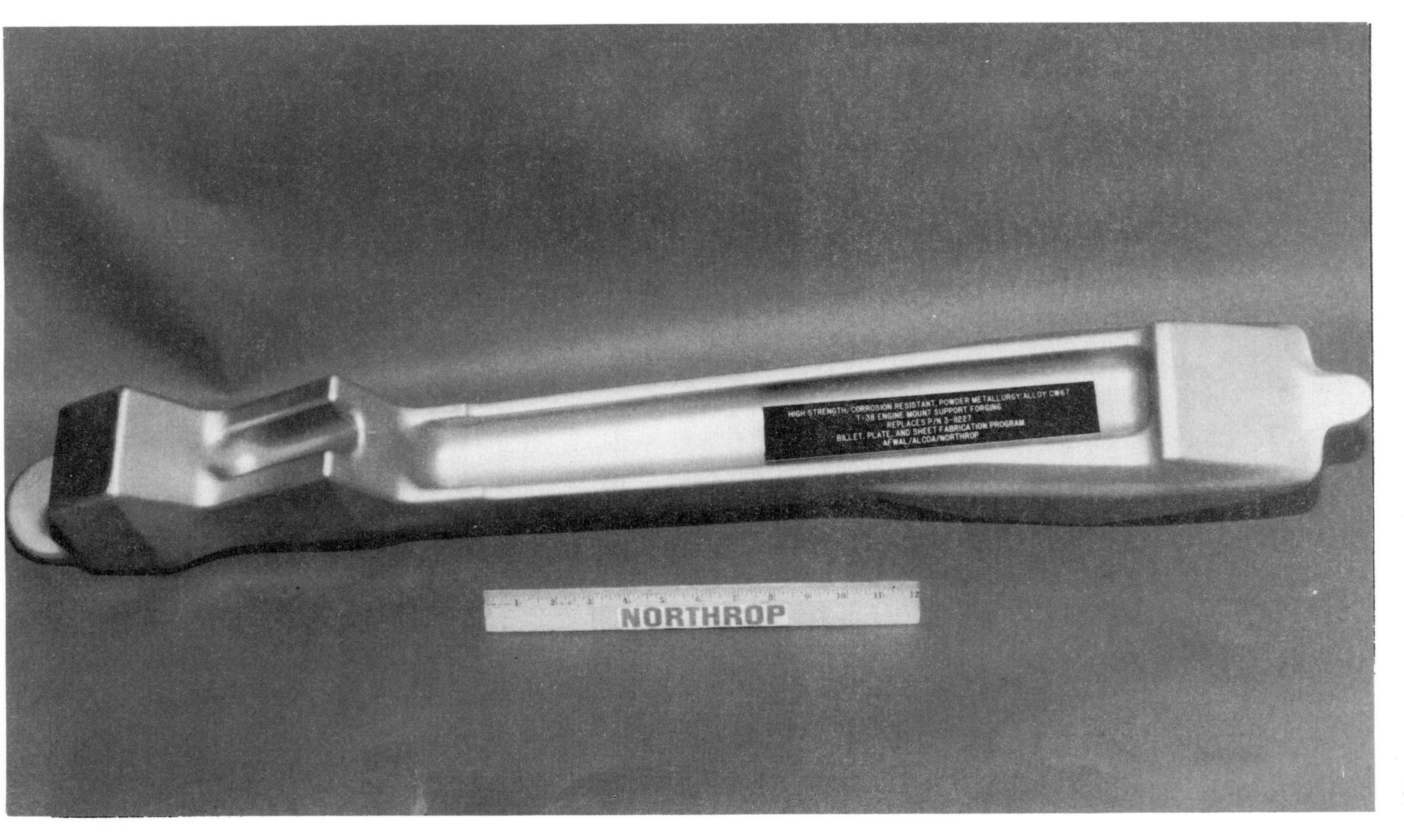

Figure 2 T-38 outboard engine mount support beam, forged from CW67 (courtesy of Eric Pohlenz, WRDC/MTPM, Wright-Patterson AFB, OH).

Table 3 Typical longitudinal properties of IM209 and IM7075 plates [59,62]

Alloy	Temper	YS, MPa	UTS, MPa	El., %	K_{Ic}, MPa$\sqrt{m}$
2090[a]	T8E41	517	552	8	38
	T8E50	441	490	9	47
7075[b]	T6	510	505	11	28
	T73	435	505	13	32

[a]Density = 2.60.
[b]Density = 2.80.

Table 4 Room temperature properties of RS/PM Al-Li alloys [35-38, 44]

Alloys	Density, g/cm³	Temper, °C/hr	YS, MPa	UTS, MPa	El., %	K_Q, MPa$\sqrt{m}$
Atomized alloys						
Al-2.5Li-11Be	2.42	--	432	504	5.2	
Al-3Li-1Mg-1.5Cu-0.2Zr	2.49	190/8	509	596	3.1	
Al-3Li-0.2Zr	-	160/32	454	492	10.5	
Al-4Li-0.2Zr	2.42	160/32	449	509	6.0	
Al-4Li-1Cu-0.2Zr	2.43	160/32	473	510	3.8	
Al-4Li-1Mg-0.2Zr	2.41	160/32	408	514	4.9	
Jet cast ribbon alloys						
Al-3.2Li	2.46	--	462	600	8	24
Al-3.1Li-2.1Cu-1.0Mg-0.45Zr	-	170/4 + 190/16	531	607	6.1	
Al-3.1Li-2.1Cu-1.5Mg-0.51Zr	-	160/4 + 180/16	554	632	5.5	
Al-2.8Li-1Cu-0.5Mg-0.5Zr	2.43	--	442	534	5.2	

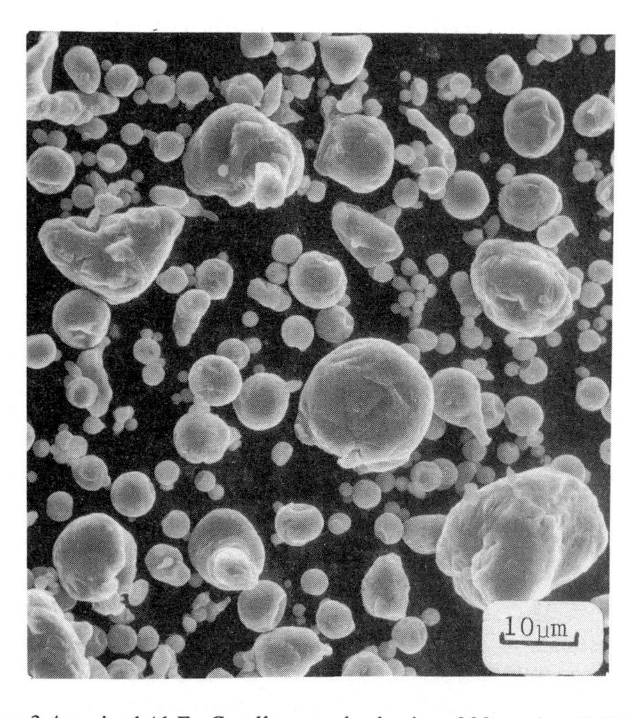

Figure 3 Atomized Al-Fe-Ce alloy powder having -200 mesh or 2-74 μm size.

powders exhibit all three types of microstructures. The PS regions have very high hardness values around 300 VHN and coarse DS regions show hardness values as low as 90 VHN [23,63,67,72]. In addition, powder particles are covered by an amorphous surface hydroxide film similar to that on 7091 powder. The alloy powders are consolidated by a PM process that consists of atomization/rapid solidification, compact, degassing, hot pressing, and hot working. The alloys are strengthened by 25-35 vol% of intermetallic compound dispersoids that are stable up to 350°C (660°F) but sensitive to processing conditions because of dispersoid coarsening. This limits processing temperature to below 400°C (750°F), resulting in relatively high residual hydrogen content.

Remarkable improvements in processing have been made recently to reduce such detrimental effects as nonuniform microstructure and residual hydrogen [23,40,41,43,52-54,63,67-71,73]. These improvements have resulted in tensile strengths of 470 MPa to 570 MPa (68 ksi to 83 ksi) at RT, and as high as 270 MPa (39 ksi) at 316°C (600°F), RT elongation as high as 12%, and RT toughness as high as $K_{Ic} = 26$ MPa $\sqrt{m}$. Table 5 lists mechanical properties of typical Al-Fe-Ce RS/PM alloys processed under optimum processing conditions [40].

2.3.2 Al-Fe-V-Si Alloys

These alloys are produced from rapidly solidified planar flow cast [72] ribbon having fairly uniform microstructure. The alloy ribbons are first mechanically comminuted into -30 mesh powders, degassed as loose powder, hot pressed in vacuum, and then either extruded or forged [74]. These ribbon powder alloys have two advantages over atomized powder alloys; the first is a relatively uniform microstructure and the other is reduced oxygen and hydrogen content. These alloys are strengthened by 27 vol% to 36 vol% of $Al_{12}(Fe,V)_3Si$ dispersoids that are finely (<0.1 μm) and fairly uniformly dispersed throughout the fine grained aluminium matrix. These fine dispersoids are reported to resist coarsening even at 425°C (800°F) and maintain their spherical shape and metastability up to 500°C (930°F). Mechanical properties of two typical Al-Fe-V-Si RS/PM alloys are listed in Table 6.

The high volume fractions of silicides increase the modulus of the Al-Fe-V-Si alloys compared with conventional aluminium alloys. The specific stiffness (modulus/density) of alloy FVS1212 is greater than that of competing titanium alloys and precipitation strengthened steels. In fact, at temperatures above 420K (300°F) the specific stiffness of alloy FVS1212 is greater than that of aluminium 6061 containing 20 vol% particulate silicon carbide. The Al-Fe-V-Si alloys do not require special machining tools and techniques for fabrication, nor complex heat treatments to achieve optimum strength levels, as is the case for aluminium based silicon carbide reinforced composites.

The Al-Fe-V-Si aluminium alloys have excellent resistance to corrosion. Alloy FVS0812 has been found to be extremely resistant to weight loss after exposure to a salt fog. Alloy FVS0812 is especially resistant to stress corrosion cracking. The alloy survives up to 40 days of alternate immersion in a 3.5% sodium chloride solution, stressed in the short transverse direction to 95% of the alloy's tensile yield strength (360 MPa [52.5 ksi]), without cracking.

2.3.3 Al-Cr Alloys

Al-Cr-Zr-(Mn) alloys are of hyperperitectic composition and are normally processed by atomizing in nitrogen followed by PM process as described above.

As-atomized powder is characterized by fine to coarse Cr-rich dispersoids ($Al_{13}Cr_2$), fine cellular microstructure and featureless regions formed by planar solidification [41,42]. When heat treated, fine Al_3Zr dispersoids are precipitated out at 270-400°C (518-750°F). Additionally, $Al_{13}Cr_2$ starts to coarsen or is newly formed at temperatures of 475°C (890°F) or higher. Because of this, these alloys can be consolidated at relatively low temperatures and hot worked at the relatively high temperatures of 400-425°C (750-800°F). Table 7 lists typical properties of two Al-Cr-Zr-(Mn) RS/PM alloys.

2.3.4 Applications

Elevated temperature RS/PM aluminium alloys are initially destined for a wide range of aerospace applications including gas turbine engines (low temperature fan and compressor cases, vanes and blades), missile applications (fins, winglets, rocket motor cases), airframe structures, and landing wheels. Some of the parts made of these alloys are shown in Figs. 4 and 5. In these applications, the improvements in both temperature capability and elastic modulus can be utilized. Perhaps the chief attribute of high temperature aluminium alloys over their chief competitor, titanium, is cost. The flyaway cost of titanium is normally very high, between $200 and $400 per pound, which is twice the flyaway cost of IM aluminium alloys [5,6]. Although the costs of the high temperature PM aluminium alloys have not yet been fully established, they are expected to be between 30% to 50% less than titanium alloys in actual use. The cost and weight benefits identified for these alloys in comparison with competitive titanium alloys assure a bright future for these PM aluminium alloys, assuming no problems are encountered [45].

2.4 Intermetallic Compound Alloys

Aluminium-rich intermetallic alloys such as Al_3Ti base [45], cubic $Al_{10}V$ base [48], and cubic Al_3Sc [76] compounds have drawn attention recently because of their low density and potential for good elevated temperature properties including strength and oxidation resistance [45]. It has been shown that replacing a certain amount of the aluminium in Al_3Ti with Cu, Ni or Fe transforms the tetragonal Al_3Ti to an ordered cubic structure. This has triggered research and development activities in these alloys using both ingot metallurgy [77] and powder metallurgy [46-49] routes. The results

Table 5 Properties of typical Al-Fe-Ce alloys

Alloy	Density, g/cm^3	Temperature, °C	YS, MPa	UTS, MPa	El., %	E, GPa	K_{Ic}, MPa$\sqrt{m}$
Al-7.1Fe-6.1Ce (extruded)	2.96	RT	387	493	11.7	78.9	12.2
		149	362	425	6.0		
		232	332	367	4.7		
		316	212	237	10.7		
Al-7.1Fe-6.1Ce (extruded and rolled	2.96	RT	524	567	5.7	78.9	26
		149	403	427	5.0		
		232	278	304	5.0		
		316	150	176	6.5		
Al-8.1Fe-6.8Ce (extruded)	3.00	RT	440	525	9.0	85.6	6.2
		149	364	433	4.0	71.8	
		232	297	351	6.0	70.4	
		316	211	246	9.0	63.5	
Al-8.4Fe-7.2Ce (extruded)	3.01	RT	458	563	8.0	80.1	6.4
		149	448	484	4.0		
		232	391	424	4.0		
		316	225	271	7.3		

Table 6 Longitudinal mechanical properties of RS/PM Al-Fe-V-Si alloy extrusions/sheet[a] [44]

Alloy	Density, g/cm^3	Temperature, °C	YS, MPa	UTS, MPa	El., %	E, GPa	K_{Ic}, MPa$\sqrt{m}$
FVS0812 Al-8.5Fe-1.3V-1.7Si	3.02	RT	390	437	10	88.4	31
		150	340	372	7	83.2	
		204	312	341	8		
		280	280	308	9		
		315	244	261	9	73.1	
FVS1212 Al-12.4Fe-1.2V-2.3Si	3.07	RT	605	636	8.7	95.5	
		345	276	286	6.7		

[a] Allied-Signal, Inc., Morristown, NJ, Engineering Data Sheet, 1987.

Table 7 Mechanical properties of RS/PM Al-Cr-Zr(Mn) alloys [75]

Alloy	Density, g/cm^3	Temperature, °C	YS, MPa	UTS, MPa	El., %	E, GPa	K_Q, MPa$\sqrt{m}$
Al-5Cr-2Zr	2.82	RT				80.8	
Al-4.8Cr-1.4Zr-1.3Mn	2.86	RT	484	515	8.0	86.5	35
		150	392	415			
		233	315	355			
		315	230	260			

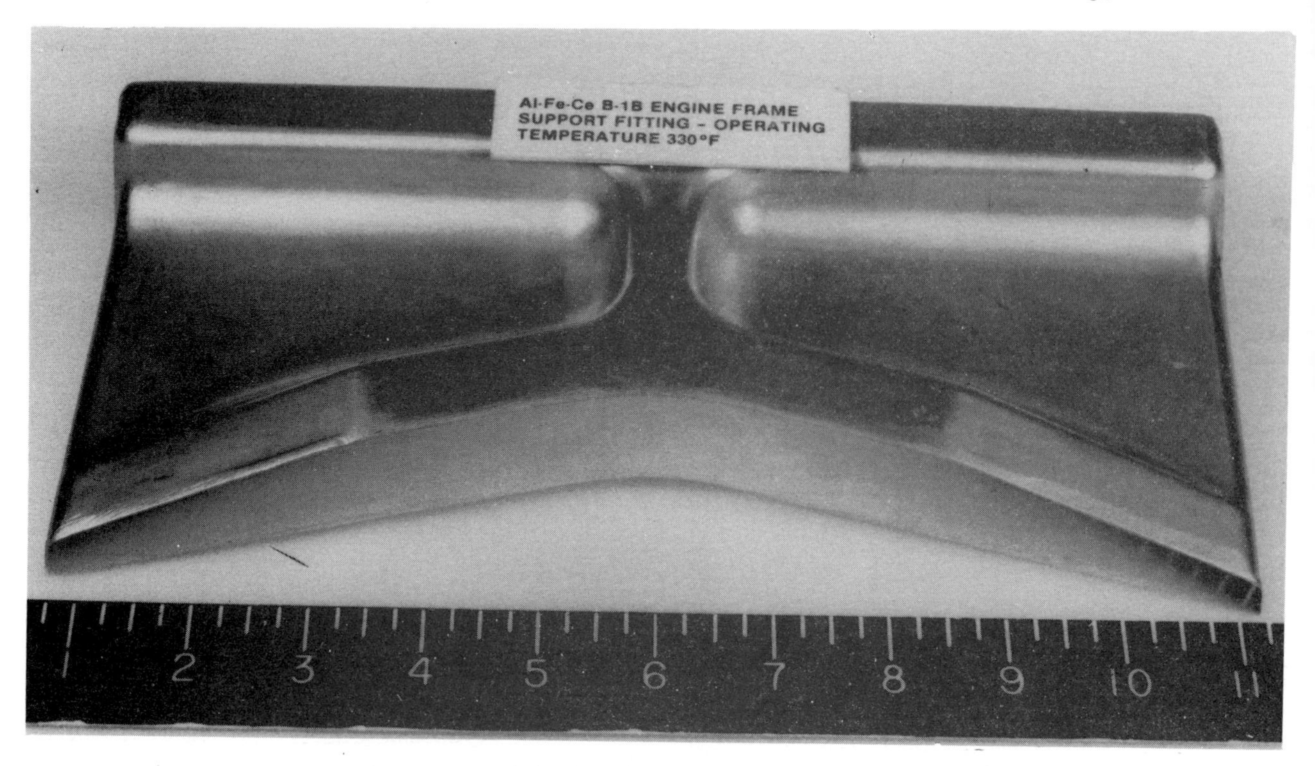

Figure 4 B-1B engine frame support fitting forged from an Al-Fe-Ce alloy (courtesy of D. Denzer, Alcoa).

Figure 5 Aircraft landing wheel forged from alloy FVS0812.

from both RS/PM and IM aluminium intermetallic alloys are rather disappointing because the cubic structure does not yield any ductility at room temperature. It appears that future work should be concentrated more on the understanding of the deformation mechanism, while characterizing much wider ranges of compositions.

2.5 Defects in PM Aluminium Alloy Products

One of the most serious challenges to the successful implementation of any high integrity PM technology is metallurgical defects [7], and this has proved to be no exception for the aluminium system [10,67]. The specific challenges are:

- That defects in the final PM aluminium alloy product will not exceed maximum allowable size (~100 μm) and level.
- End item quality control methods for identifying defects (unique to these products) must be successfully developed.

However, as always, quality assurance ("built-in" quality) is to be preferred over inspection for quality [4-6]. Experience has shown that defects in aluminium PM products are both persistent and difficult. Defects in PM aluminium alloys can be divided into four groups:

(i) Off-chemistry powder.
(ii) Excessive quantities of gas or gas generating constituents.
(iii) Unconsolidated or inadequately bonded powder particles.
(iv) Ceramic, metallic, and non-metallic impurities.

The origin of these defects stems from the many extra steps generally required in PM processing as compared to normal IM procedures and the many novel aluminium powder making and consolidation methods. The first three groups of defects are being satisfactorily dealt with in most cases, but defects involving inclusions have not been satisfactorily or consistently eliminated and still present problems. Facilities dedicated to the manufacture of aluminium PM alloys are expected to eliminate many of these deleterious particles; particularly facilities involving quality control measures to eliminate contamination associated with powder processing steps (including sizing, blending, compacting). However, some of the defects peculiar to PM aluminium alloy products are not readily detectable by inspection techniques normally applied to aluminium. Hence the need for improved and innovative quality control (assurance) techniques.

2.6 Alternative Processing

An alternative to RS/PM for production of aluminium alloys, not possible by the IM method, is to produce alloys directly from the vapour phase using electron beam melting methods [78]. The process can now produce aluminium alloy slabs on the 50-100 kg scale up to 44 mm thick at a deposition rate of 6 mm/hr. The strength of vapour-deposited Al-7.5Cr-1.5Fe alloy, up to 250°C (480°F), is significantly higher than the strength of RS/PM alloys thanks to the fine microstructure produced by the vapour deposition technique. Above about 300°C (570°F), both types of alloys have similar strength levels. Alloy development using this vapour technique is still in its early stages but potentially could result in alloys which significantly outperform the RS/PM alloys.

2.7 Metal Matrix Composites

Aluminium base MMC's processed by the PM method employ aluminium alloy powders as the matrix and generally SiC particulate or whiskers as the reinforcement [2,55]. Generally, the whiskers have an average diameter of 0.6 μm (2.5×10^{-5} in.) and lengths ranging from 10-80 μm ($4\text{-}30 \times 10^{-4}$ in.) with tensile strengths up to 7000 MPa (1000 ksi). Depending on the intended application, aluminium MMC contains from 10-40 vol% SiC reinforcement.

These aluminium MMC's may be divided into two groups depending on the matrix alloy used. The first group of aluminium MMC's are based on age hardenable aluminium alloys such as 2000-, 6000-, and 7000-series alloys. These MMC's have significantly improved stiffness and strength over the matrix or non-reinforced alloys for all temperatures. Table 8 lists properties of some of these MMC's. Because of precipitate coarsening, however, these MMC's lose their strength rapidly with temperature in a similar manner to reinforced material, limiting their use to approximately 250°C (480°F).

The second group of aluminium MMC's are based on HT aluminium alloy powder such as RS HT aluminium alloys. Several independent experiments indicate that these MMC's have problems such as fabrication difficulties and extremely poor ductility. However, with chemistry control of the matrix alloy and process optimization, the HT aluminium alloy based MMC's should be viable for HT structural applications.

3. Titanium Rapid Solidification

Titanium alloying behaviour depends in large part on the ability of elements to stabilize either the low temperature alpha (HCP) or high temperature beta phase (BCC) which relates to the number of bonding electrons [80,81]. Titanium phase diagrams are conveniently categorized into the two major subdivisions of alpha and beta stabilizers [82-86]. The alpha stabilizers are divided into those having complete stability, a simple peritectic reaction (e.g., Ti-O and Ti-N), and those which have limited alpha stability, with a peritectoid reaction into beta plus a compound (e.g., Ti-B, Ti-C, and Ti-Al). The beta stabilizers are divided into two categories; beta isomorphous and beta eutectoid. In the former system an extreme beta solubility range exists with only a restricted alpha solubility range (e.g., Ti-Mo, Ti-Ta, and Ti-V). For the beta eutectoid systems (e.g., Ti-Cr and Ti-Cu) the beta phase has a limited solubility range and decomposes into alpha and a compound.

Investigation of the RS of titanium-base alloys has trailed that of aluminium-base materials for a number of reasons including the extreme reactivity of molten titanium [85]. However, in recent years the potential for use of RS to enhance the behaviour of titanium systems has been recognized and extensive studies have been conducted [85,87-94]. However, the vast majority of this work has been in the research-base rather than applications oriented, as in the case recently with aluminium alloys, and the following review will therefore attempt to catch this flavour of the current RS titanium scenario.

A summary of the solubility extensions determined in titanium alloy systems is shown in Table 9. Here the maximum equilibrium solubility is the solubility in the high temperature beta phase. The extended solubility level indicated is the value judged [85,90] to be the level to which the matrix in question did not exhibit second phase particles involving the solute in the as-RS condition. As a transition metal titanium is characterized by its ability to provide greater than 1 at% solid state solubility for more than 50 elements [7]. Thus the potential for extended alloying is not as great as for the aluminium system where the corresponding number of elements is only eight [13].

The classes of alloys for which RS offer the potential for improved behaviour are:

- Conventional, particularly the "workhorse" Ti-6Al-4V alloy.
- Rare earth containing (including Y) which normally exhibit low solubility.
- Metalloids (B, C, Si) which exhibit low solubility.
- Eutectoid formers, which although they exhibit quite high solubility, are very segregation prone.
- Beta alloys.
- Intermetallic titanium aluminides.
- Various other systems including low density alloys.

Metal containment complicates the production of RS titanium alloys [85], although considerable attention has been given recently to circumventing this problem [14,85,90,92]. Even the most chemically stable ceramic compounds are dissolved in liquid titanium [85] and to date no scheme to minimize the reaction has become commercially viable, although significant progress is being made. Thus the melt processing for this system is different to that which has been used in other metal systems [14]. However melt spinning, melt extraction, and atomization processes have been developed at the laboratory and pilot plant scale [14,85,90,92,93].

3.1 Conventional Titanium Alloys

The most comprehensive study done to date on RS conventional titanium alloys has been on the Ti-6Al-4V alloy [7,85,87,90] using cooling rates varying from 10^4-10^7°C/s. From this work [94] two significant effects were noted; the beta grain size decreased as the cooling rate was increased (Fig. 6), and at the higher cooling rate studied, equiaxed alpha was produced on subsequent high temperature annealing (Fig. 7).

Other work on conventional alloys has shown the advantage to be gained from RS processing. In an alloy similar to Ti-6Al-4V, but containing Mo instead of V, both strength and ductility were increased as the cooling rate was increased from $10^3/10^4$ to $10^6/10^7$ K/s (Table 10) [95].

A study of the Beta III alloy (Ti-11.5Mo-6Zr-4.5Sn) indicated that the beta grain size decreased as the cooling rate increased ~40 µm in PREP, 2-4 µm in EBSQ [96]. These grain sizes were significantly smaller than in corresponding Ti-6Al-4V product (~135 µm and ~7 µm, respectively), indicating the effect of increasing solute content in decreasing grain size for a given cooling rate [13]. A further observation in Beta III was that upon aging alpha precipitated interdendritically in PREP material and at grain boundaries in the EBSQ alloy.

3.2 Rare Earth Containing Titanium Alloys

The RS of titanium alloys containing a rare earth element produced an ultrafine dispersion of stable rare earth oxides with particle sizes much finer than that produced in earlier work on ingot metallurgy (IM) material [97-99], even when mechanical alloying was used [100]. Both Er_2O_3 [101-105] and Y_2O_3 [101,106] were found to produce a fine dispersoid in rapidly solidified titanium alloys. Virtually no dispersoid was present in the RS condition [101,105,107] provided solid state cooling was sufficiently fast. Aging at various temperatures produced extremely fine dispersoids, with aging at 500°C (930°F) producing the finest dispersoid [105]. Fabrication of components requires exposure to consolidation cycles at temperatures up to 950°C (1740°F). The stability of the dispersoid is a critical issue to temperatures considerably higher than the service temperature.

A comparison of the dispersoid distribution in a series of ternary titanium- and rare earth-oxygen alloys indicated that alloys containing La, Nd, Dy, and Er had the highest resistance to coarsening at 800°C (1470°F) [101]. In these alloys, dispersoids had an average diameter range from 650-980Å with interparticle spacings from 0.3-0.8 micron. This translated to an Orowan strengthening increase of up to 85 MPa (12 ksi).

Table 8 Typical longitudinal properties of PM aluminum MMC's [2,79]

Material	Process	YS, MPa	UTS, MPa	El., %	E, GPa	K_Q, MPa$\sqrt{m}$
6061-T6	Extruded	255	290	17	70	-
6061-T6 (20% SiC)	Extruded	440	585	4	120	-
6061-T6 (30% SiC)	Extruded	570	795	2	140	-
2014-T6 (10% SiC)	Extruded	330	365	1.8	92	18
2124-T6 (15% SiC)	Rolled	570	710	5	114	59

Table 9 Solubility extensions in titanium alloy [85]

Element	Matrix	Maximum Equilibrium Solubility[a], at%	Extended Solubility[b], at%	Technique
B	Binary	0.5	6.0	Splat
C	Binary	3.1	10.0	Splat
Si	-10at%Zr	5.0	6.0	Splat
Ge	-5Al-2.5Sn	8.2	>5.0	EBSQ[c]
Nd	Binary	~0.3	>1.0	EBSQ
Y	-8Al	~0.5	>2.0	Laser
Gd	Binary	~0.3	>0.5	EBSQ
La	-5Al, -5Sn	~1.5	≥1.5	Splat
Ce	Binary	~1.5	>1.0	Splat
Er	-5Al	~0.3	≥1.5	Splat
Dy	Binary	~0.3	>0.6	EBSQ

[a]In beta titanium.

[b]As-rapidly solidified product.

[c]Electron beam splat quench.

The stability of the Er_2O_3 dispersoid has been shown to be very good in alpha Ti-Al alloys as long as the alloy remains in the alpha titanium phase field. The alloy Ti-6Al-2Sn-4Zr-2Er (wt%), rapidly solidified by melt extraction, was annealed for 1 hr at 950°C (1740°F) [108,109]. A dispersoid 200-400Å diameter, with an interparticle spacing of ~0.3-0.4 micron, was maintained in the grain interiors of this alloy. The rare earth sulphides and oxysulphides form a fine dispersion in alpha titanium alloys and exhibit good resistance to coarsening even after exposure to temperatures as high as 1000°C (1830°F) [109-112].

The rare earth dispersoid in titanium alloys appears to effect the behaviour of the material in a number of ways. First, the dispersoid produces hardening due to the Orowan effect [113]. There are also indirect effects due to a grain size reduction [101,109,110,111,114] and scavenging of interstitial elements thereby reducing interstitial strengthening. A fine dispersion of erbia particles is shown in Fig. 8 [115]. The net result in ternary Ti-RE-O alloys was a strengthening of the order of 40 MPa (6 ksi) at 700°C (1290°F) [114] and a significant creep rate reduction. An example of the creep behaviour of a dispersion strengthened alloy is shown in Fig. 9.

3.3 Metalloid Containing Titanium Alloys

The metalloid elements C, B, Si, and Ge have the potential for good chemical stability exceeded only by the rare earths. However, the stability of these elements is only moderate in titanium alloys, and their useful temperature range is below that of rare earth oxide dispersion strengthened alloys [116], i.e., below 600°C (1110°F). Fine, needle-like TiB [108] precipitates develop at temperatures above 500°C (930°F) (Fig. 10) [93,116,117,118] and coarsen dramatically at 800°C (1470°F) [108,117,118].

Rapidly solidified titanium alloys containing carbon have been shown to have good strength and ductility as-quenched. However, they lose both strength and ductility after aging at 700°C (1290°F) due to precipitation and rapid coarsening of titanium carbide [118]. Annealing carbon-bearing alloys at 900°C (1650°F) produced one-half micron diameter carbides.

The inherent instability of the metalloid compounds in titanium alloys limits their upper temperature use to about 500-600°C (930-1110°F). However, RS is an effective means of introducing relatively high volume fractions of fine titanium-metalloid compounds. The potential for alloys based upon metalloid additions is yet to be fully identified.

3.4 Eutectoid Former Titanium Alloys

The eutectoid formers (e.g., Ni, Cu, Co, Fe, Cr, Si, and W) are characterized by a phase diagram at the titanium-rich end, which involves both the eutectoid reaction ($\beta \rightarrow \alpha$ + compound) and often at higher temperatures and solute content, an eutectic decomposition ($L \rightarrow \beta$ + compound) [85]. This class of alloy has been studied in detail using IM [119]. It is normally possible to obtain relatively high levels of solute in the beta phase and lower though still significant levels in solution in the alpha phase.

A problem which arises, however, is that the eutectoid formers normally show a high partitioning coefficient between the liquid and the solid phase and a large temperature difference between the liquidus and solidus meaning that they are very segregation prone. Thus though there have been a number of attempts to develop titanium alloys containing eutectoid formers using conventional IM, only limited success has been achieved due to ingot segregation and also poor workability [120].

In one comprehensive study a few per cent of Be, Cu, Co, Fe, Ni, and Si were added to stable beta base compositions, but with a large adverse effect on hot workability and rapid aging kinetics in the case of the Ni and Si containing alloys [121]. From this study ultimately developed the commercial alloy Ti-10V-2Fe-3Al which is now seeing use in high strength applications. The addition of Si is used to enhance high temperature behaviour of commercial alloys, and 2.5Cu is used in the British moderate strength alloy IMI-230. Apart from these isolated instances, use of the eutectoid formers has not been realized.

By far the most comprehensive study of the RS characteristics of titanium eutectoid former has been carried out by Krishnamurthy and co-workers [122-126]. The eutectoid former system for which RS showed the most significant advantage was the Ti/Cr system with the addition of 4 wt%Al. Here a 15 wt%Cr alloy with the Al addition was evaluated under RS (chill block melt spun) and conventional conditions. The RS material showed a much finer microstructure than the beta grain size conventional material (30 μm compared to 300 μm) and more rapid hardening which was maintained for a considerably longer time. This enhanced behaviour was related to the finer grain size and much more uniform precipitation, the latter effect perhaps due to a higher vacancy concentration. Preliminary results on a Ti-8V-5Fe-1Al [127] PREP powder product indicate that strengths in excess of 1380 MPa (200 ksi) can be obtained with ductility levels close to 10% elongation [128,129], Table 11. These alloys would have applications in components such as landing gears.

3.5 Beta Alloys

Work on the beta alloys has included studies of the conventional alloys such as Beta III, Ti-15V-3Al-3Sn-3Cr and Ti-6Al-15V-2Er [130], microstructural studies of Ti-25V-4Ce-0.6S [131], and a heavily stabilized Ti-24V-10Cr alloy with and without Er additions [132,133]. The Ti-25V-4Ce-0.6S alloy showed a very fine dispersion leading to a predicted strength in excess of 1725 MPa (250 ksi). The Ti-6Al-15V-2Er alloy showed no dispersoid coarsening to temperatures as high as 760°C (1400°F) (beta transus estimated at 830°C [1525°F]). Second phase particles occurred in relatively large grains (~15 μm) in the Ti-24V-10Cr alloy suggesting that this system may have potential for high strength applications.

Table 10 Effect of cooling rate on tensile properties of a Ti-Mo-Al alloy [95]

Production Method	UTS, MPa	RA, %
Casting	895	12
Cast + Hot Worked	1070	22
$10^3/10^4$ Cooling + HIP	1070	22
Flake $10^6/10^7$ Cooling + HIP	1135	30

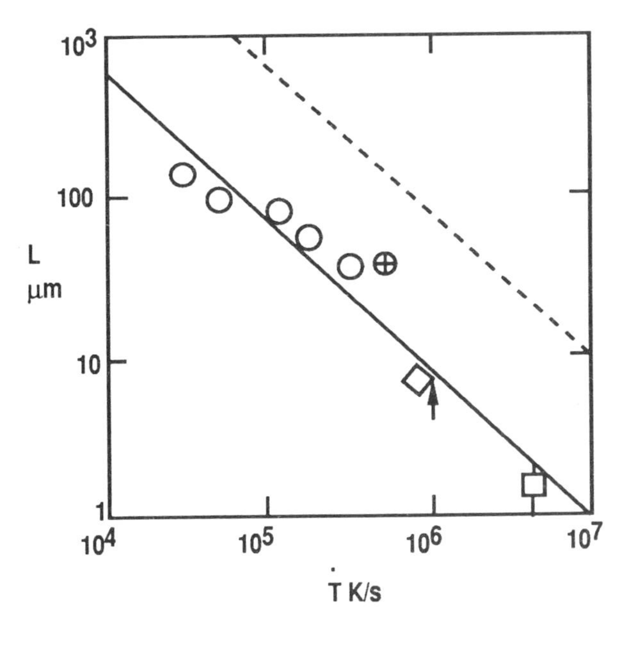

Figure 6 Variation of beta grain size in Ti-6Al-4V with cooling rate [94]. Dashed line is predicted for aluminum.

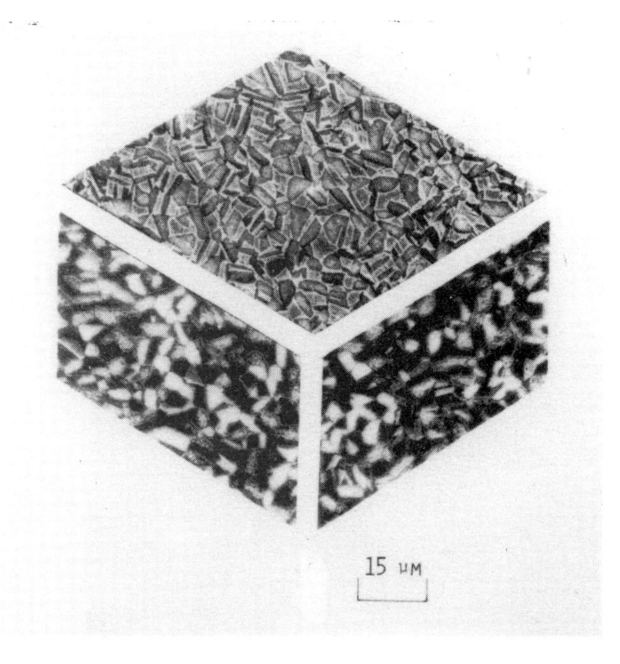

Figure 7 Equiaxed alpha morphology in Ti-6Al-4V produced after annealing material with a small beta grain size which resulted from rapid solidification [94].

Table 11 Tensile date from experimental PM/RS high strength Ti-1Al-8V-5Fe alloy [129]

Alloy	Tensile Properties YS, MPa	UTS, MPa	El., %
Ti-185	1390	1480	8
Ti-6Al-4V[a]	895	965	14

[a]Conventional.

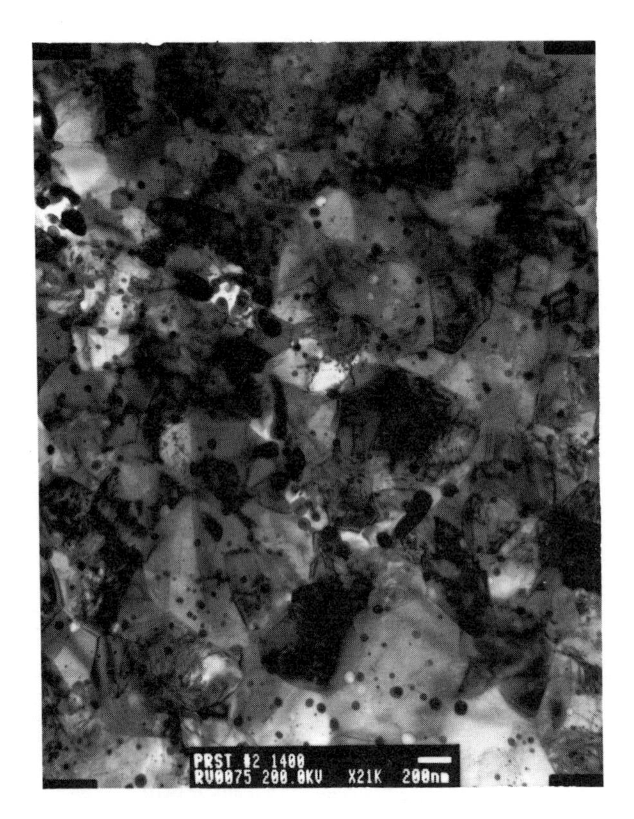

Figure 8 Fine dispersion of erbia particles in a Ti-6Al-2Sn-4Zr-2Mo-0.1Si-2Er alloy after rapid solidification (PREP≅10^4 °C/s cooling rate) and extrusion.

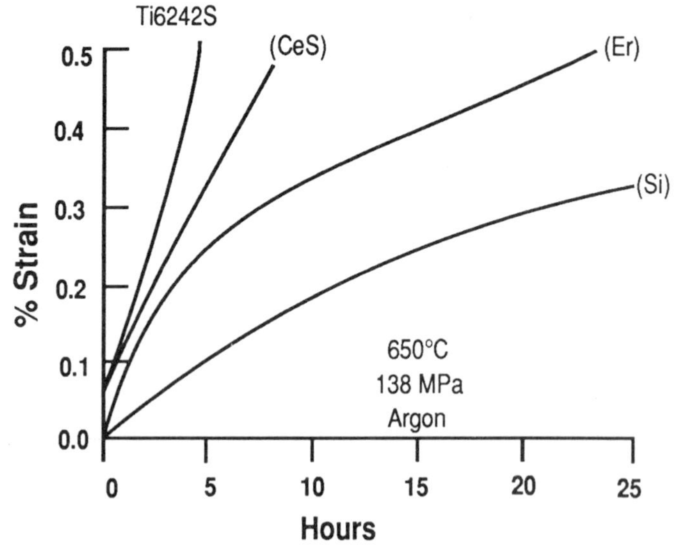

Figure 9 Creep behaviour of dispersion strengthened titanium alloys (courtesy General Electric Company) [1,2].

3.6 Aluminides

The titanium aluminides because of their ordered structure are potentially useful alloys for high temperature application. However, the planarity of slip results in low ductility at room temperature. Rapid solidification offers the potential for improved ductility in these intermetallics by disordering [13], grain refinement, and development of fine dispersoid particles. When these fine dispersoids are oxides an additional gain in ductility can result from deoxidation of the matrix.

Early work on a RS of this alloy class utilized TiAl-W alloys atomized by the PREP process. This material exhibited good creep resistance due to a fine precipitate of tungsten rich beta phase in the alpha-2+gamma lath structure of the TiAl composition [134].

Rapid solidification of alloys with a base composition of Ti_3Al with the addition of 0.4 at%Er has been shown to produce a fine dispersoid of Er_2O_3 (Fig. 11) [111,112]. This dispersoid does not coarsen in grain interiors to temperatures as high as 1000°C (1830°F) [135]. Under HIP conditions Er_2O_3 dispersoid coarsening in a Ti_3Al-Nb alloy was comparable to that observed during an anneal at the same temperature. However, extrusion caused rapid coarsening which was contrasted with cerium sulphide or cerium oxysulphide containing alloys where virtually no coarsening occurred [111,112]. The presence of a dispersoid in Ti_3Al-Nb alloys also produced a refinement in grain size. The ductility and fracture toughness of dispersoid containing alloys was dependent on the test used. Hardness indentation ductility was improved by grain refinement but fracture toughness was not [111]. Much further work is required in this generic system to optimize the microstructure and hence mechanical properties.

3.7 Other Systems

Other alloy elements which could be added to titanium to give attractive characteristics are the low density elements Li, Mg (Al and Si have a similar effect). The challenge here is to successfully add these elements which have a boiling point below the melting point of titanium; further discussion of this alloy class is contained in the MA section of this chapter.

3.8 Metal Matrix Composites

To date the majority of titanium metal matrix composites have been fabricated by pressing sheet/foil product around continuous fibres such as SiC or B_4C/B [136]. However, one study [137] investigated the production of Ti-6Al-4V fibre reinforced with continuous SiC or B_4C/B using a powder metallurgy approach. Blended elemental and prealloyed powders were used with cold pressing and hot pressing, or hot isostatic pressing techniques. Full density was achieved but poor fibre spacing occurred resulting in little longitudinal strength enhancement, but a significant (50%) modulus increase over monolithic material.

4. Magnesium Rapid Solidification

Magnesium is the lightest, nontoxic structural metal available for engineering applications. With an attractively low density of 1.74 g/cm^3, magnesium is 35.6% lighter than aluminium and 61.3% lighter than titanium. However, the use of magnesium and its alloys in structural applications has been limited because of their poor corrosion resistance and marginal mechanical properties. The corrosive behaviour of magnesium stems from its strong electropositive character leading to high chemical reactivity. Consequently, magnesium and its alloys occur at the most active ends of the electromotive force series and the galvanic series. Further, exposure of magnesium based alloys to moisture results in the formation of $Mg(OH)_2$ surface film. This hydroxide, brucite, has a hexagonal crystal structure and fractures easily by basal cleavage [138]. Since the density of this hydroxide is lower than that of magnesium, compressive rupture of the brucite film occurs easily and results in continued exposure of fresh metal surface for further corrosion [139]. In addition to the aqueous and galvanic corrosion characteristics of magnesium, its high vapour pressure, rapid oxidation and pyrophoric behaviour are considered to be the limiting factors. The marginal mechanical properties of magnesium such as its low strength and poor formability are a result of its hcp crystal structure with a limited number of slip systems and difficulties in the operation of additional strengthening mechanisms.

The alloying behaviour of magnesium is directly influenced by its electropositive nature due to two s-orbit electrons. Binary magnesium base alloy systems can be classified into three groups; eutectic, peritectic, and isomorphous (Table 12). Generally, the eutectic solidification produces an intermetallic compound except in the binaries containing Fe, Li or Pu. The formation of compound suppresses the maximum terminal solid solubility (TSS) values to below 1 at% for about 20 elements, most of which satisfy the Hume-Rothery size factor criterion for extensive solid solubility [140]. Five elements (In, Mn, Sc, Ti, and Zr) are known to form peritectic-type systems with magnesium. One element, Cd, shows complete solid solubility in the temperature range 255°C to 320°C (485°F to 610°F) and is a lone member of the isomorphous category. Just as in the case of aluminium base alloy systems, the maximum TSS values observed in magnesium alloys at the eutectic or peritectic temperatures decrease rapidly with decreasing temperature. Consequently, the room temperature TSS is nearly zero in about 35 binary magnesium systems. Pronounced TSS (5-25 at%) is retained at ambient temperature in systems containing Cd, In, Li, Sc, and Ti. The elements Er, Ho, Lu, and Tm show moderate room temperature TSS's (1-5 at%) while lower values (0.1-1 at%) are observed for Ag, Al, Y, and Zn.

The development of RS magnesium alloys is at a much earlier stage compared to that of aluminium base alloys so only

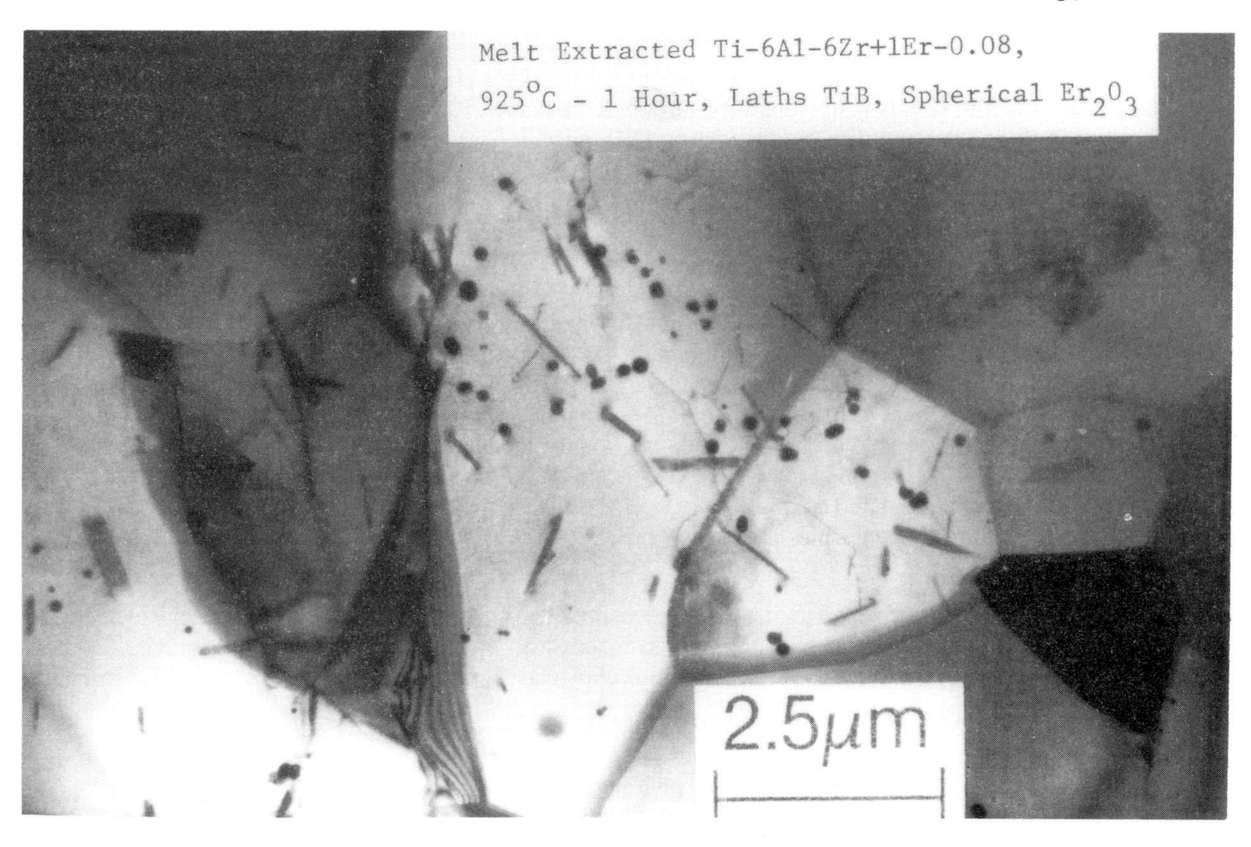

Figure 10 Fine needle-like TiB precipitates formed at temperatures above 500°C (930°F).

Figure 11 Fine dispersion of erbia particles formed in a Ti_3Al-11Nb-0.5Er (at%) after HIPing at 950°C (1740°F).

Table 12 Ranges of the maximum equilibrium solid solubilites of X in Al-X, Mg-X and Ti-X binaries

System	Type	Solubility range, at%				
		<0.1	0.1–1	1–5	5–25	>25
Al-X	E	Be, Y, B, Re, Acti, VIIIa, Ca, Vb, VIb, Sr, Ba, Sn	Be, Sc, Mn	Si, Cu, Ge	Ga, Li, Mg, Ag	Zn (66.4)
	P	Mo, Nb, Ta, W, Zr	Cr, Hf, V, Ti			
	M	Bi, In, K, Na, Pb, Tl, Cs				
Mg-X	E	As, Ba, Ce, Co, Cu, Eu, Fe, Ge, La, Na, Ni, Pd, Pr, Sb, Si, Sr	Au, Ca, Ir, Nd, Th	Ag, Bi, Dy, Ga, Gd, Hg, Pu, Sn, Y, Yb, Zn, Zr	Al, Er, Ho, Li, Lu, Pb, Tl, Tm	
	P		Mn, Ti	Zr	In, Sc	
	I					Cd (100)
Ti-X[a]	BI					V, Nb, Ta, Mo
	Eid				Fe, Co, Ni, Cu, Si, B	Cr
	Pid		B	C	Ge, Sn, O, H	Al, Ag
	M-P		Y, La, Ce, Nd, Er, Gd			

[a] In beta phase.

E = eutectic
P = peritectic
M = monotectic
I = isomorphous
BI = beta isomorphous
Eid = eutectoid
Pid = peritectoid
M-P = monotectic/peritectic

a minimal amount of application related information is currently available. However, considerable research has been conducted on RS magnesium base alloys to investigate general effects such as solubility extension, formation of metastable phases, and microstructural refinement.

The TSSE data are available for only a dozen or so systems (Table 13). These results clearly indicate that significant solid solubility extensions are possible in magnesium systems via RS. In some cases (La, Mn, and Y), the TSSE's are close to or even exceed the corresponding eutectic or peritectic composition. Although these TSSE effects are thermally unstable and decomposition can occur at ~250°C (480°F) [141], they provide a very useful means of achieving precipitation strengthening. The TSSE of rare earth metals and yttrium in magnesium decreases the c/a ratio [141-144]. These results suggest that RS could pave the way for a new generation of ductile magnesium alloys .

The formation of metastable fcc phases has been reported in rapidly solidified Mg-Pb and Mg-Sn systems [145-147]. However, the high densities and thermal instabilities of these fcc $Mg_{84}Pb_{16}$ and $Mg_{86}Sn_{14}$ phases preclude their use in structural applications. Other non-equilibrium structures which have been reported include a metastable variant of the Mg_2Si phase in the Mg-Si system [148]. In addition, metallic glass formation has been observed in the Mg-Zn [149] and Mg-Al [150] systems as a result of chill block melt spinning. The amorphous alloys are ductile in many cases and exhibit high strength levels. For example, the $Mg_{70}Zn_{30}$ amorphous alloy ribbon showed an ultimate tensile strength as high as 830 MPa (120 ksi) and a hardness of 220 kg/mm^2 [149].

However, all Mg-glasses reported so far form at relatively high alloying levels [149] and crystallize between 100°C (212°F) and 200°C (392°F) to form brittle intermediate phases of complex crystal structure [144,151].

The microstructural refinement due to RS of magnesium alloys has been studied by a number of investigators [151-155]. The grain refinement achieved by RS should suppress deformation twinning and consequently decrease the anisotropy of mechanical properties. Early work on atomized powders [156-158] showed dendrite arm spacings of ~5-8 μm leading to as-extruded grain sizes between 3 μm and 10 μm and refined size (~0.1 μm) of second phase particles. Work on RS of Mg-8.5 wt%Li with and without 1 wt%Si or 0.5 wt%Ce [155] showed microstructural refinement in the RS flakes of the binary alloy by a factor of 10 compared with chill-cast material, and an additional refinement by a further factor of 3 when silicon was present. Recent work on melt-spun ribbons of several Mg-Al-Zn alloys containing rare earth additions reported highly refined microstructures (matrix grain size = 0.3-0.7 μm, and cell size = 0.1-0.3 μm) [152, 153]. The refinement of matrix grain size is generally accompanied by a refinement of the second phase in two-phase alloys as illustrated in Fig. 12 [159]. Similarly, TEM work on melt-spun ribbons of Mg-1.7 wt%Si alloy showed extremely fine (10-20 nm diameter) Mg_2Si particles [150].

Magnesium alloys in which the potential of RS has been studied may be classified as follows:

- High strength/corrosion resistant alloys.
- High formability/low density alloys.
- Intermetallic compound alloys.

4.1 High Strength/Corrosion Resistant Magnesium Alloys

Most of the work to date on RS magnesium alloys has focused on the improvement of their strength levels and corrosion resistance. In the early work of the 1950s, extrusions of atomized powders of three commercial alloys (AZ31, M1, and ZK60) and one experimental alloy (AMZ111) resulted in the doubling of tensile and compressive yield strengths and up to 30% larger ultimate tensile strength as compared to extrusions of ingots of the same alloys [160]. Similar results were reported for a Mg-1Zn-1.6Si alloy [161]. "Interference" hardening was produced by blending Zr-bearing magnesium alloy powders with aluminium or eutectic Mg-Al alloy powders [160-163]. Precipitation resulting from interdiffusion during extrusion and heat treatment nearly doubled the yield strength of Mg-0.4Zr and increased its ultimate strength by 30%. More recently, consolidation of melt-spun ribbons has resulted in attractive tensile properties (Table 14). In particular, RS Mg-Al-Zn alloys containing Si and/or rare earths (Y, Nd, Ce or Pr) prepared by comminution and extrusion of planar flow cast ribbons exhibit ultimate strengths as high as 515 MPa (75 ksi) [152-154]. Tensile data reported for consolidated ribbons of various RS magnesium alloys range from 345 MPa (50 ksi) to 460 MPa (67 ksi) yield strength, 385 MPa (56 ksi) to 515 MPa (75 ksi) ultimate tensile strength, and 5% to 20% elongation [152-154,164-171].

As mentioned earlier, the poor corrosion resistance of magnesium has seriously limited its wider use as a structural metal. It is well known that heavy metal impurities such as Fe, Ni, Cu, and Co greatly accelerate the corrosion rate of magnesium and its alloys in salt water. The use of improved melting techniques has led to the recent development of high purity/corrosion resistant alloys such as AZ91-HP. Further increase in corrosion resistance arising from uniformity of microstructure and modification of surface film has been reported for several RS magnesium alloys. In particular, a corrosion rate of ~10mil/yr in 3%NaCl aqueous solution exhibited by RS Mg-Al-Zn alloys containing Mn and small amounts of Ce, Pr, Nd or Y is the lowest ever reported for structural magnesium alloys [154,166]. Recent polarization studies of binary magnesium alloys containing rare earths reported significantly lower current densities indicating improved corrosion resistance [159]. Preliminary results showed that the surface film was enriched in the rare earth addition [172].

Table 13 Solubility limits (at.%) of solutes in binary magnesium alloys under equilibrium and rapid solidfication conditions [141-144]

Solute	Equilibrium Maximum Solubility	Equilibrium Ce or Cp[a]	Maximum Extended Solubility
Eutectic			
Al	11.5	31.0	22.60
Ca	0.82	10.5	7.17
Ce	0.09	4.3	2.64
Eu	∿0	6.0	1.21
Ga	∿3.0	19.13	6-10
La	∿0	2.2	2.10
Nd	0.63	7.7	2.80
Y	∿3.5	9.1	9.67
Yb	1.2	10.7	3.39
Peritectic			
Mn	1.0	1.0	2.46

[a] Ce = eutectic composition and Cp = peritectic composition.

4.2 High Formability/Low density Magnesium Alloys

It is notable that elongations to fracture in the superplastic range (~200%) were observed in tensile testing at 150°C (300°F) of extrusions of comminuted melt-spun ribbon of Mg-5at%Al-2at%Zn with additions of either Y or Si and Pr [153]. This behaviour is attributable to the fine grain size stabilized by a fine dispersion of Mg-rare earth intermetallics or Mg_2Si in these alloys as a result of RS. Similar superplastic behaviour has been projected for the fine two-phase microstructure stabilized in rapidly solidified Mg-9wt%Li by dispersoid forming additions of B, Si or Ce [173]. Another study involving RS low density/high modulus Mg-9Li, Mg-9Li-1Si, and Mg-9Li-1Ce alloys suggested that the grain refinement produced by RS of the dispersoid containing ternary alloys can be preserved or further refined during subsequent consolidation [155]. The main challenge in these alloys is to improve yield strength as well as creep resistance.

4.3 Magnesium Intermetallic Compound Alloys

Recently, RS processing of alloys based on magnesium intermetallic compounds has received some attention. Rapid solidification of the intermetallic compound $Mg_{17}Al_{12}$ failed to impart any ductility to this brittle material. RS processing of $Mg_{17}Al_{12}$ after macroalloying with Cu, Ni, Fe, Zn, Sn, etc. and microalloying with B was also unsuccessful in ductilizing the compound [174]. Similar results were obtained in RS Mg_2Si and Mg_2Ge compounds [175].

5. Mechanical Alloying

5.1 Aluminium

In mechanical alloying (MA) of aluminium alloys the adherent surface oxide film on the powder particles [53] is incorporated into the interior of the processed powders by the repeated fracture and cold welding. In the case of aluminium alloys, process control additives (PCA), typically organic compounds, are necessary for control of the balance between fracture and welding [176,177]. This addition results in the formation of Al_2O_3 and Al_4C_3 [178] which are incorporated into the interiors of the powder particles as dispersoids.

5.1.1 Commercially Pure Aluminium

The dispersion produced by MA is much more efficient than the oxide dispersions in sintered aluminium powder (SAP), and results in better room temperature and elevated temperature mechanical properties [176,179]. The improved properties result directly from the novel structure developed by MA processing; very fine equiaxed grains (0.2 μm to 0.5

μm), and well distributed dispersoids 30 nm to 40 nm in size (Fig. 13). This structure has a very high thermal stability, and retains the structure, strength, and ductility on thermal exposures as high as 490°C (915°F) for 100 hr [180,181] well beyond conventional aluminium alloys and in excess of the RS elevated temperature alloys already discussed. Applications as high temperature, high strength electrical conductor wires are foreseen for the alloy.

5.2 High Strength Aluminium Alloys

MA aluminum alloy development has been built around combining ambient temperature strength from solid solution strengthening and precipitation hardening, with elevated temperature strength from dispersoid content and composition [179,182]. The alloys IncoMAP AL-9052 and AL-9021 based on magnesium [183] as an alloying addition have been well characterized for tensile strength, fatigue strength, toughness, corrosion resistance, etc. [178,184,185]. The properties of these alloys are listed in Table 15. AL-9052 is currently considered for application as torpedo hulls and other marine applications [186]. AL-9021 was developed for aircraft structural applications, but is currently being projected for application as a matrix for SiC particle reinforced composites. AL-9021 also exhibits extended ductility, over 500%, at high strain rates of 1-10 s^{-1} at 475°C (885°F) [187] and further applications based on this are foreseen.

5.1.3 Low Density Aluminium Alloys

IncoMAP alloy AL-905XL is an Al-Mg-Li MA alloy developed as a low density forging alloy for airframe applications [188]. One of the first among the Al-Li alloys to reach commercial production stage, the alloy has been well characterized [188-191]. Some properties of the alloy are shown in Table 15, and in Figure 14. Extensive evaluation of the current Al-Li alloys indicates that the MA alloy has an edge over other PM routes [192]. However, careful allowing is required, for example limiting the amount of conventional alloying additions used for precipitation hardening and for grain refinement, and also limiting the Li content to 1.3 wt% to avoid Al_3Li (δ') precipitates. Further, the MA alloy as-consolidated is considerably stronger than corresponding IM or conventional PM alloys, making fabrication steps more difficult [192].

5.1.4 High Temperature Aluminium Alloys

DISPAL alloys, developed in West Germany, employ dispersion strengthening by Al_2O_3 and Al_4C_3 as in MA, using carbon black and controlled additions of oxygen from the milling atmosphere [193,194]. Such a processing route enables independent control of dispersoid content, thereby allowing alloy tailoring for the application. With the alloy design specifically aimed at high temperature properties and low cost [194-196], automotive and energy field applications are foreseen. Forged automotive pistons, high temperature electrical conductors in Na-S battery, and aerospace cryogenic applications have also been proposed.

One of the current developments in MA alloys is the combination of MA with rapid solidification. In the Al-Fe-X (X = Ni, Ce) alloys there have been some reports indicating strength and creep rate advantages (Table 16) using this combined route [197-199]. The strength is retained in this case up to as high as 500°C (930°F) probably because the oxide and carbide dispersions are much more effective in retaining their strength and fineness than the second phase particles produced by RS.

Various Al-Ti alloys are also being explored by MA to realize the benefits of dispersion strengthening from both oxides/carbides and intermetallics. Both MA of elemental/master alloy blends [200], and MA of rapidly solidified alloy powders [201] have been investigated. With good elevated temperature strength, high stiffness, and encouraging levels of ductility [200] this alloy system appears very attractive for development.

5.1.5 Metal Matrix Composites

Aluminium-base metal matrix composites (MMC) are under development with MA as a processing route for SiC particulate reinforcements [202]. Both AL-9021 and AL-9052 have been used as the matrix material, with SiC particulate reinforcement up to 15 vol%. Typical strength and stiffness properties are shown in Table 17 [203]. AL-905XL has also been evaluated as a matrix material for MMC [190]. The elevated temperature strength of the composite AL-905XL-5 vol%SiC is also listed in Table 17. While processing details for MA composites are restricted, extensive property evaluation details are available [202-204]. Unlike other processing routes for similar composites, the particulate dispersions do not significantly enhance the strength of the matrix that has been already strengthened by the oxide and carbide dispersions. Further coarse particles of SiC break during fast fracture, and the fracture toughness of the composite is low [204]. However, at high temperatures in the 425-450°C (795-840°F) range, at high strain rates (>0.7/sec) the MA composite material shows extended ductility up to 300% total elongation [205].

5.2 Titanium

Although there has been some early work on the application of MA to titanium systems [206,207], experimental difficulties curtailed further research. Currently, several aspects of the application of MA to titanium systems are under investigation. These include extension of solid solubility limits possible by MA, the formation of amorphous phases by MA in favourable systems as a possible route to novel crystalline structures, and the microstructural engineering offered by MA.

Magnesium is an attractive alloying addition to titanium on account of its very low density, but it is very difficult to add to titanium because the boiling point of magnesium is below the melting point of titanium. It has been possible to

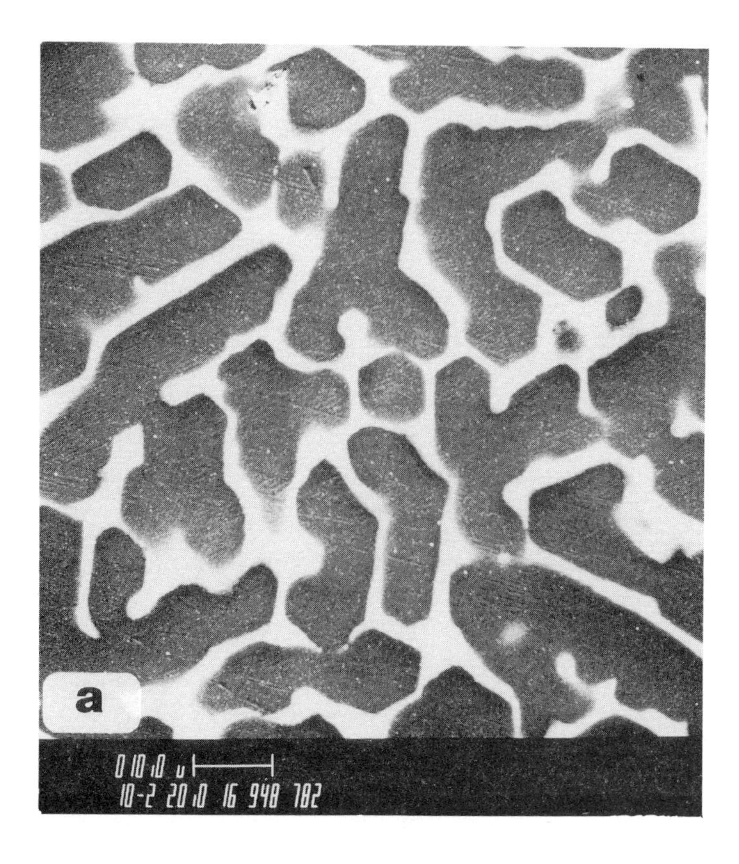

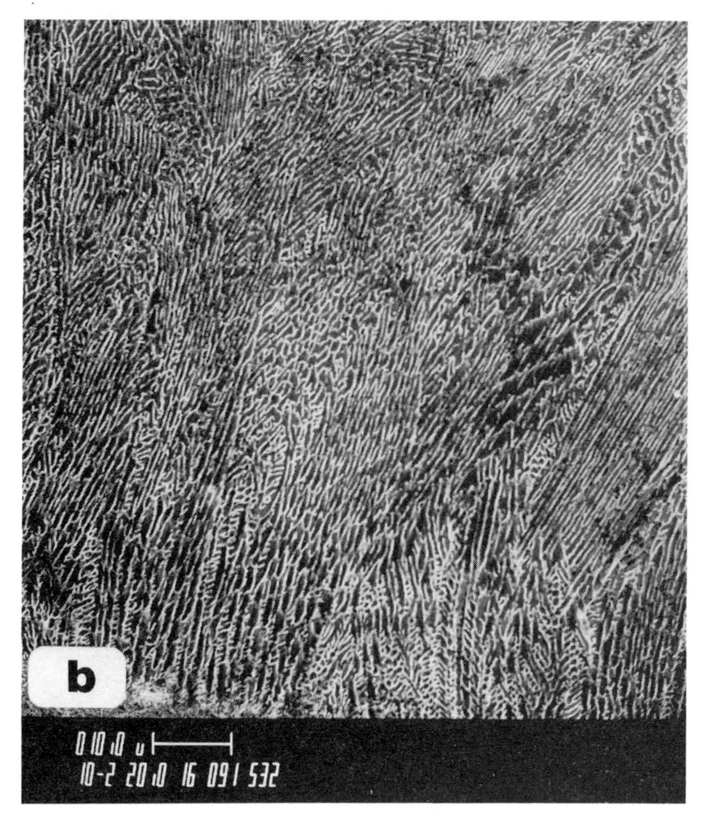

Figure 12 Scanning electron micrographs of (a) ingot and (b) splat specimens of Mg-10wt%Nd alloy showing refinement of the a+Mg_{11}Nd two-phase structure due to RS.

Table 14 Longitudinal tensile properties of RS PM magnesium alloys

Alloy	Composition	Processing	UTS, MPa (ksi)	YS, MPa (ksi)	El., %	Reference
ZK60	Mg-6Zn-0.7Zr-0.5Ca	Powder extrusion	385 (56)	365 (53)	6	156
-	Mg-1Zn-1.6Si	Powder extrusion	338 (49)	310 (45)	4	161
AK11	Mg-0.64Zr-1Al	Extrusion of Al-coated Mg-Zr pellets	351 (51)	331 (48)	10	161
ZK60	Mg-6Zn-0.45Zr	Powder extrusion + aging	420 (61)	396 (57.5)	14	168, 169
ZK60	Mg-5.5Zn-0.4Zr	Ribbon extrusion	388 (56.3)	365 (53.0)	19.6	164
-	Mg-10Al	Ribbon extrusion + solution treatment and aging	427 (62.0)	353 (51.2)	4.7	165
EA65RS	Mg-5Al-5Zn-5.9Y	Ribbon extrusion	515 (75)	460 (67)	5	166
EA65RS-B1	Mg-5Al-5Zn-4.9Nd	Ribbon extrusion	475 (69)	425 (62)	14	166
-	Mg-11Al-2.4Zn-3.2Y	Ribbon extrusion	443 (64.3)	379 (55.0)	11.5	167
-	Mg-10Al-3.2Zn-2.7Mn-5.8Ce	Ribbon extrusion	468 (67.9)	431 (62.6)	14.9	167
-	Mg-3Nd-1Pr-2Mn	Powder extrusion	427 (62.0)	420 (61.0)	5.1	170
AZ91HP	Mg-9Al-1Zn	Ribbon extrusion	457 (66.3)	391 (56.7)	12.1	171

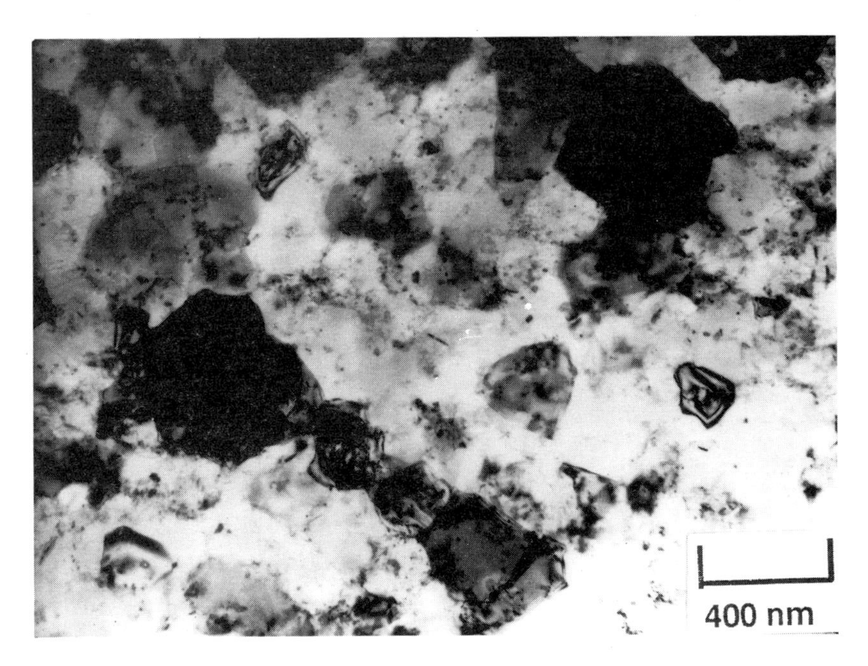

Figure 13 Transmission electron micrograph of MA aluminum.

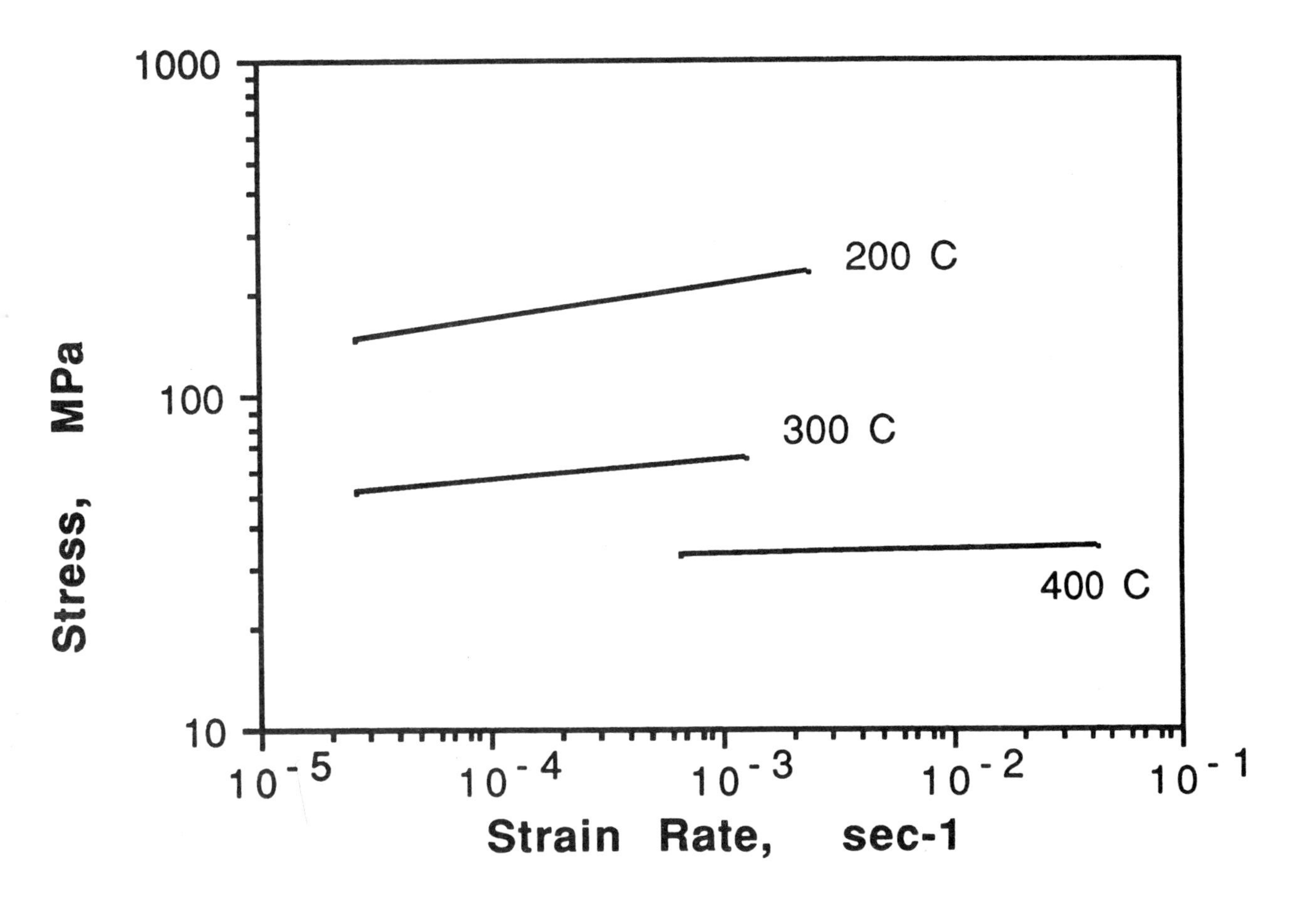

Figure 14 Creep properties of MA aluminum-lithium alloy AL-905XL.

Table 15 Mechanically alloyed aluminum-base alloys

	IncoMAP AL-9052	IncoMAP AL-9021	IncoMAP AL-905XL
Composition	Al-4Mg-1.1C-0.80	Al-1.5Mg-4Cu-1.1C-0.80	Al-4Mg-1.5Li-1.2C-0.40
Density	2.68 mg/m^3		2.58 mg/m^3
Mechanical Properties,			
YS	540MPa	590MPa	460MPa
UTS	570MPa	610MPa	490MPa
El	8%	12%	6%
Fracture Toughness, K_{Ic}	44MPa$\sqrt{m}$		32MPa$\sqrt{m}$
Elastic Modulus	76 GPa		80 GPa

Table 16 Steady state creep rate of Al-8Fe-4Ce alloy

Temperature, °C (°F)	Stress, MPa	Creep Rate in RS Alloy, s^{-1}	Creep Rate in RS+MA Alloy, s^{-1}
350 (660)	103	4.1×10^{-7}	6.3×10^{-10}
380 (715)	83	2.6×10^{-7}	1.6×10^{-9}
380 (715)	103	4.9×10^{-5}	1.4×10^{-9}

Table 17 Mechanical properties of Ma MMC

Temperature, °C	YS[a], MPa	UTS[a], MPa	El[a]., %	E[a], GPa
SiC/AL-9021				
25	420 (370)	546 (440)	6 (10)	81.8 (66.1)
100	424 (347)	483 (394)	4 (14)	
200	274 (124)	286 (137)	14 (54)	
300	77 (47)	81 (54)	12 (21)	
400	40 (30)	43 (32)	3 (10)	
SiC/AL-905XL				
25	--- (562)	581 (584)	0 (7)	
150	334 (258)	423 (388)	1 (17)	
200	197 (128)	266 (233)	3 (27)	
300	97 (55)	113 (82)	8 (29)	
400	68 (42)	68 (44)	0 (3)	

[a]Matrix material properties in parentheses.

Table 18 Knoop hardness values on teh MA processed alloys

Alloy	As-Consolidated	Solution Treated	Solution Treated and Aged
Ti-185			
Atomized	372		
MA	501	487	526
Ti-24/10			
Atomized	304	301	292
MA	437	435	440

achieve solid solution of magnesium in titanium up to 3.1 wt% (6.0 at%) by MA which reduces the density significantly [208,209].

Formation of an amorphous phase in favourable titanium systems such as Ti-Ni [200,211], Ti-Fe [212], Ti-Co [213], Ti-Cu [213], Ti-Pd [214], Ti-Cu-Pd [215], and Ti-Ni-Cu [216] can occur as a result of MA. Studies are currently in progress on low temperature consolidation of the alloy and its characterization [216,217].

Two advanced beta alloys, Ti-1Al-8V-5Fe-1Er and Ti-24V-10Cr-5Er, both with extensive rare earth addition for subsequent oxide dispersions, have been investigated using a combination RS and MA approach. An extremely fine grained (sub-micron) structure and 30-50 nm dispersions were achieved by MA in these alloys (Fig. 15). Excellent thermal stability of the structure and substantial hardening, which responds further to heat treatments, have been reported (Table 18) [218]. MA of gas atomized Ti_3Al based alloy Ti-25Al-10Nb-3V-2Mo (at%) leads to considerable grain refinement in the consolidated alloy [219]. The MA process results in a substantial hardness increase (720 Knoop as against 340 in the alloy consolidated from as-atomized powder), a thermally stable grain structure, and greater retention of the more ductile phase B2 (Fig. 16) in the alloy. Both in the beta titanium alloys and in the Ti_3Al alloy, the microstructural features that can be achieved by MA of the atomized powders suggest that the dual approach may lead to enhanced mechanical property combinations.

5.3 Magnesium

MA has been employed in the development of "supercorroding" magnesium alloys [220], which operate as short circuit galvanic cells used for making lines with precisely controllable corrosion rates for releasing deep-sea equipment at specific depths. The advantage of MA here is the extent of control and optimum electrical contact, achieved by the closeness of the anode/cathode [221]. Such supercorroding alloys have also been used in other submarine applications, such as a heat source in diving suits and as a hydrogen gas generator.

The system Mg-M is of great interest as an alloy for hydrogen storage system for energy applications [222]. By alloying with catalyst elements/intermetallic compounds the hydrogenation capability of magnesium is substantially enhanced [222-225]. MA is often the only processing technique available at present for preparing such alloys that can act as a "hydrogen pump" by absorption and desorption of hydrogen. Considerable work appears to be in progress on these systems and currently a hydrogen pump for energy conversion is being developed jointly between the USSR and France.

6. Concluding Remarks

The progress which has been made in the rapid solidification and mechanical alloying of the light metals aluminium, titanium, and magnesium, including metal matrix composites fabricated using the PM approach, has been reviewed. As with other technologies the transition from the research laboratory to production status has been slow; especially for those intimately involved in the area. However using both techniques, aluminium alloys are at the early production stage. Titanium and magnesium trail behind, but both show the potential for enhanced behaviour using the RS and MA processing methods. The future appears bright for both processing techniques, and in some cases a combined RS plus MA approach may further enhance behaviour. The increased chemistry and processing window offered by these techniques should allow the metallurgist to further improve behaviour characteristics and therefore compete more effectively with polymer matrix composite materials in the lower temperature regime, and with ceramic-based materials at more elevated temperatures.

Acknowledgements

The authors would like to acknowledge helpful discussions with H. B. Bomberger, S. K. Das, W. E. Quist, C. Suryanarayana, and G. Venkataraman. In addition the assistance of Miss Karen A. Sitzman in manuscript preparation is greatly appreciated.

References

1. F.H. Froes, Trends in High Performance Lightweight Metals, Materials Edge, 5, pp.19-31, May/June 1988.
2. F.H. Froes, Space Age Metals Technology, (eds.) F.H. Froes and R.A. Cull, SAMPE, Covina, CA, 2, pp.1-19, 1988.
3. J. Wadsworth and F.H. Froes, Trends in the Development of Structural Metallic Materials for Aerospace Applications, JOM, 41,5, 1989.
4. F.H. Froes, Dept. of Interior, Bureau of Mines Meeting Proceedings, Washington, DC, November 1988.
5. F.H. Froes, presented at the Fourth Israel Materials Engineering Conference, Beer-Sheba, Israel, December 1988, and to be published in the proceedings.
6. F.H. Froes, Aerospace Materials for the Twenty-First Century, Materials and Design, June 1989.

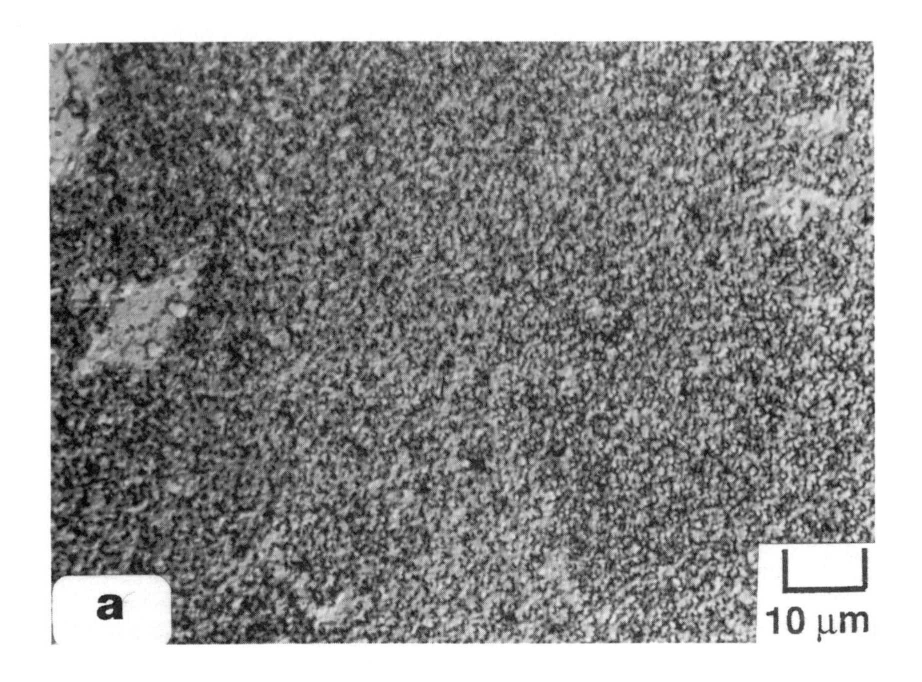

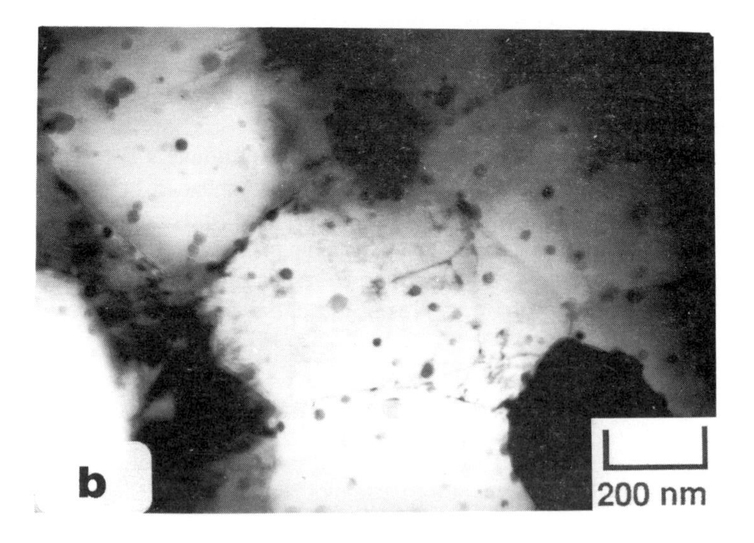

Figure 15 (a) Optical and (b) electron micrographs of MA beta-titanium alloy Ti-24V-10Cr-5Er.

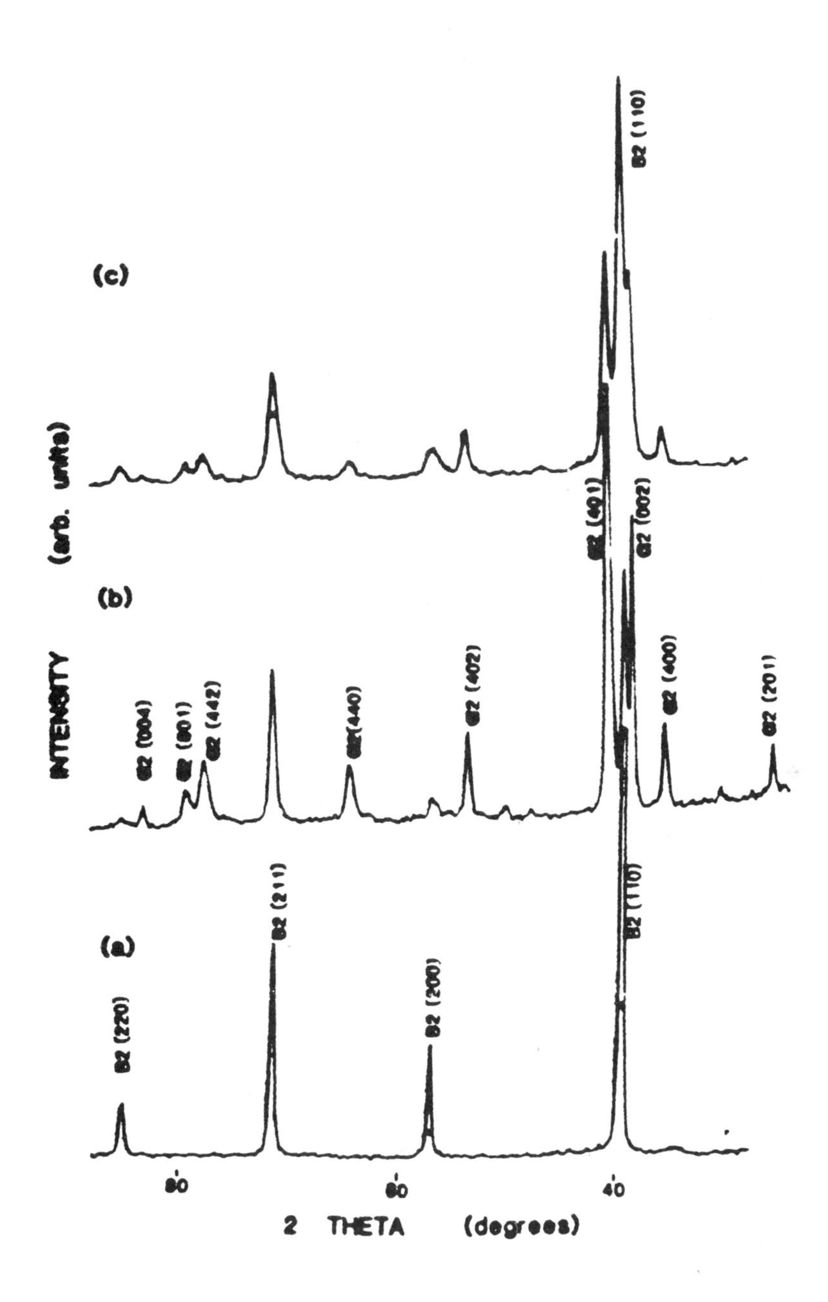

Figure 16 XRD of MA Ti_3Al alloy (Ti-25Al-10Nb-3V-1Mo at%) (a) gas atomized powder, (b) consolidated from gas atomized powder, and (c) consolidated from MA powder.

7. F.H. Froes and J.R. Pickens, Powder Metallurgy of Light Metal Alloys for Demanding Applications, J. of Metals, 36, 1, pp.14-28, January 1984.
8. H. Jones, A. Joshi, R.G. Rowe, and F.H. Froes, The Current Status of Rapid Solidification of Magnesium-Base and Titanium-Base Alloys, International Journal of Powder Metallurgy, 23, 1, pp.13-24, January 1987.
9. F.H. Froes, Y-W. Kim, and F. Hehmann, Rapid Solidification of Al, Ng and Ti, J. of Metals, 39, 8, pp.14-21, August 1987.
10. F.H. Froes, W.E. Quist, and S.R. Das, Advanced Lightweight Metals Using Rapid Solidification, Proceedings of International Conference on PM Aerospace Materials, MPR Publishing Services, Shrewsbury, England, pp.16.1-16.29, 1988; also in Metal Powder Report, 43, 6, pp. 392-404, June 1988.
11. R. Sundaresan and F.H. Froes, Mechanlcal Alloying, J. of Metals, 39, 8, pp.22-27, August 1987.
12. R. Sundaresan and F.H. Froes, Metal Powder Report, 44, 1989.
13. H. Jones, Rapid Solidification of Metals and Alloys, Monograph no. 8, The Institution of Metallurgists, London, 1982.
14. S.J. Savage and F.H. Froes, Production of Rapidly Solidified Metals and Alloys, J. of Metals, 36, 4, pp.20-33, April 1984.
15. P.S. Gilman and J.S. Benjamin, Ann. Rev. Mater. Sci., 13, p. 279, 1983.
16. R.D. Schelleng, Metal Powder Report, 43, p.239, 1988.
17. Aluminum: Properties and Physical Metallurgy, (ed.) J.E. Hatch, ASM, Metals Park, OH, 1984.
18. L.F. Mondolfo, Aluminum Alloys: Structure and Properties, Butterworth's, London, 1976.
19. T.R. Anantharaman *et al.*, Transactions of Indian Institute of Metals, 30, 6, pp.1-13, December 1977.
20. H. Jones, Aluminum, 54, pp.274-281, 1978.
21. Y-W. Kim and A.G. Jackson, Scripta Met., 20, 6, pp.777-782, 1986.
22. Y-W. Kim and F.H. Froes, Undercooled Alloy Phases, (eds) E.W. Collings and C.C. Koch, TMS, Warrendale, PA, pp.395-412, 1987.
23. Y-W. Kim and F.H. Froes, Processing of Structural Metals by Rapid Solidification, (eds) P.H. Froes and S.J. Savage, ASM International, Metals Park, OH, pp.309-320, 1987.
24. H. Jones, Materials Science Engineering, 5, pp.1-18, 1969.
25. M.H. Jacobs, A.G. Doggett, and M.J. Stowell, J. Materials Science, 9, pp.1631-1643, 1974.
26. H. Jones, Materials Science Engineering, 65, pp.145-156, 1984.
27. D.J. Skinner, R. Okazaki, and C.M. Adam, Rapidly Solidified Powder Aluminum Alloys, ASTM STP 890, (eds) M.E. Fine and E.A. Starke, Jr., ASTM, Philadelphia, PA, pp.211-236, 1986.
28. W.J. Boettinger, L. Bendersky, and J.G. Early, Met. Trans., 17A, pp.781-790, 1986.
29. Y.R. Mahajan, Y-W. Kim, and F.H. Froes, J. Materials Science, 22, 1, pp.202-206, 1987.
30. J.P. Lyle and W.S. Cebulak, Met. Trans., 6A, pp. 685-699, 1975.
31. W.L. Otto, Jr., Metallurgical Factors Controlling Structure in High Strength Aluminum P/M Products, US Air Force Materials Laboratory Technical Report, AFML-TR-76-60, May 1976.
32. F.R. Billman, J.C. Kuli, Jr., G.J. Hildeman, J.I. Petit, and J.A. Walker, Rapid Solidification Processing: Principles and Technologies III, (ed.) R. Mehrabian, NBS, Gaithersburg, MD, pp.532-546, 1982.
33. P.R. Bridenbaugh, W.S. Cebulak, F.R. Billman, and G.J. Hildeman, Light Metal Age, pp.18-26, October 1985.
34. G. Brooks, Osprey Metals, Neath, Wales, UK, private communication, June 1988.
35. I.G. Palmer, R.E. Lewis, D.D. Crooks, E.A. Starke, Jr., and R.E. Crooks, Aluminum-Lithium Alloys II, (eds) E.A. Starke, Jr. and T.H. Sanders, Jr., TMS, Warrendale, PA, pp.91-110, 1984.
36. C.M. Adam and R.E. Lewis, Rapidly Solidified Crystalline Alloys, (eds) S.R. Das, B.H. Rear, and C.M. Adam, TMS, Warrendale, PA, pp.157-183, 1985.
37. P.J. Meschter, R.J. Lederich, and J.E. O'Neal, Aluminum-Lithium Alloys II, The Institute of Metals, London, pp.85-96, 1986.
38. N.J. Kim, D.J. Skinner, K. Okazaki, and C.M. Adam, Aluminum-Lithium Alloys III, The Institute of Metals, London, pp.78-84, 1986.
39. W.M. Griffith, R.E. Sanders, Jr., and G.J. Hildeman, High-Strength Powder Metallurgy Aluminum Alloys, (eds) M.J. Koczak and G.J. Hildeman, TMS, Warrendale, PA, pp. 209-224, 1982.
40. S.L. Langenbeck, R.A. Rainen, *et al.*, Elevated Temperature Aluminum Alloys, US Air Force Wright Aeronautical Laboratories Technical Report, AFWAL-TR-86-4027, May 1986.
41. W.S. Miller, I.R. Hughes, I.G. Palmer, M.P. Thomas, T.S. Saini, and J. White, High Strength Powder Metallurgy Aluminum Alloys II, (eds) G.J. Hildeman and M.J. Koczak, TMS, Warrendale, PA, pp.311-331, 1986.
42. I.G. Palmer, M.P. Thomas, and G.T. Marshall, Dispersion Strengthened Aluminum Alloys, (eds) Y-W. Kim and W.M. Griffith, TMS, Warrendale, PA, pp.217-242, 1988.
43. M.J. Couper and R.F. Singer, High Strength Powder Metallurgy Aluminum Alloys II, (eds) G.J. Hildeman and M.J. Koczak, TMS, Warrendale, PA, pp.199-211, 1986.
44. S.K. Das, International Journal of Powder Metallurgy, 24, 2, pp.175-183, 1988.

45. M. Yamaguchi, Y. Umakoshi, and T. Yamane, Philosophical Magazine A, 55, 3, pp.301-315, 1987.
46. J. Tarnacki and Y-W. Kim, Scripta Met., 22, pp.329-334, 1988.
47. J. Tarnacki and Y-W. Kim, Dispersion Strengthened Aluminum Alloys, (eds) Y-W. Kim and W.M. Griffith, TMS, Warrendale, PA, pp. 741-762, 1988.
48. Y-W. Kim and F.H. Froes, THERMEC-88, 2, pp.588-595, 1988.
49. Y-W. Kim and F.H. Froes, Materials Science and Engineering, 98, pp.207-211, 1988.
50. R.E. Sanders, Jr., W.L. Otto, Jr., and R.J. Bucci, Fatigue Resistant Aluminum P/M Alloy Development, US Air Force Materials Laboratory Technical Report, AFML-TR-79-4131, September 1979.
51. Y-W. Kim and W.M. Griffith, Rapidly Solidified Powder Aluminum Alloys, ASTM STP 890, (ed.) M.E. Fine and E.A. Starke, Jr., ASTM, Philadelphia, PA, pp.485-511, 1986.
52. Y-W. Kim, W.M. Griffith, and F.H. Froes, ASM Metals/Materials Technology Series #8305-048, ASM Metals Congress, Philadelphia, PA, 1983.
53. Y-W. Kim, W.M. Griffith, and F.H. Froes, J. of Metals, 37, 8, pp. 27-33, 1985.
54. W.M. Griffith, Y-W. Kim, and F.H. Froes, Rapidly Solidified Powder Aluminum Alloys, ASTM STP 890, (eds) M.E. Fine and E.A. Starke, Jr., ASTM, Philadelphia, PA, pp.283-303, 1986.
55. H. Rack, Dispersion Strengthened Aluminum Alloys, (ed.) Y-W. Kim and W.M. Griffith, TMS, Warrendale, PA, pp.649-679, 1988.
56. S.G. Roberts, Powder Metallurgy, (ed.) W. Leszynski, Interscience, p. 799, 1961.
57. M. Lebo and N.J. Grant, Met. Trans., 5, pp.1547-1555, 1974.
58. Y-W. Kim and L.R. Bidwell, High-Strength Powder Metallurgy Aluminum Alloys, (eds) M.J. Koczak and G.J. Hildeman, TMS, Warrendale, PA, pp.107-124, 1982.
59. G.J. Hildeman, L.C. Labarre, A. Hafeez, and L.M. Angers, High Strength Powder Metallurgy Aluminum Alloys II, (eds) G.J. Hildeman and M.J. Koczak, TMS, Warrendale, PA, pp.25-43, 1986.
60. S.W. Schwenker, Y-W. Kim, W.M. Griffith, and F.H. Froes, Aluminum Alloys - Their Physical and Mechanical Properties, Vol. III, (eds) E.A. Starke, Jr. and T.H. Sanders, Jr., EMAS, Warley, England, pp.1837-1851, 1986.
61. J.J. Kleek, Y-W. Kim, and R. B. Nethercott, Aluminum Alloys - Their Physical and Mechanical Properties, Vol. III, (eds) E.A. Starke, Jr. and T.H. Sanders, Jr., EMAS, Warley, England, pp.1709-1723, 1986.
62. Alcoa Alloy 2090, Alcoa Green Letter No.226, Alcoa, 1988.
63. Y-W. Kim, Dispersion Strengthened Aluminum Alloys, (eds) Y-W. Kim and W.M. Griffith, TMS, Warrendale, PA, pp.157-180, 1988.
64. C.M. Adam, Rapidly Solidified Amorphous and Crystalline Alloys, 8, (eds) B.H. Kear, B.C. Giessen, and M. Cohen, Elsevier Science, New York, pp.411-422, 1982.
65. Y.R. Mahajan, Y-W. Kim, and F.H. Froes, High Strength Powder Metallurgy Aluminum Alloys II, (eds) G.J. Hildeman and M.J. Koczak, TMS, Warrendale, PA, pp.333-351, 1986.
66. D.J. Skinner, R.L. Bye, D. Raybould, and A.M. Brown, Scripta Met., 20, pp.867-872, 1986.
67. Y-W. Kim, Progress in Powder Metallurgy, compiled by C. L. Freeby and H. Hjort, MPIF, Princeton, NJ, 43, pp.13-32, 1987.
68. S.D. Kirchoff, R.M. Young, W.M. Griffith, and Y-N. Kim, High-Strength Powder Metallurgy Aluminum Alloys, (eds) M.J. Koczak and G.J. Hildeman, TMS, Warrendale, PA, pp.237-248, 1982.
69. S.D. Kirchoff and Y-W. Kim, Processing of Structural Metals by Rapid Solidification, (eds) F.H. Froes and S.J. Savage, ASM International, Metals Park, OH, pp.297-308, 1987.
70. Y-W. Kim and P.H. Froes, Proceedings of 8th International Light Metals Congress, Leoben-Wien, Austria, pp.756-759, 1987.
71. Y-W. Kim, M.M. Cook, and W.M. Griffith, Rapid Solidification Processing: Principles and Technologies III, (eds) R. Mehrabian, NBS, Gaithersburg, MD, pp.609-614, 1982.
72. H.H. Liebermann and R.L. Bye, Jr., Rapidly Solidified Crystalline Alloys, (eds) S.K. Das, B.H. Kear, and C.M. Adam, TMS, Warrendale, PA, pp.61-76, 1985.
73. Y-W. Kim and F.H. Froes, Aluminium, 64, 5, pp.1035-1038, 1988.
74. D. Raybould, Dispersion Strengthened Aluminum Alloys, (eds) Y-W. Kim and W.M. Griffith, TMS, Warrendale, PA, pp.199-216, 1988.
75. W.S. Miller and I.G. Palmer, Metal Powder Report, 41, 10, pp.761-767, October 1986.
76. J. Tarnacki and Y-W. Kim, Scripta Met., 22, pp.329-334, 1988.
77. K.S. Kumar and J.R. Pickens, Dispersion Strengthened Aluminum Alloys, (eds) Y-W. Kim and W.M. Griffith, TMS, Warrendale, PA, pp.763-788, 1988.
78. R.L. Bickerdike, D. Clark, J.N. Eastabrook, G. Hughes, W.N. Mair, P.G. Partridge, and H.C. Ranson, Rapidly Solidified Materials, (eds) P.W. Lee and R.S. Carbonara, ASM, Metals Park, OH, pp.137-144, 1985.
79. P. Niskanen and W.R. Mohn, Advanced Materials and Processes, 133, 3, pp.39-41, March 1988.
80. H. Margolin and J.P. Neilson, Modern Materials, Advances in Development and Application, (ed.) H.H. Hauser, Academic Press, New York, 2, pp.225-325, 1960.

81. S.S. Joseph and F.H. Froes, Titanium Metallurgy and Applications, Light Metal Age, 46, Nos.11, 12, pp.5-12, December 1988.
82. I.I. Kornilov, Interaction of Titanium with Elements of the Periodic System, The Science, Technology and Application of Titanium, (eds) R.I. Jaffee and N.E. Promisel, Pergamon Press, Oxford, pp.407-418, 1970.
83. I.I. Kornilov, Critical Review - Equilibrium Diagrams, Electronic and Crystalline Structures and Physical Properties of Titanium Alloys, Titanium and Titanium Alloys, (eds) J.C. Williams and A.F. Belov, Plenum Press, New York, 2, pp.1281-1305, 1982.
84. E.K. Molchanova, Phase Diagrams of Titanium Alloys, Israel Program for Scientific Translations, Jerusalem, 1965.
85. F.H. Froes and R.G. Rowe, Rapidly Solidified TItanium, Rapidly Solidified Alloys and Their Mechanical and Magnetic Properties, (eds) B.C. Giessen, D.E. Polk, and A.I. Taub, Materials Research Society, Pittsburgh, PA, 58, pp.309-334, 1986.
86. D.E. Polk and B.C. Giessen, Metallic Glasses, ASM, Metals Park, OH, pp.1-35, 1978.
87. F.H. Froes and D. Eylon, Powder Metallurgy of Titanium Alloys - A Review, Titanium, Science and Technology, (eds) G. Lutjering, U. Zwicker, and W. Bunk, DGM, Oberursel, West Germany, 1, pp.267-286, 1985.
88. S.H. Whang, Polytechnic University, New York, private communication, 1985.
89. Titanium, Rapid Solidification Technology, (eds) F.H. Froes and D. Eylon, TMS, Warrendale, PA, 1986.
90. F.H. Froes and R.G. Rowe, Rapidly Solidified Titanium - A Review, Titanium, Rapid Solidification Technology, (eds) F.H. Froes and D. Eylon, TMS, Warrendale, PA, pp.1-20, 1986.
91. H.B. Bomberger and F. H. Froes, Prospects for Developing Novel Titanium Alloys Using Rapid Solidification, Titanium, Rapid Solidification Technology, (eds) F.H. Froes and D. Eylon, TMS, Warrendale, PA, pp.21-43, 1986.
92. R.G. Rowe and F.H. Froes, Titanium Alloy Rapid Solidification, Rapid Solidification of Crystalline Alloys, (ed.) J.V. Wood, The Institute of Metals, London, 1989.
93. S.M.L. Sastry, T.C. Peng, P.J. Meschter, and J.E. O'Neal, J. of Metals, 35, 9, pp.21-28, September 1983.
94. T.F. Broderick, A.G. Jackson, H. Jones, and F.H. Froes, The Effect of Cooling Conditions on the Microstructure of Rapidly Solidified Ti-6Al-4V, Met. Trans., 16A, pp.1951-1959, November 1985.
95. A.F. Belov and I.S. Polkin, Modern Trends in Titanium Production and Processing, Germany Metallurgical Society Workshop, University of Nuremberg, Erlangen, July 1982.
96. T.F. Broderick, F.H. Froes, and A.G. Jackson, Cooling Rate Effects on Ti-6Al-4V and Beta III Titanium Alloys, Rapidly Solidified Metastable Materials, (eds) B.H. Kear and B.C. Glessen, Elsevier Science Publishing, New York, 28, pp.345-351, 1984.
97. T.C. Peng, S.M.L. Sastry, and J.E. O'Neal, Lasers in Metallurgy, (eds) R. Mukherjee and J. Mazumder, TMS, Warrendale, PA, pp.279-292, 1981.
98. J.E. O'Neal, T.C. Peng, and S.M.L. Sastry, Proceedings of 39th Annual EMSA Meeting, (ed.) G.W. Balley, Claitor's Publishing Division, Baton Rouge, LA, pp.66-67, 1981.
99. B.C. Muddle, D.G. Konitzer, and H.L. Fraser, Strength of Metals and Alloys, (ed.) R.C. Gifkins, Pergamon Press, Oxford, 1, pp.313-318, 1983.
100. I.G. Wright and B.A. Wilcox, Battelle Columbus Laboratories Report no. AD-781-133, 1974.
101. S.M.L. Sastry, P.J. Meschter, and J.E. O'Neal, Met. Trans., 15A, 7, pp.1451-1463, July 1984.
102. D.B. Snow, Laser Processing of Materials, (eds) K. Mukherjee and J. Mazumder, TMS, Warrendale, PA, pp. 83-98, 1984.
103. D.G. Konitzer, B.C. Muddle, R. Rirchheim, and H.L. Fraser, Rapidly Quenched Metals, (eds) S. Steeb and H. Warlimont, Elsevier Science Publishers, Amsterdam, 1, pp.953-956, 1985.
104. D.G. Konitzer, B.C. Muddle and H.L. Fraser, Scripta Met., 17, p.963, 1983.
105. D.G. Konitzer, B.C. Muddle, H. L. Fraser, and R. Kirchheim, Titanium, Science and Technology, (eds) G. Lutjering, U. Zwicker, and W. Bunk, DGM, Oberursel, West Germany, 1, pp.405-410, 1985.
106. D.G. Konitzer, B.C. Muddle, and H.L. Fraser, Met. Trans., 14A, p.1979, 1983.
107. D.G. Konitzer and H.L. Fraser, The Production of Powders of Titanium Alloys with Dispersion of Rare Earth Oxides by Laser Spin Atomization, Modern Developments in Powder Metallurgy, (eds) E.N. Aqua and C.I. Whitman, MPIF, Princeton, NJ, 16, p.607, 1984.
108. R.G. Rowe, T.F. Broderick, E.F. Koch, and F.H. Froes, Microstructural Study of Rapidly Solidified Titanium Alloys Containing Erbium and Boron, Rapidly Solidified Materials, (eds) P.W. Lee and R.S. Carbonara, ASM, Metals Park, OH, pp.107-114, 1985.
109. M.F.X. Gigliotti, R.G. Rowe, G.E. Wasielewski, G.R. Scarr, and J.C. Williams, Rapidly Solidified Alloys and Their Mechanical and Magnetic Properties, (eds) B.C. Giessen, D.E. Polk, and A.I. Taub, Materials Research Society, Pittsburgh, PA, 58, pp.343-351, 1986.
110. R.G. Rowe and E.F. Koch, Rapidly Solidified Materials, (eds.) P.W. Lee and R.S. Carbonara, ASM, Metals Park, OH, pp.115-120, 1985.
111. R.G. Rowe, J.A. Sutliff, and E.F. Koch, Rapidly Solidified Alloys and Their Mechanical and Magnetic Properties, (eds) B.C. Giessen, D.E. Polk, and A.I. Taub, Materials Research Society, Pittsburgh, PA, 58, pp.359-364, 1986.

112. J.A. Sutliff and R.G. Rowe, Rapidly Solidified Alloys and Their Mechanical and Magnetic Properties, (eds) B.C. Giessen, D.E. Polk, and A.I. Taub, Materials Research Society, Pittsburgh, PA, 58, pp.371-376, 1986.
113. L.M. Brown and R.K. Ham, Strengthening Methods in Crystals, (eds) A. Kelly and R.B. Nicholson, Elsevier, New York, p.9, 1971.
114. S.M.L. Sastry, T.C. Peng, and L.P. Beckerman, Met. Trans., 15A, 7, pp. 465-1474, July 1984.
115. R.G. Vogt, D. Eylon, and F.H. Froes, unreported work, 1983-1986.
116. S.H. Whang, J. of Metals, 36, 4, pp.34-40, April 1984.
117. S.H. Whang, Polytechnic University, New York, private communication, 1985.
118. S.M.L. Sastry, T.C. Peng, and J.E. O'Neal, Titanium, Science and Technology, (eds) G. Lutjering, U. Zwicker, and W. Bunk, DGM, Oberursel, West Germany, 1, pp.397-404, 1985.
119. G.W. Franti, J.C. Williams, and H.I. Aaronson, Met. Trans., 9A, 11, p.1641, 1978.
120. H.B. Bomberger and F.H. Froes, AFWAL Report no. TR-84-4164, 1985.
121. D.B. Hunter, AFWAL Report TR-405/2-15, Part II, 1966.
122. W.A. Baeslack III, S. Krishnamurthy, and F.H. Froes, Strength of Metals and Alloys, (eds) H.J. McQueen, J.-P. Bailon, J.I. Dickson, J.J. Jonas, and M.G. Akben, Pergamon Press, Oxford, 2, pp.1633-1638, 1985.
123. S. Krishnamurthy, A.G. Jackson, D. Eylon, R.R. Boyer, and F.H. Froes, Rapidly Quenched Metals V, (eds) S. Steeb and H. Warlimont, Elsevier Science Publishers, Amsterdam, 1, pp.945-948, 1985.
124. S. Krishnamurthy, I. Weiss, D. Eylon, and F.H. Froes, Strength of Metals and Alloys, (eds) H.J. McQueen, J.-P. Bailon, J.I. Dickson, J.J. Jonas, and M.G. Akben, Pergamon Press, Oxford, 2, pp.1627-1632, 1985.
125. W.A. Baeslack III, L. Weeter, S. Krishnamurthy, P.R. Smith, and F.H. Froes, Rapidly Solidified Metastable Materials, (eds) B.H. Kear and B.C. Giessen, Elsevier Science Publishers, New York, 28, pp.375-379, 1984.
126. S. Krishnamurthy, R.G. Vogt, D. Eylon, and F.H. Froes, Rapidly Solidified Metastable Materials, (eds) B.H. Kear and B.C. Giessen, Elsevier Science Publishers, New York, 28, pp.361-366, 1984.
127. J. Gross, Mallory-Sharon Titanium Corporation Internal Report No.1000R164, 1956.
128. R.G. Vogt, P.R. Smith, D. Eylon, and F.H. Froes, unreported work, 1983-1986.
129. F.H. Froes and R.G. Rowe, Engineered Titanium-Base Structural Materials, presented at the Sixth World Conference on Titanium, Cannes, France, June 1988, and to be published in the proceedings.
130. T.F. Broderick, F.H. Froes, and J.A. Snide, unreported work, 1985-1986.
131. T.F. Broderick and F.H. Froes, unpublished work, 1985.
132. F.H, Froes and P.R. Smith, unpublished work, 1985.
133. M. Gutierrez, I.A. Martorell, and F.H. Froes, unpublished work, 1986.
134. P.L. Martin, M.G. Mendiratta, and H.A. Lipsitt, Met. Trans., 14A, p.2170, 1983.
135. D.G. Konitzer, Alcoa, and H.L. Fraser, University of Illinois-Urbana, private communication, 1985.
136. P.R. Smith and F.H. Froes, Developments in Titanium Metal Matrix Composites, J. of Metals, 36, 3, pp.19-26, March 1984.
137. T. Egerer, Selective Reinforcement of P/M Titanium Compacts, US Air Force Wright Aeronautical Laboratories Technical Report, AFWAL-TR-84-4175, March 1985.
138. H.P. Godard, W.B. Jepson, M.R. Bothwell, and R.L. Kane, The Corrosion of Light Metals, John Wiley and Sons, Inc., New York, p.260, 1967.
139. P.S. Frederick, Corrosion and Protection of Magnesium, presented at the Annual Congress for Automotive Engineers, Detroit, Michigan, February 1980.
140. F. Hehmann and H. Jones, Magnesium Technology, compiled by C. Baker, G.W. Lorimer, and W. Unsworth, The Institute of Metals, London, pp.83-96, 1987.
141. N.I. Varich and B.N. Litvin, Fiz. Metal. Metalloved., 16, 4, pp.526-529, 1963.
142. F. Hehmann, F. Sommer, and H. Jones, Processing of Structural Metals by Rapid Solidification, (eds) F.H. Froes and S.J. Savage, ASM International, Metals Park, OH, pp.379-398, 1987.
143. H.L. Luo, C.C. Chao, and P. Duwez, Trans. Met. Soc. AIME, 230, pp.1488-1490, 1964.
144. B. Predel and K. Hulse, J. Less Comm. Metals, 63, pp.45-56, 1979.
145. H. Abe, R. Ito, and T. Suzuki, Acta Met., 18, pp.991-994, 1970.
146. H. Abe, R. Ito, and T. Suzuki, Trans. Jap. Inst. Met., 11, pp.368-370, 1970.
147. H. Abe, R. Ito, and T. Suzuki, J. Pac. Eng. University of Tokyo, A10, pp.60-61, 1970.
148. A.F. Belyanin, N.A. Bulenkov, V.R. Martovitsky, and T.A. Toporenskaya, T. Moskov. Inst. Toukoi Khim. Tekhnol., 4, 1, pp.3-7, 1974.
149. A. Calka, M. Madhava, D.E. Polk, B.C. Giessen, H. Matyja, and J. Vandersande, Scripta Met., 11, pp.65-70, 1977.
150. A. Joshi, Rapidly Solidified Magnesium Alloys, Lockheed Missiles and Space Company Interim Technical Report, LMSC-F156889, January 1987.
151. F. Hehmann and H. Jones, Rapidly Solidified Alloys and Their Mechanical and Magnetic Properties, (eds) B.C. Giessen, D.E. Polk, and A.I. Taub, Materials Research Society, Pittsburgh, PA, 58, pp.259-274, 1986.

152. S.K. Kas and C.F. Chang, Rapidly Solidified Crystalline Alloys, (eds) S.K. Das, B.H. Kear, and C.M. Adam, TMS, Warrendale, PA, pp.137-156, 1985.
153. C.F. Chang, S.R. Das, and D. Raybould, Rapidly Solidified Materials, (eds) P.W. Lee and R.S. Carbonara, ASM, Metals Park, OH, pp.129-135, 1985.
154. C.F. Chang, S.K. Das, D. Raybould, and A. Brown, Metal Powder Report, 41, 4, pp.302-305, 308, 1986.
155. P.J. Meschter and J.E. O'Neal, Met. Trans., 15A, pp.237-240, 1984.
156. R.S. Busk, J. Met. Trans. AIME, 188, pp.1460-1464, 1950.
157. T.E. Leontis and R.S. Busk, US Patent No. 2,659,131 (1953), British Patent No. 690783, 1953.
158. R.S. Busk, Light Metals, 23, 266, pp.197-200, 1960.
159. S. Krishnamurthy, E. Robertson, and F.H. Froes, Processing of Structural Metals by Rapid Solidification, (eds) F.H. Froes and S.J. Savage, ASM International, Metals Park, OH, pp.399-408, 1987.
160. R.S. Busk and T.E. Leontis, Trans. AIME, 188, pp.297-306, 1950.
161. G.S. Foester, Met. Eng. Quart., 12, 1, pp.22-26, 1972.
162. G.S. Foester, US Patent No. 3,067,028, 1962.
163. G.S. Foester, US Patent No. 3,219,490, 1965.
164. M.C. Flemings and A. Mortensen, Rapid Solidification Processing of Magnesium Alloys, US Army Materials and Mechanics Research Center Technical Report, AMMRC TR-84-37, September 1984.
165. P.J. Meschter, Met. Trans. A, 18A, pp.347-350, February 1987.
166. Allied-Signal, Inc., Morristown, NJ, Engineering Data Sheet, 1987.
167. A. Joshi, R.E. Lewis, and H. Jones, work in progress, 1988-1989.
168. S. Isserow and F.J. Rizzitano, International Journal of Powder Metallurgy, 10, 3, pp.217-227, 1974.
169. S. Isserow, Investigation of Microquenched Magnesium ZK60A Alloy, DTIC Technical Report, AD780799, Alexandria, VA, 1974.
170. S. Krishnamurthy, I. Weiss, and F.H. Froes, Powder Metallurgy and Related High Temperature Materials, (ed.) P. Ramakrishnan, Trans Tech Publications, Rheinfelden, Switzerland, pp.135-146, 1989.
171. G. Nussbaum, H. Gjestland, and G. Regazzoni, Magnesium: A Strategic Material, Int. Magnesium Association, McLean, VA, pp.19-24, 1988.
172. S. Krishnamurthy, M. Khobaib, E. Robertson, and F.H. Froes, Materials Science and Engineering, 99, pp. 507-511, 1988; also Rapidly Quenched Metals 6, (eds) R.W. Cochrane and J.O. Strom-Olsen, Elsevier Sequoia, The Netherlands, 3, pp.507-511, 1988.
173. P.J. Meschter, R.J. Lederich, and J.E. O'Neal, McDonnell-Douglas Research Laboratories, St. Louis, MO, unpublished research, 1987.
174. S. Guha, Structure and Properties of $Mg_{17}Al_{12}$, M.S. Thesis, Dartmouth College, Hanover, New Hampshire, October 1987.
175. S. Krishnamurthy, unpublished research, 1986.
176. J.S. Benjamin and M.J. Bomford, Met. Trans., 8A, p.1301, 1977.
177. W.D. Nix, A Fundamental Stuty of the Processing of Oxide Dispersion Strengthened Metals, AFOSR-TR-81-0013, 1980.
178. R.F. Singer, W.C. Oliver, and W.D. Nix, Met. Trans., 11A, p.1895, 1980.
179. J.H. Weber and R.D. Schelleng, Dispersion Strengthened Aluminum Alloys, (eds) Y-W. Kim and W.M. Griffith, TMS, Warrendale, PA, pp.467-482, 1988.
180. J.A. Hawk and H.G.F. Wilsdorf, Scripta Met., 22, p.561, 1988.
181. J.A. Hawk, P.R. Mirchandani, R.C. Benn, and H.G.F. Wilsdorf, Dispersion Strengthened Aluminum Alloys, (eds) Y-W. Kim and W.M. Griffith, TMS, Warrendale, PA, pp.517-538, 1988.
182. D.L. Erich, Development of a Mechanically Alloyed Aluminum Alloy for 450-650°F Service, US Air Force Materials Laboratory Technical Report, AFML-TR-79-4210, 1979.
183. J.S. Benjamin and R.D. Schelleng, Met. Trans., 12A, p.1827, 1981.
184. A. Renard, A.S. Cheng, R. de la Veaux, and C. Laird, Materials Science and Engineering, 60, p.113, 1983.
185. D.L. Erich and S.J. Donachie, Metal Progress, 121, p.22, 1982.
186. R. D. Schelleng, Metal Powder Report, 43, p.239, 1988.
187. T.R. Bieler, T.G. Nieh, J. Wadsworth, and A.R. Mukherjee, Scripta Met., 22, p.81, 1988.
188. R.D. Schelleng, A.I. Kemppinen, and J.H. Weber, Space Age Metals Technology, (eds) F.H. Froes and R.A. Cull, SAMPE, Covina, CA, 2, pp.177-187, 1988.
189. R.T. Chen and E.A. Starke, Jr., Materials Science and Engineering, 67, p.229, 1984.
190. P.S. Gilman, J.W. Brooks, and P.J. Bridges, Aluminum-Lithium Alloys III, (eds) C. Baker, P.J. Gregson, S.J. Harris, and C.J. Peel, The Institute of Metals, London, p.112, 1986.
191. W. Ruch and E.A. Starke, Jr., Aluminum-Lithium Alloys III, (eds) C. Baker, P.J. Gregson, S.J. Harris, and C.J. Peel, The Institute of Metals, London, p.121, 1986.

192. W.E. Quist, G.H. Narayanan, A.L. Wingert, and T.M.F. Ronald, Aluminum-Lithium Alloys III, (eds) C. Baker, P.J. Gregson, S.J. Harris, and C.J. Peel, The Institute of Metals, London, p.625, 1986.
193. G. Jangg, F. Kutner, and G. Korb, Powder Metallurgy Institute, 9, p.24, 1977.
194. V. Arnhold and K. Hummert, Dispersion Strengthened Aluminum Alloys, (eds) Y-W. Kim and W.M. Griffith, TMS, Warrendale, PA, pp.483-500, 1988.
195. G. Jangg, H. Vasgyura, R. Schroder, M. Slesar, and M. Besterci, Horizons of Powder Metallurgy, Part II, (eds) W.A. Kaysser and W.J. Huppmann, Verlag Schmid GmbH, Freiburg, West Germany, pp.989-992, 1986.
196. E. Arzt and J. Rosler, Dispersion Strengthened Aluminum Alloys, (eds) Y-W. Rim and W.M. Griffith, TKS, Warrendale, PA, pp.31-56, 1988.
197. S. Ezz, M.J. Koczak, A. Lawley, and M.K. Premkumar, High Strength Powder Metallurgy Aluminum Alloys II, (eds.) G.J. Hildeman and M.J. Koczak, TMS, Warrendale, PA, pp.287-307, 1986.
198. D.L. Yaney, M.L. Ovecoglu, and W.D. Nix, Dispersion Strengthened Aluminum Alloys, (eds) Y-W. Kim and W.M. Griffith, TMS, Warrendale, PA, pp.619-630, 1988.
199. S.S. Ezz, A. Lawley, and M.J. Koczak, Dispersion Strengthened Aluminum Alloys, (eds) Y-W. Kim and W.M. Griffith, TMS, Warrendale, PA, pp.243-264, 1988.
200. P.R. Mirchandani and R.C. Benn, Space Age Metals Technology, (eds) F.H. Froes and R.A. Cull, SAMPE, Covina, CA, 2, pp.188-201, 1988.
201. R.E. Frazier and M.J. Koczak, Scripta Met., 21, p.129, 1987.
202. A.D. Jatkar, R.D. Schelleng, and S.J. Donachie, Metal-Matrix, Carbon and Ceramic-Matrix Composites, (ed.) J.D. Buckley, NASA, CP 2406, p.119, 1985.
203. T.G. Nieh, C.M. McNally, J. Wadsworth, D.L. Yaney, and P.S. Gilman, Dispersion Strengthened Aluminum Alloys, (eds) Y-W. Kim and W.M. Griffith, TMS, Warrendale, PA, pp.681-692, 1988.
204. D.L. Davidson, Met. Trans., 18A, p.2115, 1987.
205. T.G. Nieh, C.A. Henshal, and J. Wadsworth, Scripta Met., 18, p.1405, 1984.
206. T.K. Wassel and L. Himmel, A Study of Mechanical Alloying of Metal Powders, TACOM-TR-12571, 1981.
207. I.G. Wright and A.H. Clauer, Study of Intermetallic Compounds: Dispersion Hardened TiAl, US Air Force Materials Laboratory Technical Report, APML-TR-76-107, 1976.
208. R. Sundaresan and F.H. Froeg, Powder Metallurgy and Related High Temperature Materials, (ed.) P. Ramakrishnan, Trans Tech Publications, Rheinfelden, Switzerland, pp.199-206, 1989.
209. R. Sundaresan and F.H. Froes, Development of the Titanium-Magnesium Alloy System Through Mechanical Alloying, presented at the Sixth World Conference on Titanium, Cannes, France, June 1988, and to be published in the proceedings.
210. R.B. Schwarz, R.R. Petrich, and C.K. Saw, J. Non-Crystalline Solids, 76, p.281, 1985.
211. R.B. Schwarz and C.C. Koch, Appl. Phys. Let., 49, p.146, 1986.
212. G. Cocco, S. Enzo, L. Schiffine, and L. Battezzatti, Materials Science and Engineering, 97, pp.43, 121, 1988.
213. B.P. Dolgin, M.A. Vanek, T. McGory, and D.J. Ham, J. Non-Crystalline Solids, 87, p.281, 1986.
214. C. Politis and W.L. Johnson, J. Appl. Phys., 60, p.1147, 1986.
215. C. Politis and J.R. Thompson, Science and Technology of Rapidly Quenched Alloys, (eds) M. Tenhover, W.L. Johnson, and L.E. Tanner, Materials Research Society, Pittsburgh, PA, 80, p.91, 1987.
216. R. Sundaresan, A.G. Jackson, S. Krishnamurthy, and F.H. Froes, Materials Science and Engineering, 97, pp.115-119, 1988.
217. C. Politis and J.R. Thompson, Horizons of Powder Metallurgy, Part I, (eds.) W.A. Kaysser and W.J. Huppmann, Verlag Schmid GmbH, Freiburg, West Germany, p.141, 1986.
218. R. Sundaresan, A.G. Jackson, and F.H. Froes, Dispersion Strengthened Titanium Alloys Through Mechanical Alloying, presented at the Sixth World Conference on Titanium, Cannes, France, June 1988, and to be published in the proceedings.
219. R. Sundaresan and F.H. Froes, Application of Mechanical Alloying to Titanium Systems, Modern Developments in Powder Metallurgy, compiled by P.U. Gummeson and D.A. Gustafson, MPIF, Princeton, NJ, 21, 1988.
220. S.S. Sergev, S.A. Black, and J.F. Jenkins, US Patent 4,264,362, April 28, 1981.
221. S.A. Black, Development of Supercorroding Alloys for Use as Timed Releases, CEL-TN-1550, 1979.
222. M.Y. Song, E.I. Ivanov, B. Darriet, M. Pezat, and P. Hageumuller, Int. J. Hydrogen Energy, 10, p.169, 1985.
223. E. Ivanov, I. Konstanchuk, A. Stepanov, and V. Boldyrev, J. Less Common Metals, 131, pp.25, 89, 1987.
224. M.Y. Song, E. Ivanov, B. Darriet, M. Pezat, and P. Hagenmuller, J. Less Common Metals, 131, pp.71, 181, 1987.
225. M. Khrussanova, M. Terzieva, P. Peshev, and E. Yu. Ivanov, Materials Research Bulletin, 22, p.405, 1987.

17

The Microstructure and Properties of Al-SiC_p Composite Materials Produced by Spray Deposition

T.C. WILLIS, J. WHITE, R.M. JORDAN AND I.R. HUGHES

Alcan International Limited, Banbury, UK

Abstract

Production of metal matrix composite (MMC's) using the Osprey [1] spray deposition process is described. Fabrication and machining practices are discussed and potential product forms are shown, along with microstructures and mechanical properties.

It is concluded that spray deposition has considerable potential for the production of bulk quantities of high quality, low cost aluminium based SiC particulate (SiC_p) MMC's.

1. Introduction

For over two decades metal matrix composites (MMC's) have attracted considerable academic and industrial attention but have not yet found significant commercial application. The potential advantages of these materials are well documented. In aluminium alloys reinforced with a percentage volume fraction (Vf) of 40% SiC_p increases in modulus of greater than 100% have been reported. Associated with the modulus improvement are increases in proof and ultimate strength of up to 60% [2,3,4]. Table 1 shows the relative specific modulus for several competing metals. As can be seen, the specific modulus for most alloys is ≈ 26: Al-Li and MMC (15% SiC_p) are comparable, with values ≈ 32, with Al-Li MMC (12% SiC_p) ≈ 38, which is approaching an increase in specific modulus of 50% over that for all the conventional metal alloys.

Table 1 Specific modulus for a number of metal alloys and MMC's

Material	Modulus (GPa)	Density (Mg/m^3)	Specific Modulus (GPa m^3/Mg)	Improvement Relative to Steel (%)
Steel	200	7.7	26.0	0
Ti - IMI 318	106	4.4	24.1	-7
Mg - AZ80A	45	1.8	25.0	-4
Al - 2014	75	2.8	26.8	3
Al-Li - 8090	80	2.54	31.5	21
Al + 15% SiC_p	95	2.84	33.5	29
Al-Li + 12% SiC_p	100	2.61	38.3	47

Other reported advantages of these materials, over their non-composite counterparts, include a potential for high abrasion resistance [5], improved fatigue crack initiation characteristics, increased elevated temperature strength [3,6], improved creep rupture properties [7], and good micro creep performance [8].

The presence of reinforcement can also modify the physical properties of the host metal. For example, the coefficient of thermal expansion (CTE) of the resultant composite can be controlled by altering the level of reinforcement. Figure 1 shows that a Vf of 15-20% SiC_p in aluminium results in a CTE equivalent to stainless steel whereas a Vf of 34% produces a CTE equivalent to beryllium and plain carbon steel.

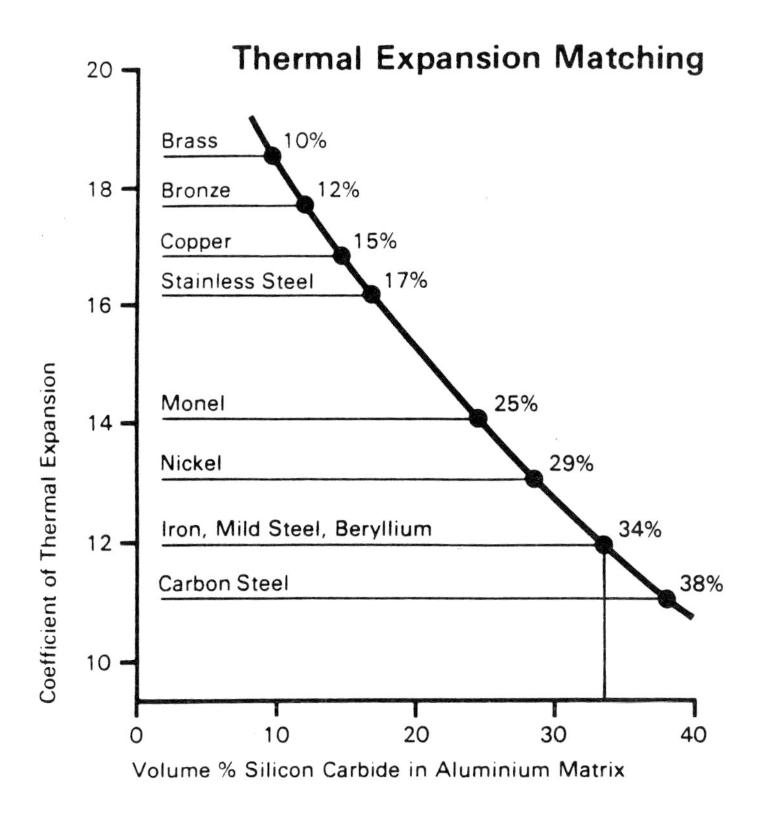

Figure 1 Graph of CTE versus Vf of SiC for particulate reinforced 6061.

These improvements in properties make MMC's attractive for structural applications in the aerospace, automotive and leisure industries [9]. Limited use has been made of these materials to date, primarily because of the current cost of production and the limited availability of evaluation material. Part of the high cost of MMC's arises from the use of expensive reinforcements, such as SiC fibres or whiskers, which cost in the range £50 to £1000/kg. These also result in anisotropy in the final product which, in specific applications can be advantageous, but for a multi-purpose material is undesirable. The cheapest type of reinforcement available with suitable specific properties is SiC$_p$ which has the additional advantage of producing isotropic properties in the product.

SiC$_p$ also has the advantage of being biologically inert [10]. For pathogenic reasons (such as mesothelioma induction, fibrogenic and lung cancer potential) it has been found that it is the size and shape of any given particle that causes problems, as well as its chemistry, and it is hence advisable to use particulate with diameters > 5μm, and to avoid all contact with whiskers/fibres in the range 0.1-3μm, with lengths > 5μm [11,12]. Successful bulk composite systems are therefore most likely to be based on SiC$_p$ as the reinforcement.

There are numerous methods that have been used to produce MMC's. These can essentially be reduced to those involving incorporation of reinforcement in either the liquid or the solid state. The most commonly used route for the preparation of particulate and whisker reinforced MMC's is the powder route, for details see Ref. 3 and Fig. 3. In the liquid state incorporation is usually achieved through infiltration of a preform of reinforcement by liquid metal under pressure [e.g. Ref. 13]. This method, however, is primarily only suitable for the incorporation of fibres or whiskers and the choice of matrix alloy is rather limited.

An alternative liquid metal route is that of mixing the particulate into the melt prior to casting. This method is being used successfully by Dural Aluminum Composites Corporation (DACC), San Diego, a wholly owned subsidiary of Alcan.

Additionally, Alcan has installed a laboratory scale Osprey spray deposition unit at its Banbury Laboratories. The process is being developed as an additional low cost production route for particulate reinforced MMC's, with specific advantages for some alloy systems and certain products. This paper describes the progress made during the initial phase of the study.

2. The Spray Deposition Process

A schematic of the spray deposition process is shown in Fig. 2. The aluminium alloy to be sprayed is melted in a crucible, ejected through a nozzle into an atomizer, and atomized. A solid deposit is built up on the collector; atomized powder that is not deposited is separated from the exhaust gas by a cyclone and collected. Metal flow rates are typically 6 kg/min. MMC is produced by mixing particulate into the metal spray, and co-depositing it onto a substrate.

Typical efficiency is in the range 60-90%, depending on the product form. The shape of the final product depends on the atomizing conditions, and the shape and motion of the collector. The equipment can be easily modified to produce hollow tube, near net shape forging stock, extrusion ingot or plate.

Precise control of gas pressures and particulate feed rates are required to ensure that a uniform distribution of particulate is produced within a typically 95-98% dense aluminium matrix. A range of matrix alloys from which MMC's have been successfully made by this technique is listed in Table 2. Alloys with a wide variety of freezing ranges have been made, from commercial purity aluminium to 2618 and 8090 [14].

Figure 3 compares schematically the production routes for MMC's manufactured by the powder blending and spray deposition routes. As can be seen the spray deposition route contains considerably less process steps, as well as eliminating the potentially hazardous blending step. The elimination of the powder handling steps also leads to a cleaner end product with no opportunity for undesirable inclusions to be incorporated.

A typical MMC microstructure for 2014 + 15% SiC_p is shown in Fig. 4. The average particle diameter in this instance is ≈ 13μm. For comparison the as-cast microstructure for chill cast 2014 of the same composition, without SiC_p is shown in Fig. 5. The considerably refined microstructure of the composite, combined with the lack of precipitation at the interface between matrix and SiC_p is reflected in the ease of homogenization, (Fig. 6) of the material and the fine as-cast grain size of ≈ 15 μm.

Table 2 A range of matrix alloys from which MMC's have been made.

Alloy	Comment
1100	Commercial purity aluminium
LM13	Al-Si casting alloy
2014	Al-Cu-Mg alloy
2618	Al-Cu-Mg-Ni alloy
6061	Al-Mg-Si alloy
7475	Al-Zn-Mg high purity base alloy
8090	Al-Li-Cu-Mg (Lital A) alloy

3. Fabrication of As-sprayed MMC's

MMC ingots containing SiC_p have been processed by extrusion, rolling, forging and remelting. The emphasis has been placed on the first two for the purpose of evaluating the characteristic properties of this type of material. Examples of all four product forms are shown in Fig. 7.

3.1 Extrusion

Billet has been extruded, with minor variations in extrusion practice, and a variety of cross-sections produced (Fig. 7). The main difference between these MMC's and conventional aluminium alloys is an increase in breakthrough pressure of ≈ 25% and marginally higher running loads. Both tapered and shear dies have been used to produce satisfactory extrusions. With the limited amount of material processed to date there has been no observable extrusion die wear.

A microstructure of extruded 2014 MMC is shown in Fig. 8. The SiC_p is evenly distributed throughout the section. Since the SiC_p used, although particulate, has an aspect ratio of ≈ 2:1, there is some alignment of the SiC_p but not to the extent observed in whisker reinforced MMC's. No evidence of break-up of the SiC_p during extrusion or decohesion at the matrix/SiC_p interface has been observed.

Longitudinal tensile results from 2014 + 10% SiC_p and 6061 + 13% SiC_p extrusions, aged to peak strength, are shown in Fig. 9, and compared with the relevant base alloys processed under identical conditions. Figure 10 shows longitudinal tensile results from dilute 8090 + 12% SiC_p extrusions, aged to peak strength [see Ref. 14 for details]. In all cases there is a clear increase in modulus. The ductility of the matrix alloy is reduced by the presence of the SiC but only a limited amount of work has been done so far with regard to optimization of downstream fabricating practices, and mechanical properties. In the Discussion section below it will be demonstrated that it is possible to alter the balance of strength and ductility, by heat treatment, while retaining the modulus improvements.

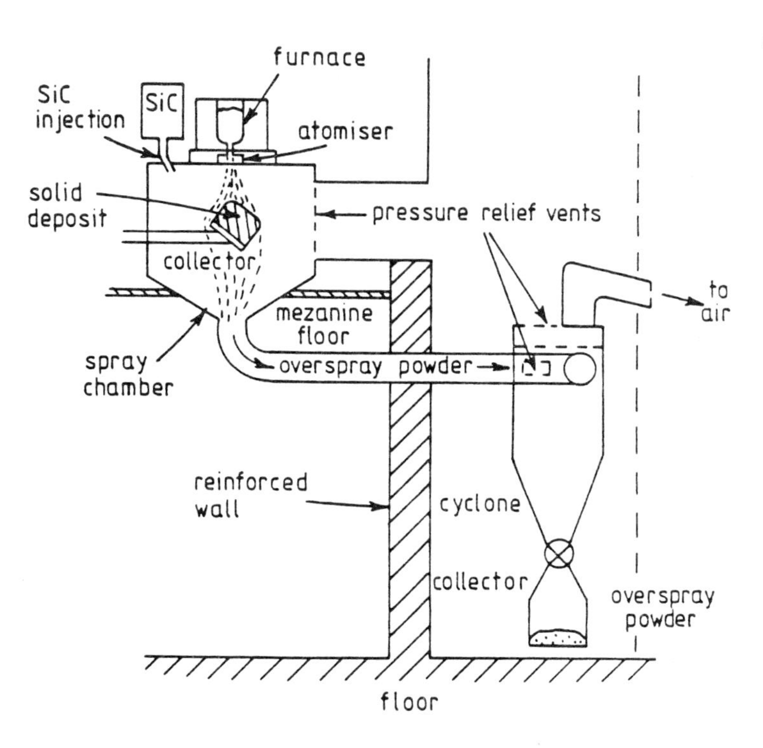

Figure 2 Schematic drawing of the Osprey process.

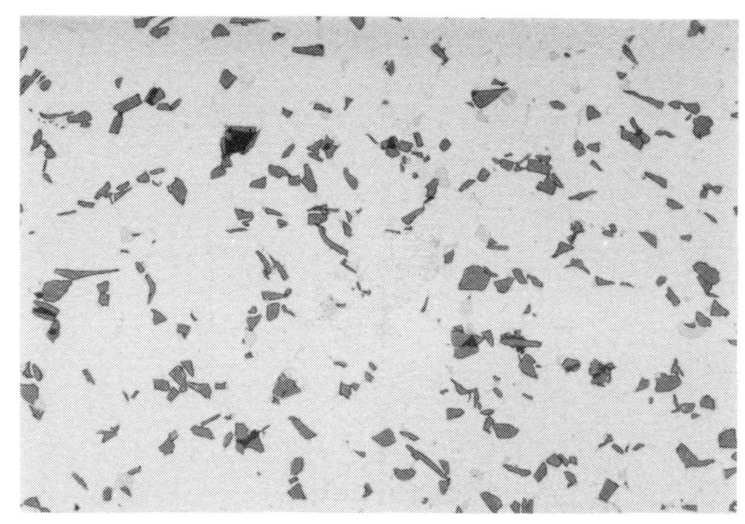

Figure 4 Microstructure of as-produced spray deposited 2014 + SiC.

Figure 5 Microstructure of chill cast 2014 - No SiC.

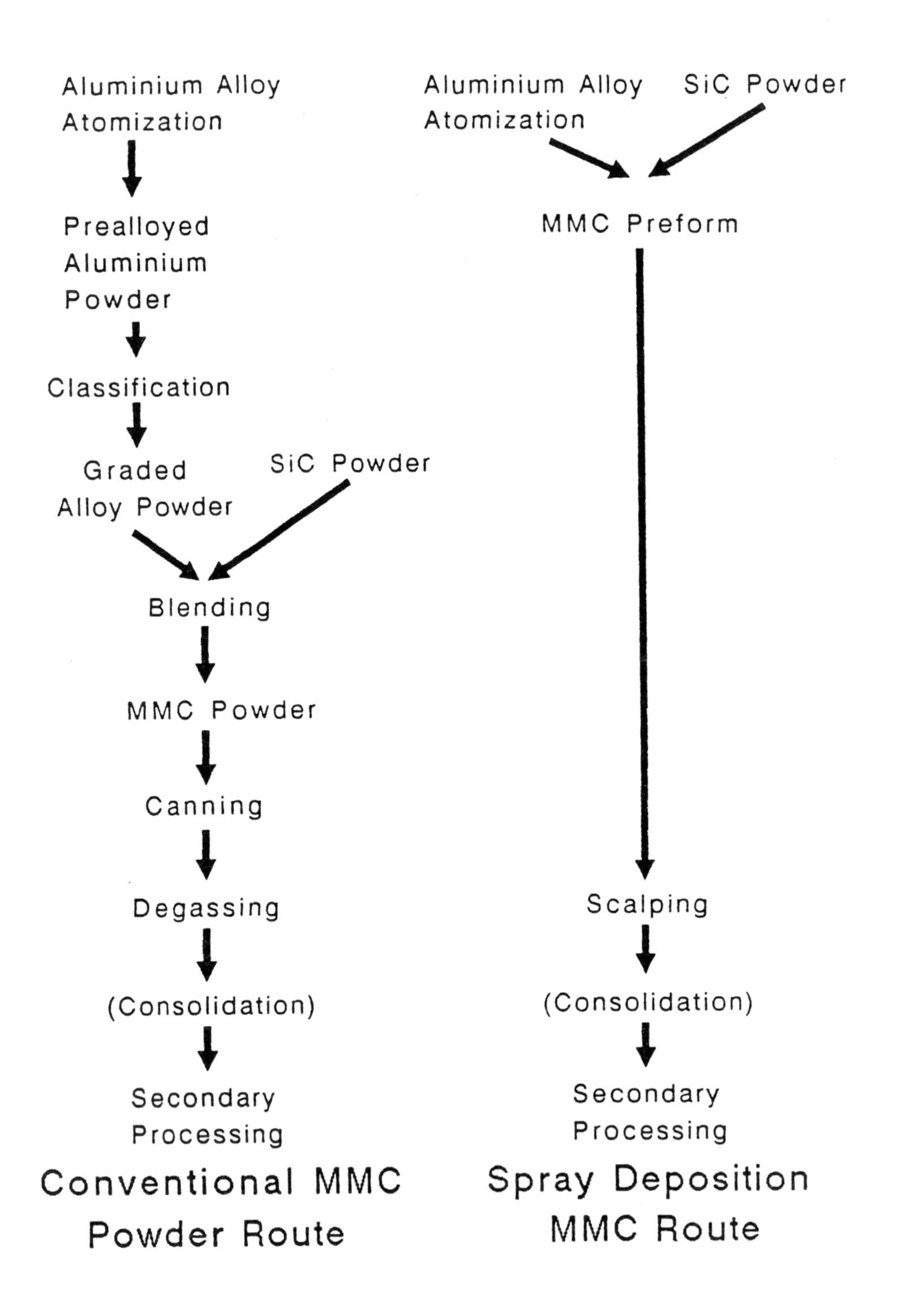

Figure 3 Schematic flow diagram comparing conventional powder blending route versus spray deposition route for MMC production.

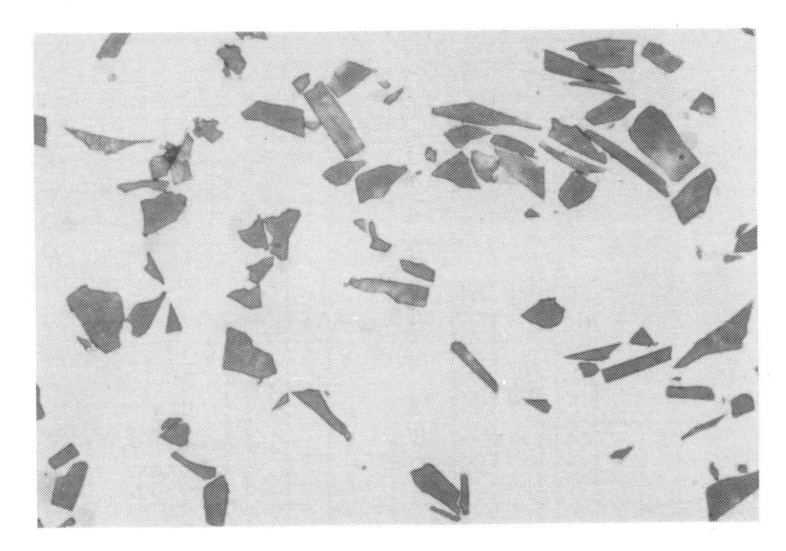

Figure 6 Microstructure of homogenised spray deposited 2014 + SiC.

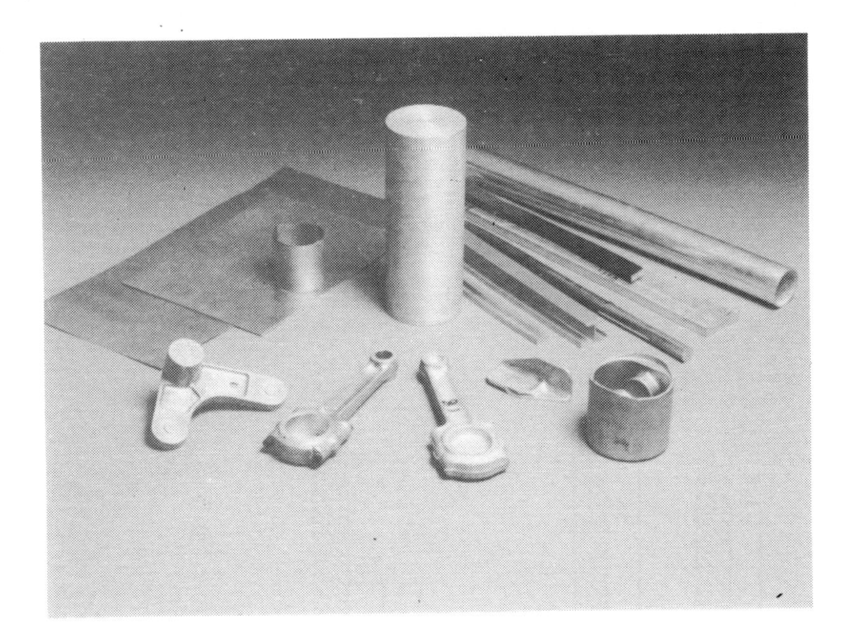

Figure 7 Collectlon of four MMC product forms: Extrusion, sheet, forging and pressure die casting.

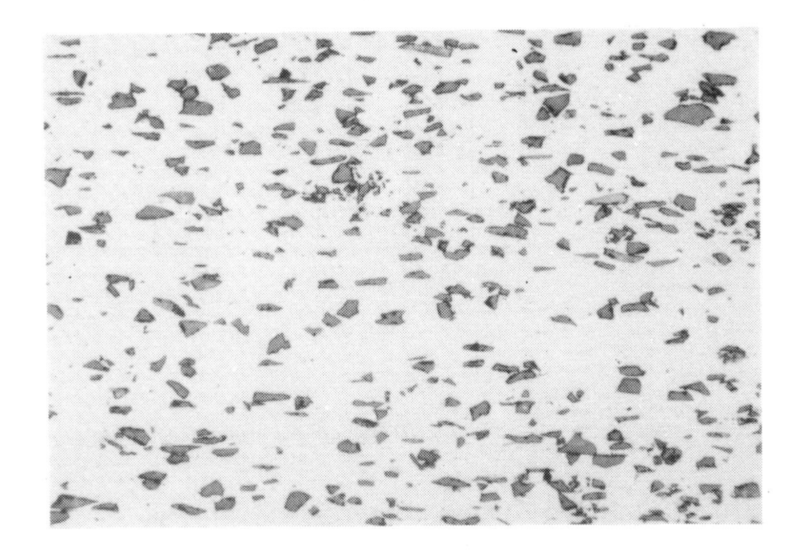

Figure 8 Microstructure of extruded spray deposited 2014 + SiC.

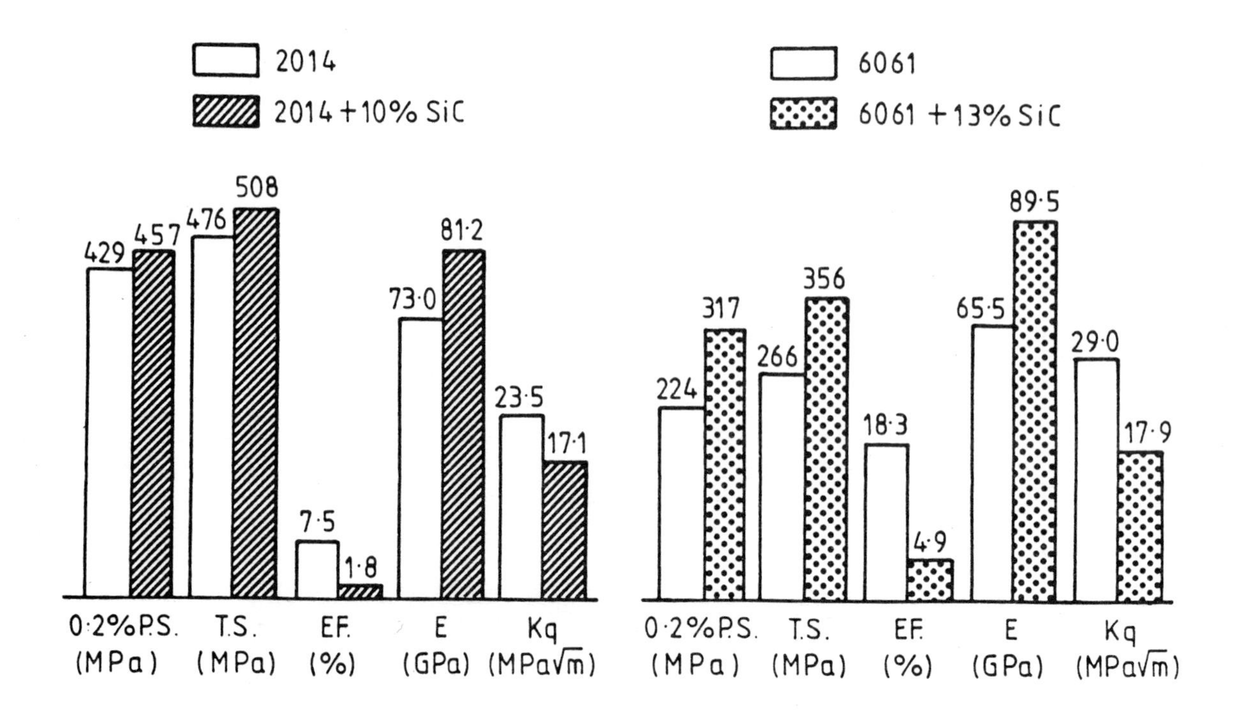

Figure 9 Mechanical properties for 6061 and 2014 MMC's, with base alloy comparisons.

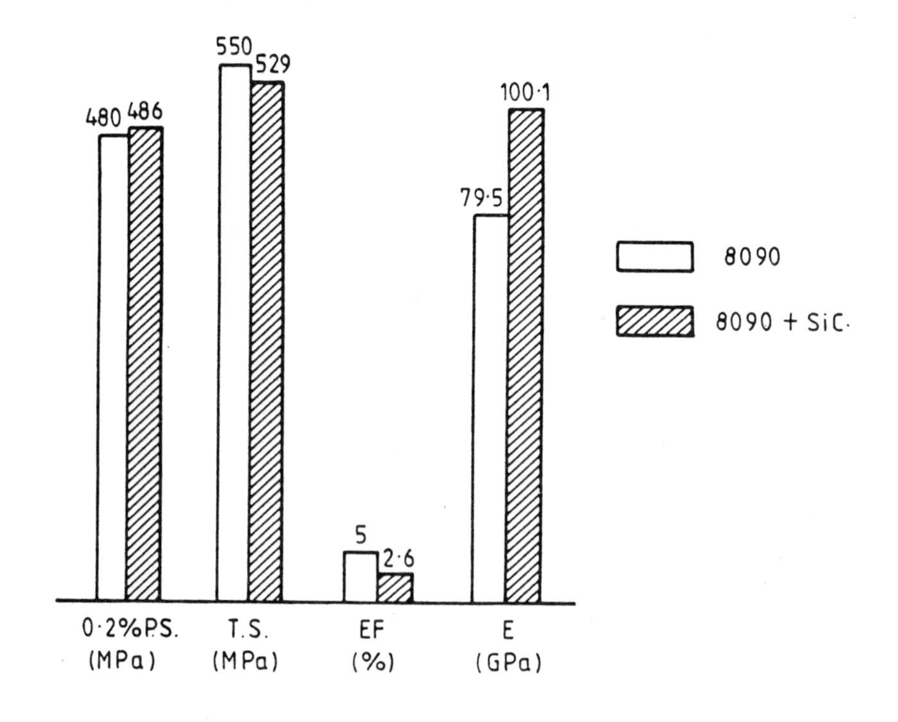

Figure 10 Mechanical properties for dilute spray deposited 8090 MMC.

3.2 Rolled Products

The hot and cold working characteristics of this type of MMC are similar to those of the base alloys with some alterations to account for the increased elastic modulus and work hardening rates. To demonstrate the workability of the material it has been cold rolled to 100μm thick foil. Figure 11 shows transverse and longitudinal sections from this foil indicating that the SiC_p is still uniformly distributed throughout the product, with no sign of break-up of the SiC_p or decohesion at the SiC_p/aluminium interface.

Typical properties from 2618 sheet containing 12% SiC_p are shown in Fig. 12. A comparison of longitudinal and transverse results illustrates the good isotropy found in this MMC sheet (Fig. 13).

4. Discussion

The foregoing description of process and properties demonstrates the feasibility of spray deposition as a method for the production of MMC's. The properties of the MMC's produced so far show an attractive combination of strength, modulus and ductility/fracture toughness. Furthermore the combination of properties in the final heat treated product can be controlled to a large extent by the selection of the heat treatment schedule. For example in the 2618 MMC sheet described earlier (Fig. 12) an improvement in modulus of ≈ 30% can be obtained for equivalent strength and ductility, compared with the conventionally produced base alloy, by deliberate underaging: Further improvements in strength alone can be achieved by peak aging. The properties desired can thus be tailored.

One of the primary reasons for the attractive properties of spray deposited MMC's is the integrity of the matrix/SiC_p interface. Transmission electron microscopy (TEM) investigations [15] of the interface show that precipitation and reaction products in this region are negligible (Fig. 14) in direct contrast to observations of the matrix/fibre interface in squeeze cast MMC's containing SiC_f (Fig. 15). This difference can be directly related to the relative contact time between molten metal and the SiC in the two processes; this can be in the order of minutes during squeeze casting but a few seconds during spray deposition.

An attractive combination of properties, however, will not guarantee successful introduction of a new material. In real applications ease of fabrication is important. In this and other work with Vf's up to 25% SiC_p, fabrication of the MMC by either hot or cold working appears to present no great difficulty. One important aspect of fabrication is machining. In this study both large (extrusion and forging billet) and small (e.g. test pieces) cross-section MMC has been machined using conventional techniques, allied with carbide and diamond tooling. Machine shops used to machining high silicon containing aluminium alloys report similar machining characteristics for the high silicon alloys and aluminium based MMC's with ≈ 15% SiC_p particulate.

5. Conclusion

In conclusion the spray deposition route for MMC production has both microstructural and property advantages over competitive routes. This, combined with low starting material costs, indicates that the spray deposition process has considerable potential for the production of large quantities of high quality MMC's at a relatively low cost.

Acknowledgements

The authors would like to thank Alcan International Limited for permission to publish this work, and to T.J. Warner (Cambridge University) and P. Withers (Cambridge University) for the TEM micrographs.

Nomenclature

CTE = Coefficient
MMC = Metal Matrix Composite
SiC_f = Silicon Carbide Fibre
SiC_p = Silicon Carbide Particulate
TEM = Transmission Electron Microscopy

Figure 11 Microstructure of 100 mm foil, longitudinal and transverse sections.

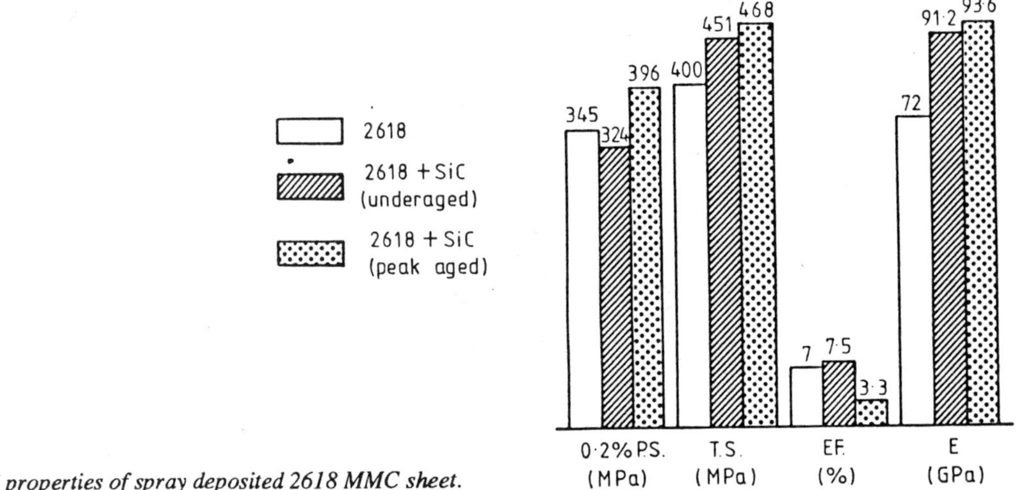

Figure 12 Mechanical properties of spray deposited 2618 MMC sheet.

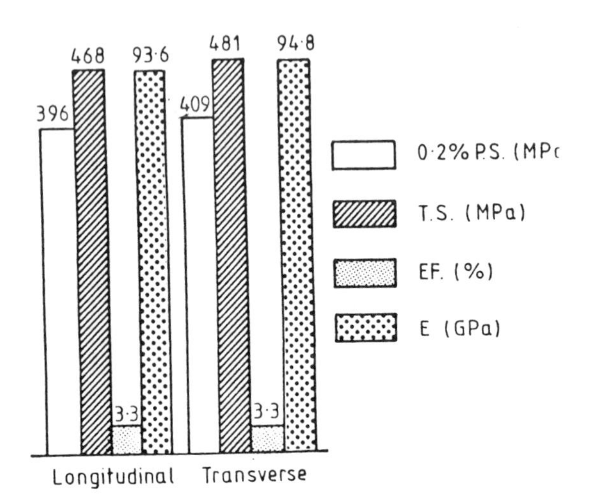

Figure 13 Longitudinal and transverse tensile properties of spray deposited 2618 MMC sheet, showing isotropy.

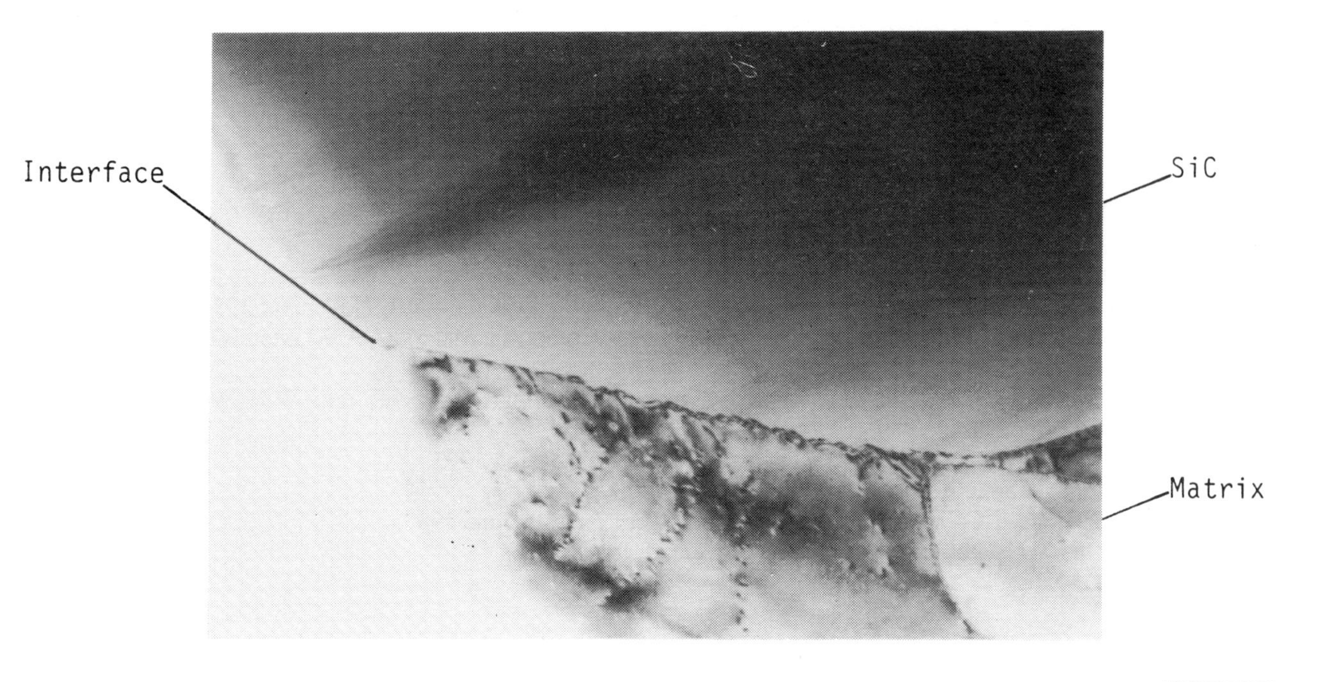

Figure 14 TEM micrograph of spray deposited SiC$_p$/aluminium interface - T. J. Warner.

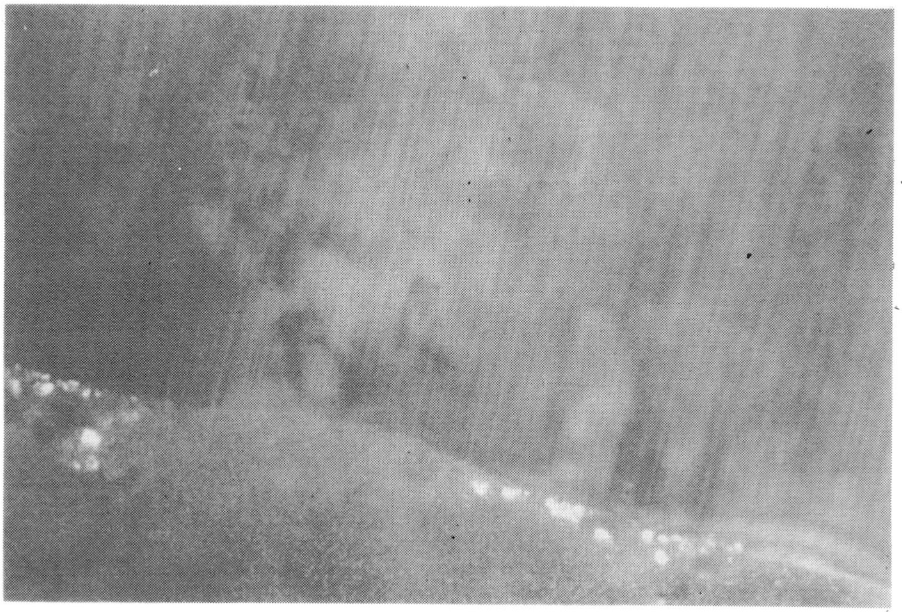

Sic

Interface

Matrix

Figure 15 TEM micrograph of squeeze cast SiC$_p$/aluminium interface - P. Withers.

50 nm

References

1. UK Patent No. 1,479,239.
2. S.V. Nair, J.K. Tien, R.C. Bates, Int. Metals rev., 30, 6, p.275, 1985.
3. D.L. McDanels, Met. Trans. A., vol.16A, p.1105, June 1985.
4. S. Dermarkar, Met. & Mat., 145, March 1986.
5. S. Norose, T. Sasada and M. Okabe, 28th Japan Congress On Materials Research, p.231, March 1985.
6. W.L. Phillips, Proc. of 1978 Conference on Composite Materials, (ed.) B. Noton *et al.*, New York, Met. Soc. of AIME, 567, 1978.
7. T.G. Nieh, Met. Trans. A., vol.15A, p.139, January 1984.
8. G. Gould, Proc. 3rd International Conference on Isostatic Pressing, London, vol.1, November 1986.
9. J.W Weeton, D.N. Peters and K.L. Thomas, Engineers' Guide to Composite Materials, ASM, Metals Park, Ohio, 1987.
10. Norton Co., Material Safety Sheet, November 1985.
11. G.H. Pigott and J. Ishmael, Proc. 5th International Symposium on Inhaled Particles, (ed.) W.H. Walton, Pergamon Press, Cardiff, 8-12 September 1980.
12. M.F. Stanton and C. Wrench, J. Nat. Cancer Inst., 48, 3, p. 797, March 1972.
13. J.A. Cornie *et al.*, Ceramic Bull, 65, 2, 1986.
14. J. White, I.R. Hughes, T.C. Willis and R.M. Jordan, Al-Li IV, Paris, June 1987.
15. T.J. Warner *et al.*, To be published at ICCI II, Cleveland, June 1988.

18

Powder Metallurgy Composites

E.A. FEEST AND J.H. TWEED
Metals Technology Centre, Harwell Laboratory, UK

1. Introduction

It is arguable that almost all powder metallurgy (PM) materials are composites in view of the near impossibility of excluding inclusions or naturally forming oxides in powder compacts. If the definition of a composite is limited to materials whose microstructures include second phases which have been deliberately added to enhance performance, then there are still many examples of the successful exploitation of the concept. Some of the established families of PM composite materials are:

Thoriated tungsten - Here the ThO_2 additions of up to a few percent increase the hot strength of wires and allow high electron emission rates.

Electrical contact materials - Here appropriate combinations of electrical and thermal conductivity and resistance to abrasion and arcing are achieved by combinations of powders, such as Ag-C, Cu-W and W-Ag.

Mechanically alloyed nickel based alloys - Here the added fine oxide is further broken down in processing to give dispersion strengthening of the alloy.

Dispersion nuclear fuels - Here fissile uranium compounds are dispersed in, for example, an aluminium based matrix, whose role is to facilitate handleability, heat extraction, corrosion resistance, etc.

Hardmetals - Here the relatively low volume fraction of matrix metal acts as a bonding agent and heat transfer path for tooling materials.

In addition to powder based composites used as bulk materials, there are also established examples of composite surface materials involving powder processing, such as in the field of hardfacing alloys.

There is thus considerable diversity of type of PM composite ranging from predominantly metal matrix, as in the case of the various dispersion strengthened families, through to predominantly filler, as in the case of the hardmetals. Whilst exploitation has tended to be in "niche" applications, the cumulative wealth of experience arising from both successful and unsuccessful PM composite development over the years has provided us with a versatile armoury of processing techniques on which to base further composite materials initiatives. The past decade has witnessed a concerted drive towards the development of metal matrix composites (MMC) aimed at high performance, structural applications and which therefore appear to offer the possibility of markedly influencing the global pattern of materials selection and use. These composites present us with one of the most exciting fields of materials development activity for the 1990s, and powder metallurgy based materials feature prominently in the vanguard of this MMC activity. This review will concentrate on this emerging family of composites both because of the topicality of the technology and because many of the more established powder based composite families mentioned earlier in this introduction are reviewed elsewhere in this publication.

The exploitation of MMC materials, together with the associated R&D activity, is advancing with such speed at present that a detailed review would rapidly become outdated. The range of developments associated with even PM MMC materials is sufficiently extensive and uneven in terms of position on the development stream that a book rather than a chapter would be required to do justice to a comprehensive review. The aim of this chapter is therefore to point out the main factors associated with PM MMC development and exploitation, to illustrate some of these points by considering one family of such materials in some detail and to summarize the main materials of interest at present.

2. Comparative Processing Options

Powder metallurgy techniques can be used for producing all of the main families of MMC materials. In the case of continuously reinforced composites, powder (slurry) impregnation routes have been used to produce multifilament reinforced precursor "prepreg" for subsequent consolidation [1] and the concept is still being pursued with vigour for MMC development [2]. Good fibre distributions and fully dense matrix microstructures can be achieved (Fig. 1), but it is difficult to the conflicting requirements of avoiding fibre damage in processing and of achieving matrix property optimization. In these respects, PM based routes are at a disadvantage compared with melt infiltration routes which can produce composite material with shorter high temperature exposure histories and with less reliance on deformation processing. This case is discussed in greater detail in section 4.1.

In the case of discontinuously reinforced MMC there is far more direct competition between PM and other processing routes. For example, in the case of particle reinforced aluminium alloy MMC, PM is only one of four feedstock preparation routes already forming the basis of commercial supply. The others are co-spray, melt stirring and melt infiltration. The four competing processing routes are represented schematically in Fig. 2. The relative advantages and disadvantages of the PM processing route over these other three families of "ingot" preparation are summarized in Table 1 and representative microstructures produced by these routes are shown in Fig. 3.

This comparison highlights the versatility of the PM approach in terms of the range of matrix and reinforcement types that it can encompass. The PM route is also relatively easy to experiment with and has been the starting point for many a research group's foray into the MMC field. It therefore can provide samples relatively quickly but this can in some cases be counter-productive because of the risk that conclusions are drawn on the basis of unoptimized processing. In terms of ultimate commercial exploitation, expense could lead to the other routes pushing out PM processing in all but those applications for which it is the only route which can provide the required MMC microstructure.

3. Particle or Whisker Reinforced Aluminium Alloys Produced by Powder Metallurgy

This family of MMC materials is reviewed here in some detail because of its topicality in terms of commercial exploitation and because many of the technical and economic factors influencing its development have at least qualitative relevance to other MMC systems.

3.1 Historical Context

The present range of particle or whisker reinforced MMC materials is part of a wider range of MMC developments which have in recent decades achieved varying degrees of industrial take up based on the asset that they can in principle be used in wrought form having undergone "conventional" secondary processing. Prominent examples were SAP [6] which exploited thermodynamically stable dispersion strengthening at temperatures approaching the melting point of aluminium, and mechanically alloyed composites which, in the early nickel base alloy examples, achieved the beneficial combination of high temperature strengthening by both dispersion and precipitation hardening [7]. A simplified comparison between these examples is given in Table 2.

The logic behind the strong interest in particle reinforced MMC is that some useful property enhancement can be achieved relatively cheaply and in a relatively "user friendly" way. The modesty of the achievable property enhancement in the context of MMC materials in general is illustrated in Fig. 4, but this is offset by, amongst other factors, the near isotropy of the material and its ease of processing to semi-finished products such as rolled sheet or extruded section. This has led to MMC comprising relatively coarse (~10μ) ceramic particles distributed in near conventional aluminium alloy matrices being the subject of nationally funded alloy development programmes and deployment in appreciable production runs [8].

3.2 Selection of MMC Formulations

The main parameters defining the MMC formulation are the matrix specification (composition, powder form and size) and the reinforcement specification (type, form and size).

The matrix specification is usually based on an iterative consideration of factors such as the required ensemble of properties (see section 3.4), compatibility with processing requirements (see section 3.3) and availability. Other more psychological factors often come into play, such as familiarity with the matrix material as an unreinforced conventional alloy. To date, a combination of these factors has led to aluminium alloy matrices tending to be centred on the more common conventional alloys within the 2XXX, 6XXX, 7XXX series and the more advanced lightweight alloys based the Al-Li system. Particular variations of these alloys are generally only used in the case of MMC materials that have undergone exhaustive optimization programmes as would be the case for a conventional alloy development. Both pre-mixed (blends of powders of the elemental constituents of the alloy) and pre-alloyed (solidified from the alloy) matrix powders have been used in the preparation of particle reinforced MMC. Because of improved cleanliness and the avoidance of problems of incomplete solute dissolution, most formulations are currently based on pre-alloyed atomized powders. The atomizing medium has an influence on particle size distribution and the degree of oxidation of the as-atomized powder. Inert gas atomization helps in both these respects and commonly provides the matrix alloy powder

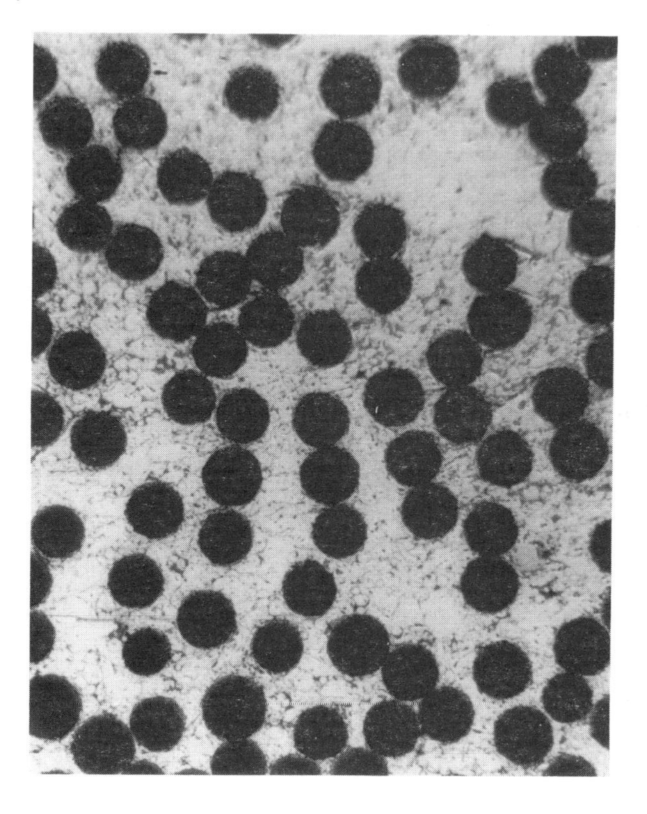

a

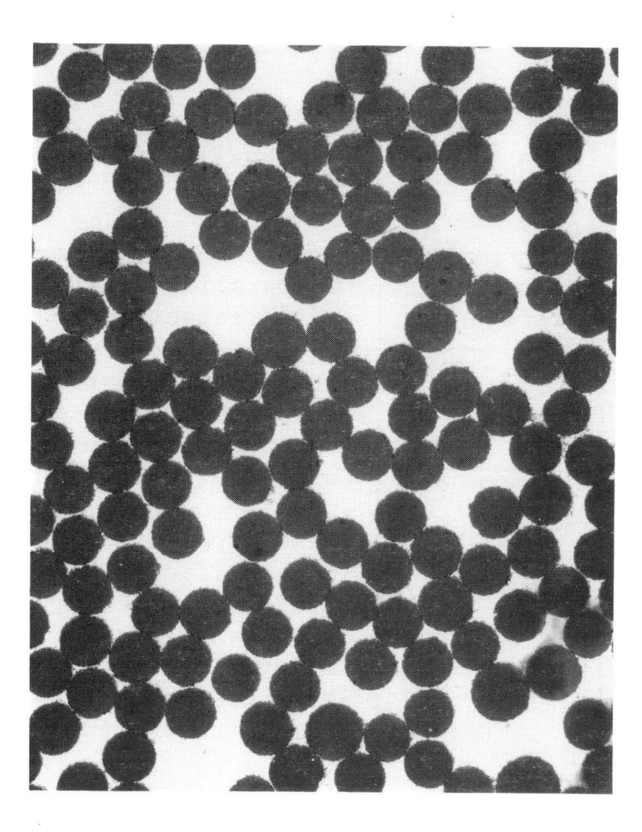

b

Figure 1 Alumina multifilament reinforced 5083 aluminium alloy composites produced by (a) powder impregnation and hot pressing and (b) low pressure melt infiltration.

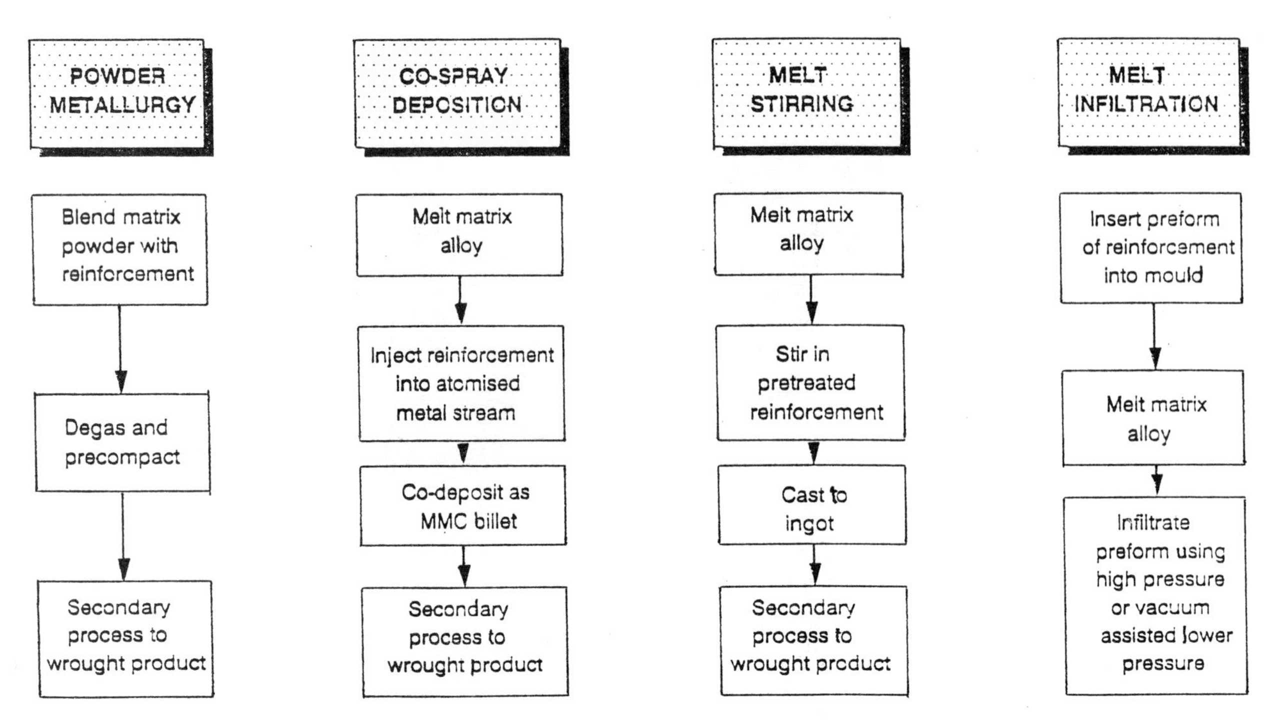

Figure 2 Key steps in alternative processing routes to particulate reinforced MMC.

Table 1

Competing route	Example reference	Comparative strengths and weaknesses with respect to PM processing route
Co-spray deposition	3	Less expensive processing; Lower volume fraction achievable; Narrower range of reinforcement size and shape possible
Melt stirring	4	Less expensive processing; More difficult to maintain uniformity of dispersion; More restrictive range of matrix alloys; More restrictive range of reinforcement forms handleable.
Melt infiltration	5	Less expensive processing; Narrower range of volume fractions achievable; Less risk of whisker damage.

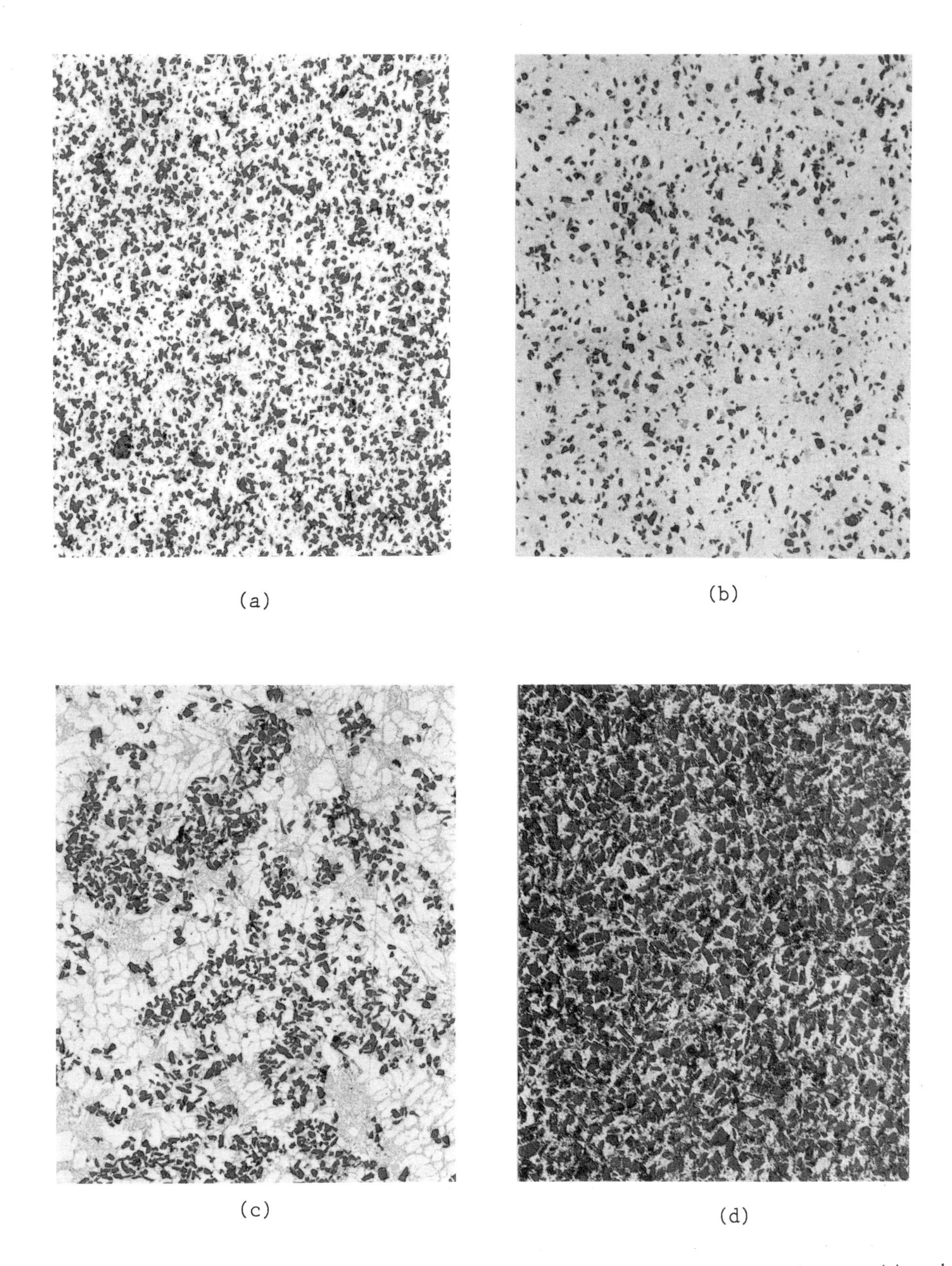

Figure 3 Microstructures of SiC particulate reinforced aluminium MMC produced by different processing routes; (a) powder metallurgy, (b) co-spray, (c) melt stirring, (d) melt infiltration.

source. The degree of rapid solidification achieved in for example inert gas atomized powder might be considered advantageous in terms of reducing matrix homogenization times in reactive MMC systems, but the main advantage of the rapid solidification is in the finer particle sizes that can be achieved and hence the improved microhomogeneity of the MMC microstructure.

The selection of reinforcements is based primarily on the trade off of achieved property enhancement versus cost. This is not a simple matter to optimize since, for example, apparent gains on a primary property such as specific modulus might be accompanied by a disadvantage on an equally relevant property such as toughness. Similarly, costs are not uniquely related to reinforcement material cost but also to factors such as differences in processing schedules. These inter-relationships are not always clear at the outset of an MMC development and major changes in formulations may have to take place during the course of an alloy development programme. In terms of enhancement of specific modulus and of acceptable cost, the Al_2O_3 and SiC families of reinforcements have been pursued in the majority of aluminium based PM MMC developments. Intrinsic specific moduli (modulus ÷ specific gravity) for these two compounds are respectively ~100 and 150 GPa. The choice between whiskers or particles has remained relatively open over the last decade, particularly as there has been an ever increasing multiplicity of SiC whisker sources - notably in Japan [9]. The emphasis is currently very much on particle rather than whisker reinforcement mainly for the following reasons:

- There is concern over the possible health hazards associated with handling SiC whiskers [10].
- Now that MMC processing costs have been reduced, the material cost penalty of whiskers over particles is a significant disincentive to use.
- There is insufficient difference between whiskers and particles in the mechanical property enhancement that they have achieved.

The latter point is illustrated by Table 3 which compares properties of examples of currently available whisker and particle aluminium alloy MMC.

It is notoriously difficult to make rigorous comparisons from data such as given in Table 3 since they do not necessarily compare materials at the same level of development, particularly in terms of processing scale and optimization. This is further complicated by inconsistencies in testing techniques particularly in the case of fracture toughness measurements. A recent survey [13] of published fracture toughness data reported that values in excess of 60 MPa $\sqrt{m}$ have been claimed for thin sheet of both 20% SiC particulate in 2124 and 15% SiC whisker in 2124. However the test techniques employed were not valid for providing plane strain toughness data and lack of consistency in test techniques has inhibited assessment of real gains in toughness achieved through improvements in processing.

It should be noted, however, that there is a spectrum of forms between high integrity whisker and irregular shaped particles. Indeed some whisker products have been made available as a mixture of fibrous whiskers and irregular particulate matter. Whisker products are continually being improved and comparisons between products have not always taken account of this evolution. Some of the available discontinuous forms of SiC reinforcements are given in Table 4.

Now that major particle reinforced MMC alloy development programmes are under way, considerable attention is being focussed on the morphology of the particulate matter. Thus the relative reinforcement efficiency of irregular grit, regular platelets and other morphologies are being compared [14]. Section 3.4 discusses further relationships between reinforcement structure/dispersion and properties.

A current theme in MMC formation is the incorporation of more than one second phase to achieve desired combinations of properties. A frequently tried combination for bearing applications is that of a hardening phase, such as alumina, with a lubrication enhancing phase, such as graphite.

3.3 Processing

The essence of PM MMC processing is that all of the care and precision associated with "conventional" PM processing has to be employed together with additional measures specific to the MMC system in question.

For particle reinforced MMC it is desirable to select the matrix alloy particle size distribution to minimize microsegregation in the MMC. Since the aim is usually to reinforce with particles which are as fine as practicable it follows that matrix particle sizes should be at least fine enough to avoid "necklacing" effects. Examples of microstructures of MMC formulations with severe and minimum necklacing are shown in Fig. 5. Failure to take into account these constraints can lead to inappropriate microstructures even in the heavily wrought state, particularly at high reinforcement volume loading. The trend, therefore, is to use as fine a matrix powder as is practicable in terms of handleability, economy and safety. Attempts have also been made to optimize the distribution of matrix particle sizes to maximize geometric packing. In general, the rigour required in matrix particle size selection increases with decreasing degree of deformation subsequently introduced into the wrought product.

The blending of the MMC constituents has been satisfactorily achieved with both wet and dry blending.

In general dry blending has been found to be adequate for blends of particulate down to 1200 grit with aluminium powders. However, for finer powders and for chopped fibres and whiskers wet mixing may be required to avoid agglomeration of the reinforcement during blending. The wetting agent should be reasonably volatile to aid subsequent

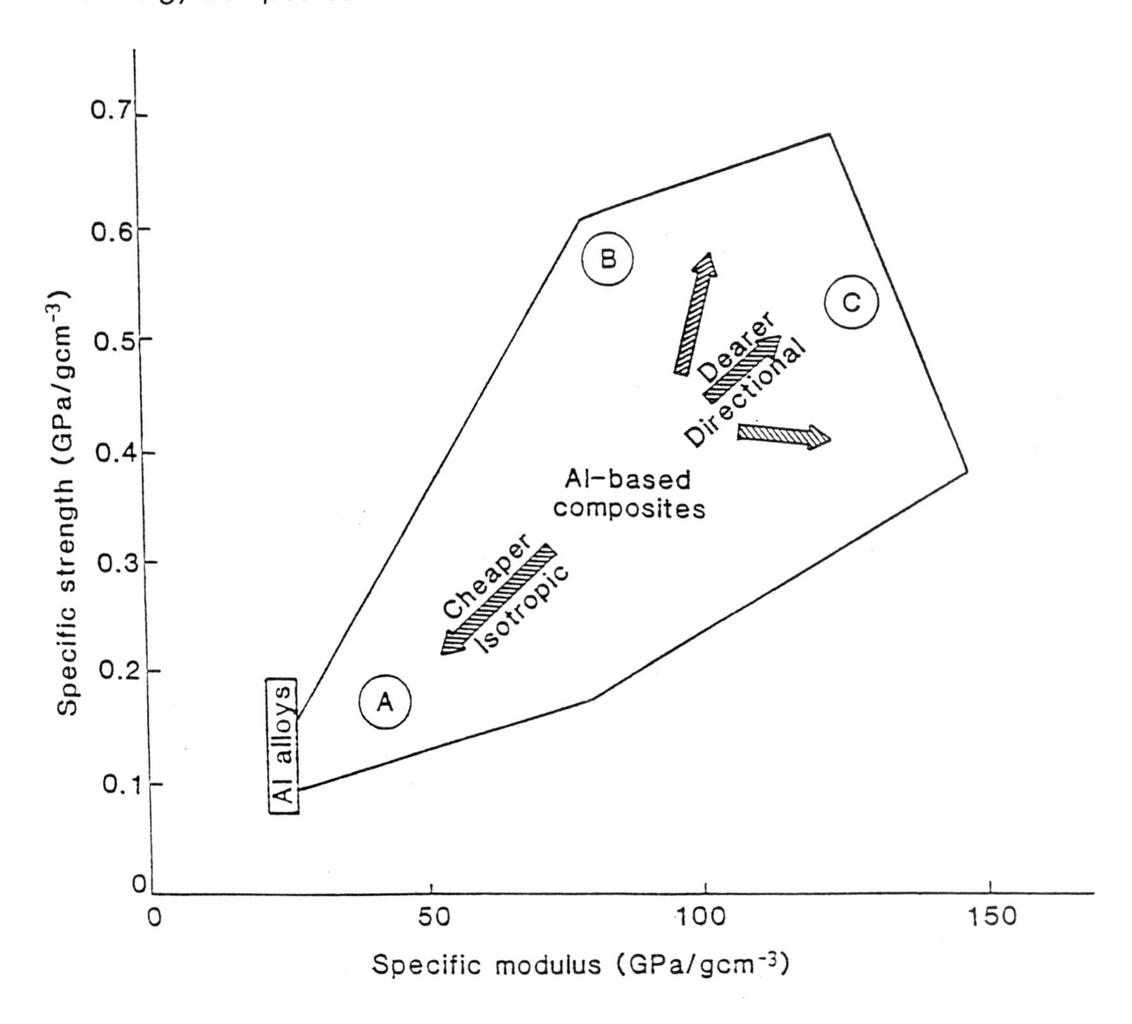

Figure 4 Bounds of combinations of longitudinal tensile specific strengths and stiffnesses achieved in aluminium based metal matrix composites. Typical reinforcements: 'A' = SiC particulate or whisker, 'B' = SiC or B monofilament, 'C' = high modulus carbon fibre.

Table 2

MMC Type	advantages	disadvantages
SAP	High temperature stability; simplicity of processing	Industrial development discontinued
Mechanically alloyed MMC	Good high temperature performance	Expensive
Whisker reinforced MMC	Potential for exploiting fibre reinforcement concept	Expensive
Particle reinforced MMC	Cheap; reasonable specific modulus gain	Modest degree of general property enhancement; toughness problem

removal and should not oxidize the matrix powder. Thus oxygen free organics or aqueous solutions with additives to suppress oxidation have been used. Dispersion of the reinforcement and matrix then generally requires assistance either from ultrasonic or mechanical agitation.

In common with PM practice for unreinforced materials a de-gassing stage may be required. This is particularly so for aluminium (alloy) based composites where the surface oxide around the powder particles may become hydrated. This hydrated alumina layer decomposes with temperature to yield water vapour (typically from 100 to 300°C) and then hydrogen (predominant above 300-350°C). This is a strong function of temperature and extended degassing at low temperature will not prevent evolution of gases at higher temperatures [15]. (See also Fig. 6 which illustrates the evolution of surface chemistry of 2014 powder). Thus degassing is required if the thermal exposure during consolidation is lower than for some subsequent treatment (e.g. solution treatment of a heat treatable alloy). For 2014 aluminium alloy/SiC particulate composites the gas evolution scales with the amount of matrix powder present (Fig. 7).

It is customary to include a pre-compaction step prior to deformation processing via conventional routes (e.g. extrusion or rolling) to final wrought form. Batch compaction processes such as vacuum hot pressing and hot isostatic pressing are commonly employed and those involving vacuum processing offer the opportunity to degas during a programmed pre-heating cycle. In special cases where near-net-shape sintering is adopted as a consolidation step, reaction assisted bonding has been used to achieve near full densification [16].

Deformation processing of PM MMC is in principle no more difficult than for the equivalent multiphase conventional alloys. They have even been reported to be more tolerant to variations in extrusion ratio and die aspect ratio than equivalent unreinforced alloys [17]. Added complications arise, however, from the differences in constitutive data (the MMCs being in general markedly less ductile and also stiffer and stronger at processing temperatures). Other constraints include the need to avoid degradation either by physical damage of reinforcement (e.g. break-up of whiskers) or chemical reaction (e.g. modification of matrix alloy microchemistry by reinforcement/matrix reaction). In well developed MMC systems these obstacles have been surmounted. They do however add to the R&D investment and associated time delay associated with bridging the gap from initial MMC formulations to optimized alloy development.

3.4 Structure/Property Relationships

One of the major attractions of particulate reinforced aluminium is the significant increase in material stiffness which is attainable. In cases where there is only limited deformation after compaction this increase in stiffness is near isotropic (even for short fibre and whisker reinforcements). Increased deformation processing leads to texturing of the matrix and alignment of reinforcement particles giving anisotropic properties. Data from a range of sources for elastic modulus of aluminium/SiC powder route particulate reinforced materials are presented in Fig. 8. These are compared with three criteria:

- Voigt estimate (rule of mixtures). This effectively assumes equal strains in matrix and reinforcement.
- Reuss estimate. This represents the case of equal stresses in matrix and reinforcement.
- Hashin-Shtrikman bounds. These are held to be the narrowest bounds which could be specified for a homogenous, isotopic two phase system based only on knowledge of elastic properties and volume fractions of the matrix and reinforcement.

In this figure all estimates assume elastic moduli of 70 GPa and 400 GPa for matrix and reinforcement respectively. The Hashin-Shtrikman bounds are generally appropriate with some deviations for whisker reinforced materials which tend to be anisotropic.

The use of SiC reinforcement in aluminium also leads to increased density and hence the performance benefit which varies as E/ρ (for tensile or compressive deflection), $E^{1/2}/\rho$ (for buckling of rods) or $E^{1/3}/\rho$ (for buckling of plates) is less than the stiffness benefit. This can be counteracted by the use of Al-Li alloys as the matrix material [18] in that incorporation of up to about 20 vol% SiC still gives lower densities than for conventional aluminium alloys.

MMC produced through powder routes can also give some increase in strength over unreinforced materials. A broad spread of behaviour is shown ranging from little strengthening to a proof strength increase around 100 MPa for a 20 vol% increase in ceramic content. No strong trends are noted with matrix alloy strength, chemistry or heat treatment conditions.

Possible mechanisms for this observed strengthening effect with increasing ceramic content include the following:

- Particle strengthening effect (very small as particles are large).
- Refinement of grain/subgrain size of the matrix through retardation of grain boundary migration.
- A substantial increase in matrix dislocation density to accommodate the thermal expansion mismatch strains generated around ceramic particles during cooling from processing temperatures.
- Composite strengthening effect.
- Matrix constraint between particles.

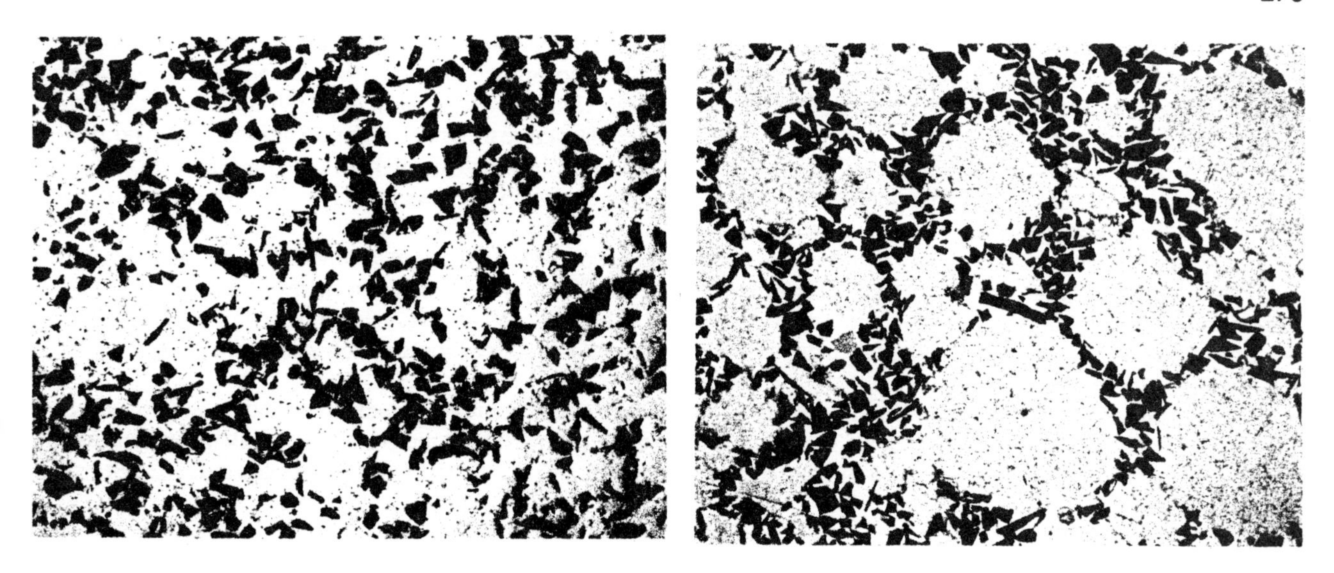

Figure 5 Effect of aluminium alloy matrix powder size on 'necklacing' of 600 grit SiC particulate reinforcement in hot pressed composite (a) 10-15 μm matrix powder, (b) 40-50 μm matrix powder.

Table 3

Longitudinal (L) and long transverse (LT) tensile properties of 2124 sheet					
Ref	Reinforcement	Modulus (GPa) E_L	E_{LT}	Proof stress (MPa) $0.2\%_L$	$0.2\%_{LT}$
11	25 vol% SiC_p	113	116	385	385
12	15 vol% SiC_w	114	95	573	385
	None	71		420	
Tensile properties of wrought 6061, T-6 condition					
Ref	Reinforcement	Modulus (GPa)		Proof stress (MPa)	
11	20 vol%SiC_p	103		415	
11	30 vol%SiC_p	121		435	
9	20 vol%SiC_w	115		475	
	None	69			

Evidence for the composite strengthening effect of ceramic particles is given in a study of load sharing in 2014/20 vol%SiC material using a neutron diffraction technique [19]. This demonstrated lower matrix average strains in the MMC, compared with unreinforced 2014, at a given macroscopic stress. Although use of particulate reinforcement can lead to benefits in modulus and strength it also leads to loss of ductility and fracture toughness. Optimization of these properties is the major current thrust of research in Al/SiC materials. Particulate reinforced metal matrix composites with sufficient ceramic to give a useful increase in modulus (say at least 12 vol%) show tensile elongations (Fig. 9) which are generally substantially less than specified for aluminium ingot-route alloys of similar strength. In addition the elongations fall still further with increasing ceramic content. These low elongations for particulate reinforced MMC are a continuation of the trend for mechanically alloyed, dispersion strengthened materials [20] and are considerably lower than elongations for unreinforced powder metallurgy route aluminium alloys [21]. This indicates that it is the presence of the high volume fraction of non-deforming ceramic particles, rather than the powder metallurgy route which is primarily responsible for the low elongations observed.

The micromechanism of fracture in aluminium based material is generally by nucleation and growth of voids at particles in the crack tip stress/strain field. These voids eventually coalesce with the crack tip by linkage either directly between adjacent major voids or by shear linkage between smaller scale particles between the major voids [22]. Potential void nucleation sites include:

- Relatively large (~1μm) second phase particles in ingot alloys.
- Ceramic reinforcing particles in MMC.
- Relatively large grain boundary aging precipitates in over-aged alloys.
- Al_2O_3 particles originally from aluminium powder boundaries in powder-route materials.

Shear linkage between major voids typically involves intermediate sized particles (0.1 - 0.5μm diameter) which may be:

- Dispersoids intended as grain boundary pinning agents.
- Al_2O_3 particles in powder route materials.

A model of this void growth and coalescence fracture process [23] predicts the criterion for void coalescence with the crack tip as a function of inclusion diameter (d) centre-to-centre spacing (r) and the crack tip opening displacement (δ) at coalescence (Fig. 10). This model does not include void nucleation strain or (explicitly) shear linkage between major voids. Void nucleation strains for iron and silicon-rich second phase particles in ingot alloys are reported to be low [22]. For particulate MMC, void nucleation strains have not been reported but may be higher than for second phase (Fe, Si rich) particles . This is a result of compressive stresses acting on the MMC reinforcement due to thermal expansion mismatch between the particles and matrix. Shear linkage between major voids will lead to experimental fracture initiation toughness data lying below the model predictions as is indeed observed from results on ingot alloys [22,24] and aluminium MMC [25].

A variety of SiC reinforcement forms has been incorporated in aluminium base matrices by powder routes:

- particulate
- whiskers
- whisker/particulate mixtures
- short (or chopped) fibres

More recently hexagonal platelets (~2μm thick by 2.5 to 50μm diameter) have become available [14] but only limited work has incorporated this reinforcement. In principle, high aspect ratio reinforcements should give enhanced strength properties with whiskers and short fibres providing uniaxial improvements for extrudes and platelets providing biaxial improvements for plates. However, to date, the difficulty in maintaining high aspect ratios during processing, has mitigated against the full realization of these benefits.

The technological importance of the alumina film surrounding aluminium powder particles has been highlighted above in relation to degassing and fracture properties. If an unreinforced aluminium alloy is hot pressed to full density these films remain virtually continuous (Fig. 11a) and can provide a low energy fracture path. For unreinforced aluminium PM alloys these films are disrupted by deformation (Fig. 11b) and an extrusion ratio of at least 20 has been quoted to provide sufficient dispersion of the oxide films [26] to avoid reduced ductility/toughness. The presence of non-deforming ceramic particles in PM MMC may be expected to assist break-up and dispersion of oxide films and "full" ductility has been obtained for extrusion ratios as low as 10 [17]. However, at these low extrusion ratios it has also been observed that fatigue cracks may still follow prior particle boundaries [27].

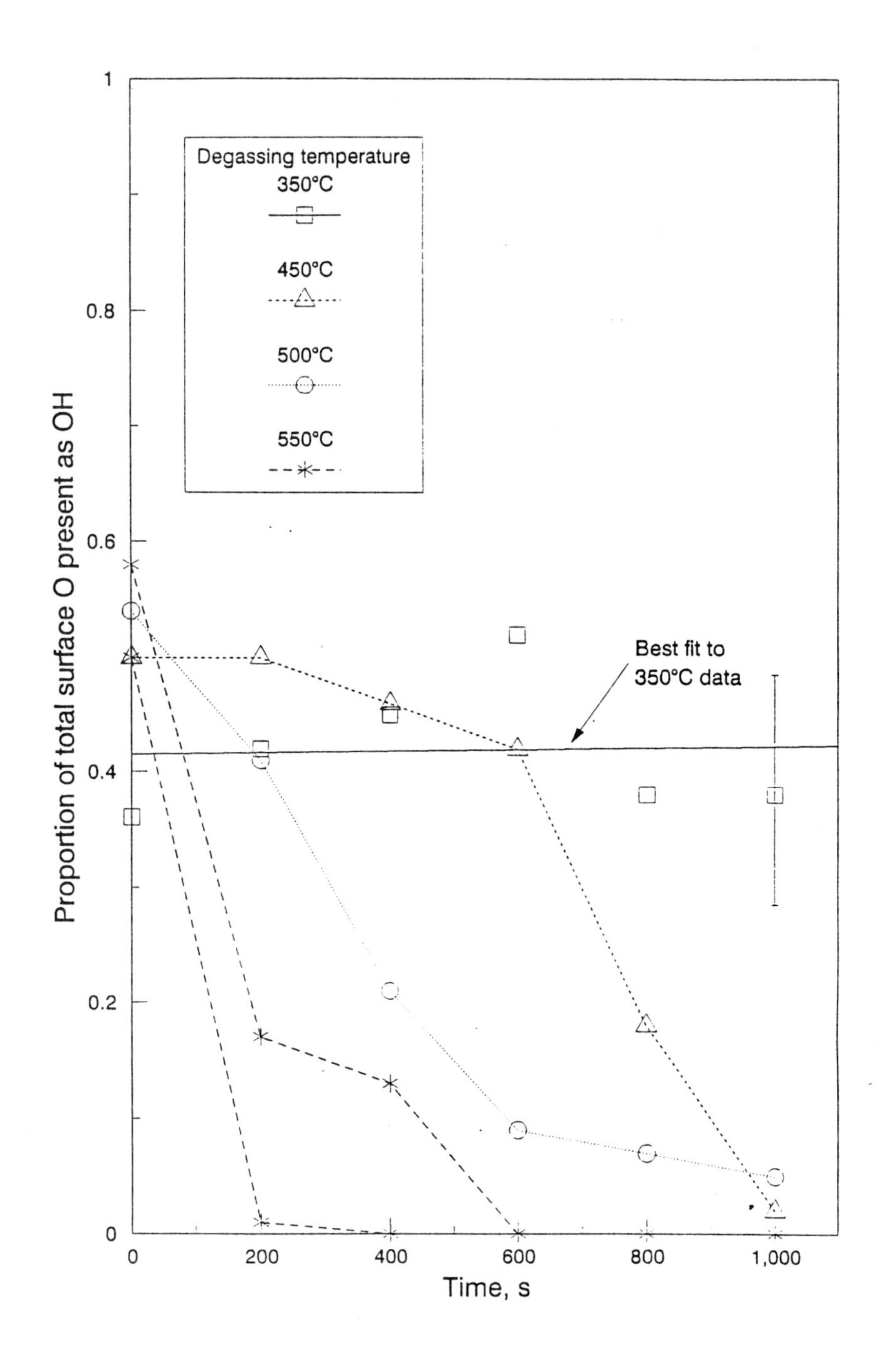

Figure 6 Degassing of 2014 powder as a function of time and temperature determined in situ by x-ray photoelectron spectroscopy. (Ordinate represents proportion of total surface oxide which is present as OH species.) [15]

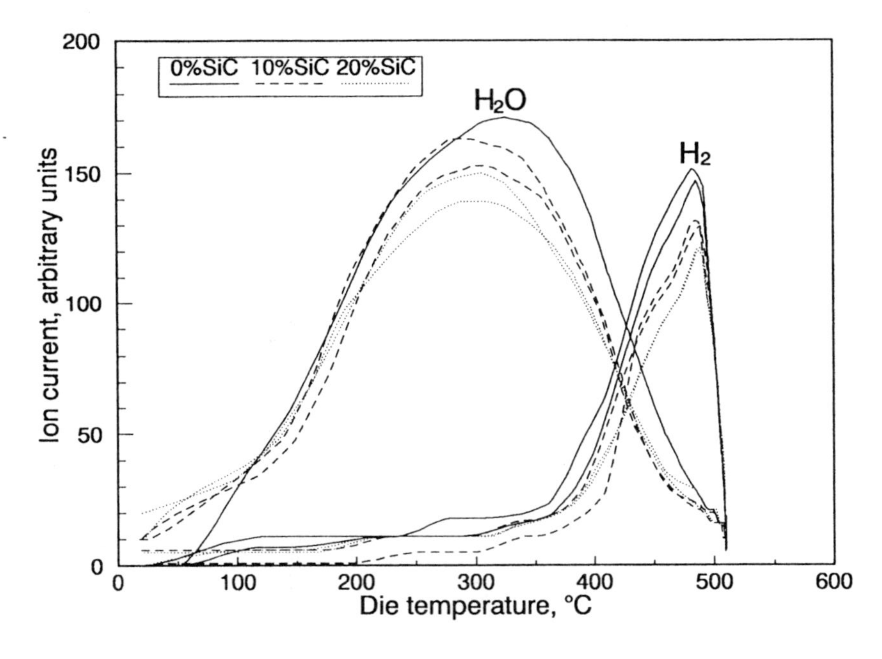

Figure 7 Gas species evolved during vacuum degassing of 2014/SiC particulate mixtures as a function of time and container temperature.

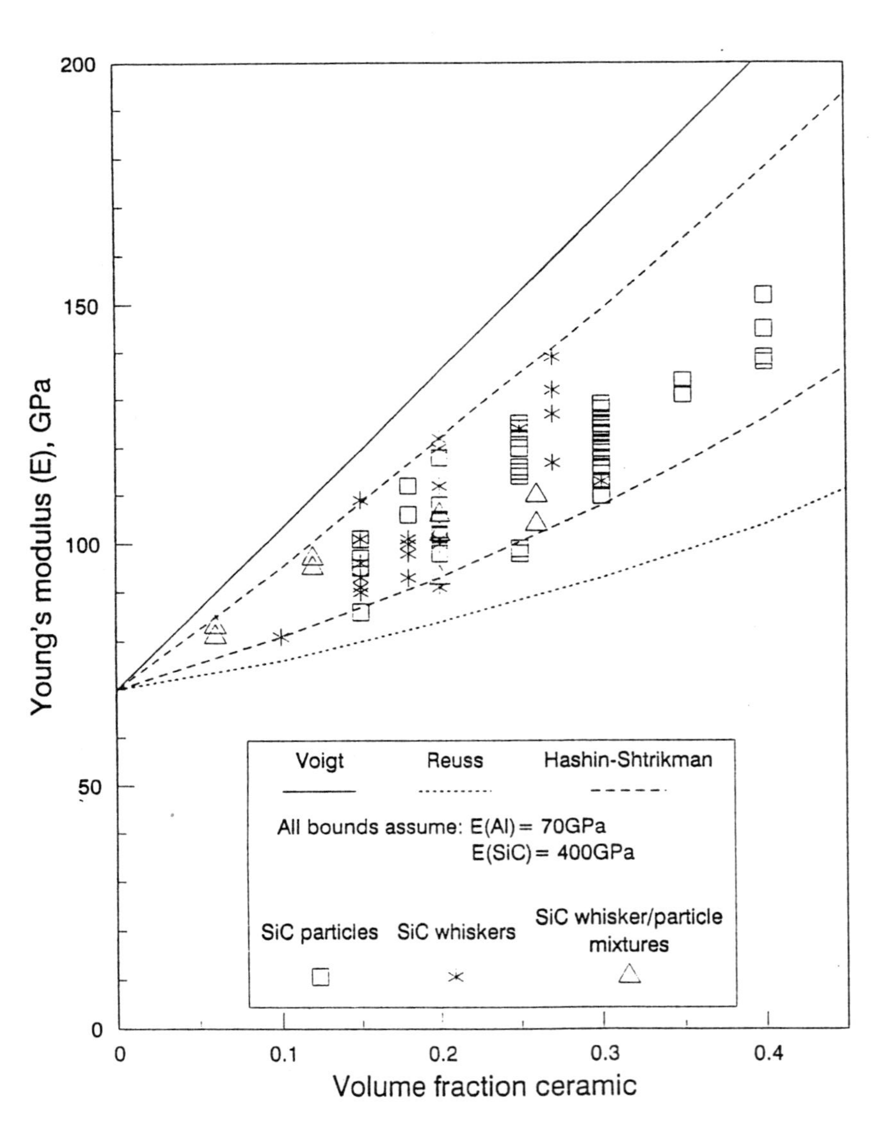

Figure 8 Comparison of predicted and achieved Young's Modulus improvements in SiC reinforced aluminium alloys as a function of reinforcement type and volume fraction (Compendium of results published up to 1986).

3.5 Aspects of Microstructural Stability

Almost all metallurgical microstructures are thermodynamically unstable, if only because of the excess interfacial energy associated with interdispersed phases. MMC systems have in general an additional degree of instability associated with the non-equilibrium relationship between the reinforcement and matrix and this can give rise to dissolution of the reinforcement into the matrix or the formation of reaction products at the reinforcement/matrix interface. This can be minimized by taking account of thermodynamic activities in the formulation of the MMC system, e.g. in the case of alumina reinforcement in aluminium or in the use of high silicon alloys to stabilize silicon carbide reinforcements. However such measures are seldom viable because they compromise too greatly the potential of the system for property enhancement. Nevertheless they are always adopted to the point of at least ensuring that the reinforcing phase is not significantly dissolved during processing or use of the MMC. In some cases the reactive reinforcement is coated with a material which is relatively stable to both the reinforcement and matrix. This is equivalent to the approach sometimes adopted for the protection of fibres in reactive matrices and is generally much more easy to achieve in the case of particulate reinforced MMC since the performance of the reinforcement is far less sensitive to surface degradation than is the case for most high performance fibres.

Perhaps the most significant aspect of this general instability is that it makes irreversible some of the reactions which in conventional metallurgical processing are considered as reversible. Thus the heat treatment response of an MMC sample may not be a function only of its initial matrix alloy composition, but also of the cumulative thermomechanical history that it has undergone up to the point of heat treatment. This means that within development programmes and research investigations into heat treatment response, care has to be exercised in the interpretation of the reproducibility of results. Similarly, recommended heat treatment schedules have to be linked to specified prior processing thermomechanical histories.

The degree to which the heat treatment response of MMC formulations differs from those of the unreinforced alloy understandably differs with both the reinforcement type and the influence that the reinforcement has on the behaviour, within the matrix, of the key solute elements controlling the heat treatment response. In the MMC formulations currently studied there has been evidence of solute enrichment or denudation at the reinforcement/matrix interface. For example in a 7XXX matrix material, particle growth at SiC interfaces is associated with segregation to the interface coupled with some solute depletion up to 50 nm away. For the over-aged case this segregation is restricted to a narrow band ~ 2 nm wide and the loss of solute in the rest of the precipitate free zone is sufficient to change the fracture mode from particle fracture to decohesion in the depleted zone [28] (Fig. 12). Such segregation influences the distribution of precipitation within the matrix both in terms of its amount (due to changes in the quantity of available solute) and its inhomogeneity (due to solute activity gradients). Similarly the presence of coarse particles of reinforcement with thermal expansion properties markedly different from the matrix alloy influences the dislocation substructure of the matrix to a marked extent and this also influences heat treatment response. MMC properties are therefore far more sensitive to cumulative irreversible compositional changes and to the detail of heat treatment cycles (quench rates etc.) than are the equivalent unreinforced alloys.

3.6 Applications and User Acceptance

To date most "production" PM MMC material has been used in "demonstrator" programmes, particularly in the aerospace and defence sector. This is, in part, due to the pattern of national R&D funding for materials developments which links both the investment in materials development and the opportunity for applications assessment to major national goals. Examples from recent US experience with PM MMC are:

- Tank components (track links and mirror components).
- An electronics rack for the Advanced Technical Fighter.
- Vertical tail demonstrators for the Advanced Technical Fighter.
- An electronics mounting for a missile system.

The last example would appear to correspond to an appreciable production run. Bearing in mind the time required for a near-conventional material, such as an Al-Li based alloy, to achieve applications acceptance, the surprise is not that there have been relatively few PM MMC applications established but that such a wide variety of applications trials has been carried out on as yet unoptimized formulations. Companies investing early in MMC materials development have to look to reasonably big tonnage markets in the short term to achieve an acceptable return on investment and this inevitably points to the automotive, general engineering and leisure goods market sectors. Of these, only the latter affords the possibility of anything but cost competitiveness being the prime criterion for acceptance. Thus PM MMC materials, which in general terms have an intrinsic processing cost disadvantage over those produced by some competitive routes, are likely to find profitable exploitation only in niche applications which capitalize on unique technical strengths. These strengths are basically those which give powder metallurgy products their edge in non-composite materials competition; for example the exploitation of cost savings arising from the near-net-shape processing of components. Additional benefits for PM composites are the breadth of reinforcement shapes, sizes and volume fractions that can be incorporated. These combine

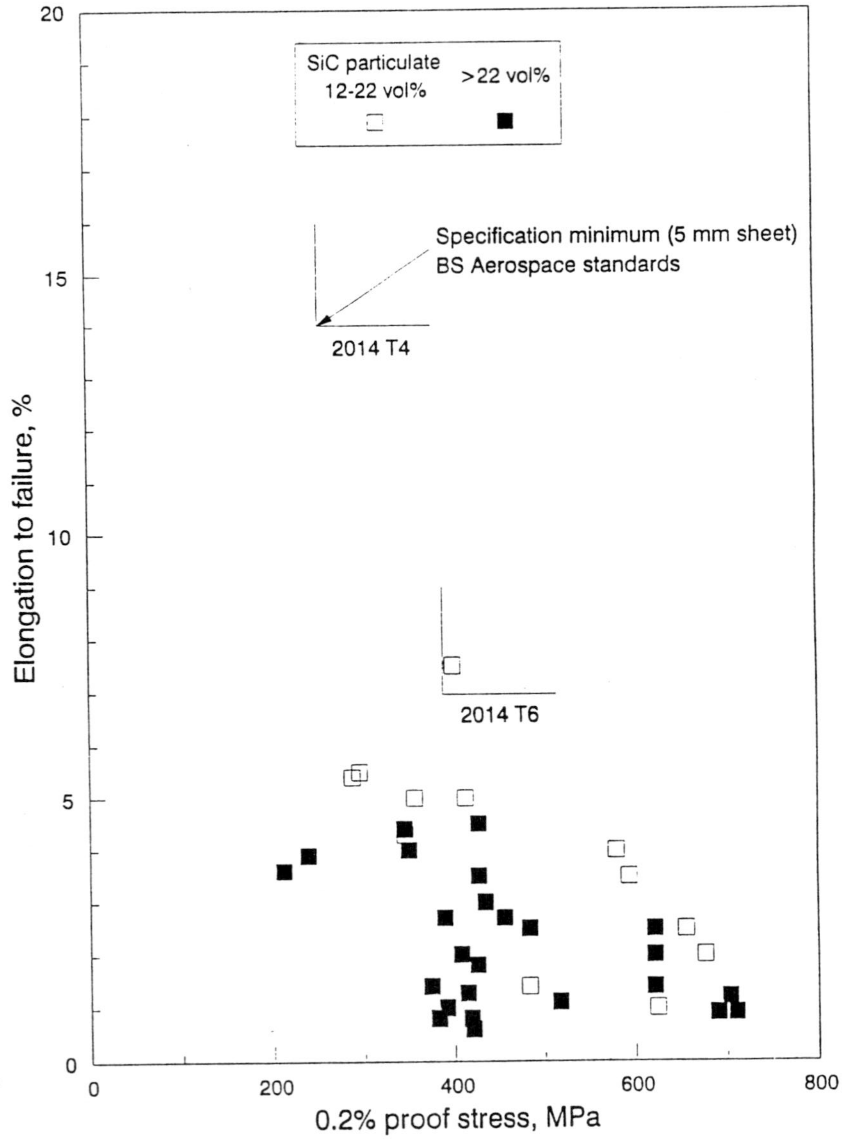

Figure 9 Combinations of elongation to failure and proof stress achieved in powder metallurgy SiC particulate reinforced aluminium alloy MMC materials (Compendium of results published up to 1986).

Table 4

Type	Examples of source	Morphologies	Comments
'VLS' powder catalysed	Los Alamos	0.1–1μm dia x < 10 cm	Lab scale. Expensive. 6.4 GPa mean strength
Cereal chaff or mineral source Si	Tateho Silag/Exxon Tokai Carbon	0.05–1μm dia x 50μm. High whisker proportions can be screened out from whisker/particulate mixes	In production. Cost: 10^2–10^3 S/kg. Products being improved and optimised
Platelet	American Matrix	0.5–20μm thick hexagonal α SiC platelets	
Crushed grit	Carborundum	Irregular, angular. Graded down to - 1μm	In production Cost: 10° - 10 S/kg.

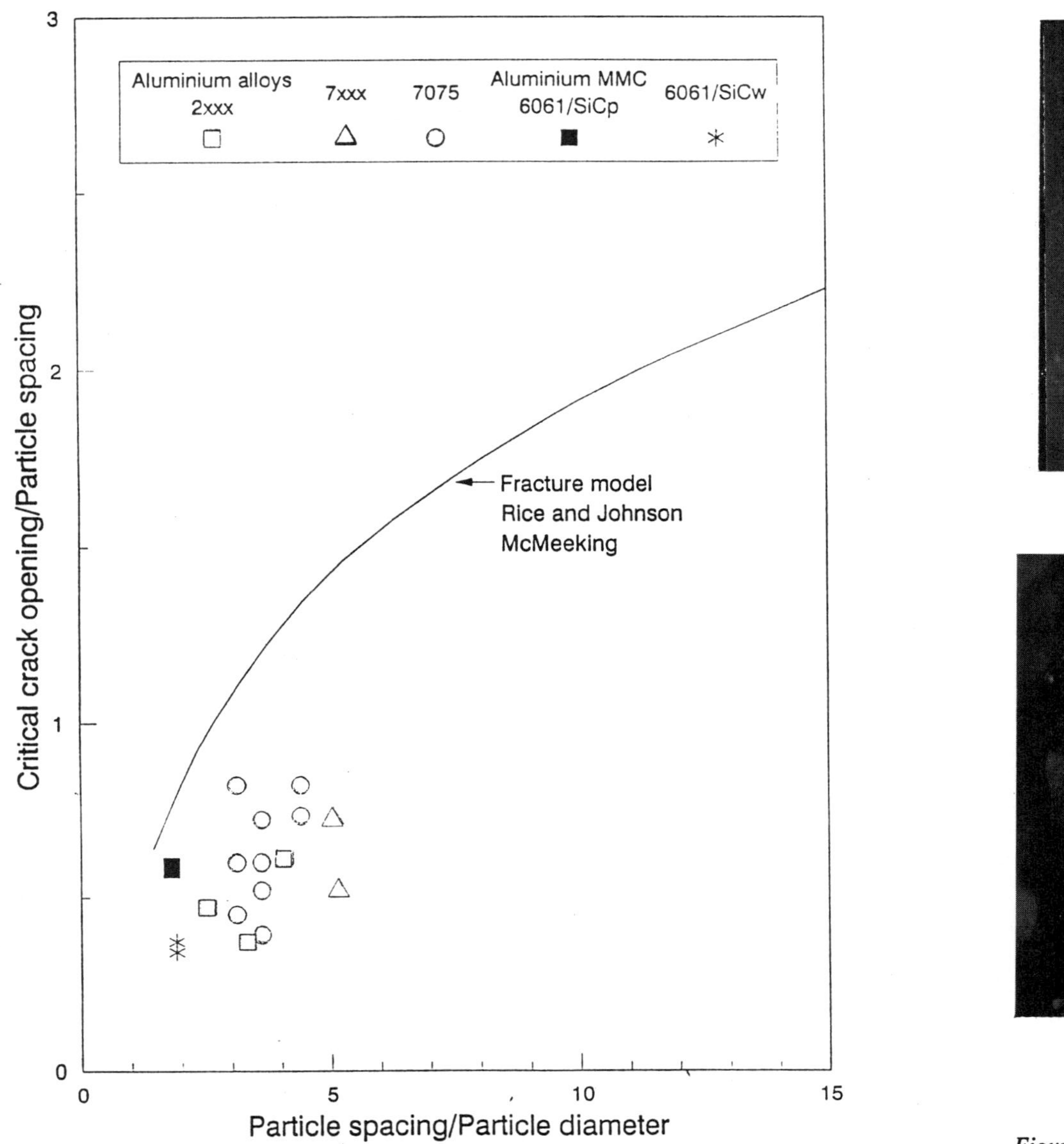

Figure 10 Comparison of experimental structure property data with predictions of void growth and coalescence fracture process model.

a

b

Figure 11 TEM dark field images of oxide distributions in (a) hot pressed aluminium alloy powder compact (continuous oxide films) and (b) extruded compact (discrete oxide fragments).

to provide the freedom to satisfy combinations of properties, such as say elastic modulus, expansion coefficient and tribological performance, optimized for specific applications. It is noteworthy that in the examples quoted earlier in this section, the applications have generally been dependent on an appropriate combination of mechanical and thermal properties. Powder route composites may also find temporary niche markets where other, potentially cheaper, processing routes may take longer to develop. These are likely still to be at the higher performance end of the PM MMC applications field since in less demanding applications the cheaper competitive processing routes can be readily mobilized. Thus melt route particulate reinforced MMC has already been successfully used for horseshoes [29] and this will probably rule out the use of more expensive PM route composites in this application.

In all but the least demanding of applications, user acceptance will be dependent on the demonstration of appropriate quality control procedures. In addition to the physical defects which have to be screened out in conventional alloys, PM MMC materials could exhibit macro and micro inhomogeneities in structure due to inadequate process control. Thus collaborative R&D programmes have been addressing the development of new nondestructive examination techniques for checking the local particulate filler content in MMC material [30]. This represents an additional barrier to user acceptance since a new NDE technology has to be developed and brought into use. Fortunately the economic logic demands that this quality assurance step be carried out at the earliest possible stage of processing that thus concerns most directly the supplier rather than the users.

In addition to acceptance barriers, such as the above, which are specific to a given family of MMC, there are the general barriers to innovation in the following categories:

- Powder metallurgy versus ingot metallurgy. Here the general suspicion with respect to PM materials for critical components will have to be overcome.
- MMC versus conventional alloys. Here the general requirement to change design criteria and any specific requirements such as consistency of repainting procedures for aircraft structures, need to be met.
- Materials substitution. Here the general reluctance to change has to be overcome and the normal time delays for the qualification of new materials have to be accommodated.

All of these barriers will ensure that PM MMC materials will have to have stronger technical and economic justification for use than would be required for say an incrementally improved conventional alloy. This reinforces the conclusion that PM structural composites are only likely to be exploited in "niche" applications and that the future for PM MMC materials in general may well be more promising in fields where tailored physical properties at least complement any mechanical property advantage.

4. Other Powder Metallurgy Composite Families

In addition to the relatively long established families of PM MMC referred to in section 1 and the main example chosen on the subject in section 3, there are other families of PM composite which merit some discussion on the basis of the degree of R&D effort being expended on them worldwide. These include multifilament reinforced metals and composites being developed primarily for their tribological properties. These two families will therefore be briefly reviewed here to illustrate some of the relevant technical factors which were not covered or emphasized in the review of the particle or whisker-reinforced aluminium alloys.

4.1 Multifilament Reinforced Composites

The availability since the 1960s of carbon fibre and, more recently, of multifilament ceramic reinforcements based particularly on silicon carbide and alumina, has resulted in continuing developments aimed at exploiting them as reinforcements for metals. Most process developments have been based on pressure assisted melt infiltration of the multifilament tows. However, powder metallurgy processing in the form of powder impregnation to produce tape "prepreg" has also been pursued because the "prepreg" processing route is more easily applied to the production of essentially two dimensional structures, such as panels, than are the infiltration casting routes. There is also the successful precedent for such a processing route in the preparation of multifilament SiC reinforced glass composites [1]. Most attention has been directed to aluminium base matrices and the process has been pursued at least in the UK and Japan.

This process is capable of producing good distributions of fibres - indeed better than can currently be achieved, without hybridization by melt infiltration techniques. However the properties achieved in the consolidated composites, after hot pressing or roll bonding in the solid state, have been inferior to those achieved after super solidus processing and therefore process development within for example MITI's BTFI programme in Japan has evolved to be focused on plasma spray infiltrated [31] or melt infiltrated (wire) [32], rather than powder impregnated, precursors for consolidation. Thus the powder metallurgy route has been superceded by routes which require less processing steps, and thermomechanical histories which are both more controllable and less aggressive in terms of fibre degradation. This evolution is perhaps not surprising in the case of aluminium based MMC, bearing in mind the degree of deformation processing that has been found to be necessary to achieve reasonable toughness in particulate reinforced MMC (see section 3.4) and which could

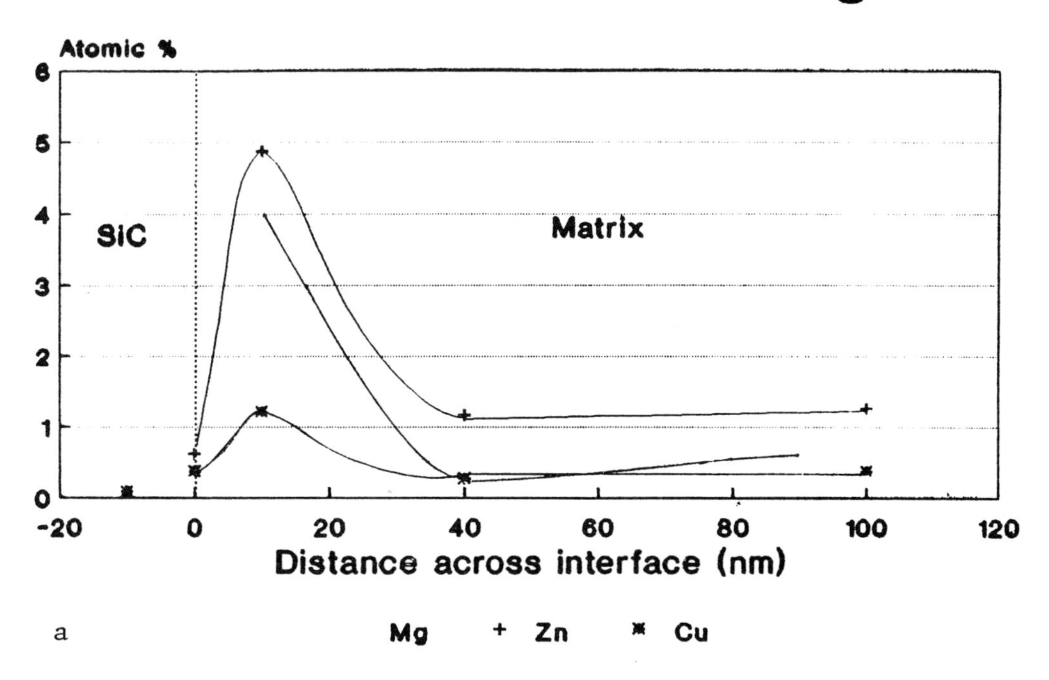

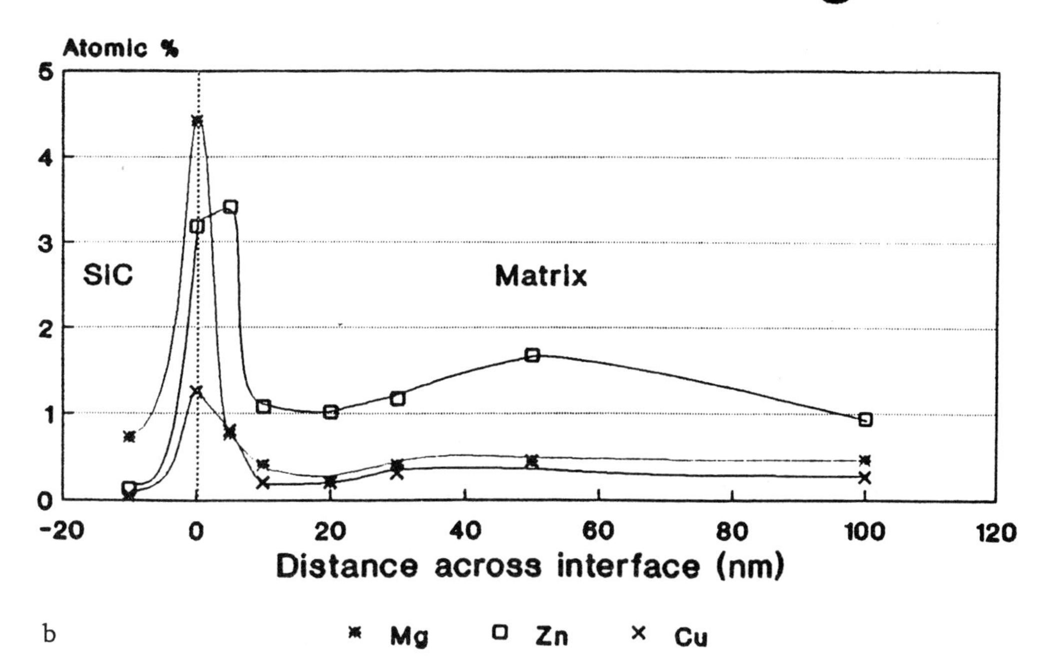

Figure 12 Solute concentration profiles in viscinity of SiC particle/7XXX alloy matrix interface in (a) underaged and (b) overaged condition [28].

not be realistically applied to a fibre reinforced matrix during consolidation without damage to the reinforcing fibres.

Here then we have an example of a powder metallurgy route apparently losing out to simpler, more appropriate routes to achieving the same end.

4.2 Composites for tribological applications

One of the first product developments for carbon-fibre reinforced metals was in the field of bearing materials where load bearing strength combined with low wear rate and coefficient of friction was to be exploited (33). Powder metallurgy produced MMC offer the flexibility of tailoring mechanical and tribological performance through multiple reinforcement dispersions. In view of this and the historical role that powder metallurgy has played in the evolution of material for electrical contacts, hard metals and hard facing materials, it is perhaps surprising that powder metallurgy MMC have not been pre-eminent in present developments in applications fields for which tribological performance is paramount. In practice it is the melt processed MMC formulations which have been most evident in recent MMC developments in the tribological field, whereas tribological behaviour in powder produced formulations has tended to be treated as secondary to mechanical property optimisation. There are a number of possible reasons for this apparent inconsistency. For example, melt routes, either because of their relatively low maximum reinforcement volume fraction loadings or because of their lack of potential for thermo-mechanical processing, are less capable of achieving major mechanical property enhancement over unreinforced matrix materials than are PM MMC. Attention therefore has to be turned to properties other than strength, stiffness and toughness as development objectives. In addition, the principles governing predictive microstructural design for tribological performance are far less well developed than those concerning front line mechanical properties. This is true even for conventional alloys, never mind MMC, primarily because of the complexity of individual service regimes and the difficulty in devising test techniques which simulate service environments or which have basic significance from which to draw general conclusions. Furthermore, near net shape processing is a central requirement for many of the candidate applications and this tends to mitigate against the achievement of a satisfactory balance between acceptable mechanical properties and the required tribological characteristics in unworked PM MMC.

There are however indications that some of these possible barriers are being attacked. Systematic studies on the tribological behaviour of established MMC systems are under way. For example, the wear rates and mechanisms of SiCp reinforced 2014 alloy have been investigated under representative conditions of dry sliding, abrasive and lubricated sliding wear against steel counterfaces (34). In all cases the wear rates were lower than those of a control sample of unreinforced 2024 Al. The comparative effects of various soft and hard particulate additions (including graphite) on the mechanical properties of sintered 6061 alloy have been determined (35). The data base on the tribological performance of multiple filler composites is increasing as a result of development activities mainly involving other MMC processing routes (36). Process developments are also being pursued which could lead to near-net-shape sintered materials with improved mechanical performance (16,37). Thus the diverse knowledge base on which to develop PM composites for tribological application is growing to the point where serious product developments may soon be contemplated. As is generally the case for advanced MMC applications, exploitation will depend on the full versatility of the composite materials approach being marshalled in niche applications.

5. Conclusions

For well over half a century powder metallurgy processing has had a significant role to play in a wide range of MMC developments and powder metallurgy MMC materials have proved their worth in applications exploiting electrical, mechanical, thermal and nuclear properties. This is mainly due to the versatility and conceptual simplicity of the powder metallurgy approach - particularly when applied to particle or whisker reinforced composites. In the present drive towards developing high performance MMC materials for structural components there has been significant commercial exploitation of particulate reinforced PM composites but there is growing competition from composites produced by potentially cheaper processing routes. This reflects the complexities arising from the superimposition of the composite materials concept on the already demanding problem of high integrity PM processing. Future applications are likely to be those which combine the strengths of both technologies; for example exploiting microstructural versatility coupled with near-net-shape processing. This will require creativity on the part of both materials producers and users.

Acknowledgements

The authors thank their collaborators from industry and academia and their colleagues from many disciplines at the Harwell Laboratory who have contributed to the body of knowledge drawn upon in this review. Particular thanks are due to those in industry who have been prepared to invest patiently in the exploitation of the PM MMC concept, thereby providing the raison d'etre for this exciting field of materials development. The review has been prepared with the support of the UKAEA's Underlying Research Programme.

References

1. D.H. Bowen, N.J. Mattingley and R.A.J. Sambell, Continuous coating of fibres, UK Pat. No. 1279252, 1972.
2. S. Kohara and N. Muto, Fabrication of SiC fibre-aluminium composite materials, Proc. ICCM-IV, Tokyo, p.1451, 1982.
3. E.A. Feest, Metal matrix composite manufacture, UK Pat. No. 2172825, 1988.
4. M.D. Skibo and D.M. Schuster, Cast reinforced composite materials, Int. Pat. WO 87/06624, 1987.
5. S-Y. Oh, J.A. Cornie and K.C. Russell, Wetting of ceramic particulates with liquid aluminium alloys: Pt 1. Experimental techniques, Met. Trans, 20A, p.527, 1989.
6. E.A. Bloch, Dispersion-strengthened aluminium alloys, Met. Rev. 6, p.193, 1961.
7. J.S. Benjamin, Dispersion strengthened superalloys by mechanical alloying, Met. Trans. 1, p.2943, 1970.
8. US Congress, Office of Technology Assessment, Advanced Materials by Design, DTA E 351 Series, 1988.
9. E.A. Feest, M.J. Ball, A.R. Begg and D.A. Biggs, Metal matrix composites developments in Japan, Report on "OSTEM" visit to Japan, Harwell Laboratory, 1986.
10. M.F. Stanton and M. Layard, The carcenogenicity of fibrous minerals, NBS Special Pub. 506, p.143, 1978.
11. S.E. Booth, M.J. Ball, A.J. Clegg, N.J. Hurd and R. Savery, Metal matrix composites developments in the USA, Report on "OSTEM" visit to USA, BNF Metals Technology Centre, 1987.
12. P. Niskanen and W.R. Mohn, Versatile metal-matrix composites, Adv. Mats. & Processes, 133, 3, p.39, 1988.
13. R.D. Goolsby and L.K. Austin, Fracture toughness of discontinuous SiC reinforced aluminium alloys, Adv. in Fracture Research, (ed.) K Salama *et al.*, Pergamon, 4, 1989.
14. J.A. Black, Shaping reinforcements for composites, Adv. Mats. & Processes, 133, 3, p.51, 1988.
15. P.R. Chalker, J.A.A. Crossley, V.J. Moore, J.H. Tweed and R.M.K. Young, Degassing of rapidly solidified Al-4Cu powders studied by XP5, Harwell Report AERE R13747, 1990.
16. M.S. Maclean, Sintered aluminium alloys, UK Pat. GB 2 179 369, 1988.
17. T.F. Bryant, S.T. Durham, A.E.J. Forno and W.S. Miller, Extrusion processing of aluminium alloy base metal matrix composites, Metal Matrix Composites: Property Optimisation & Applications, The Inst. of Metals, London, 1989.
18. C.J. Peel and R. Moreton, Design studies for optimisation of metal matrix composites in aerospace applications, Metal Matrix Composites: Structure and Property Assessments, The Inst. of Metals, 1987.
19. A.J. Allen, M. Bourke, M.T. Hutchings, A.D. Krawitz and C.G. Windsor, Neutron diffraction measurements of internal stress in bulk materials:- metal matrix composites, Residual stresses in Sci. & Tech., (ed.) E Macherauch & V. Hauk, DGM Inf. Verlag, Oberursel, 1, p.151, 1987.
20. V. Arnold and J. Baumgarten, Dispersion strengthened aluminium extrusions, Powder Metal Int., 17, p.168, 1985.
21. E. Lavernia, G. Rai and N.J. Grant, Rapid solidification processing of 7XXX aluminium alloys: A review, Mat. Sci. Eng., 79, p.211, 1986.
22. R.H. Van Stone, R.H. Merchant and J.R. Low Jr., Investigation of the plastic fracture of high-strength aluminium alloys, ASTM STP 556, p.93, 1974.
23. R.M. McMeeking, Finite deformation analysis of crack-tip opening in elastic-plastic materials and implications for fracture, J.Mech. Phys. Sol. 25, p.357, 1977.
24. T. Ohira and T. Kishi, Effect of iron content on the fracture toughness and cracking processes in high strength Al-Zn-Mg-Cu alloy, Mat. Sci. & Eng., 78, p.9, 1986.
25. C.R. Crowe and R.A. Gray, The effect of Notch Root Radius on Crack Initiation in SiC/Al, in 'Failure Mechanisms in High Performance Materials', Ed. J G Early *et al.* CUP, 157, 1985.
26. Y-W. Kim, W.M. Griffiths and F.H. Froes, Surface oxides in aluminium alloys, J. of Metals, 37, p.27, August 1985.
27. M. Strangwood, Corrosion fatigue behaviour of powder processed 2014 and 2014-20 vol.% SiCp material, Harwell Report AERE R13665, 1989.
28. M. Strangwood, Interfacial structure in under- and over-aged 7XXX - 20 volume % SiCp MMCs, Harwell report AERE R13664, 1989.
29. J.B. Borradaile (Hydro Aluminium), private communication, 1989.
30. QT News, National NDT Centre, Harwell Laboratory, Issue 37, November 1989.
31. A Okura, S Sakai "Development of fibre reinforced composite materials. The properties of SiC-fibre reinforced aluminium composites" Rept. of the Res. Group for Fibre Reinf. Aluminium Matrix Composites, Light Metal Educational Foundation Inc. Osaka, (1986).
32. A Kohyama, N Igita, Y Imai, H Teranishi & T Ishikawa, "Microstructures and mechanical properties of silicon carbide fibre reinforced aluminium composite materials and their preform wires", Proc. ICCM-V, ed Harrigan et al. TMS p . 609 (1985) .
33. B W Howlett, C F Old & D C C Minty, "Improvements in or relating to composite bearing materials", UK Pat. GB 1403862 (1975).
34. A T Alpas & J D Embury, "Sliding and abrasive wear behaviour of an aluminium (2014) - SiC particle reinforced composite", Scripta Met. 24 p . 93 1 (1990) .

35. A K Jha, S V Prasad & G S Upadhyaya, "Mechanical behaviour of sintered 6061 aluminium alloy and its composites containing soft or hard particles", Z . Metallkde 8I p . 457 (1990) .
36. P K Rohatgi, R Asthana & S Das, "Solidification structures and properties of cast metal-ceramic particle composites", Int. Met. Rev. 31 p. 115 (1986) .
37. L Christodoulou, D C Nagle & J M Brupbacher, "Aluminium-ceramic composites", Int . Pat . WO 86/06366 (1986)

19

Applications and Developments of Sintered Ferrous Materials

G.F. BOCCHINI AND P.F. LINDSKOG*
Höganäs Italia srl, Italy
*Consultant**

1. First Industrial Developments

Powder metallurgy entered mass production less than 70 years ago. The first industrial applications of sintered ferrous PM parts were developed in the United States, bearings being the first examples, followed very soon by gears and cams. The oil pump gear for an Oldsmobile car, in 1937/1938, was the "pioneer" of ferrous powder metallurgy parts mass production. The starting material was an iron-graphite mix, while the sintered structure was quite unusual: a hypereutectoid matrix with frequent inclusions of free graphite. The final properties were similar to those of a rather poor cast iron, but the sintered (and sized) gears offered different and significant advantages, in comparison with analogous components produced by machining of cast iron [1]. The European approach towards mass production by powder metallurgy before the war was quite different from that in America. The most advanced European country was Germany where, apart from cemented carbides, developments were mainly devoted to military applications. The components selected were shell driving bands. Studies on this subject began in 1934 and the first samples were ready in 1935. The official patents were granted in 1937. After having developed a new domestic process of iron powder making, shell production began in 1938-1939. Material properties, at first, were not as high as required for the application, but the production process was steadily improved; new presses and new furnaces were also developed. The total tonnage of shell driving bands produced up to 1945 was more than 100,000 tons. In the same period more than 200 millions of sintered parts were produced in Germany. Even if the mechanical properties of the materials used were quite low, many technological problems were dealt with and brilliantly solved. The basis for the post-war peaceful development of powder metallurgy was established.

2. The Growth of PM Ferrous Technology

The availability of statistical data covering the production of iron powders and sintering applications allows the reconstruction of the development of powder metallurgy in quantitative terms. As far as material characteristics are concerned, however, the data is quite scanty. For about 15 years after the Second World War, the great majority of sintered parts was mainly competitive with parts made of grey cast iron, or high zinc alloys. The highest mechanical properties (obtained on tensile test bars) reached until the mid 1970s, are believed to be those mentioned in Table 1, partially modified with respect to the indications of Hulthèn and Smedstam [2,1]. It should be pointed out, however, that the properties of current applications at the time were generally lower than the values given in Table 1. In fact, before introducing a new material, especially for highly stressed services, the potential users considered long term practical tests necessary to confirm the reliability of the new materials.

A substantial contribution to the improvement of mechanical properties of sintered steels was made by powder producers, through the introduction of better grades of powder. Improvements were achieved by following two main lines:

(i) Enhancement of iron powder properties.
(ii) Introduction of partially prealloyed powders [3,4,5].

Figure 1, drawn from a paper by Lindskog and Arbstedt [6], shows the substantial increase in availability of iron powders having high properties.

Other factors, of course, contribute to final strength characteristics, in addition to the properties of the iron powders. The mastering of alloying techniques, for instance, is fundamental.

Among the various factors that have played a decisive role towards improvement of sintered materials, it is appropriate to mention:

Table 1 Characteristics of sintered steels up to the mid-1970s

Period	Material Types	Production Cycle	Characteristics: U.T.S. N/mm^2	Characteristics: Elongation, %	Competitive Materials
From 1940	Fe	$P_1 + S_1$	200 - 300	5 - 25	Common Cast Iron
	Fe-C	$P_1 + S_1 + P_2 + S_2$	250 - 400	0 - 5	
From 1945	Fe-Cu Fe-Ni Fe-Cu-Ni	$P_1 + S_1$	350 - 450	1 - 5	Special Cast Irons, Unalloyed Steels
About 1950	Fe-C Fe-Cu-C Fe-Ni-C	$P_1 + S_1$	400 - 500	0 - 3	Unalloyed Steels
From 1950	High Carbon Steels	$P_1 + S_1$	450 - 600	0 - 2	Low Alloy Steels
From 1965	Ternary Diffusion Bonded Steels	$P_1 + S_1$	450 - 600	2 - 4	Medium Alloy Steels
From 1970	Ternary Diffusion	$P_1 + S_1$	500 - 800	3 - 5	Quality Construction Steels
	Steels with very high Compressibility Powders	$P_1 + S_1 + P_2 + S_2$	600 - 1000	3 - 6	

P_1 = Pressing; S_1 = Sintering (or First Sintering); P_2 = Repressing; S_2 = Second Sintering

- Progressive improvement of sintering equipment;
- Increased knowledge and technological progress of the powder metallurgy industry, achieved by means of a remarkable build up of studies and experiences, increased process control and quality assurance and an increasing awareness of the users' different needs for improved technical performance and upgrading;
- Ever-growing use of heat treatments, correctly made, i.e. recognising and mastering fully the peculiar properties of sintered steels.

3. Typical Market Sectors

After the pioneer attempts of introducing powder metallurgy structural parts in varied mechanical mass productions, the most significant advantages of the technology were, identified and positively exploited. The typical features of any sintered component, which has to compete effectively with similar parts produced by other mass-producing technologies were, and are:

- Easy-to-manufacture complicated shapes, provided that they are suitable for compaction in rigid dies.
- High precision.
- Medium or high strength.
- Good surface finish.
- Good tribological features.

Progressively, the automotive industry, all over the world, became the biggest user of sintered parts.

A typical distribution among different markets for sintered components is illustrated in Fig. 2 [7].The distribution has remained practically the same for several years.

According to D.G. White [8], the typical car manufactured in the USA contains today about 9.5 kilos of sintered parts. Table 2 shows the growth in application of powder metallurgy parts in US cars up to 1983.

Table 2 Increase of PM parts consumption in US cars (after K.H. Roll)

Period	Average weight in kilos
From 1955 to 1960	2.0
1965	3.8 *
1970	4.5
1975	7.0
1983	9.0

(*)Introduction of automatic gearbox

In terms of tonnage, the application of P.M. parts in the automative industry has been steadily increasing. On the other hand, from the point of view of the commercial importance of the applications, the growth has occurred in stages. From a technical point of view, the more arduous applications have involved onlyu after many drawbacks of the process and its products have been overcome. The important stages in the introduction of P.M. parts in the various automotive component groups are as follows:-

- Reserve devices (windscreen wipers).
- Body (door locking teeth).
- Suspension (shock absorbers).
- Electrical equipment (distributor).
- Steering gear (steering rod).
- Automatic gearbox.
- Brakes (pistons).
- Distribution (pulleys).
- Gearbox (synchronizing rings and hubs, shifting levers, fingers).
- Engine (valve seats).

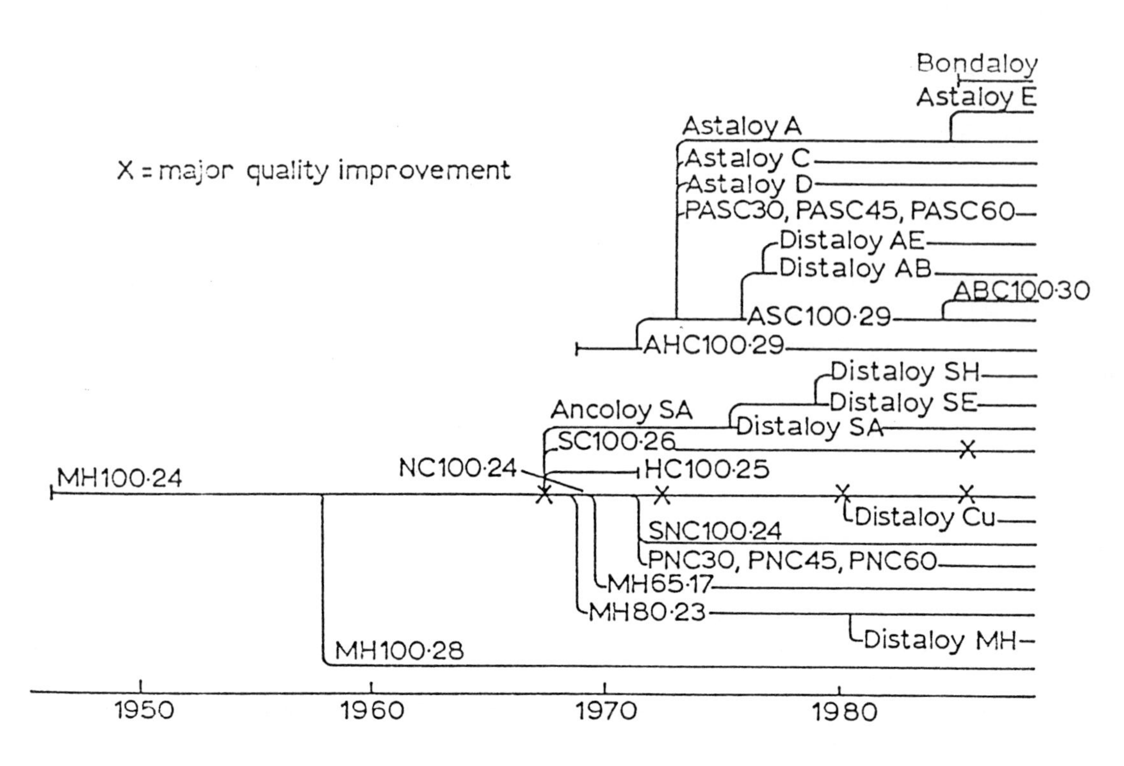

Figure 1 The growth of Höganäs P/M powder grades.

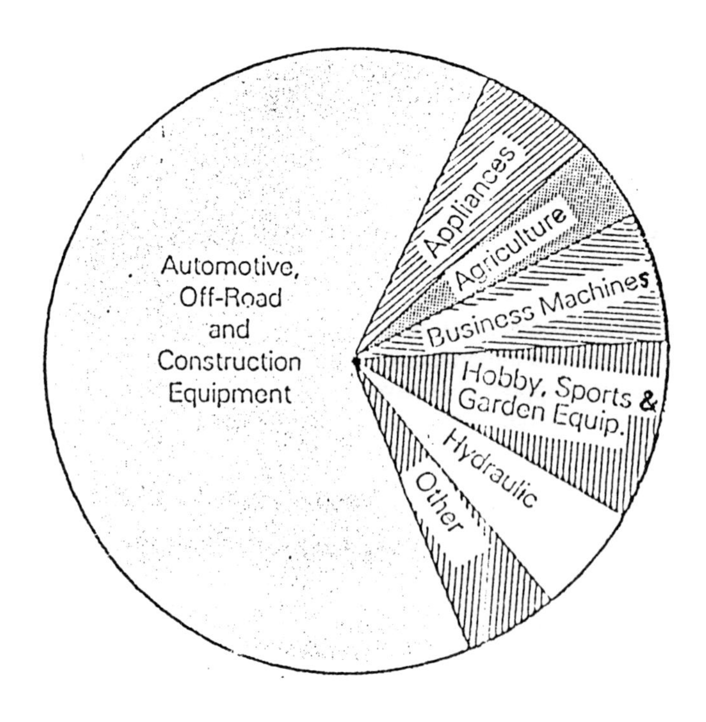

Figure 2 Typical markets of iron and steel sintered components.

The primary reasons for the growth of powder metallurgy applications in the motor industry are listed below:-

Basic reason: Economy

Basic factors of economy:
- Initial investments rather modest.
- High equipment flexibility.
- High productivity.
- Possibility of easily - and rapidly - producing complex shapes.
- Wide materials choice.
- Possibility of producing materials with unique properties.
- No (or very little) machining required.
- Very small quantity of swarf.
- High surface finishing.
- Good or high dimensional precision.
- Possibility of "modulating" density (and reducing weights).

The same reasons, of course, are valid for other important market sectors such as: household appliances, agricultural machines, business machines, sports and garden equipment, hydraulic and other applications. Mass production is a constant, common characteristic. An exhaustive survey on the automobile industry and powder metallurgy, with special attention to France, has been made, recently, by Lochon [9]. A very complete list of sintered components, currently used in typical US cars, is reported in the relevant section of Metals Handbook [10]. Other detailed papers, including technical descriptions of sintered parts commonly produced for US automobiles, have been prepared in recent years by S. Mokarski and D.W. Hall *et alia*. [11,12,13]. In the latter report, the survey is extended even to some applications typical of cars made in the Far East.

4. The North American Powder Metallurgy Industry

The very effective promotional activity of the US Metal Powder Industries Federation has included the collection and distribution of statistical data and many publications issued periodically. Recent up-dated information has been given by D. G White [8] who notes that in 1988 the North American PM market consumed more than 200,000 tons of iron powder, which is about one half of the total world consumption . Since the automotive interest in PM is so strong, the unit wieght of PM parts per car could exceed 13.5 Kgs within a few years.

Figure 3, based on official data published by the Metal Powder Industries Federation, shows the course of iron powders shipments in North America within the last two decades. But, according to White [8], "To set things in their proper place, we must remember that 1978 shipments of iron powder were almost 217,000 tons and the part sector of this figure, 178,000 tons. A lot of things have happened since then, including 30% of the automotive market being taken over by foreign made cars, downsizing of cars, all types of products going from mechanical functions to electronic functioning, and sharply rising imports of appliances and business machines". In spite of all these adverse factors, the positive growth of ferrous PM - with some cyclical difficult periods - is evident.

5. The Growth of Powder Metallurgy in Japan

In the last two decades, the continuous formidable expansion of Japanese industries, in different mass-production sectors, has been astonishing. It goes without saying that activities in powder metallurgy followed the same course. The Japan Powder Metallurgy Association, at intervals, issues statistical data on industrial production. The information is extremely detailed, with precise figures for every branch of powder metallurgy applications. The annual amounts of sintered products in the last 15 years have been collated in Table 3, based on papers of Watanabe and Sakurai [14,15], updated for 1987.

The powerful growth of many different sectors of the powder metallurgy industry is evident. It is necessary to underline, however, that the increase in consumption of sintered parts exceeded other mass-production technologies. This is clearly shown in Fig. 4, taken fromNagane [16]. It is also interesting to note that Japan is a strong exporter of sintered parts. The export is mainly indirect: that is to say in cars, motorcycles, etc. This statement is supported by the relationship between the tonnage of sintered parts and GNP, as illustrated in Ref.17.

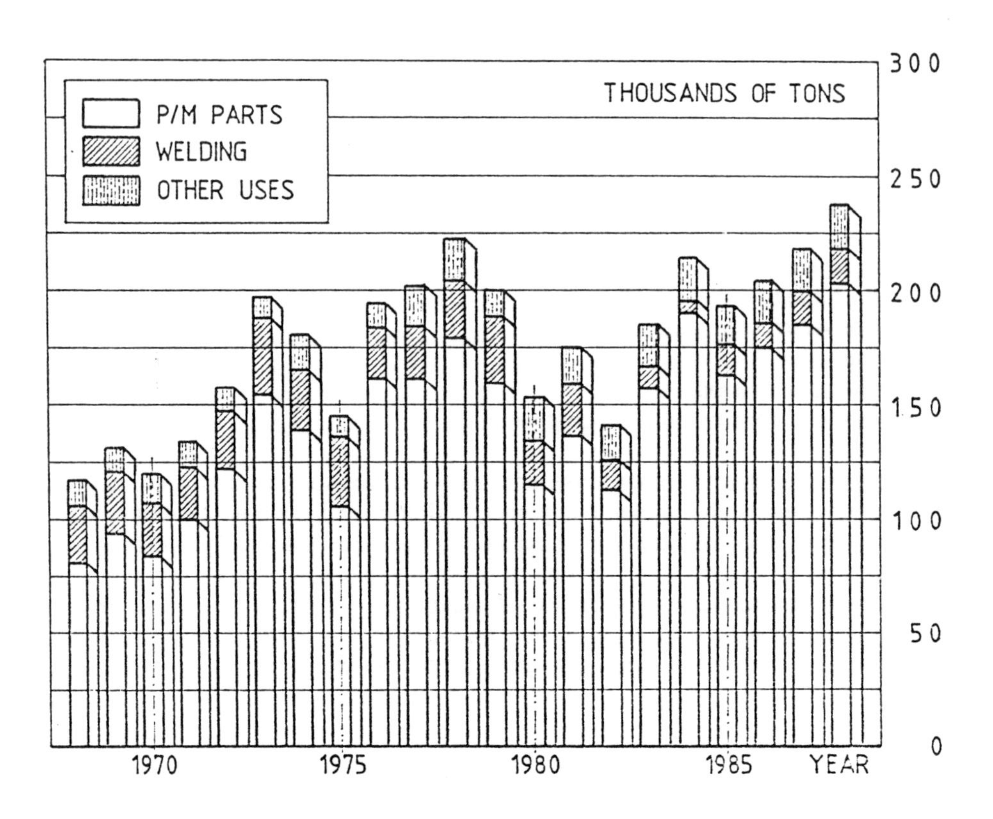

Figure 3 Iron powder shipments in North America. Source: Metal Powder Industries Federation

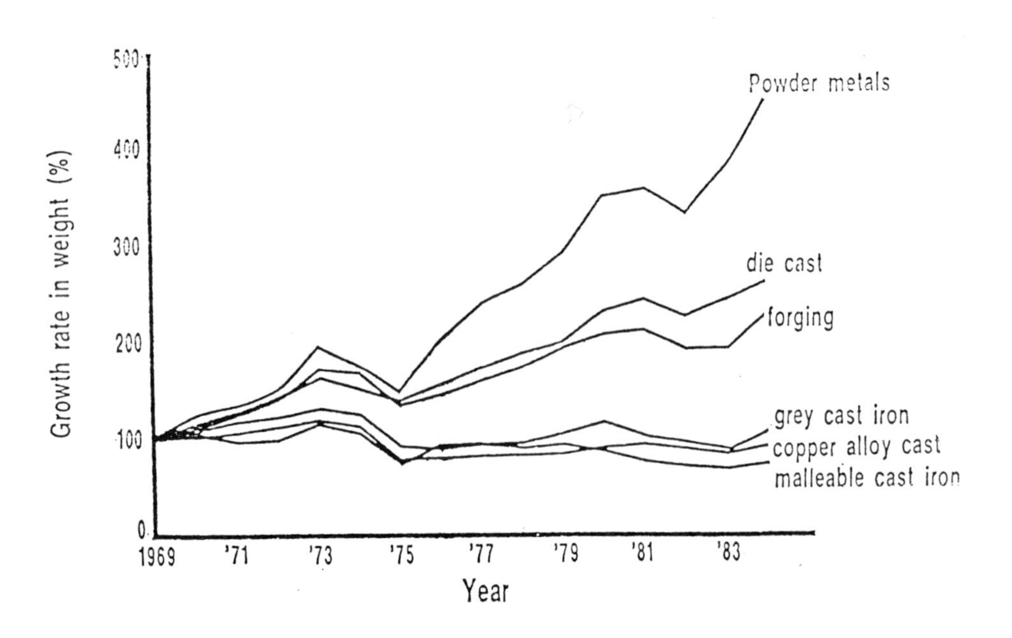

Figure 4 Growth rate of metal working industries in Japan.

Table 3 Annual production amounts of powder metallurgy products in the last 15 years in Japan (tons) Source: Japan Powder Metallurgy Association

PRODUCTS	1972	1973	1974	1975	1976	1977	1978	1979	1980	1981	1982	1983	1984	1985	1986	1987
SINTERED POROUS BEARINGS	4,181	5,232	4,764	3,419	5,310	5,831	6,009	6,444	6,850	6,665	5,929	5,951	6,943	6,826	6,715	7,116
SINTERED MACHINE PARTS	14,473	19,113	19,976	18,016	22,959	26,158	28,819	33,406	41,551	44,401	44,488	48,094	53,538	55,278	58,544	59,461
SINTERED FRICTION MATERIALS	694	988	1,059	567	489	505	607	668	641	547	672	653	581	559	487	449
SINTERED ELECTRIC CONTACTS	182	275	209	133	192	212	228	228	233	202	176	189	209	230	180	189
MATERIALS (W & Mo) FOR BULBS AND TUBES	470	606	501	256	408	457	434	511	550	523	528	551	700	695	567	616
SINTERED COLLECTOR BRUSHES	235	287	337	249	286	310	247	275	266	278	274	273	288	230	264	239
SINTERED MAGNETIC MATERIALS (HARD)	9,887	13,271	12,234	10,487	17,766	24,235	26,774	29,960	37,038	36,276	32,830	38,869	45,310	47,762	48,494	42,645
SINTERED MAGNETIC MATERIALS (SOFT)	18,509	23,022	16,121	12,743	18,644	16,358	19,710	21,526	26,395	25,593	19,934	24,474	34,464	35,200	34,535	37,962
SINTERED HARD ALLOYS (CEMENTED CARBIDES)	1,233	1,701	1,657	948	1,087	1,159	1,207	1,915	2,194	2,136	2,320	2,357	3,222	3,505	3,499	3,240
OTHERS	48	54	59	46	88	103	92	113	117	157	232	277	368	362	386	438
(TOTAL)	49,912	64,549	56,917	46,864	67,229	75,328	84,127	95,046	115,835	116,778	107,383	121,688	145,623	150,647	153,671	152,355

6. Prospects for the Future

The analysis of the evolution of powder metallurgy during the last decades justifies some well-founded optimism for the years to come. In fact, the technical cooperation between manufacturers and automotive engineers, which was not so open at the beginning, has, by now, become more commonplace and constructive. This change has contributed to improving and refining the design of sintered structural components and the choice of sintered materials. The problem of the direct transfer of parts, originally designed for other forming technologies, better known and familiar to design engineers, to PM, has now definitely arrived.

Today, in the main automotive industries, design engineers study new components keeping in mind, from the outset, that they will be produced by powder metallurgy. They are now convinced of the high strength, ductility and precision characteristics which can be consistently achieved. The present confidence in sintered materials, is the result of the very many past positive experiences already acheived, of steady, continuous and painstaking work in collecting and distributing technical information followed with great perserverance by far-seeing iron powder producers and parts manufacturers, firmly convinced of the validity of the powder metallurgy processes. Another positive element has been the amount of research and development activities carried out by the main industries involved aiming at providing the underpinning science to the technology and broadening the applications for sintered materials. A simple list of the main themes on which PM researchers and engineers are working gives an idea of the remarkable potential for growth. Some fundamental subjects are:

- Higher densities in cold compaction.
- Improvements in alloying techniques and processes.
- Further reduction of furnace energy consumption.
- Design of efficient continuous furnaces for vacuum sintering.
- Design of new forming techniques.

On the latter point, injection moulding appears more than a promise. The possible analogy with the considerable development potential of sinter-forging, only partially exploited, after substantial technical and research studies, however, introduces some caution about the possible rate of growth. According to L.F. Pease III [18], research into metal injection moulding started nearly twenty years ago. In spite of this long period, the growth, in terms of application to parts making, has been relatively slow. Emerging factors, however, suggest that metal injection moulding will become a viable branch of the powder metallurgy industry within few years [18].

The well-established production processes, however, are also benefitting from improved presses and furnaces. It can be expected that manufacturing equipment will become more and more efficient, providing better process control, improved product reliability, and an increase in productivity. As far as pollution and environmental protection are concerned, it is recognised that powder metallurgy activities are characterized by very negligible environmental effects. Therefore, the industry is destined to grow, with respect to other "less clean" technologies.

Despite the advantages of improved mechanical properties listed earlier, however, some technical and economic drawbacks will limit possible developments in the use of sintered components. The drawbacks are listed in Table 4.

Table 4 Technical and economical drawbacks to possible developments in the extension of uses of P/M components

Main Features of the Components	Causes of Drawbacks
Heavy weight parts (>1kg)	Cost of raw materials, equipment
Parts with elongated shapes	Technical difficulties in manufacturing
"Two-dimensional" parts (sheet metal parts)	Experimental equipment, cost of powders, rolling
Special shapes, with undercuts (carburettor body)	Technical difficulties or inability to manufacture
Helical gears, highly loaded (gearbox)	Technical difficulties in manufacturing, poor materials strength
Parts subjected to high fatigue (connecting rods, crankshaft)	Technical difficulties in manufacturing, negative porosity effects

It must be remembered, however, sintering technology has acquired a strong reputation for reliability within the existing forming possibilities and established techniques. This positive situation will always stimulate new applications, while consolidating the existing ones.

For this reason, it is believed that the stage of maturity of the ferrous powder metallurgy is still far ahead. A significant example of this is that car engineers are today considering seriously the development of a porous sintered connecting rod [19,20] and have already actively initiated extensive research work and bench tests. Only fifteen years ago, such an idea would have been considered complete madness.

References

1. P.F. Lindskog and G.F. Bocchini, Development of high Strength PM Precision Components in Europe, The Int. J. of Powder Metallurgy & Powder Technology, APMI, 15, 3, pp.199-230, 1979.
2. S.I. Hulten and J.A. Smedstam, An Elaboration in the Art of Making and Utilizing Sponge Base and Atomized Iron Powders, Annual Meeting of PM Joint Group of the Iron and Steel Institute, Swansea, 1968.
3. P. Lindskog and G. Skoglund, Alloying Practice in the Production of Sintered Steels, Powder Metallurgy, pp.375-396, 1971.
4. G. Wastenson, Methods of Extending the Applicability of Sintered Steels, PM Group Meeting, Eastbourne, October 1974.
5. P. Lindskog, Dimensional Accuracy, Microstructure and Mechanical Properties of Parts Made from Some High Strength Sintered Steel, 4th European Symposium on PM, Grenoble, 13-15 May, 1975. Published in Materiaux et Technique, Numero Hors Serie, Paris, 1976.
6. P. Lindskog and P. Arbstedt, Iron Powder Manufacturing Techniques: a Brief Review, Powder Metallurgy, 29, 1, 1986.
7. A.G. Dowson, Powder Metallurgy: Its Products and Markets, in Int. Powder Metallurgy Directory 1988/1989, MPR Publishing Service Ltd., Old Bank Buildings, Shrewsbury, England.
8. D.G. White, PM Developments in North America, International Conference on Evolution of Advanced Materials, Associazione Italiana di Metallurgia and American Society for Materials, Milan, May 31-June 2nd, 1989.
9. M. Lochon, The Automobile Industry and Powder Metallurgy, Metal Powder Report, pp.107-109, February 1989.
10. A.S.M., Powder Metallurgy, Volume 7 of Metals Handbook, Ninth Edition (Automotive Applications), ASM, Metals Park, Ohio, 1984.
11. D.W. Hall and S. Mocarski, Update on PM Automotive Applications, The Int. J. of Powder Metallurgy and Powder Technology, APMI, 21, 1, pp.79-109, 1985.
12. S. Mocarski and D.H. Hall, PM Parts for Automotive Applications, The Int. J. of Powder Metallurgy, APMI, 23, 2, pp.109-125, 1987.
13. S. Mocarski, D.V. Hall, J. Khanuja and S.K. Suh, Parts for Automotive Applications - Part III, APMI, The Int. J. of Powder Metallurgy, 25, 2, pp.103-124, 1989.
14. T. Watanabe and T. Sakurai, Present Status of Powder Metallurgy in Japan, APMI, The Int. J. of Powder Metallurgy, Powder Technology, 20, 3, pp.225-231, 1984.
15. T. Watanabe and T. Sakurai, Global PM Industry Trends, Japan, APMI, The Int. J. of Powder Metallurgy, 24, 2, pp.154-158, 1988.
16. I. Nagare, The Recent Trend of Sintered Components for Automobiles (Japan), Modern Developments in Powder Metallurgy, MPIF, APMI, Princeton, N.J., 21, pp.591-603, 1988.
17. G.F. Bocchini, Global PM Industry Trend Italy, APMI, The Int. J. of Powder Metallurgy, 24, 2, pp.143-151, 1988.
18. L.F. Pease III, Metal Injection Molding: The Incubation is Over!, APMI, The Int. J. of Powder Metallurgy, 24, 2, pp.123-127.
19. 0. Morocutti and R. Stickler, Concerted Efforts in Powder Metallurgy within the European Community, Powder Metallurgy Int. Verlag Schmid, 5 pp.40- 43, 1987.
20. S. Suzuki, K. Toyama and N. Konda, I'Development of As-Sintered Connecting Rods for Automobiles, APMI, The

20

PM High Speed Steels

P.R.BREWIN
Powdrex Ltd, Tonbridge, Kent, UK

1. Historical

The development of powder processing routes for the production of high speed steels was stimulated by the demand of tool manufacturers and users for improved microstructures and mechanical properties. Compared with conventional ingot wrought material, PM tool steels exhibit superior homogeneity together with finer, more uniform grain size and carbide distributions. This improved structure can lead to a marked reduction in distortion and cracking during heat treatment and provide better properties, including superior strength and toughness, improved grindability and increased tool life.

In the late 1960s and early 1970s, therefore, several powder metallurgy processes were developed the most significant of which are ~~probably the following:~~

Table 1 PM High Speed Steel Processes

Developer (Country)	Process	Product
Aerojet General (USA)	Cold Isostatic Compaction and Vacuum Sintering of Impact Pulverized Powders	Components
Amsted Research Laboratories (USA)	Mechanical Compaction and Vacuum Sintering of Water Atomized Powders	Components
Asea Stora (Sweden)	Hot Isostatic Pressing of Gas Atomized Powders	Bars
Brico Engineering (UK)	Mechanical Compaction and Copper Infiltration of Water Atomized Powders in Gas Sintering Furnaces	Components
Crucible Steels (USA)	Hot Isostatic Pressing of Gas Atomized Powders	Bars
Davy Loewy (UK)	Extrusion of Ingots produced by Cold Isostatic Compaction and Vacuum Sintering of Water Atomized Powders	Bars
Edgar Allen Tools (UK)	Mechanical Compaction and Vacuum Sintering of Water Atomized Powders	Components
Powdrex (UK)	Mechanical Compaction and Vacuum Sintering of Water Atomized Powders	Components
Powdrex (UK)	Forging of Ingots produced by Cold Isostatic Compaction and Vacuum Sintering of Water Atomized Powders	Bars

1.1 Processes for the Production of High Speed Steel Bars
In the **Asea Stora Process** [Ref. 1] a melt of high speed steel is atomized using high pressure nitrogen, and cooled by free fall in a horizontal chamber. The powder is then loaded into steel cans and hot pressed to full density. The resultant ingots are then forged or rolled to size as required.

Because the powders are nitrogen atomized they are free of the surface oxides which are characteristic of water atomization. The particles are also spherical in shape, and therefore, unlike water atomized powders, only deformable at high temperature. The resultant product is of very high quality, clean and fine grained. Hot Isostatic Pressing is an expensive slow operation, for the process to be economic it requires large ingots (of the order of 1000kg), and some of the isotropy of the powder metallurgical route is lost by the time these ingots are reduced in diameter to the final size.

The Crucible Process [Ref. 2] is similar to the Asea Process.

In the **Davy Process** [Ref. 3] high speed steels are water atomized and annealed in vacuo to an oxygen content in the region of 2000 ppm, These powders are then cold isostatically pressed into ingots and sintered in hydrogen atmosphere to chemically bond the powder particles together. The sintered ingots are heated to about 1100°C and extruded to the required diameter for sale. The extrude is coiled or cut to length, annealed, and peeled or bright drawn prior to sale.

The process produces a fine grained product but fails to reduce the powder surface oxides produced during water atomization - these oxides are broken up and uniformly distributed throughout the product by the extrusion step. Fully dense product is only achieved after extrusion through an area reduction of 16:1. This means in practice that to produce an extrude of 50mm diameter requires a starting ingot of 200mm diameter. As extrusion forces are in the order of 5 ton/sq cm this requires extrusion forces of the order of 1570 tonnes. Therefore to cover the entire range of bar sizes required by the market would require a very large extrusion press.

In the **Powdrex Process** [Ref. 4] high speed steels are water atomized and annealed in vacuo to an oxygen content in the region of 400ppm. These powders are then cold isostatically pressed into ingots 200-600kg in weight, and sintered in vacuo to 95-98% of full theoretical density (FTD). The sintered ingots are then open die forged to close all residual porosity, and then rolled to the required diameter for sale.

Because the production route incorporates vacuum sintering, the process produces a product which is free of residual oxides. However, the product is coarser grained than the Davy Process. This is a result of the need to raise the density of the sintered ingot by introducing an incipient liquid phase during sintering. It was found that below 95% density the sintered ingots had insufficient hot strength to withstand the tensile stresses introduced during the forging step. Efforts were made to carry out rotary hammer forging (GFM) of the as sintered ingot, in order to eliminate one process step. However, it was found that this process caused surface densification only, and that severe cracking occurred during subsequent rolling initiated at the residual centre porosity.

1.2 Processes for the Production of Fully Dense High Speed Steel Components
In the **Aerojet Process** [Ref. 5] coarse prealloyed shot is impacted at subzero temperatures against a hard target. The resulting pulverized material is sub-sieve size and of relatively low oxygen content. This process was operated for some time by Consolidated Metallurgical Industries (CMI).

The impact pulverized powder was too fine to flow freely, and therefore was not suitable for automatic mechanical presses. The process was therefore used to produce larger complex shaped parts such as milling cutters. This process was discontinued when processes were developed which produced low oxygen water atomized powders which flowed freely.

In the **Amsted Process** [Ref. 6] the powder is atomized with zero prealloyed carbon in order to avoid the need for a separate annealing operation. High silicon contents (up to 1.0%) are used in the melt. These have the dual purpose of reducing the powder surface oxides and of supressing the formation of undesirable carbide structures during vacuum sintering to full density. This process was operated by HTM.

The process produces a high quality product, and for some time this was of sufficient quality to be used on a large scale for the production of stator rings for diesel fuel injection pumps.

In the **Edgar Allen Tools Process** [Ref. 7] water atomized powders were hydrogen reduced to soften, mechanically compacted, and then sintered in a liquid phase to full density,

Partly because this process was developed by a major UK high speed steel bar manufacturer, it was limited to the standard high speed steel compositions such as T1 and M2. Carbide modifiers such as silicon were not used, and therefore the sintering process required temperature controls of the highest possible standard. However, process yields were hampered by the relatively high oxygen content of the powders - of the order of 2000 ppm. It was found that such high oxygen contents caused a significant change in carbon content during sintering, such changes varying not only within a furnace charge but also on larger sections even within a component from outside to inside; in turn this change affected the solidus of the steel and therefore the degree to which densification occurred.

The **Powdrex Process** [Ref. 8] was similar to that developed by Edgar Allen except that the powder was reduced in vacuo. This process was also operated by the Valeron Corporation.

This produced powders of below 1000 ppm oxygen, and overcame the problem of controlling the sintering response.

1.3 Processes for the Production of Partly Dense High Speed Steel Components

In the **Brico Process** [Ref. 9] water atomized high speed steel powders are mechanically compacted and placed in close contact with copper during sintering. The sintering temperature is controlled to melt the copper which then flows into the high speed steel by capillary action, thus filling the residual porosity. Following cooling to ambient the parts are then tempered.

Because of the presence of 25% copper, the resultant component has enhanced thermal conductivity while also retaining some of the abrasion resistant properties of high speed steels.

This process was developed for valve seat inserts for special high duty engines. It shows that on smaller components (below 100 grammes) PM high speed steels are cost effective even in the mass production automotive market. The process produces a 2-phase material which could not be produced by any other technique. The sintering process is designed to effect the infiltration, and not to densify the material.

1.4 Alloy Development

1.4.1 Bar

Processes for the production of high speed steel bar had a dilemma: was it necessary to stay with the alloy compositions which had been developed over the years for production by casting and forging, or could one successfully market new compositions tailor-made for production by powder metallurgy? The advantage of the former approach would be that the material could be sold in the open market with the minimum of market resistance; the advantage of the latter approach would be that higher grade compositions could be introduced without the drawbacks associated with the conventional process.

Asea Stora decided to follow the latter route, and marketed 3 compositions in Europe examples of which are given in Table 2.

Table 2 Asea Stora High Speed Steel Compositions

Element	ASP 23	ASP 30	ASP 60
C	1.28	1.28	2.30
Cr	4.20	4.2	4.0
Mo	5.0	5.0	7.0
W	6.4	6.4	6.5
V	3.1	3.1	6.5
Co	—	8.5	10.5

Although the introduction of these non-standard compositions took longer, it gradually became clear in the marketplace that powder metallurgy bar was more easily machined and more stable in heat treatment than a conventional bar of the same composition. Increasingly these grades (especially ASP30) were specified for larger cutting tools such as gear hobs and also for larger hard wearing components such as metalworking dies.

1.4.2 Components

PM component processes did not have the same problem as the bar producers, as individual components were generally developed in close cooperation between producer and customer.

The main Amsted grade was a high silicon M2: although it was found that increased vanadium gave improved sintering response, high vanadium led to a serious falloff in grindability. In Europe this drawback limited the wider use of vacuum sintered T15 high speed steels, although US toolmakers had more experience of this 5% vanadium grade and were therefore able to accept it in vacuum sintered form.

Powdrex grades also included carbide modifiers such as silicon; Powdrex also developed sintering data on a range of different standard high speed steels, showing that different alloys sintered to different structures [Ref. 10]. In particular Powdrex showed that M2 high speed steel, by far and away the grade the most widely used in the marketplace, was one of the most difficult alloys to sinter to full density by cold pressing and vacuum sintering.

In 1987 Wood *et al.* investigated the fundamental reasons underlying the sintering response of different high speed steel types, and established that this was controlled by the alloy carbides present at the sintering temperature. This work included the development and use of computer generated equilibrium data [Ref. 11].

2. Current Situation

2.1 Bar

Largely through the marketing efforts of Crucible and Asea Stora, PM HSS bar is now established in Europe, USA and Japan as the main route to the production of gear hobs, end mills, cold pressing dies and other high quality tools. Titanium nitride and titanium carbide coating processes have significantly increased tool life in many applications; while this has brought significant savings to the final customer, it has resulted in a reduction in steel volume manufactured.

The largest consumers of high speed steels are twist drills and hacksaw blades. In both cases isotropy of microstructure is of less importance, and therefore PM high speed steels have made little inroad.

2.2 Fully Dense PM Components

The development of the PM high speed steel component took place against a moving scene, as in the 1970s and early 1980s massive rationalization took place worldwide in the high speed steel industries, and competitive pressures prevented high speed steel bar prices matching inflation. Furthermore, tool users became more cost conscious and premature tool failures were no longer tolerated. It was several years before it was realized by PM HSS component makers that, with advances in conventional machining processes and the reduction in prices of conventional bar, components would only sell which had sufficient dimensional accuracy to avoid machining, and which were totally repeatable in metallurgical quality.

In the early days the problem of maintaining high dimensional accuracy was seriously underestimated. Undeniably impressive shapes were produced incorporating the most intricate features, but the passage of time showed that very often these features were insufficiently accurate to use without rectificatory machining,

The problem of maintaining consistent metallurgical quality was twofold. First, the powders had to be consistent in quality, predicating the requirement for blends sufficiently large to cover several complete sinter loads. Second, the sintering furnace had to be in top-class condition in order to ensure that the total sinter charge experienced the same temperature, to an accuracy as close as ±1°C. Efforts were made to sinter charges containing simultaneously large and small components, but even this was found to affect sinter temperature distribution and therefore yield.

Increasingly, the trend on PM high speed steels followed that of hard metals, away from the large complex shaped component towards the indexable insert. Certain larger components continued to be made from PM HSS where this had undoubted advantages. For example, square section toolbits in high vanadium and high cobalt high speed steels were successfully produced because square sections in high alloys are specially difficult to forge and roll.

2.3 Part Dense HSS Components

Developments in composite materials are an exciting new area, and the incorporation of more than one powder in a sintered component is an example of this. As yet the potential of this in the area of PM HSS has yet to be established. It will however depend to some extent on the producers coming to some understanding of the role of the constituent materials, and also of the residual porosity.

However, as it would appear that bulk material properties are sufficient to withstand the arduous conditions of engine exhaust valves, it would appear that it is not necessary to have full HSS density in order to achieve advanced properties. This means that marketable components can be made which are sintered with zero size change from pressed to sintered component. This fact means that, in contrast to fully dense HSS parts, HSS components can be produced by this route the dimensional accuracy of which lies within the requirements of computer controlled machine tools.

References

1. (Asea Stora Process description), Method of Manufacturing Billets from Powder. US Patent 3 728 111, 1971.
2. (Crucible Steel Process description), E. J. Dulls and T. A. Neumeyer, New and Improved High Speed Tool Steels by Particle Metallurgy. Progress In Powder Metallurgy, 1972, Powder Metallurgy Conference Proceedings. Published by Metal Powder Industries Federation.
3 (Davy Process Description), R. J. Causton and J. J. Dunkley, "Extrusion of powder metallurgy high speed steel", in Proceedings of International Conference on Hot Working and Forming Processes, Sheffield, July 1979. Published by The Metals Society, London.
4. (Powdrex Billet Process Description), P. Brewin, P. Nurthen and B. Toloui, "The Powdrex high speed steel billet process". Steel Times (London) Oct. 1987.
5. (Aerojet Process Description), "Powder Metallurgy Process". Canadian Patent 823 134, 1967.
6. (Amsted Process Description), J. A. Rassenfoss, "Production of Full Density M2 automotive diesel parts via the HTM Powder Metal Process". SAE Technical Paper Series 800 309, 1980.
7. (Edgar Allen Tools Process description), E. A. Dickenson and P. I. Walker, "P/M High Speed Steels Production at Edgar Allen Tools". Metal Powder Report, pp.14-16, Jan. 1980.

8. (Powdrex Preform process description), P. Brewin, P. Nurthen and B. Toloui, "Developments in hard wearing components. Metals and Materials, pp.470-3, August 1988.
9. (Brico Process description), M. S. Lane and P. Smith, "Developments in Sintered Valve Seat Inserts". Metal Powder Report, pp.474-480, Sept. 1982.
10. P. Brewin, B. Reed and H. Maurer, "The Influence of Chemical Composition and Production Conditions on the Metallurgical Properties of Sintered High Speed Steels". Proceedings of the 10th Plansee Seminar, Reutte, Austria, 1981.
11. J. Bee, J. Wood, P. Nurthen et al. "Phase Distributions during the sintering of high speed steel powders". Proceedings of the Powder Metallurgy Conference, San Francisco, MPIF, July 1985.

21
Hard Phase Containing Alloys - Wear and Corrosion Resistant Coatings

E. ESCHNAUER
Plasma Technik AG, Wohlen, Switzerland

AA p 309

1. Introduction

If a metallic hard phase is combined with a metal, mostly with elements from the iron group, it is called hard alloy or hard metal depending on the ratio.

Common sintered hardmetals consist of a hard-phase matrix with small amounts of a metallic binder. A well known example is tungsten carbide, using cobalt as the binder material.

Hard alloys are usually produced in an atomizing process from the liquid state. They are composed of a metal matrix, e.g. iron,cobalt or nickel, in which amounts of hard phases (carbides, borides, silicides) are embedded to reduce wear.

Often the above mentioned matrix-metals are alloyed with copper or other metals like manganese, molybdenum, or with the metalloids boron, carbon, and silicon. Hence, borides, carbides and silicides are formed, which are precipitated in the form of binary or ternary eutectics or in the form of primary or secondary solidified phases.

However, hard alloys are multiphase alloys, usually consisting of three or four components. The tough corrosion resistant matrix metals as well as the brittle and hard character of the hard phases influence the properties of the hard alloys, especially resistance against abrasive and erosive wear, corrosion resistance against aqueous chemicals, molten salts and in special cases resistance against hot gases and molten metals.

2. Classification of hard alloys

Hard alloys are classified into three different classes, depending on the base matrix metal used: there are iron based, cobalt based or nickel based hard alloys as well as pseudo hard alloys, where further hard phases than those alloyed are added.

Table 1 shows the historical development of hard alloys.

2.1 Iron based hard alloys

This group of hard alloys have been as rods in welding, and as cast material against wear since 1930. Pseudo iron based hard alloys in the form of wires or small tubes are of great importance in wear protection.

Since 1970 self-fluxing iron based hard alloys gain more and more importance as centrifugal alloys to protect the inner surface of bimetal cylinders in plastic processing machines [17]. In recent last years modified iron based hard alloys are used for special thermal spraying applications.

2.2 Cobalt based hard alloys

Since the 1920s these alloys gained considerable importance as wear resistant hard surfacing materials. As early as 1895 E. Haynes discovered the high values of hardness of cobalt - chromium alloys, when he was searching for cutting materials for tools. The addition of tungsten and other metals resulted in the classic cobalt based hard alloys, which are commonly known as "Stellites". In the following years these alloys were developed and patented in 1913. "Stellite" - powders are gaining use in plasma hardsurfacing [9, 27].

Important new advanced "Stellite" alloys have rapidly gained importance, particularly Co-Ni-Cr-W-C "Stellite" as well as Ni-Cr-W-C-"Stellite", the so called cobalt free "Stellites" [8 ,10, 29] .

In general, self fluxing boron containing "Stellite" alloys are seldom used for thermal spraying processes.

2.3 Nickel based hard alloys

This kind of hard alloy is most important for thermal spraying. The standard composition of such alloys was patented in 1937 in the USA [28], soon they alloys were "discovered" for thermal spraying. Nowadays these are the "classical" self fluxing nickel base alloys, which are of outstanding importance are widely used for flame spraying. (world-wide consumption of 3000 tons a year).

With additions of the metalloids boron and silicon, these nickel base hard alloys gain self fluxing properties. Thus an undisturbed metallurgical bonding (diffusion) between coatings and base metal can be expected, due to the fact that the

Table 1

Year	Matrix	Refractory Metals	Metalloides	Matrix	Hardphases	Hard Materials
1920	Co	Cr, W (Mo, V, Nb, Ta)	C (B, Si)	Ni, Fe, Cu, Mn	M_7C_3, $M_{23}C_6$, M_6C, (M-Si, M-B)	
1930	Fe	Cr (W, Mo, V)	C (Si, B)	Mn, Ni, Co	M_7C_3, $M_{23}C_6$, M_6C, M_3C, M-Si, M-B	
1940	Ni	Cr (W, Mo)	B, Si (C)	Fe, Co, Cu, Mn	Ni_3B, CrB, Ni_3Si, (M-C)	
1980	Ni	Cr, W (Mo)	C (B, Si)	Fe	M_7C_3, $M_{23}C_6$, M_6C, (M-B, M-Si)	
1982	Ni	Cr	B, C (Si)		CrB, Ni_3B, Ni_3Si, M-C	
1985	Ni	Ti, Zr, Hf V, Nb, Ta	B (C, Si)	(Fe)	τ-$Ni_{21}RM_2B_6$, CrB, Ni_3B, (M-C, M-Si)	
Pseudo Hard Alloys						
~1950	Ni	Cr (W, Mo)	B, Si (C)	Fe, Co Cu, Mn	Ni_3B, CrB Ni_3Si, (M-C)	FTC WC, WC-Co, Cr_3C_2, Cr_3C_2-Ni
~1960	Co	Cr, W (Mo, V, Nb, Ta)	C (B, Si)	Ni, Fe Cu, Mn	M_7C_3, $M_{23}C_6$, M_6C, (M-Si, M-B)	WC-Co
~1985	Ni	Cr (W, Mo)	B, Si (C)	Fe, Co Cu, Mn	Ni B, CrB Ni_3Si (M-C)	NbC, TaC, VC

substrate is completely wetted, converting the oxide films on the base metal into boron oxides and silicon oxides. If self fluxing nickel base hard alloys are used, no bond coat or intermediate layer is necessary.

Two new outstanding developments concerning nickel base hard alloys must be mentioned. These are carbide boride nickel hard alloys, developed in 1982 [14, 30] and τ - boride hard alloys developed in 1985 [7, 11, 12, 19, 21, 31]. More details are given later.

2.4 Pseudo hard alloys

Pseudo hard alloys consist of a mixture of different powders, generally nickel base hard alloys with further additions of hard phases to improve their wear behaviour. Originally sintered hard metal granulates were used, but after a short time fused tungsten carbides with suitable elastic structures were substituted (Fig. 1). Other WC - Co powders produced by different processes such as alloying, agglomeration, spray drying or coating as well as chromium carbide or chromium boride powders, or nickel coated chromium carbide and chromium boride, are blended with suitable nickel hard alloys.

New pseudo hard alloys with refractory hard phases such as vanadium carbide, niobium carbide or tantalum carbide increase resistance against wear and corrosion [3].

Approximately 1000 tons a year of pseudo hard alloys are used for thermal spray processes.

3. Spray powder production

The process of atomizing molten metals has become the most important process for powder production due to its comparatively low cost.

Atomizing with variations of cooling media such as gas, water or oil, or water atomizing with water cooling are well established. Nitrogen, argon (helium) are used as atomizing gases. The alloys are melted in induction furnaces in vacuum or in protective inert gas atmospheres. The atomizing is carried out either in a free fall process or in a confined stream.

The morphology and surface of atomized powder particles as well as their gas content or their fine crystalline, homogeneous and liquation free microstructure depends on the atomizing parameters. Controlled processing of atomizing parameters is necessary, if constant powder qualities are required [4, 22]. This is most important in order to produce sprayed coatings with reproducible qualities.

Mechanical comminution of "sinter-alloyed cakes" has lost its importance, whilst other techniques such as agglomeration or hydrometallurgical processes are insignificant.

The microstructure of atomized powders particles, as shown in Fig. 2, underlines the presence of homogeneously distributed hard phases in the powder particles produced using this process.

4. Properties

The type of use for coatings depends on the chemical composition of the alloys and their mechanical or physical properties [2, 5, 14, 15, 20, 32].

With high contents of the metalloids the formation of hard-phases in *boride nickel hard alloys* increases the brittleness, due to decreased tough matrix content. Wear resistant hard nickel alloys can be used up to elevated temperatures of 700°C. and the hot hardness behaviour is similar to that of comparable "Stellites". The non-scaling property is guaranteed up to 950°C.

Characteristic properties of conventional τ - boride hard alloys are based on the τ-phase $Ni_{21}RM_2B_6$, which is stabilized by a refractory metal (RM). Concentrations of 50 at -% of tantalum or niobium produce the outstanding chemical wear resistance and corrosion behaviour of the τ – phase, which is similar to those of pure tantalum and niobium.

Carbidic cobalt hard alloys ("Stellites") are resistant against corrosion, erosion, abrasion and fretting and they keep these properties even at elevated temperatures, thereby being also resistant against oxidation. Newly developed *carbidic nickel hard alloys* of the type Ni-Cr-W-C have hardness values of 35-50 HRC and exhibit both excellent hot hardness behaviour up to 1000°C and all the usual good mechanical and physical properties of "Stellites".

With chromium contents up to 30 percent by weight the new Ni-Cr-B-C-Si *carbidic-boridic-hard alloys* posses high corrosion resistance. Because of maximum boridic-and carbidic hard phases in coatings the wear resistance is very high. The hot hardness reaches the upper level of hard alloys and the hard matrix increases the good mechanical properties.

In Table 2 the chemical composition of important hard alloys and their mechanical and physical properties are listed.

5. Processing of hard alloys

During recent years different applications for hard alloys have been identified. They include the automatic industrial production as well as the most advanced technologies such as vacuum-, HIP- or laser-processing. These processes are extended not only to the production of thermally sprayed coatings but also include the production of engineering parts and bimetals.

Table 2

YEAR	TYPE	MATRIX	TYPICAL CHEMICAL COMPOSITION Wt.-%								
			C	B	Si	Cr	W	Nb,Ta,Ti	Others	Co	Ni
	Co - Cr - W - C		CARBIDIC COBALT HARD ALLOYS/STELLITES								
(1895)	1	Co	2.4-2.8	-	0-1	29 -33	11 -14	-	(5)	Bal.	0- 3
(1913)	6	Co	0.9-1.4	-	0-1.5	26 -31	3.5-5.5	-	(5)	Bal.	0- 3
1920	12	Co	1.1-1.8	-	0-1	28 -32	7 -9.5	-	(5)	Bal.	0- 3
1970	F	Co-Ni	1.7	1.5	1.5	25	12	-	(5)	39	22
	Ni - Cr - W - C		CARBIDIC NICKEL HARD ALLOYS								
1980	40	Ni	0.9-1.2	0.7-0.9	2.2-2.5	29 -31	6.5-8.5	-	2-5	0-2	Bal.
	50	Ni	1.2-1.4	0.9-1.2	2.2-2.5	30 -32	8.5-10	-	2-5	0-2	Bal.
	Ni - Cr - B - Si		BORIDIC-SILICIDIC NICKEL HARD ALLOYS/SELF-FLUXING								
(1937)	20	Ni	0-0.25	1.0-1.5	3.0-3.5	0 - 5.0	-	-	(2)	-	Bal.
1940	30	Ni	0-0.45	1.5-2.3	2.3-4.0	7.5-10	-	-	(3)	-	Bal.
	40	Ni	0.3-0.6	1.5-2.5	3.0-4.0	10 -12	-	-	(3)	-	Bal.
	50	Ni	0.3-1.0	3.0-3.5	4.2-4.5	13.5-17	-	-	(4)	-	Bal.
	60	Ni	0.9-1.0	3.5-4.0	3.5-4.0	26 -27	-	-	(2)	-	Bal.
	Ni - Cr - B - C		BORIDIC-CARBIDIC NICKEL HARD ALLOYS								
1982	65	Ni	2.5-3.5	2.5-2.7	3.0-4.0	22 -27	-	-	(5)	-	Bal.
	Ni - Cr - B - RM		TAU-BORIDIC NICKEL HARD ALLOYS								
1985	Nb-τ	Ni(Nb)		2.0-3.0		17.0-20.0	-	4.0- 5.0	1-2	-	Bal.
	Nb-τ	Ni(Nb)		1.5-2.0		10.0-13.0	-	25.0-30.0	1-2	-	Bal.
	Ta-τ	Ni(Ta)		4.0-4.5		11.5-14.5	-	3.0- 5.0	1-2	-	Bal.
	Ta-τ	Ni(Ta)		2.0-2.5		9.5-13.0	-	35.0-40.0	1-2	-	Bal.
	Ti-τ	Ni(Ti)		1.5-3.5		11.0-14.0	-	3.5- 6.5	1-2	-	Bal.
	Ti-τ	Ni(Ti)		2.0-3.0		7.5-10.5	-	20.0-24.0	1-2	-	Bal.

HARDPHASIS

	Hardness Rockwell C	Melting Range [°C]	Density [g/cm^3]	Tensile Strength [N/mm^2]	Compression Strength [N/mm^2]	Elong-ation [%]	Elastic Modulus [N/mm^2]	Thermal Expansion Coefficient 10 x 10^6 20-800°C
	51-58	1255-1290	8.09	620	1600-2000	<1	24-25	12.0
M_7C,$M_{23}C_6$,	39-43	1285-1395	8.46	890	1500-1700	1	20-21	14.5
(M_6C),	47-51	1280-1315	8.58	830	1500-1700	<1	21-22	13.8
(M-Si,M-B)	40-45	1210-1300	8.60	450		<1		12-15
M_7C_3,$M_{23}C_6$,	36-42	1180-1230	8.2-8.3	700-750	1880-2000			14.5
M_6C,	42-50	1160-1220	8.2-8.3	700-750	1880-2000			14.2
(M-B,M-Si)								
	20-30	1020-1150	8.4	335	>1800	>6	19	15.2
CrB	30-40	990-1150	8.3					
Ni_3B,Ni_3Si	40-50	980-1070	8.2	>320	>2000	>4	19.2	13.9
(M-C)	50-60	970-1040	8.0	>300	>2000	>4	19.8	13.0
	55-62	1000-1050	7.8	>290	>2450	3.4	20	12.1
CrB,Ni_3Si,Ni_3B (M-C)	60-65	970-1050	7.7					14.0
τ-$Ni_{21}Nb_2B_6$	55-60	1070-1140						
CrB,Ni_3B	55-62	1080-1190						
(M-C,M-Si)								
τ-$Ni_{21}Ta_2B_6$	55-60	1050-1100						
CrB,Ni_3B	55-62	1160-1240						
(M-C,M-Si)								
τ-$Ni_{21}TiB_6$	55-62	1040-1200						
CrB,Ni_3B	55-62	1090-1220						
(M-C,M-Si)								

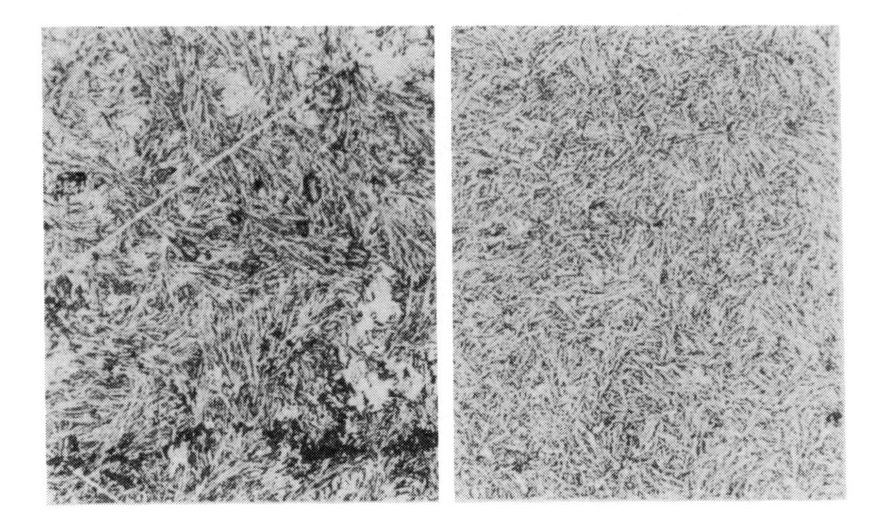

Figure 1

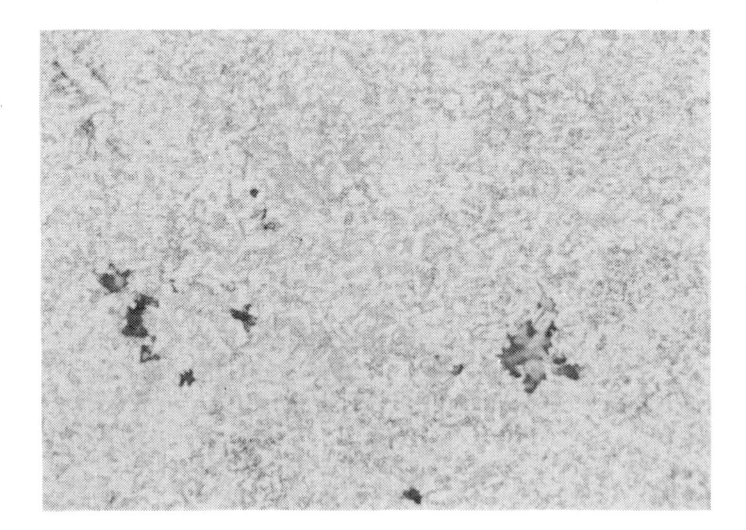

Figure 2 Phase distribution of an atomized particle (Boridic-carbidic Nickel Hard Alloy) - etched magnification x500.

5.1 Thermally sprayed coatings of nickel hard alloys

Substrate materials to be coated have to be of the right dimensions and free of contaminations such as rust, grease, carburization, nitrations or any kind of plating. In order to roughen the surface, sand blasting with corundum or grit blasting is required immediately before the parts are being coated.

Adaptation of base metal and coating material is most important for the adhesion and the quality of sprayed coatings. Numerous construction materials, e.g. cast iron, steel, nickel, copper etc. can be used for thermal spraying processes. Base metals with melting points lower than those of the hard alloys can not be coated. Standards and regulations for coating on steel substrates have to be taken into account, especially if base- and coating-material differ in their coefficient of thermal expansion.

Nickel hard alloys are widely used for repair coatings (60%) but nowadays coatings on new parts gain more and more importance. Single layer coatings with thicknesses from a few microns up to 1 mm are possible as well as double or multi-layered coatings of 10 mm or even more.

For manufacturing, two different processes have been established:

- Using the *"one-step-process"* low energy spraying and fusing of the coating with the flame is done simultaneously. Edges, sides and other defects are coated and repaired with layers of 0.1-0.8mm thickness.

- Using the so called *"two-step-process"*. In the first step the deposition of the coating is carried out by flame spraying and in the second separate step the coating is fused to a homogeneous layer of about 1-3mm thickness.

High energy processes like plasma spraying and high velocity flame spraying (e.g. Jet-Kote®) are also widely used. For fusing and sintering of sprayed coatings different methods are available namely:-

5.1.1 Fusing of coatings by burner flame

An oxygen-acetylen flame is usually for spraying and fusing , but sometimes a softer propane flame is used for fusing.

5.1.2 Fusing of coatings by induction

This excellent process is preferred for mass production of rotating symmetrical parts. By means of medium frequency equipment with electrical power up to 100 kW, dense homogeneous coatings with smooth surfaces and good metallurgical diffusion zones are obtained.

5.1.3 Fusing by furnaces

Fusing (sintering) processes are carried out in gas, electron, induction and vacuum furnaces for mass and single production. Remelting in vacuum furnaces usually with pressures of 10^{-3} bar, leads to very dense and homogeneous coatings with homogeneously distributed hard phases and excellent diffusion zone at the interface between base material and coating [6,23,32].

Plasma sprayed turbine blades using robots followed by vacuum fusing of the sprayed coatings were developed for mass production 15 years ago [25]. The quality of the coating and its microstructure is shown in Fig. 3.

5.1.4 Electron beam treatment

Fusing and densification of flame sprayed coatings using the electron beam causes no improvement of coatings qualities [18].

5.1.5 Laser treatment

Laser treatment of surfaces includes remelting, alloying, and glazing. Flame sprayed and laser remelted coatings generate improved wear properties due to extremely homogeneous, fine crystalline and partially amorphous microstructures, which can be realized, depending on the composition of the treated alloy (Fig. 4). The metallurgical bonding between coating and substrate can be regulated using the laser process so that the behaviour against traverse loads and impact stresses can be improved significantly [13,16].

5.2 Manufacturing of engineering parts

Hard alloys are used for thermal spraying processes achieving composite structures, as well as for the manufacture of complete engineering parts. The following are examples of the different processes which are used:-

5.2.1 Casting

Cast engineering parts with highest resistance against heat, abrasion and corrosion are produced using chilled casting and sand casting. Small parts with tolerances are manufactured by precision casting.

5.2.2 Hot isostatic pressing

With HIP-technology now well-established powder densification of hard alloys, especially "Stellite" [6, 21] and others, is a well known process, whereby the properties of the alloys are completely preserved [24]. Based on experienced data regarding HIP-encapsulation techniques both the new boridic-carbidic nickel hard alloys and the τ - nickel hard alloys

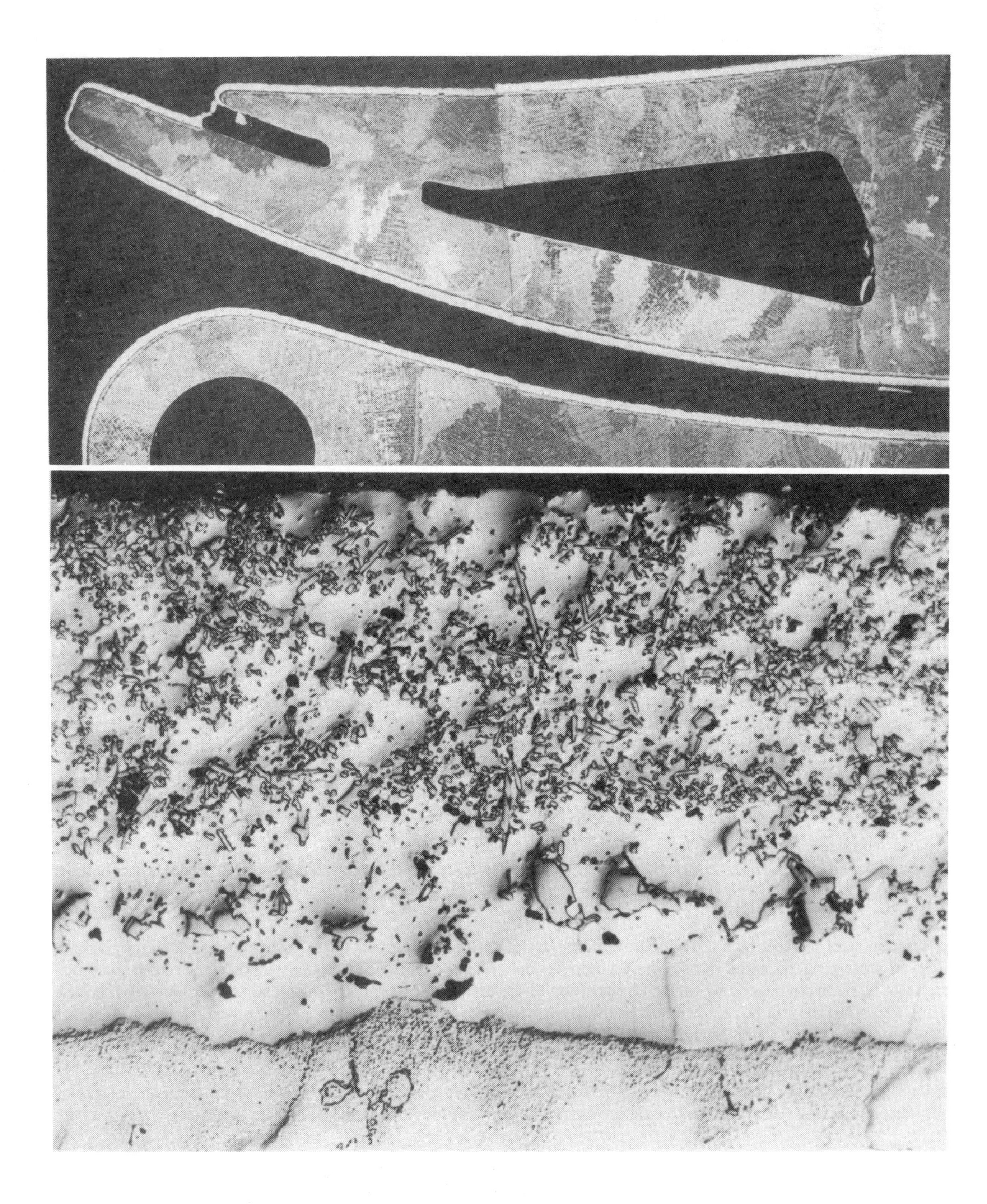

Figure 3 Plasma-sprayed and vacuum-fused Nickel Hard Alloy.

can be manufactured to parts with homogeneous, liquation-free and pore free microstructures, with improved quality-free (Fig. 5). Thermally sprayed and subsequently HIPped hard alloy coatings are the subject of present research.

5.2.3 Bimetal cylinders
The protection of the inner surface of bimetal cylinders (tubes, etc.) using the centrifugal zone melting process generates thick coatings with high wear resistance and long lasting resistance against abrasive and erosive filler materials. Usually two different alloys are chosen, nickel-cobalt based and iron based hard alloys. New advanced alloys of this type exhibit increased wear and corrosion resistance [1, 17, 29].

5.2.4 Composite elements
Composite structures have hard alloy coatings up to 10mm thickness, containing up to 50 vol-% coarse grained carbides. Composite materials posses high resistance against abrasive sliding wear as well as grain sliding wear. Composite structures also can be produced from "hard alloy pastes". This method offers numerous and promising results.

5.3 Plasma transferred arc welding
20 years plasma transferred arc welding of cobalt and nickel hard alloys has been established in repair and new part production, e.g hardsurfacing of ship engine valves, armatures and extruder rods in plastic processing machines. The protection of small automobile valves seems to be possible, due to progress with equipment, processes and alloys [9, 26, 27].

6. Future aspects

During recent years considerable development and progress concerning nickel hard alloys and cobalt hard alloys have been made. Numerous processes for the production of protective coatings and engineering parts have been successfully modified, tested and used for industrial mass production with the use of the most advanced technologies such as HIPing or laser.

Further progress can be expected with automatic atomizing technologies providing homogeneous microstructures of powder particles. As a result both nickel hard alloys and cobalt hard alloys will gain further applications resulting in increased powder consumptions.

References

1. Y. Attiyate, Innengepanzerte Extruder- und Spritzgiesszylinder erhöhen den Nutzungsgrad Kunststoffe-Plastics, Nr. 9 (1981).
2. H. Eschnauer, H. Meinhardt und E. Lugscheider, Verbundkonstruktionen aus Nickel-Hartlegierungen und Stahl über Diffusionszonen Tagung Deutsche Gesellschaft für Metallkunde Verbundwerkstoffe-Stoffverbunde Konstanz, 9. - 11. 5. 1984.
3. H. Eschnauer, H. Meinhardt, E. Lugscheider und W.-G. Burchard, Phasenbeurteilung von Oberflächenschutzschichten aus Nickelhartlegierungen durch Interferenzschicht-Metallographie Praktische Metallographie, 15, 101-104, 1984.
4. H. Eschnauer, H. Meinhardt, B. Häuser und E. Lugscheider, A new Nickel-base Alloy for Rapid Solidification,Intern. Conference on Rapidly Quenched Metals, Würzburg, Sept., 19845. O. Knotek, E.Lugscheider und H. Eschnauer, Hartlegierungen zum Verschleißschutz. Verlag Stahleisen mbH, Düsseldorf, 1975.
6. O. Knotek und R. Reimann, Vakuumeingeschmolzene Oberflächenschichten aus Carbid-Hartlegierung-Pulvermischungen. Metell, 34, 711-715, 1980.
7. A. Krautwald, Konstitution τ -Borid-haltiger Hartlegierungen und ihre Verarbeitung Dissertation, RWTH-Aachen 1986.
8. G. Kruske, Neue Nickel-.Hartlegierungen für den Verschleißschutz Werkstoffe Veredlung, Heft 4, 24 u. 25, 1979.
9. M. Kunath und E. Pfeiffer, Trennen + Fügen, 19 (1988), 9-14. Plasma-Pulver-Auftragsschweissen - ein Anlagenkonzept für neue Anwendungen.
10. E. Lugscheider und G. Kruske, Neue Hartlegierungen für Verschleiss- und Korrosionsbeständige Auftragungen, DVS-Bericht 52, 97-104, 1978.
11. E. Lugscheider, H. Reimann und R. Pankert, Untersuchungen im τ -boridhaltigen System Nickel-Chrom-Tantal-Bor bei 850°C. Z. Mettallk. 71, 654-657, 1980.
12. E. Lugscheider, H. Reimann und R. Pankert, Mit 4a- und 5a -Metallen. stabilisierte τ-Boride des Nickel Metall., 36, 247-251, 1982.
13. E. Lugscheider, H. Eschnauer, A. Krautwald und H. Meinhardt, Neuere Hartlegierungen für Verschleiss- und Korrosionsschutzschichten, Achema, Frankfurt 9-15. 6. 1985.

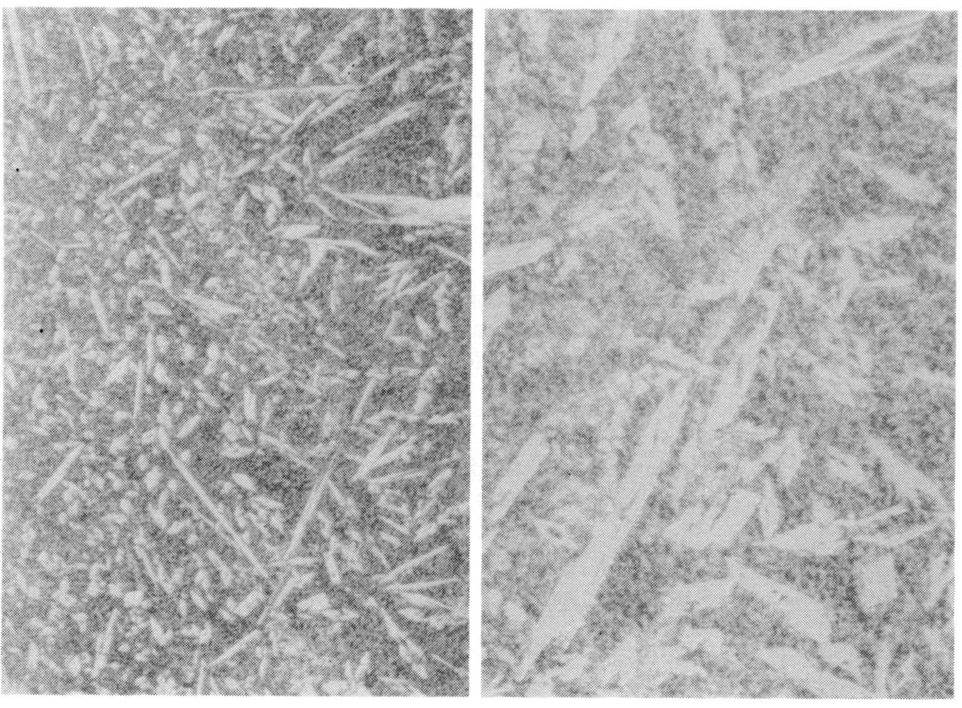

Magnification x100 *Magnification x200*

***Figure 4** Sprayed and lasered Boridic-carbidic Nickel Hard Alloy.*

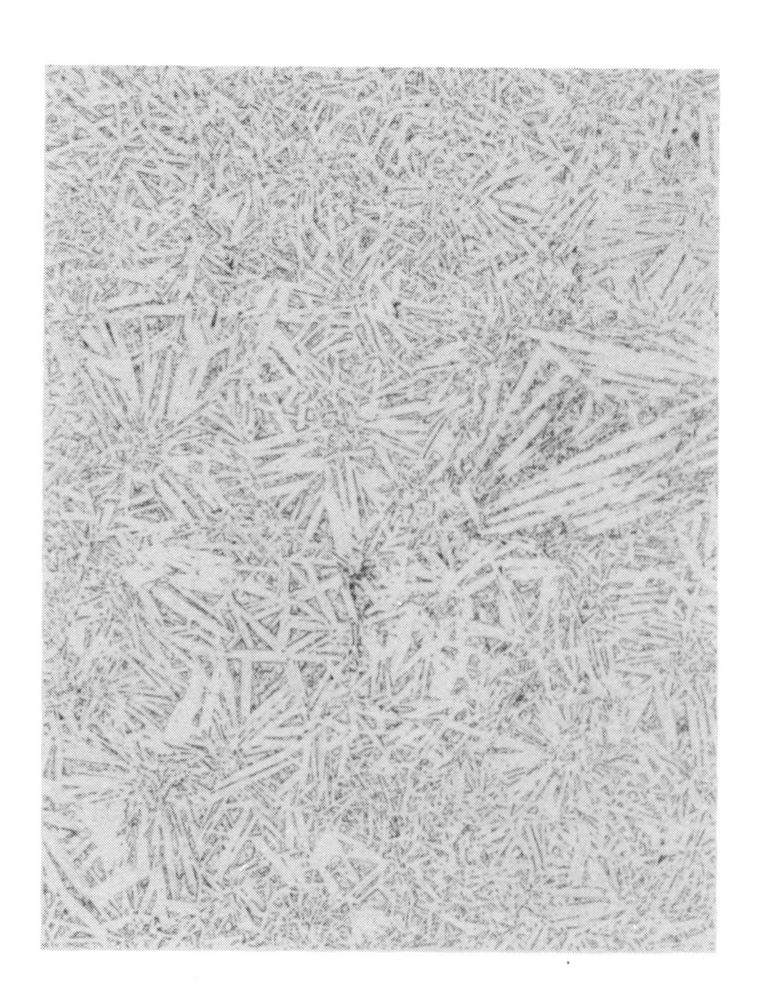

14. E. Lugscheider, H. Eschnauer, A. Krautwald, H.W. Bergmann and H. Meinhardt, Laser Remelting of Thermal Sprayed Coatings. European Conference on Laser Treatment of Materials, Bad Nauheim, 25, 9 - 26 . 9. 1986.
15. E. Lugscheider, A. Krautwald, H. Eschnauer, J.Wilden and H. Meinhardt, A new Type of Atomized Coating Powder for Protection against Wear and Corrosion. Surface CoatingTechnology 32, 273-284, 1987.
16. E. Lugscheider, A. Krautwald and J. Wilden, Laser Treatment of Thermal Spray Coatings. International Conference 9Heat Treatment '87) London, 11-15 May, 1987.
17. P. Lül sdo rf, Mit gepanzerten Schnecken und Zylindern gegen Verschleiss Reiloy Bericht Nr. 5. Tagung Verschleiss in der Kunststoffverarbeitung, Darmstadt, 10.2.1987.
18. H. G. Mayer, G. Haufler, J. Föhl, T. Weissenberg, G. Gnädic, D. Röckle, B. Krismer und J.E. Albers, Herstellung von Verschleißschutzschichten durch Diffusionsschweissen und Elektronenstrahlumschmelzen und deren Eigenschaften. BMFT-Förderungsvorhaben 01 ZT 110, 120, 130, 140, 77-156, 1984.
19. H. Meinhardt, H. Eschnauer, E. Lugscheider und A. Krautwald, Neue boridische Nickel-Hartlegierungen mit hohen Gehalten an Niob und Tantal für Oberflächenschutzschichten gegen Verschleiss und Korrosion 11 . Plansee-Seminar Reutte, 20. - 24. 5. 1985.
20. H. Meinha rdt and J. Beczkowiak, Boride Ni-Hard Alloys with Hard Phases Stabilised using Refractory Metals; 1st Plasma-Technik Symposium, Luzern, Vol. 2, 229-236, 1988.
21. H. Meinhardt and H. Eschnauer, "Development Production of a New Class of Nickel-Hardalloys for Substitution of Cobalt-Hardalloys", in European Research on Materials Substitution, edited by I. V. Mitchel and H. Nosbusch. Elsevier Applied Science, London and New York, 409-416, 1988.
22. H. S. Pacil and R. K. Upadhyay, Intelligent Processing for Metal Atomization. NTSC '88 Conference, Cincinnati, 23-27 October, 1988.
23. H. Reimann, K. Papendick und G. Kruske, Oberflächenschichten durch das pulverflammspritzverfahren aufbringen Industrie-Anzeiger, Nr. 94, 1984.
24. H. Seilstorfer, 10 Jahre Heiss-Isostat-Press (HIP) - Technik Seilstorfer GmbH & Co., 1985.
25. M. Villat and P. Felix, High-temperature Corrosion Protective Coating for Gas Turbines. Sulzer Technical Review, 3, 1-8, 1976.
26. R. Weintz, Eigenspannungen und Ausscheidungen in der Sitzpanzerung von Gaswechselventilen. Dissertation, RWTH Achen, 1987.
27. M. Wollner, Stanzeitversuche mit dem Plasma-Pulver-Auftragsschweissbrenner PPA 201 unter Einsatz Pkw-Ventilschweisstypischer Parameterkombinationen. Industrie-Anzeiger, H 66, 28-30, 1987.
28. US-Patent No 2088838, Aug. 3, 1937. N. W. Cole and W. H. Edmonds, Hardening Material, Resistant to Heat, Acid, Corrosion and Abrasion, and Method of Producing the same.
29. DP 2639325, 19. 6.1980 Metallgesellschaft AG Verwendung einer Nickel-Basis-Legierung.
30. DP 3315920, 2. 5. 1983 u. EP 124134, 2. 5. 1984. Hermann C. Starck Berlin Hartlegierung auf Nickelbasis.
US-Patent, June 9., 1987. Hermann C. Starck Berlin. Nickel-Based Hard Alloy.
31. DP 35009242, 14. 3. 1985 u. EP 1947C1, 14. 3. 1986 Hermann C. Starck. Berlin Verfahren zur Herstellung von Oberflächenschutzschichten mit Niob oder Tantal und ihre Verwendung.
32. Produkt Informationen Bernex, Colmonoy, Deloro, Gotek, Plasma-Technik, Seilstorfer, Reiloy, Starck.

22

Cemented Carbide Powders and Processing

B. ARONSSON AND H. PASTOR*

AB Sandvik Hard Materials, Stockholm
*Ugicarb Morgon and Eurotungstene Poudres, Grenoble**

1. Introduction

The excellent properties of materials obtained by sintering powders of tungsten monocarbide WC, together with cobalt, at temperatures above the melting point of this metal, were discovered in the early 1920s.

After a slow start there has been a steady flow of new products of this powder metallurgical material, called hard metal or cemented carbide, with applications in the areas of metal cutting, rock drilling and wear parts. Important steps in the development were the introduction of grades with titanium carbide (TiC) and tantalum-niobium carbide (Ta,Nb)C additions around 1930, greatly improving the performance in metal cutting, and grades with hard coatings around 1970.

Materials with TiC rather than WC as the major constituent and a nickel-rich binder phase were invented before the second World War but had moderate commercial impact. In recent times TiC/TiN-Ni/Mo grades have attracted renewed interest and are often referred to under the heading "cermets".

Although after more than 60 years a mature material, there is still a strong development of cemented carbide and its products.

The purpose of this article is to give a survey of the processing of tungsten monocarbide and of cemented carbides with some remarks on its relation to properties and microstructure. For the sake of simplicity we shall concentrate on straight WC-Co grades. Whenever possible reference is made to the many books and review articles [1-24] that are available on cemented carbides and their processing.

The article will begin with some comments on the structure and properties of cemented carbides since many details of the processing are due to their effects on structure and on properties of the final product and hence cannot be discussed without some basic understanding of processing-structure-properties relationships. We will subsequently give an overview of the processing and present its different steps: intermediate powder products from ore concentrates and by recycling, powder mixing, milling and agglomeration, consolidation and sintering.

2. Some Comments on Processing, Structure, Properties Relationships in Cemented Carbides

When discussing the effects of variations in processing of hard materials such as cemented carbide on properties, it is appropriate to distinguish between different types of properties (Table 1) which admittedly are not separated by any sharp boundaries. Some properties such as density, thermochemical properties, etc., only depend on chemical composition and hence are independent of processing or microstructure, others depend on microstructure (e.g. grain size) but not on the way in which this has been obtained. Many technological factors depend strongly on the population of three-dimensional defects (e.g. transverse rupture strength) or on more complex features such as gradients of composition, residual stresses, etc., and the reason for a particular processing route is to govern such structural details so as to achieve the best possible performance of the final product.

It is not possible to give many comments on these questions in the following and we will mainly refer to hardness and transverse rupture strength when discussing the effects of processing on properties.

3. Processing Tungsten Carbide and Cemented Carbide - An Overview

The various processing steps are summarized in Fig. 1.

The most important intermediate product is tungsten monocarbide powder. As seen from the W-C phase diagram in Fig. 2, this phase decomposes peritectically and cannot be obtained in pure form from the melt (the commercial powder product "fused carbide" consists mainly of the eutectic mixture of WC and W_2C).

The starting materials are concentrates of scheelite or wolframite ore from which ammonium paratungstate (APT) or tungstic acid is prepared by hydrometallurgical processing, today often implying solvent extraction and ion exchange.

Table 1 Classification of properties (see text) [36]

	Intrinsic properties			
	Elastic constants, thermochemical properties	Fracture toughness, hardness, creep	Transverse rupture strength	Performance properties
Chemical composition, T, p	●	●	●	●
Microstructure	...	●	●	●
Defects: porosity, inclusions, etc.	...	...	●	●
Surface region conditions Residual stresses Microgeometry, testing conditions	...	...	(●)	●

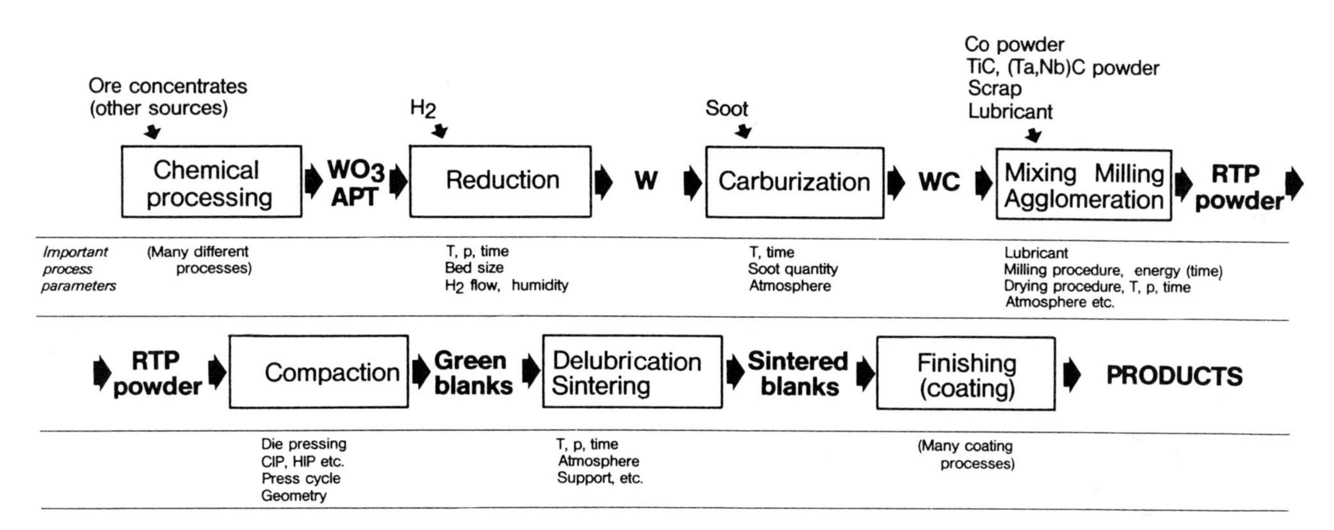

Figure 1 Processing steps in the making of cemented carbide [36].

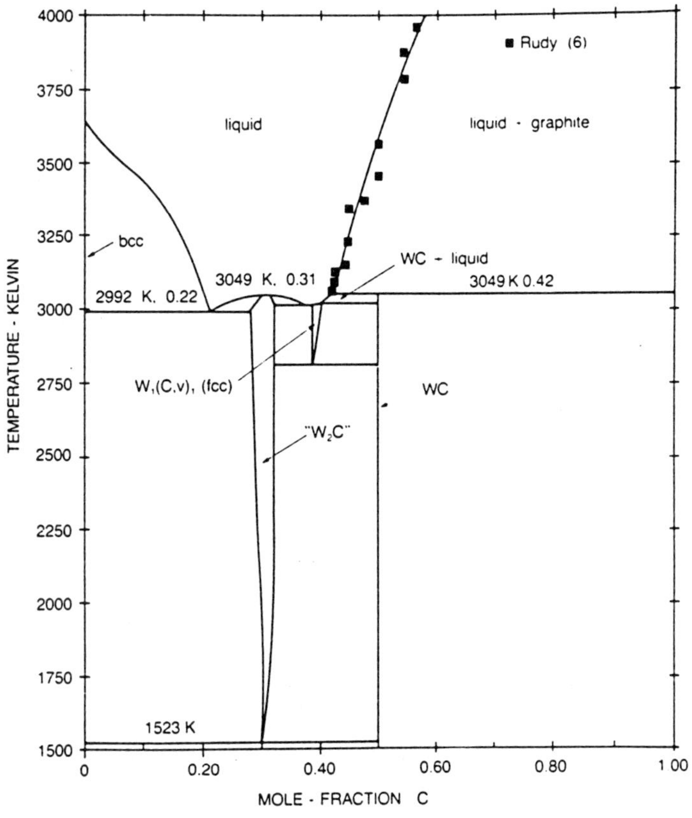

Figure 2 The W-C phase diagram [101,103].

After reduction to tungsten metal and mixing with graphite, tungsten carbide is obtained by exothermic reaction at around 1350-1600°C. Tungsten carbide powder is subsequently mixed with cobalt, and if appropriate with other carbides and comminuted cemented carbide scrap to the desired overall composition. After mixing and milling, the powder mixture is agglomerated, consolidated and sintered. For many products there will be finishing and coating operations but these will only be given a perfunctory treatment in this context.

4. Processing of Ammonium Paratungstate (APT) and Tungstic Acid

In virtually all processing of tungsten bearing minerals to tungsten products, ammonium paratungstate $(NH_4)_{10}H_2W_{12}O_{42}.4H_2O$ (also written $5(NH_4)_2O$, $12WO_3$, $5H_2O$) or APT for short, is an intermediate product. With modern techniques this compound can be made to almost any desired purity and granulometry. It has largely replaced ore concentrates as the commodity from which production starts, and can easily be converted to tungstic acid $WO_3.H_2O$.

We shall briefly recall the major steps in making APT since it is of importance for the subsequent processing of tungsten monocarbide and cemented carbide.

4.1 Minerals, Ores, Ore Concentrates

There are many techniques for making concentrates of the common tungsten minerals, scheelite ($CaWO_4$) and wolframite (Fe, Mn)WO_4: flotation, magnetic or electrostatic separation, etc. Excellent summaries of these can be found in the proceedings of the International Tungsten Symposia [20-23] and also in the 1988 Conference at Changsha, China [24] and reference books [17,18,19].

Previously the ore was commonly concentrated to 60-70%WO_3 before chemical treatment. Nowadays concentrates with 10-40%WO_3 are often fed-directly into the chemical plant where they are - for instance - treated by sodium carbonate in an autoclave. An overview of the various processes for making APT, W and WC [17-25] from ore concentrates or scrap is shown in Fig. 3.

4.2 Hydrometallurgical Processes

Scheelite concentrates are decomposed by leaching with hot concentrated hydrochloric acid. Wolframite concentrates are usually treated with sodium carbonate under pressure; the sodium tungstate solution thus obtained is sometimes converted to calcium tungstate, often referred to as artificial scheelite, which can then be fed into a scheelite processing line.

APT obtained by fractional crystallization in the acid route process has a purity which is satisfactory for many applications.

The sodium tungstate solution obtained by the alkaline route can be purified by a series of chemical treatments [17,18,19,25,91] or by modern techniques such as solvent extraction and ion-exchange [17,18,19,25,92]. The level to which impurities can be eliminated is demonstrated by typical current analyses of APT, W and WC given in Table 2.

As shown in Fig. 3, tungsten-containing scrap is also treated chemically and converted to sodium tungstate and APT [25,35,37]. This is particularly appropriate for contaminated scrap. An interesting new method for treating such starting material is electrolytic dialysis [26]. It can sometimes be advantageous to oxidize scrap and dissolve it in salt baths for subsequent leaching to sodium tungstate solution and a modification of this process route has recently been presented [35,38].

5. Reduction of APT and WO_3 to Tungsten Metal Powder

Some basic facts about the various reduction processes are collected in Table 3. The most common process is hydrogen reduction of APT or WO_3 in which intermediate oxides are being formed.

The "blue oxide" sometimes referred to as W_4O_{11} is a mixture of $W_{18}O_{49}$ and $W_{20}O_{58}$, while the "brown oxide" is WO_2. The variation of the equilibrium constants with temperature is shown in Fig. 4.

Of particular importance for the subsequent processing are the purity, the granulometry and the morphology of the tungsten powder obtained during the reduction, and this has been the subject of many recent publications, particularly concerning the influence of small amounts of impurities [53-60,95]. The reduction temperature being moderate, there is little elimination of impurities during this process which sets limits on the amount of dopants that can be added for modifying granulometry or morphology.

On regulating temperature, humidity of the hydrogen gas and the bed size, the average grain size of the tungsten powder can be varied between 0.3 and 10 microns or more [25]. The mechanisms and kinetics of reduction of WO_3 are well described in refs [29,63]. The use of an oxygen probe (yttria-stabilized zirconia) at present permits the study and soon the control of the composition of the furnace atmosphere during reduction [28].

For fine-grained materials it may be advantageous to stop the reaction at the "blue oxide" stage and mill this, before continuing the reaction [64-70,93,94]. Coarse-grained materials require the addition of particular dopants of which the alkaline metals are the most important ones.

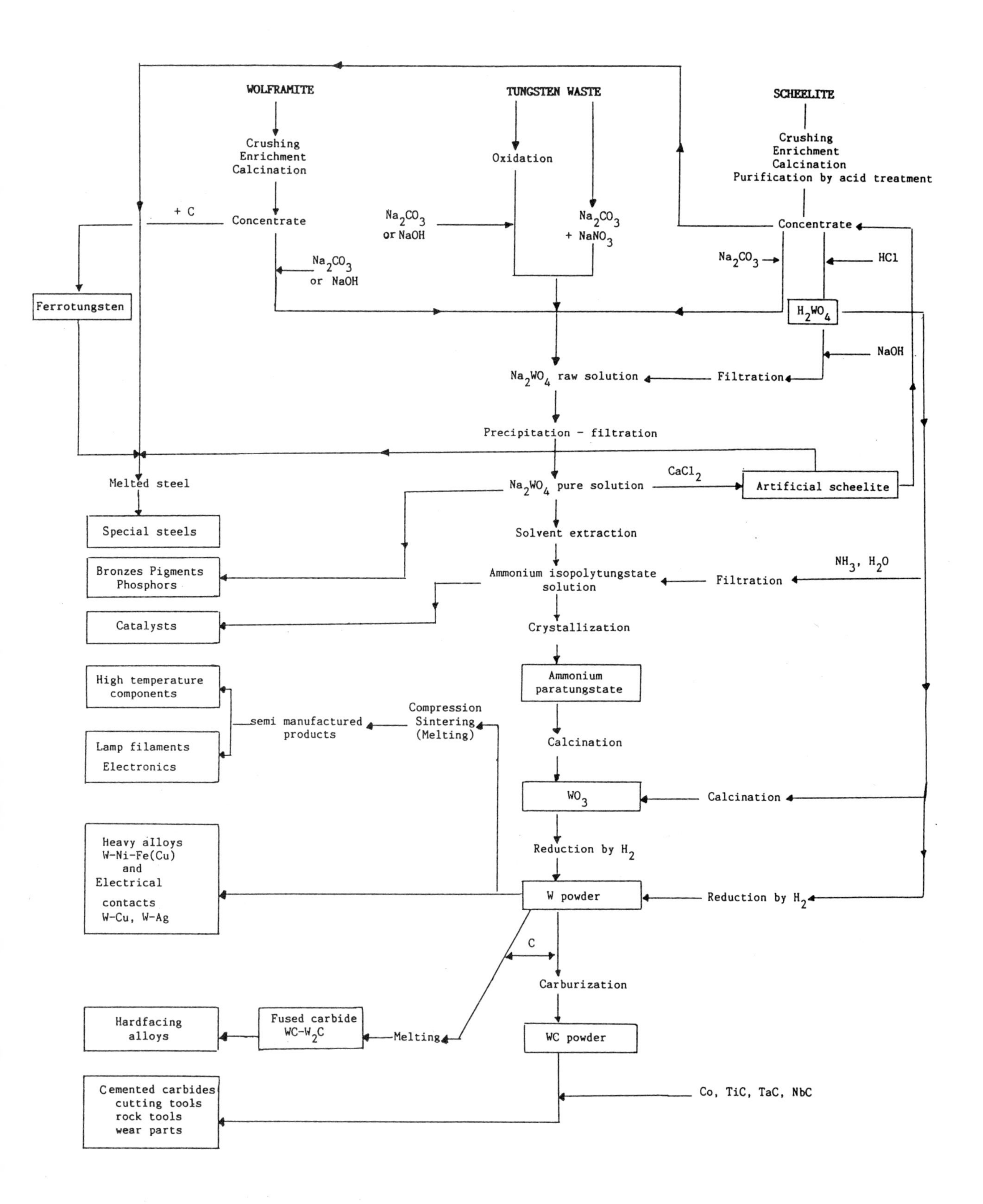

Figure 3 The industry of tungsten and its compounds and alloys [25]

Table 2 Contents of trace elements in different powders according to current specifications (in ppm) [36]

	APT				
	1976	1984	W	WC	Co
Al	...	...	...	<20	...
As	<50	10	<50	<5	...
Ca	<10	<5	<40	<20	<100
Cd	...	<1	...	...	...
Co	...	<2	...	<300	...
Cr	...	<1	<200	<30	...
Cu	...	...	...	<10	<20
Fe	30	<10	<500	<200	<100
K*	40	<10	...	<50	...
Mg	...	<1	...	<10	<100
Mn	...	...	...	<10	<30
Mo	300	15	<1000	<90	...
Na*	<10	<3	<25	<30	<50
Nb	...	...	...	...	...
Ni	<10	<1	<100	<20	<450
O*	...	...	<800	<500	<5000
P	40	12	<40	<50	...
Pb	<10	<3	...	<10	...
S	17	<10	<40	<30	<50
Si	25	20	<50	<40	<40
Sn	...	<3	...	<10	...
Ta	...	...	...	<30	...
V	...	...	...	<5	...
Zn	...	...	...	...	<50

* The contents of oxygen tend to be higher in fine grain W or WC powder; those of sodium and potassium are higher in coarse grain powder.

Table 3 Reduction processes [39]

Reduction process/ furnace type	Compound	Reducing Agent	Temperature (°C)
Batch Operations			
1. Carbon reduction	APT	Carbon or Hydrocarbons	1300-1400
2. Hydrogen reduction	Oxide Acid	H_2	700-900
Continuous Operations			
3. Walking beam or belt type	APT Oxide Acid	H_2	700-1000
4. Multitube pusher type	Oxides	H_2	700-1000
5. Rotary kiln	APT Oxides	H_2	700-1200
6. Plasma	Chlorides APT Oxide	H_2	>5000
7. Fluidised bed	Oxide	H_2	> 700
Melting Operations			
Carbothermic Silicothermic Aluminothermic	ore concentrates	C Fe Si Al	≈2500

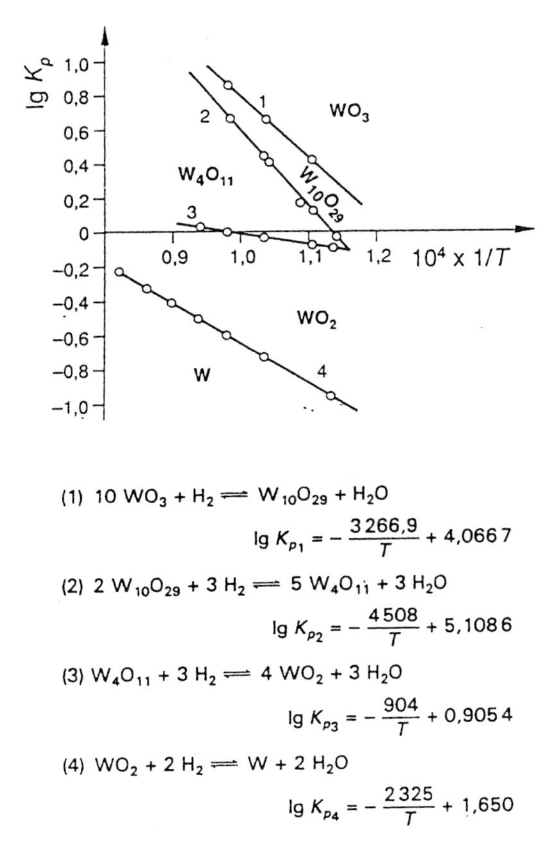

Figure 4 Reduction of tungsten oxides by hydrogen: variation of the equilibrium constant Kp with temperature T (K) [27].

Table 4 Some characteristics of commercial tungsten carbide powders [25]

Mean grain size (FISHER SSS) μm	Non taped density (Scott density) g/cm³	Taped density g/cm³	B E T Specific surface cm²/cm²	Oxygen content (reducible by hydrogen at 900°C) wt. %
0.35 ± 0.05	0.85 ± 0.15	1.5 ± 0.25	750 000	0.5
1.2 ± 0.1	2.7 ± 0.25	4.05± 0.3	300 000	0.3
1.6 ± 0.15	2.8 ± 0.25	4.3 ± 0.3	220 000	0.12
3 ± 0.3	3.1 ± 0.3	4.6 ± 0.4	110 000	0.07
6 ± 0.6	3.7 ± 0.4	5.2 ± 0.5	25 000	0.03
8.1 ± 0.8	4.2 ± 0.5	6.5 ± 0.8	–	0.02
12 ± 1.2	4.4 ± 0.5	6.8 ± 0.8	–	0.02

It is possible to obtain well-faceted tungsten single crystals which are believed to depend on chemical transport by the volatile compound $WO_2(OH)_2$.

6. The Fabrication of Tungsten Monocarbide

As seen from the W-C phase diagram (Fig. 2) the tungsten monocarbide has a very narrow homogeneity range and decomposes peritectically at 2776°C. It is made by reacting pure tungsten carbide powder with carbon black of low ash content, at 1350-1600°C. Even during this high temperature reaction there is a considerable elimination of impurity elements. The reaction is exothermic ($G°_{273}$ = -39.63 kJ) and this may give rise to temperature gradients and a concomitant scatter in grain-size in batch production. The continuous carburization process gives a more homogeneous grain size: the mixture W + C is poured into graphite boats circulating under a counter-current hydrogen flow in a pusher-type Tamman furnace - a graphite tube is used as a heating element. Coarse and polycrystalline WC powders (0-2 > μm) are obtained from coarse tungsten powders at a relatively high temperature (1600°C), while fine and nearly monocrystalline WC powders (μ < 1 um) are produced from submicron tungsten powders at 1350-1400°C. Small amounts of cobalt influence the carburization of coarse tungsten powder [96]. During the carburization process the hemicarbide W_2C is also obtained as an intermediate phase [30]. The carburization temperature is correlated with the WC phase substructure and consequently with the mechanical properties of WC-Co alloys; it influences the subgrain size, microstrain and microhardness of WC [31,44,45,51]. A good summary of the mechanisms and kinetics of tungsten powder carburization is given in ref.[46].

The temperature of synthesis being much higher than during the carbon reduction of tungsten oxides, there is as yet no commercial rotary furnace for the carburization process.

Table 4 gives the main characteristics of commercial tungsten carbide powders. The morphology of W and WC powders can be seen on SEM photographs in Fig. 5.

7. Alternative Routes for Making Tungsten Carbide

In addition to the conventional carburizing of tungsten there are several other ways of making WC.

Direct carburization of tungstic oxide WO_3 is obviously of interest since it would eliminate one processing step. Recent studies have confirmed this possible route but there is still uncertainty about its advantages over the traditional route. The tungsten trioxide reduction-carburization with CO-CO_2 mixtures leads to the production of submicron WC powders (0 - 0.4 μm) at 900°C [40].

A new fairly direct method for making WC from ore concentrates has been developed by the US Bureau of Mines [32,97]. In a system with two immiscible salt melts, the gangue is concentrated in the silicate melt while the tungsten-rich component is in the sodium chloride melt. The latter is subsequently subjected at 1070°C to a gas sparging operation using natural gas (methane) which will give precipitation of WC which is cleaned by a wet chemical treatment. Submicrometer crystals of WC are initially produced which then grow into thin plates up to 100 μm on a side or into popcorn-shaped conglomerates.

The still used "menstruum technique" [33], based on old patents, consists of dissolving tungsten in a graphite-covered melt of iron, nickel or cobalt (the so-called "auxiliary bath") and extracting the carbide crystals from the cooled and crushed material by acid treatment. Very pure (low oxygen content) and well crystallized carbide is obtained.

Alumino-thermic reduction of tungstic compounds combined with carburization is also an established procedure but with limited economic attraction, as is also the case for preparation of WC by molten salt electrolysis.

Tungsten carbide with special structure and properties (coatings for instance) can be prepared via gaseous tungsten compounds : WF_6, WCl_6, $W(CO)_6$. Plasma technology has also been used for making WC but, again, the economy of these processes (involving consideration of energy consumption and protection of the environment) has not made them widely applicable.

A mixture of cobalt and tungsten carbide can also be made by gas phase reduction/carburization of cobalt tungstate ($CoWO_4$) and there is complete information on the thermodynamics of this reaction [34] which, so far, has not been commercially exploited.

8. Processing of Cemented Carbide Scrap (35,41,42)

It is possible to feed cemented carbide scrap to the chemical processing described in paragraph 4, and this is always done when the scrap is contaminated (Fig. 5). However there are also vast amounts of "pure" scrap available, since in many applications only a small portion of a tool or a wear part is actually used. Such scrap can be recycled by less costly processes. Before any processing the scrap has to be cleared and sorted [43].

In the "COLDSTREAM" process the crushed scrap (passing through mesh 6 = 3.4 mm) is fed into a blast chamber and drawn by vacuum into a primary classifier. The full charge thus collected is pressurized and metered into a high pressure dried air system and accelerated through a venturi toward the blast chamber. The fragments are impinged

Figure 5 Scanning electron microscope photograph of tungsten (top) and tungsten carbide (bottom) powders.

Table 5 Properties of tungsten carbide [5]

α - WC

Carbon content :	6.13 wt. %
Crystal structure :	Hexagonal - Space group : $P\overline{6}m2$ (D_{3h})
Cell parameters : :	a = 2.906 Å, c = 2.837 Å; c/a = 0.976
Density :	15.67 g/cm³
Peritectic decomposition temperature :	2776 ± 4°C
Coefficient of thermal expansion :	$\alpha_a = 6.7 \times 10^{-6}\ K^{-1}$
	$\alpha_c = 4.9 \times 10^{-6}\ K^{-1}$
	$\alpha = \frac{2\alpha_a + \alpha_c}{3} = 6.1 \times 10^{-6}\ K^{-1}$
Thermal conductivity :	λ = 0.290 cal/cm.s.K
Free enthalpy of formation :	ΔG°_{298} = - 39.6 kj/mole
Electrical resistivity :	19.2 μohm.cm
Single crystal elastic constants :	C_{11} = 720 GPa
	C_{33} = 972 GPa
	C_{44} = 328 GPa
	C_{12} = 254 GPa
	C_{13} = 267 GPa
Young's modulus	E = 706 GPa
Bulk modulus	K = 443 GPa
Poisson's ratio	ν = 0.23

Microhardness (VICKERS) (kg/mm²)

	Charge (g)			
Plane	20	50	84	100
(0001)	2500	2165	1970	1950
$(00\overline{1}0)$	1450	1420	1350	1360

(at a speed of mach 2 against a fixed target of cemented carbide. The air expands as it leaves the venturi, creating an adiabatic cooling which lowers the possible oxidation of the particles. After blasting the powder is transported to a primary classifier, a secondary classifier or a fines collector, depending on the size. Oversize particles are returned to the blast chamber and the process is repeated.

The most common method for recycling cemented carbide scrap is the "zinc process" [99]. The scrap is first cleaned with particular attention to remnants of brazing alloys or coatings. Thereafter the scrap is immersed in liquid zinc which reacts with the cobalt binder phase, forming an intermetallic compound with a concomitant large increase in volume, thus resulting in a disintegration of the scrap to powder. After distillation of the zinc (at 900°C) the powder is ready to use again. At present some 30% of all cemented carbide is recycled by this route.

A third method is to make the scrap more brittle by a high temperature treatment (> 1800°C) and then make a cemented carbide powder by mechanical comminution [98].

The advantage with all three methods is that the costly chemical processing can be circumvented. The other side of the coin is that there is no purification and rather some risks of contamination. The composition and granulometry of the powder obtained by these methods may also not correspond to what is required in commercial production. Some loss of carbon is usually unavoidable and requires additions of carbon to the reclaimed powder. There is also likely to be accumulation of some impurities such as iron and aluminium. To some extent this can be compensated for by adding virgin carbide and cobalt powders of high purity but, sooner or later, scrap has to be recycled by the chemical route in order to decrease the contents of impurities.

It is more difficult to adjust for granulometry although a desired average grain size can be obtained by adding an appropriate grade of carbide but this will cause a broader spread of grain size.

9. Characteristics of WC Powder

Some properties of WC are shown in Table 5. There is no detectable variation in composition of this compound, the C/W ratio always being very close to unity.

One important reason for the excellent properties (in particular the toughness) of composites containing WC is the plasticity of this compound. Several papers [47-50,52,100] describe the dislocations and slip systems of WC. Plasticity depends naturally on the grain size, subgrain structure and other defects of the grains, which in turn depend on the processing.

10. The Mixing, Milling and Granulation of Cemented Carbide Powder

For the production of cemented carbide the tungsten carbide powder is mixed with cobalt powder (with a grain size around one micron) and, if appropriate, with powders of other carbides and of re-cycled materials. A lubricant, usually paraffin wax or polyethylene-glycol (PEG) is added to facilitate the subsequent consolidation processes.

Mixing in order to homogenize larger quantities is made by conventional methods with modifications required by the abrasive nature of the powder.

Milling for obtaining an even distribution of the various components and the desired average grain size is usually made in ball mills rotating around a horizontal axis or attritor mills with a vertical axis. As milling media, cylpebs of cemented carbide with a total weight about 4.3 (ball mills) or 3.7 (attritor mills) times that of the powder to be milled are used.

In order to protect the powder from oxidation, milling takes place in ethanol (e.g. when using PEG as lubricant) or acetone (e.g. when using paraffin wax). In spite of this there is some oxygen pick-up, increasing with milling time [71]; this can be reduced by subsequent treatment in hydrogen. Oxygen pick-up is a more difficult problem in TiC-based materials ("cermets") due to the high affinity of titanium for oxygen.

During the first stage of milling de-agglomeration takes place. There follows a decrease in the grain size, with the specific surface as measured by the B.E.T. method increasing linearly with the logarithm of milling time [51,71,72]. As recently pointed out it may be more appropriate to relate the grain size to the milling energy [73,74].

In the cobalt binder phase there is a transformation from the cubic high-temperature modification to the hexagonal one with an increasing density of lattice defects. The distribution of cobalt becomes finer with increasing milling time. In grades with high binder contents there is a tendency to form cobalt islands also containing fine fractions of carbide grains. The presence of cobalt oxide Co_3O_4 adhering to the WC grains may play an important role during the first stage of sintering. Incomplete milling leaving some coarse-grained tungsten carbide may lead to discontinuous grain growth during sintering [72,90].

For the most common cemented carbide grades with cobalt contents in the range 6-10 wt.% an even distribution of cobalt is easily obtained. On this point there appears to be more problems with nickel-rich binders [71], although corrosion-resistant grades, with chromium and nickel in the binder are now supplied by several producers. Grades with compositions similar to nickel-based superalloys have also been tested with satisfactory results.

With a grain size around one micron the cemented carbide powder has to be agglomerated in order to facilitate subsequent handling. Nowadays this usually takes place by spray-drying in a non-oxidizing atmosphere. By varying the process parameters the agglomerated size and size distribution, as well as the mechanical properties of the agglomerates, the different producers have arrived at solutions that fit their proprietary processing. In a recent publication [75] an apparatus for measuring the mechanical strength of powder agglomerates is described, with some results of the effects of reactions between lubricant and atmosphere on their mechanical strength. Such information will be of increasing importance in the future development of agglomerated or ready-to-press (RTP) powders, which are now a common commercial product (Fig. 7).

With the agglomeration procedures now developed it is possible to produce a large series of cemented carbide products with very small variations in size; for a common 12.5 mm insert the variation in linear dimensions of a sintered blank is of the order of 10 microns.

The processes of mixing, milling and agglomeration will be considerably modified when starting from compound powders already containing the ingredients intimately mixed. As yet such powders have not found significant commercial applications.

11. Consolidation

For the most common cemented carbide products, such as inserts for metal cutting or buttons for rock-tools, made in large numbers, die-pressing is the most appropriate consolidation method.

The agglomerated powder with a very reproducible flowability is fed into an automatic press equipped with a robot for handling the pressed green bodies. The high precision tools are made of cemented carbide. A fairly low pressing pressure in the range 100-200 MPa is used, giving a density of the as-pressed body around 50% of the full theoretical density.

For optimum performance modern products have a complicated geometry (e.g. for good chip breaking) and this, together with the friction between powder and tool, gives rather large (in the range of 10%) density variation in the as-pressed body. An example of the density distribution in a body only pressed from one side is shown in Fig. 8. There is intensive research in the cemented carbide industry to develop powder properties, tool geometries and pressing cycles so as to achieve density variations below the critical limit at which pressing cracks or porosity during sintering leads to rejection or inferior properties of the final product.

Cold isostatic pressing, extrusion (e.g. for rods and tubes) and injection moulding [76,77] are also consolidation techniques used in the cemented carbide industry.

12. Sintering of Cemented Carbides

The pressed bodies are sintered in furnaces in which temperature (20 - 1600°C), pressure (0 to one atmosphere or more) and chemical environment can be changed. In the early production of cemented carbide, de-lubrication at temperatures up to 350-400°C was a separate process, often followed by machining of the presintered bodies to a more exact shape. With the much improved ready-to-press powder now available this process step has been suppressed and de-lubrication and sintering take place in the same furnace.

De-lubrication is carried out at low pressure and with a temperature- time curve that ensures a slow evaporation of the lubricant so as to avoid blister in the as-pressed body. Temperature is then raised to the sintering temperature at which some impurities disappear as volatile compounds.

As seen from the isothermic section of the W-Co-C phase diagram (Fig. 9) there is a fairly broad Co(W,C)-WC two-phase region. The carbon balance has to be arranged so that the composition does not fall outside this region with concommitant formation of eta-carbide or graphite.

There is a great advantage in the production of cemented carbide that graphite can be used as a heating element, in the thermal insulation and as a support. The latter is coated by some refractory compounds in order to reduce the risk from carburization during sintering.

As seen from the phase diagram the liquid cobalt phase dissolves fairly large amounts of tungsten and carbon and, due to this, the liquid binder phase occupies a larger volume than the binder in the solidified material. There is a considerable solubility of WC also below the solidus temperature. During a normal sintering process tungsten carbide precipitates on the undissolved WC grains. There will be a tungsten gradient in the solid binder phase and the composition at the centre of this corresponds to the equilibrium at around 1000°C, the diffusion of tungsten being too slow below this temperature in relation to the thickness of the binder phase for any further tungsten carbide to precipitate [102].

There has been intensive development of the sintering processes of cemented carbide during the last few years and these have mainly concerned three topics: sintering at elevated pressure (sinter-HIP), control of inclusions during sintering, and compositional gradients in the sintered products.

The advantage of adding a period with enhanced pressure (up to 6 MPa) towards the end of the sintering cycle is the suppression of porosity [78,79]. As demonstrated in many recent publications this results in an improvement in the transverse rupture strength and, more important, less scatter in this property. This is particularly so for the low cobalt, difficult-to-sinter grades and, for products of such grades, this technique (more economic than the older one implying a

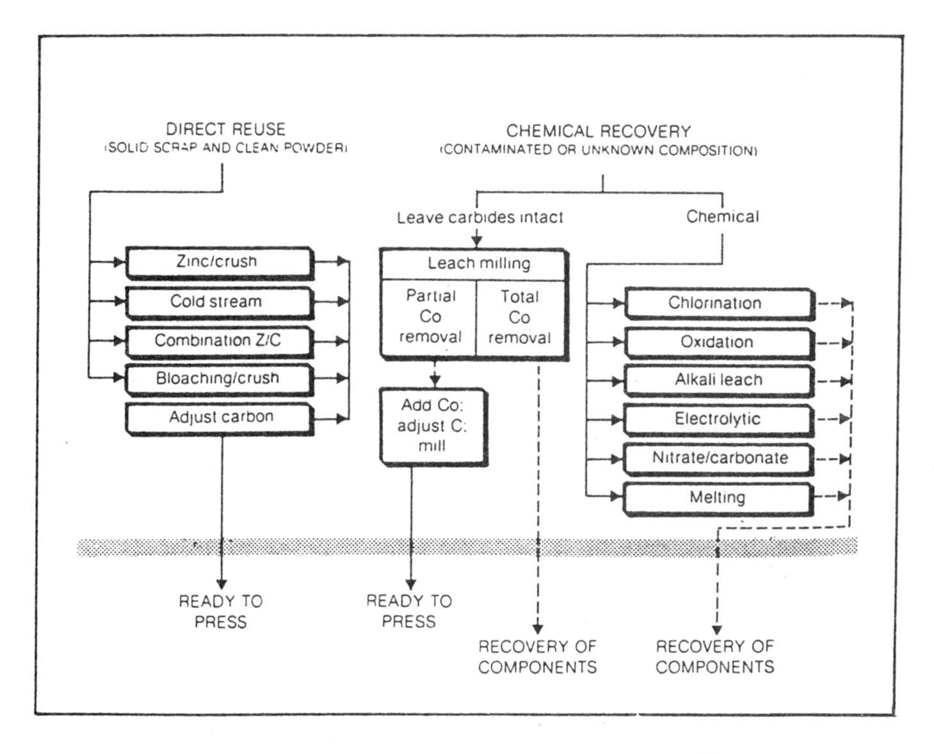

Figure 6 Treatment of cemented carbide scrap [42].

Figure 7 Agglomerated cemented carbide (ready-to-press) powder.

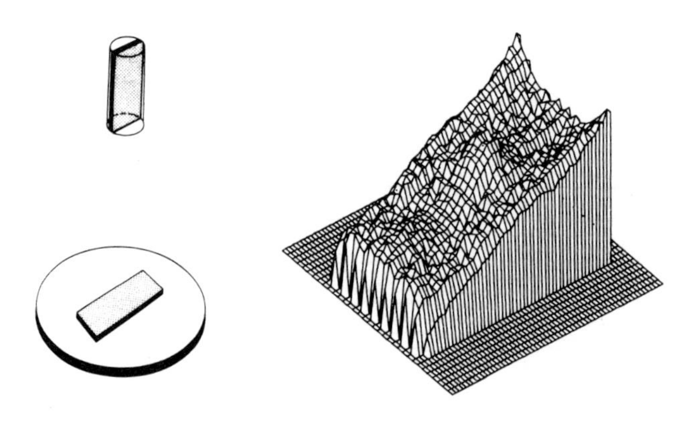

Figure 8 Density distribution in die-pressed body (pressing only from one side).

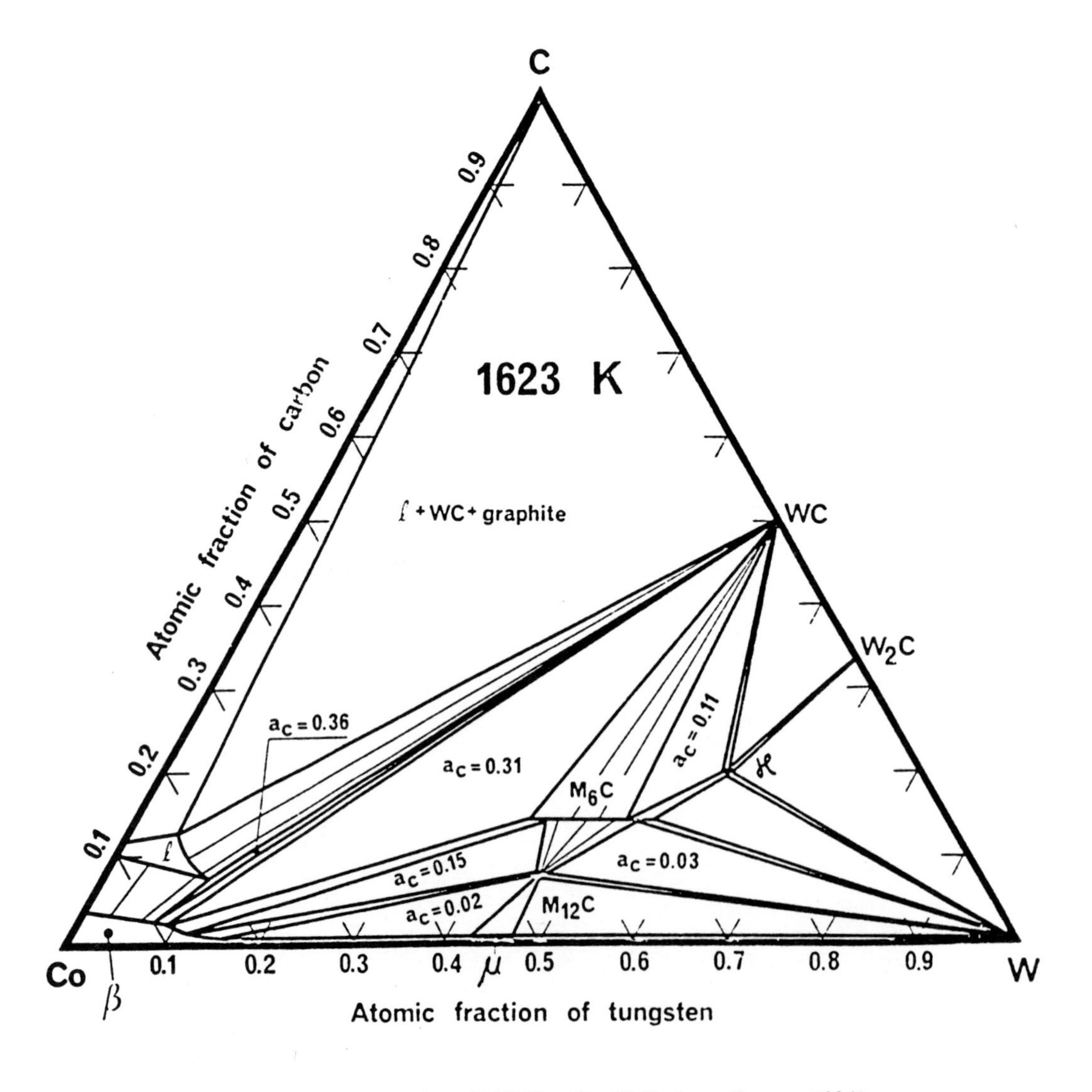

Figure 9 Isothermal section (1350°C) of Co-W-C phase diagram [104].

post-HIPing treatment [80]) is certainly justified, as for products with complicated geometries often leading to porous regions. For more common grades, refinement of the powder preparation and the pressing operations can give the same result.

A better control of inclusion contents [81] (silicates, sulphides, etc.) has the same purpose since inclusions will have a very similar effect on properties, in particular transverse rupture strength and related properties (see Table 1), as porosity.

As a general comment on this development it should be said that the important goal is to suppress the population of three dimensional defects below a critical level at which a further decrease of them does not lead to better properties. This is illustrated in Fig. 10, in which different porosity populations in a WC-10% Co grade are shown. When there are more defects than corresponding to the dotted line in the figure, there is a marked deterioration in transverse rupture strength, whereas variations within the region below this line have no effect on this property [82,83]. The line dividing the two regions depends on the cobalt content. In high cobalt grades extrinsic defects normally have little effect while the reverse is true in low cobalt grades. This could partly explain the variation of transverse rupture strength with cobalt content [84].

The third important aspect of sintering concerns gradients in chemical composition, microstructure, etc. Such gradients can be obtained by treatment in appropriate atmospheres and temperatures towards the end of the sintering cycle [85,86]. An example is the dual properties cemented carbide for rock drilling applications [82]. In this there is a hard cobalt-poor outer region, greatly enhancing the wear resistance, while a more cobalt-rich interior region gives the product an adequate toughness (Fig. 11). We are likely to see a rapid development in this area in particular as furnaces with better possibilities to control the chemical composition of the sintering atmosphere become available.

13. Some Notes on Finishing and Coating

Many cemented carbide products are subjected to grinding and polishing operations before being used. Although often representing high added value there is little information on these processes. Here we only want to remark in passing that the performance properties may depend very strongly on the finishing, for instance through the residual stress field resulting from grinding, polishing, edge-honing, etc. It is also noteworthy that virtually no studies have been published on the grindability of cemented carbide.

The coating of cemented carbide with hard (a few microns thick) layers of titanium carbide, titanium nitride and alumina, often in combination, has been very successful and to-day more than 50% of the indexable inserts are coated, in a few cases also with other hard substances than those mentioned. In the area of coated cemented carbide there is also a rapid development of substrates with gradients (obtained during the sintering process) which give improvements of the properties of the final products [87].

14. Some Notes on Health and Safety

When handled correctly there is little hazard associated with cemented carbide powder. Earlier negligence in this respect has led to a few cases of so-called hard metal disease. Several such cases have been well documented in recent publications [10,88,89] and it has been shown that the damage on the respiratory organs is mainly due to inhalation of cobalt powder. It is of particular importance to take due precautions when grinding cemented carbide.

Some persons may suffer from skin irritation when they come in contact with cemented carbide powder and such over-sensitivity should be tested before engagement in certain parts of the production.

In all industrial countries there are authorities and organizations that can give advice on how to take due precautions in the handling of cemented carbide powder.

References

1. P. Schwarzkopf, R. Kieffer, W. Leszynski and F. Benesovsky, Cemented Carbides, The Macmillan Co., New York, p.349, 1960.
2. R. Kieffer and F. Benesovsky, Hartmetalle, Springer Verlag, Wien, p.541, 1965.
3. G.S. Kreimer, Prochnost'tverdykh splavov, Izdat. Metallurgiya, Moskva, (1966) American translation: Strength of hard alloys, Consultants Bureau, Div. of Plenum Publishing Corp., New York, p.166, 1968.
4. V.I. Tumanov, Svoistva splavov sistemy karbid vol'frama karbid titana - karbid tantala - karbid niobiya - kobal't, Izdat. Metallurgiya, Moskva, p.184, 1973.
5. G.V. Samsonov, V.K. Vitryanyuk and F.I. Chaplygin, Karbidy vol'frama, Izdat. Naukova Dumska; Kiev, p.176, 1974.
6. I.N. Chaporova and K.S. Chernyavskii, Struktura spechennykh tverdykh splavov, Izdat. Metallurgiya, Moskva, p.248, 1975.
7. V.I. Tretyakov, Osnovy metallovedeniya i tekhnologii proizvodstva spechennykh tverdykh splavov, Izdat. Metallurgiya, Moskva, p.528, 1976.
8. Recent advances in hard metal production. 17-19 September 1979, Loughborough University of Technology, Loughborough, UK and Metal Powder Report, no pagination, 1979.

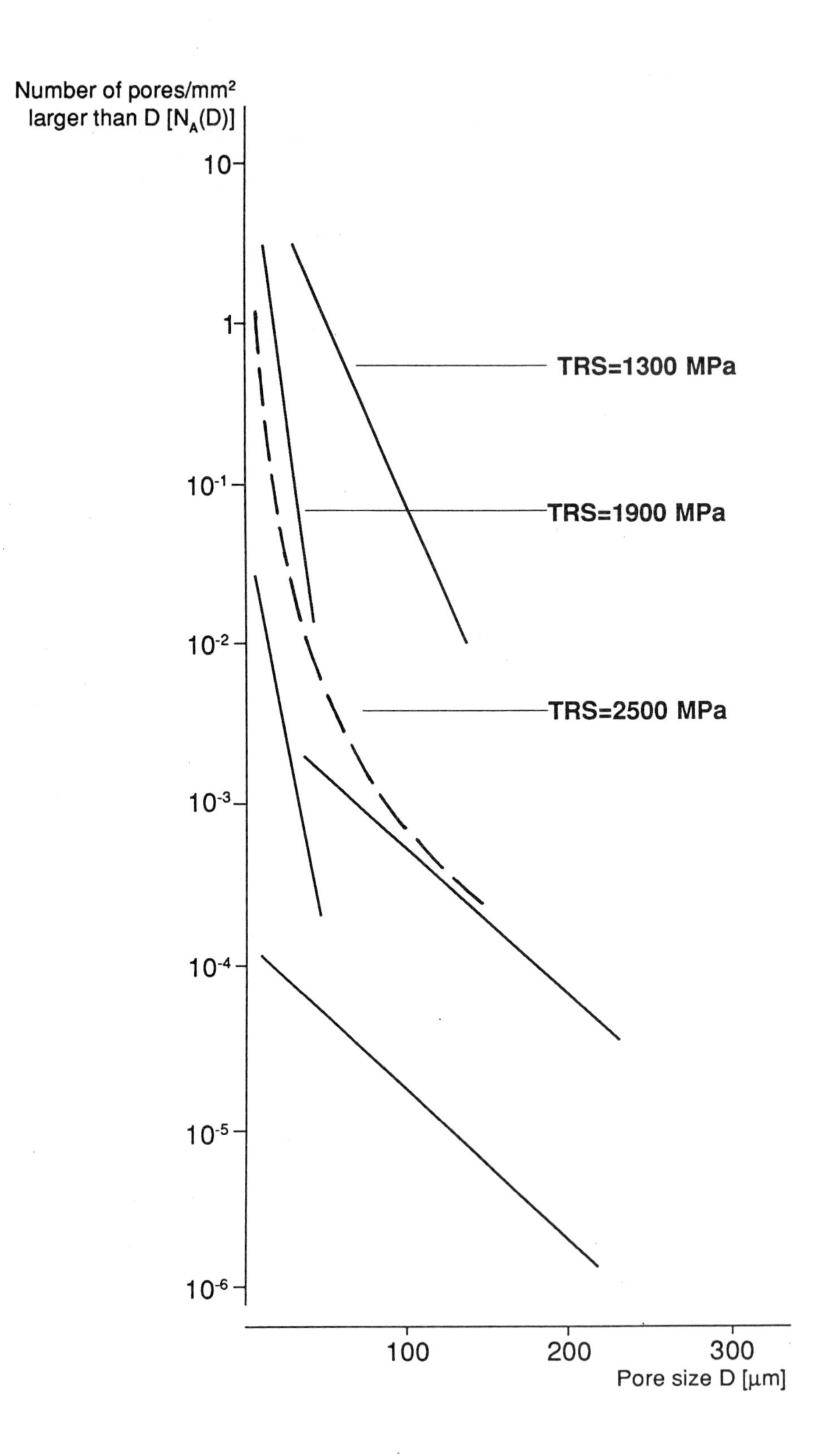

Figure 10 Cumulative size distributions for pores in different WC-Co cemented carbides.

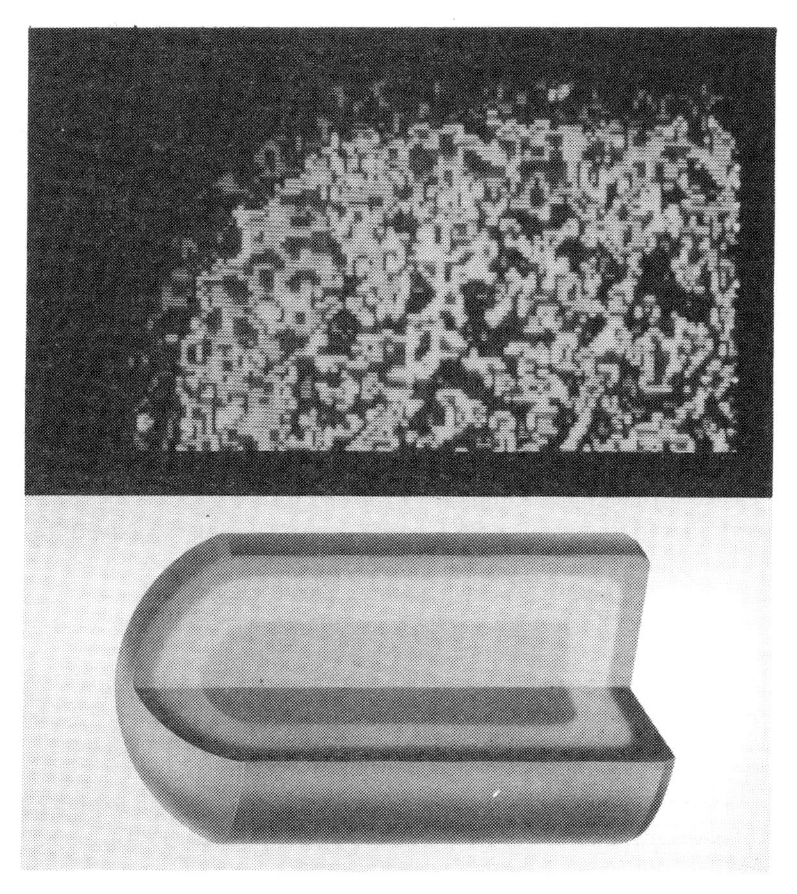

Figure 11 Button of dual properties carbide with a hard cobalt-poor outer zone.

9. Advances in hard metal production, Conference held in Luzern, CH, 7-9 November 1983 (in 2 volumes) Metal Powder Report Publishing Services Ltd, Bellstone, Shrewsbury, UK, no pagination, 1983.
10. Advances in hard metal production, Proceedings of a M.P.R. Conference held in London, UK, 11-13 April 1988. Metal Powder Report Publishing Services Ltd, Bellstone, Shrewsbury, UK, no pagination, 1988.
11. R.K. Viswanadham, D.J. Rowcliffe and J. Gurland, Science of hard materials, Proceedings of the 1st International conference on the science of hard materials held in MORAN, WY, USA, August 23-28, 1981, Plenum Press, New York, p.1011, 1983.
12. E.A. Almond, C.A. Brookes and R. Warren, Science of hard materials, Proceedings of the 2nd International conference on the science of hard materials held in RHODES, Greece, September 23-28, 1984, Institute of Physics Conference Series, Adam Hilger Ltd., Bristol, no.75, p.1091, 1986.
13. V.K. Sarin, Science of Hard Materials 3, Third International Conference on the Science of Hard Materials, Nassau, The Bahamas, November 9-13, 1987, Elsevier Applied Science, London, p.584, 1988; Materials Science & Engng. vol.A, pp.105-106.
14. H.M. Ortner, Proceedings of the 10th Plansee-Seminar 1981: Trends in refractory metals, hard metals and special materials and their technology, Metallwerk Plansee GmbH, Reutte, Tirol, Austria, 2 volumes: pp.732,1004, 1981.
15. H. Bildstein and H.M. Ortner, Proceedings of the 11th international Plansee-Seminar'85: New applications, recycling and technology of refractory metals and hard materials, Metallwerk Plansee GmbH, Reutte, Tirol, Austria, 3 volumes: pp.442, 972, 1039, 1985.
16. F. Kriener and H.M. Ortner, Proceedings of the 12th International Plansee-Seminar'89: to be published.
17. A.N. Zelikman and L.S. Nikitina, Vol'fram, (Tungsten), Izdat. Metallurgiya, Moskva, p.272, 1978.
18. F. Bensovsky, K. Swars, Wolfram, Erganzungsband Al Metall, Technologie, Gmelin Handbuch der Anorganischen Chemie, System Nr. 54 Springer Verlag, Berlin, p.241, 1979.
19. S.W.H. Yih and C.T. Wang, Tungsten Sources, Metallurgy, Properties and Applications, Plenum Press, New York, p.500, 1979.
20. Tungsten, Proceedings of the First International Tungsten Symposium. Stockholm, 1979, Mining Journal Books Ltd., London, and Primary Tungsten Association and Consumer Reporting Group, p.181, 1979.

21. Tungsten 1982, Proceedings of the second International Tungsten Symposium, San Francisco, June 1982, Mining Journal Books Ltd., London, and Primary Tungsten Association and Consumer Reporting Group, p.179, 1982.
22. Tungsten 1985, Proceedings of the third International Tungsten Symposium. Madrid, May 1985, MPR Publishing Services Ltd. Bellstone, Shrewsbury, UK and Primary Tungsten Association and Consumer Reporting Group, p.255, 1985.
23. Tungsten 1987, Proceedings of the fourth International Tungsten Symposium. Vancouver, September 1987, MPR Publishing Services Ltd. Bellstone, Shrewsbury, UK and Primary Tungsten Association and Consumer Reporting Group, p.190, 1987.
24. W-Ti-RE-Sb in '88, Proceedings of the first International Conference on the metallurgy and materials science of tungsten, titanium, rare earths and antimony, Changsha, China, 1988. International Academic Publishers (distribution Pergamon Press), 2 volumes, p.1353, 1988.
25. H. Pastor, Metallurgie du tungstene, (fascicule M 2378) Techniques de l'Ingénieur, Paris, p.20, 1988.
26. K. Vadasdi, R. Olah, I. Szylassy and A. Jeszensky, Preparation of APT, by means of electrodialysis and solvent extraction, in Ref. 15, 1, p.77-89.
27. A.I. Vasil'eva, Ya. I. Gerasimov and Ya. P. Simanov, Equilibrium constant for the reaction of reduction of tungsten oxides by hydrogen, (in Russian) Zhur. Fiz. Khim.,31, 3 , pp.682-691, 1957.
28. J. Qvick, Use of oxygen probes to study and control the atmosphere during reducing of tungsten oxide, Intern. J. Refract. & Hard Metals, 7, 4, pp.201-205, 1988.
29. D.S. Parsons, The reduction of tungsten oxides by hydrogen, Electrochem. Technol., 3, 9/10, pp.280-283, 1965.
30. A. Hara and M. Miyake, Studies on the formation process of tungsten carbide powder from tungsten powder, Planseeber. Pulvermetall., 18, pp.90-110, 1970.
31. Z. Junxi, Effect of reduction and carburization temperatures of tungsten powder on WC phase substructure and mechanical properties of WC-Co cemented carbide, Intern. J. Refract. Hard Metals, 7, 4, pp.224-228, 1988.
32. G. Kazonich, A.E. Raddatz and T.G. Carnahan, Investigation of the important parameters in a gas sparging technique for production of tungsten carbide, Metal Powder Report, 12, pp.800-810, 1988.
33. R. Kieffer and H. Rassaerts, The use of an auxiliary metal bath for the production of high purity carbide powders of the IV A - VI A Group of elements, Intern. J. Powder Metall., 2, 2, pp.15-22, 1966.
34. T.D. Halliday, F.H. Hayes and F.R. Sale, Thermodynamic considerations of the production of Co/WC mixture by direct gas phase reduction/carburization reactions, in Industrial Use of Thermochemical data, Proc. Conf. Chem. Soc. 1979; Spec. Publ. Chem. Soc. no.34, pp.291-300, 1980.
35. B. Aronsson and H. Pastor, Processes for recycling cemented carbides and some aspects of the influence of recycled materials on finished products, in Ref. 24, pp.650-656.
36. B. Aronsson, Influence of processing on properties of cemented carbide, Powder Met., 30, 3, pp.175-181, 1987.
37. B.F. Kieffer, Recycling systems with particular reference to the zinc process, in Ref. 21, pp.103-114.
38. S.V. Naik and S. Basu, New salt bath technique for recovery and extraction of tungsten from scrap and other secondary sources, in Ref. 10, Paper no. 6, p.7.
39. P. Borchers, Processing of tungsten, in Ref. [20], pp.64-77.
40. F.C. Nava Alonso, M.L. Zambrano Morales, A. Uribe Salas and J.E. Bedolla Becerril, Tungsten trioxide reduction carburization with $CO-CO_2$ mixtures: kinetics and thermodynamics, Intern. J. Mineral Processing, 20, pp.137-151, 1987.
41. B. Aronsson and H. Pastor, Processes for recycling cemented carbides and some aspects of the influence of recycled materials on finished products, in International Conference on sintering of multiphase metal and ceramic systems, New-Delhi, January 31- February 3, 1989. To be published in 1989.
42. B.F. Kieffer and E. Lassner, Reclamation of tungsten containing scrap material: economic and technical aspects, in Ref. 23: pp.59-67.
43. F. Clin and H. Pastor, Sur une technique de tri par densité de déchets de carbures cémentés en vue de leur valorisation et recyclage, Industrie Minérale, Mines et Carrières, Les Techniques, December 1987.
44. J. Li, A study of microstructures of tungsten carbide powder, Metall. Trans., 18 A, pp.753-758, 1987.
45. T. Zhengji, Production of fine quality tungsten carbide for cemented carbide, Intern. J. Refract. & Hard Metals, 7, 4, pp.215-218, 1988.
46. L.V. McCarty, R. Donelson and R.F. Hehemann, A diffusion model for tungsten powder carburization, Metall. Trans., 18A, pp.969-974, 1987.
47. R.M. Greenwood, M.H. Loretto and R.E. Smallman, The defect structure of tungsten carbide in deformed tungsten carbide-cobalt composites, Acta Metall., 30, pp.1193-1196, 1982.
48. J. Vicens, S. Hagege, F. Osterstock, G. Nouet and J.L. Chermant, Analyse des défauts dans la phase carbure des composites WC-Co déformés à température ambiante, Rev. Int. Hautes Temp. Refract., 19, pp.359-371, 1982.
49. S. Lay, P. Delavignette and J. Vicens, Analysis of subgrain boundaries in tungsten carbide deformed at high temperature, Phys. Status Solidi, A90, pp.53-68, 1985.

50. S. Lay, G. Nouet and J. Vicens, Mécanismes de déformation des carbures réfractaires : application aux composites carbure de tungstène-cobalt, J. Microsc. Spectros. Electron., 11, pp.179-194, 1986.
51. U. Seeker and H.E. Exner, Eigenschaften unterschiedlich hergestelter wolframkarbidpulver und ihr Verhalten bei der Hartmetallherstellung, Metall, 40, 12, pp.1238-1247, 1986.
52. S.A. Horton, M.B. Waldron, B. Roebuck and E.A. Almond, Characterization of powders and dislocation structures in processing of WC-Co, (W,Ti)C-Co and TiC-Co hardmetals, Powder Metallurgy, 27, 4, pp.201-211, 1984.
53. R. Haubner, W.D. Schubert, H. Hellmer, B. Lux and E. Lassner, Die Wolframreduktion, in Ref. 15, vol.2, pp.161-179.
54. R. Haubner, W.D. Schubert, E. Lassner and B. Lux, Einfluss von Alkalidotierungen auf die Reduktion von WO_3 zu W mit wasserstoff, in Ref.15 , vol.2, pp.69-97.
55. R. Haubner, W.D. Schubert, E. Lassner and B. Lux, Influence of small amounts of Fe and Ni during the production of tungsten powder, in Horizons in PM, Proc. 1986 Intern. PM Conference held in Dusseldorf, Part I, pp.147-150, July 7-11, 1986.
56. J. Qvick, Influence of sodium on the reduction of WO_3, Reactivity of solids, 4, pp.73-91, 1987.
57. R. Haubner, W.D. Schubert, E. Lassner and B. Lux, Einfluss von Al, Si, P und As auf die Stabilisierung APW-pseudomorpher Teilchen während der wolframreduktion, Intern. J. Refract. & Hard Metals, 6, 1, pp.40-45, 1987.
58. R. Haubner, W.D. Schubert, E. Lassner and B. Lux, Uber den Einfluss von P und Si auf die Reduktion von WO_3 zu W mit wasserstoff, Intern. J. Refract. & Hard Metals, 6, 2, pp.111-116, 1987.
59. R. Haubner, W.D. Schubert, E. Lassner and B. Lux, Einfluss von Aluminium auf die Reducktion von wolframoxid zu Wolfram, Intern. J. Refract. & Hard Metals, 6, 3, pp.161-167, 1987.
60. R. Haubner, W.D. Schubert, E. Lassner and B. Lux, Einfluss von Eisen und Nickel auf die Reduktion von Wolframoxid zu Wolfram, Intern. J. Refract. & Hard Metals, 7, 1, pp.47-55, 1988.
61. J. Qvick, Thermodynamical calculations on the reduction of W (VI) oxide with special emphasis on trace element behaviour, Intern. J. Refract. & Hard Metals, 3, 3, pp.121-131, 1984.
62. J. Qvick, Trace element evaporation during the reduction of W (VI) oxide - a comparative study between thermochemical calculations and experiments, in Ref. 12, pp.1047-1062.
63. H. Hellmer, W.D. Schubert, E. Lassner and B. Lux, Kinetik der Wolframoxidreduktion, in Ref. 15, vol.3, pp.43-86.
64. R. Haubner, W.D. Schubert, H. Hellmer, E. Lassner and B. Lux, Mechanism of technical reduction of tungsten, Hydrogen reduction of tungsten blue oxide to tungsten powder, Intern. J. Refractory & Hard Metals, 2, 4, pp.156-163, 1983.
65. T.Z. Ji, W.Y. Fang and L.J. Bo, Multidimensional regression equations for forecasting particle size of tungsten powder, Intern. J. Refractory & Hard Metals, 4, 4, pp.166-171, 1985.
66. T.Z. Ji, Production of submicron tungsten powder by hydrogen reduction of WO_3,in Ref. 22, pp.219-228.
67. S.Q. Zou, E. Wu, A. Tan and C.L. Qian, Formation of tungsten blue oxide and its hydrogen reduction, in Ref. 15, vol.1, pp.337-348.
68. S.J. Li and H.Y. Lai, The control of phases of tungsten blue oxide and their effect on the particle size of tungsten powder, in Horizons in PM, Proc. 1986 Intern. PM Conference, Dusseldorf, July 7-11, 1986, Part 1, pp.155-158 and: Intern. J. Refract. & Hard Metals, 6, 1, pp.35-39, 1987.
69. Z. Tao, Production of submicron tungsten powder by hydrogen reduction of WO_3, Intern. J. Refract. & Hard Metals, 5, 2, pp.108-112, 1986.
70. Z.Q. Zou, C.L. Qian, E. Wu and Y. Chang, H -reduction dynamics of different forms of tungsten oxide, Intern. J. Refract. & Hard Metals, 7, 1, pp.57-60, 1988.
71. A. Gabriel, Remplacement du cobalt dans les carbures cémentés, Thèse de Docteur-Ingénieur de l'Institut National Polytechnique de Grenoble, p.166, 27 Avril 1984.
72. U. Seeker and H.E. Exner, Exaggerated grain growth in cemented carbides due to inhomogeneous milling, J. Am. Ceram. Soc., 70, 2, C31 - C32, 1987.
73. R. Goodson, L. Sheeran and F. Larson, Correlating WC grain size analysis techniques with attritor mill monitoring in cemented carbides, pp.165-191 in E.N. Aqua and C.I. Whitman, Modern Developments in Powder Metallurgy, Volume 17, Special Materials. Proc., 1984 Intern. PM Conf. MPIF-APMI, June 17-22, 1984, Toronto, Canada MPIF - APMI, Princeton, NJ, p.921, 1985.
74. S.J. Mashl, D.W. Smith, G.H. Becking and T.E. Hale, Attritor milling of WC + 6% Co: effects of powder characteristics and compaction behaviour, pp.233-262 in C.L. Freeby, Annual PM Conference Proc., MPIF-APMI, Dallas, TX, May 17-20, 1987.
75. U. Oscarson, P. Samuelson, M. Hehenberger, C.E. Drakenberg and B. Bolin, A method of determining the strength of powder agglomerates, in Ref. 15, 2, pp.769-780.
76. Dr Poniatowski and G. Will, Injection moulding of tungsten carbide base hard metals, Metal Powder Report, pp.812-815, December 1988.
77. M.T. Martyn, P.J. James and B. Haworth, Injection moulding of hardmetal components, Metal Powder Report, pp.816-823, December 1988.

78. R.C. Lueth, Moldless hot press for sintering and HIPing, Metal Powder Report, 39, 1, p.404, 1983.
79. Sinter-HIP-Technologie: Vorträge und Firmen - Präsentationen anlässlich des Symposiums am 2. und 3. Dez. 1987 in Hagen. Verlag Schmid GmbH, Freiburg in Breisgau, 1987.
80. S. Takatsu, Effects and problems of HIP on cemented carbides, Metal Powder Report, 42, 7/8, pp.500,502-503,505, 1987.
81. B. Uhrenius, L. Akesson and M. Mikus, The role of atmosphere and impurities during sintering of cemented carbides, High Temp.-High Pressures, 18, pp.337-346, 1986.
82. B. Aronsson, T. Hartzell and J. Akerman, Structure and performance of dual properties carbide for rock drilling and similar applications, in Ref. 10, Paper no.19, 6 pages.
83. A. Nordgren and A. Melander, Influence of porosity on strength of WC-10% Co cemented carbide, Powder Metallurgy, 31, 3, pp.189-200, 1988.
84. C. Chatfield, Comments on microstructure and the transverse rupture strength of cemented carbides, Intern. J. Refract. & Hard Metals, 4, 1, p.48, 1985.
85. M. Schwarzkopf, H.E. Exner, H.F. Fischmeister and W. Schintlmeister, Kinetics of compositional modification of (W,Ti)C-WC-Co alloy surfaces, in Ref. 13, pp.225-231.
86. K. Kobori, M. Ueki, Y. Taniguchi and H. Suzuki, The binder enriched layer formed near the surface of cemented carbide, J. Jap. Soc. Powder & Powder Met., 34, 3, pp.129-132, 1987 (in Japanese).
87. H. Pastor and V. Brozek, Les dépôts chimiques en phase vapeur, Sci. Papers Prague Inst. Chem. Technol., Inorg. Chem. & Technol., B 26, pp.237-274, 1981.
88. G.E. Spriggs (Chairman), Open Forum - Health and Safety in the Hardmetal industry, in Ref. 10, D43-D73.
89. M. Hartung, Lungenfibrosen bei HartmetallschleifernBedeutung der Cobalteinwirkung, Schriftenreihe des Hauptverbandes der gewerblichen Berufsgenossenschaften e.V., Bonn, p.62, 1986.
90. M. Schreiner, T. Schmitt, E. Lassner and B. Lux, On the origins of discontinuous grain growth during liquid phase sintering of WC-Co cemented carbides, Powder Metall. Intern., 16, 4, pp.180-183, 1984.
91. E. Lassner, Modern methods of APT processing, in Ref.21, pp.71-80.
92. M.B. MacInnis and T.K. Kim, Impact of solvent extraction on the tungsten industry, in Ref. 22, pp.132-143.
93. S. Aoki, S. Yamada and M. Tsuchiya, Production of tungsten metal from blue oxide, in Ref. 24, pp.156-161.
94. T. Xinhe and C. Rongjiang, New processes for the production of blue tungsten oxide and superfine tungsten powder, in Ref. 24, pp.162-167.
95. B. Lux, W.D. Schubert, R. Haubner and E. Lassner, Behaviour of impurities during the technical hydrogen reduction of tungsten oxides, in Ref. 24, pp.4-10.
96. S. Jianxin, W. Youming and H. Conqxun, Effect of small amounts of cobalt addition on carburization of coarse tungsten powder, in Ref. 24, pp.150-155.
97. J.M. Gomes, A.E. Raddatz and T.G. Carnahan, Preparation of tungsten carbide by gas sparging tungstate melts, in Ref. 22, pp.96-112.
98. T. Kobayashi, Reclaimed powder from cemented carbide scrap by high temperature processes, in Ref. 24, pp.645-649.
99. J.L. Dassel and D.R. de Halas, Applications for zinc-reclaimed powders in the cemented carbide industry, in Ref. 10, Paper no.7, p.15.
100. D.J. Rowcliffe, V. Jayaram, M.K. Hibbs and R. Sinclair, Compressive deformation and fracture of WC materials, in Ref.13, pp.299-303.
101. E. Rudy, St. Windisch and J.R. Hoffman, Ternary phase equilibria in transition metal-boron-carbon-silicon systems. Part I. Related binary systems. Volume VI: W-C system: supplemental information on the Mo-C system, Technical Report no. AFML-TR-65-2, Part I, vol.VI, p.79, January 1966.
102. M. Hellsing, High resolution microanalysis of binder phase in as sintered WC-Co cemented carbides, Mater. Sci. Techn., 4 , pp.824-829, 1988.
103. B. Uhrenius, Calculation of the Ti-C, W-C and Ti-W-C phase diagrams, Calphad, 8, 2, pp.101-119, 1984.
104. L. Akensson, An experimental and thermodymanic study of the Co-W-C system in the temperature range 1473-1698 K, Technical Doctor Thesis, Kunglika Tekniska Hogskola, Stockholm, p.38 + appendix and figures, 1982.
105. P. Samuelson and B. Bohn, Experimental studies of frictional behaviour of hard metal powders sliding on cemented carbide wall, Scand. J. Metallurgy, 12, pp.315-322, 1983.

23

Industrial Applications of the Refractory Metals

J.G. HEYES AND R.G.R. SELLORS
Plansee Metals Limited, Slough, UK

1. Introduction

The term refractory metal is widely used to describe the relatively small number of metallic elements which have extremely high melting points and therefore can be used at very high temperatures. For the purpose of this article a melting point of 2000°C or above is assumed in arriving at a somewhat arbitrary definition of refractory. Thus the metals molybdenum, tungsten, tantalum and niobium, all of which have technically interesting and commercially important applications, feature prominently in this discussion.

The field of use of these metals is vast and still growing. In terms of tonnage consumption their most significant usage is as melting additions in both the steel industry and in the manufacture of important non-ferrous alloys, including heat resisting superalloys for jet engines. However, the refractory metals in pure form concern us here, with applications as diverse as complex components for military avionics to humble tungsten dart bodies; from guided missile propulsion systems and warheads to specially developed non-toxic fishing weights made from tungsten loaded polymers.

Although it is often the heat resisting capabilities of these metals which leads to their selection, they have other interesting and useful properties which give rise to specific applications. For example the outstanding corrosion resistance of tantalum benefits the chemical industry and the high specific gravity of tungsten is taken advantage of in some military applications. A short review paper of this nature must however be restricted to a relatively small number of interesting examples. It is appropriate to begin with the earliest practical application of a refractory metal, namely the tungsten filament in an incandescent lamp.

2. The Lighting Industry

The principle of producing a light source from a thin strip of metal, glowing at white heat, was demonstrated by Sir Humphrey Davy in 1802. More than a hundred years were to pass, however, before the first ductile tungsten wires were produced, enabling the manufacture of the electric light bulb to become a commercial reality.

Refractory metals now find a multitude of applications in the manufacture of light sources. As well as being the standard filament material for incandescent lamps, tungsten is widely used for electrodes in various discharge lamps, either in pure form or with additions of thoria to reduce the electron work function. In high power flash lamps, specially produced porous tungsten cathodes, in which the porous skeletons are impregnated with a mixture of metallic oxides, are frequently used in preference to thoriated tungsten. Examples of such cathodes are shown in Fig. 1.

Molybdenum wire is employed to support the tungsten filament in an incandescent lamp, and interestingly, is sometimes also used in the production of the filament itself. Although many standard filaments are formed around retractable steel mandrels, this is not possible for some complex designs; instead the filaments must be wound around disposable metal mandrels. Molybdenum is the only metal which both remains stable during the subsequent high temperature annealing treatment, and can be removed easily by chemical dissolution.

Molybdenum foils are used to solve problems in forming the glass-to-metal seal in certain lamps. Normally the type of glass used can be specified according to its ability to form a hermetic seal with a suitable metal, which in turn requires compatibility of thermal expansion coefficient over a wide temperature range. Thus, matched seals between soda-lime and lead-alkali silicate glasses and 50% nickel-iron coated with copper (Dumet) wires are readily achieved.

In some lamp types, however, very high operating temperatures and severe thermal shock conditions dictate the use of fused silica. Certain tungsten-halogen lamps and the arc tubes of mercury vapour lamps are good examples. Fused silica has such a low expansion coefficient that no metal is able to provide a matched seal. Special techniques have been developed to permit the use of thin molybdenum foils (typically 0.025 mm) with feather edges, which are pinch-sealed into the vitreous material. In this form molybdenum foil is sufficiently ductile to prevent cracking of the seal by deforming under the influence of the resultant tensile stresses.

Dimming shields made from molybdenum sheet are used in halogen lamps for car headlamps, along with wires made from molybdenum doped with traces of potassium and silicon. (A typical "H4" lamp is illustrated in Fig. 2). Special manufacturing methods enable such doped wires to withstand extreme temperatures without becoming embrittled, a feature which enhances the life of the lamp.

In high pressure sodium lamps precision seamless tubes made from niobium and niobium - 1% zirconium serve the dual purpose of current lead and electrode holder. Niobium is an excellent material for this application because its good ductility facilitates easy tube drawing and manipulation, and it is highly resistant to sodium vapour.

3. High Temperature Processing and Special Melting Processes

It is an unfortunate fact that the refractory metals tend to be thermodynamically unstable in the presence of oxidizing atmospheres. Thus, although molybdenum can be alloyed with transition metal carbides to achieve outstanding levels of hot strength, or doped and thermomechanically processed to obtain excellent high temperature creep resistance, its oxidation resistance is poor above 400°C.

Many attempts have been made to develop suitable protective coatings, to allow refractory metals to be used under oxidizing conditions. The use of molybdenum as a construction material in extremely efficient gas turbines is one example of such a potential application. However, none of the protective coatings evaluated to date has been able to resist thermal cycling or mechanical damage and, once the coating is penetrated, the resulting oxidation is catastrophic. Therefore, the high temperature applications of the refractory metals are limited to non-oxidizing conditions.

3.1 Vacuum and Inert Atmosphere Furnace Technology

To meet the demands of present day manufacturing technology, the thermal processing of metals, ceramics and composite materials must be carried out under increasingly stringent conditions. Whereas processes such as sintering have almost always required controlled atmospheres, the benefits of vacuum or protective atmosphere processing are today becoming ever more widely recognized in annealing, hardening and brazing treatments. The materials required to build the internal parts of such furnaces must have some rather special properties. The ability to withstand extreme temperature is of course essential, as is high purity and low vapour pressure. The refractory metals are ideally suited to meet these demands and are employed in high temperature furnaces in the following ways:

- Resistance heating (elements and supports made from wire, rod, strip, plate or mesh).
- Thermal protection of the furnace walls from radiant heat (heat shields).
- Handling and positioning of the furnace charge (boats, trays, racks, etc.).
- Temperature measurements and control (thermocouples, pyrometers).

Figure 3 shows a set of heating elements made from molybdenum rod surrounded by a series of heat shields made from molybdenum sheet. Such heat shields are essential to protect the cold outer wall of the furnace from the intense heat generated during operation.

Molybdenum is suitable for continuous operation up to around 1825°C. An alloy known as TZM (0.5% titanium, 0.08% zirconium, 0.025% carbon, balance molybdenum) has superior hot strength and resistance to recrystallization as well as improved welding properties. It is therefore used for highly-loaded radiation shields, welded assemblies and furnace boats used at very high temperatures, such as those used for sintering uranium dioxide pellets in the nuclear industry (Fig. 4).

Special grades of doped molybdenum have achieved truly outstanding levels of high-temperature creep resistance because of their highly directional interlocking grain structure. Furnace parts manufactured from such materials combine remarkable high temperature dimensional stability with good ambient temperature ductility and toughness, even after they have been subjected to many furnace cycles. Initial trials in nuclear fuel manufacturing have given some excellent results with these materials, and it is likely that they will be used increasingly in processes requiring temperatures above 1400°C.

Tungsten has the highest melting point and lowest vapour pressure of all metals. It can therefore be used to manufacture heating elements and heat shields for furnaces operating at up to 3000°C and under ultra-high vacuum. Whereas both molybdenum and tungsten can also be used with protective atmospheres of nitrogen or hydrogen, tantalum and niobium absorb large quantities of these gases and become extremely brittle. Their usage is therefore restricted to furnaces operating under vacuum or with inert gas atmosphere protection. Tantalum also has excellent getter properties and is thus a useful material for furnaces where ultra-high purity is an essential requirement.

3.2 Hot Isostatic Pressing

Many benefits can be obtained by subjecting sintered or cast components to a combination of pressures of up to 207 MPa and accurately controlled temperatures of up to 2000°C. This novel process, hot isostatic pressing, is carried out in large steel pressure vessels, into which sealed stainless steel cans containing the workpieces are loaded. After initial evacuation the operating pressure is developed by compressing pure argon gas to a point where its density approaches that of water. As in cold-wall vacuum furnaces, refractory metals are ideal materials for resistance heating elements and heat shields. In large presses several separate sets of heating elements may be used to facilitate accurate temperature control and, as illustrated in Figure 5, extremely large TZM heat shields are needed. TZM is also used to produce shafts and rotors for circulating fans, which are frequently employed in hot isostatic presses to boost the flow of quenching gases.

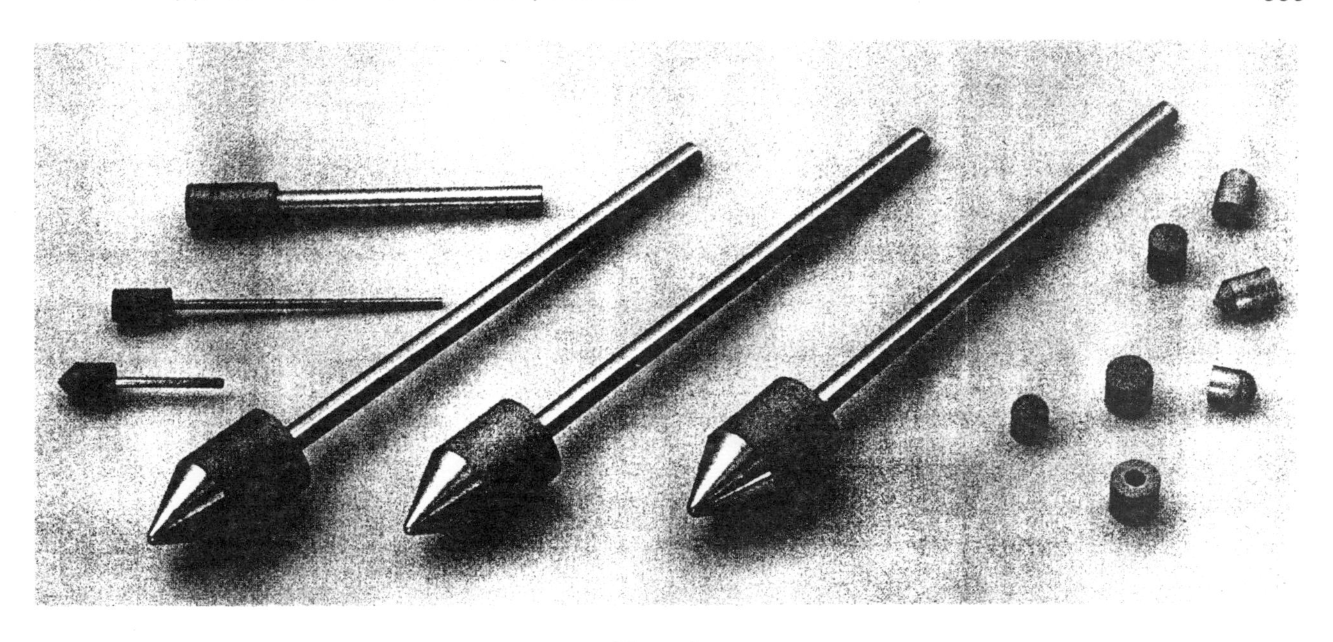

Figure 1

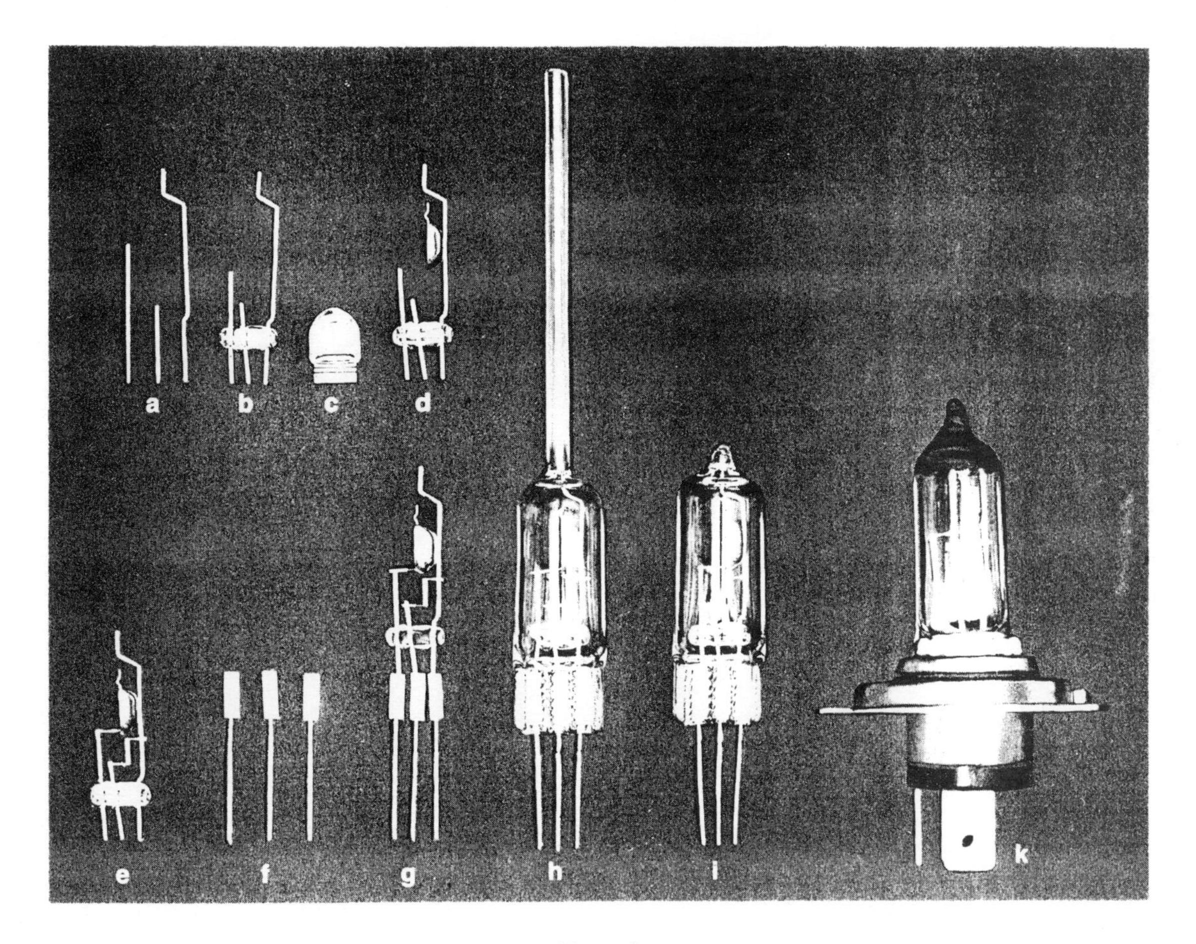

Figure 2

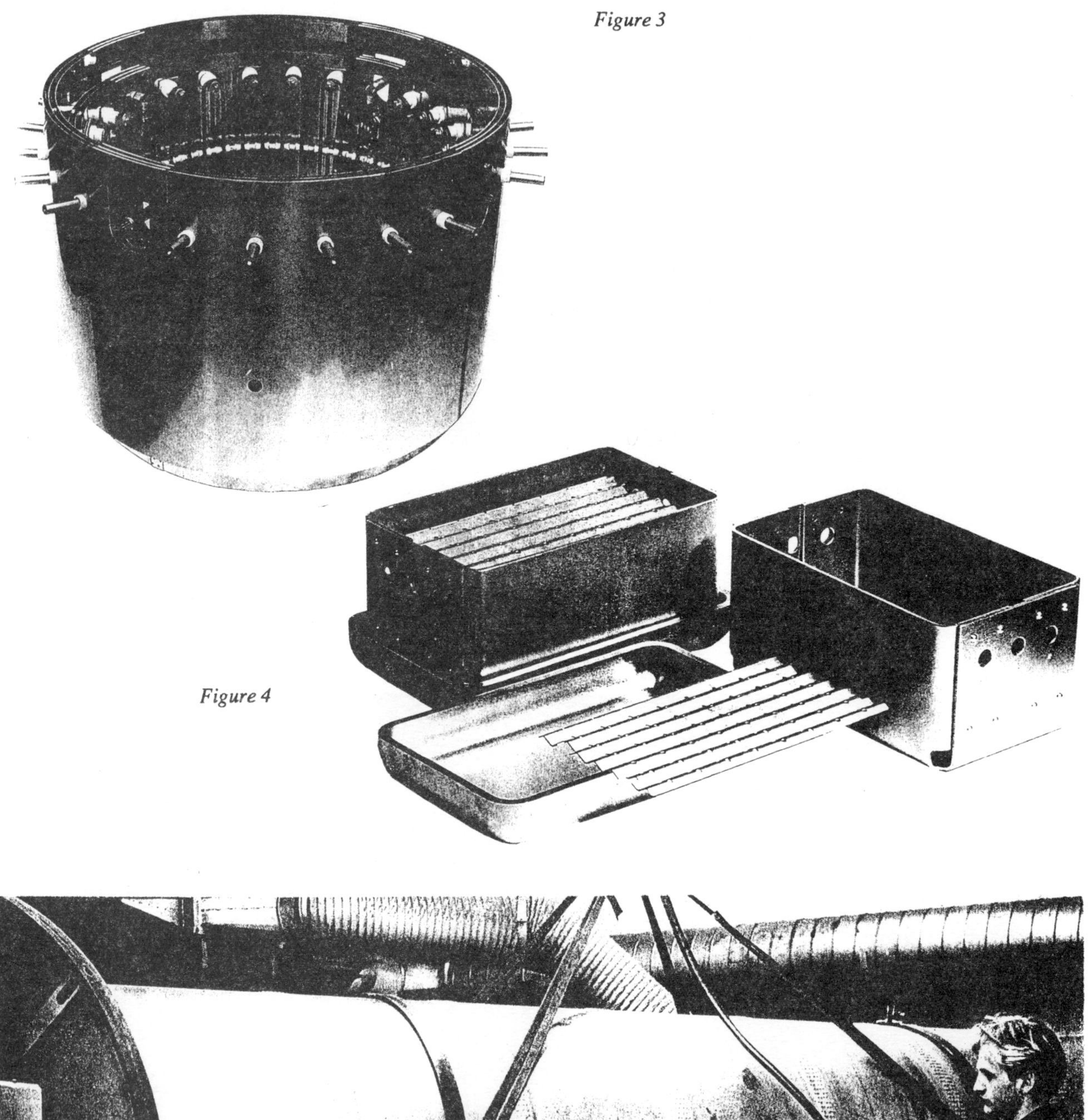

Figure 3

Figure 4

Figure 5

Because of its ability to eliminate voids and porosity, one of the earliest uses of hot isostatic pressing was the densification of sintered tungsten carbide products. It is also used to enhance the mechanical properties of cast components to a point where they approach those of wrought products. Interesting recent applications include diffusion bonding, the consolidation of surface coatings and "Powder HIPing" to produce parts in alloys which are unsuitable for casting.

3.3 High Temperature Mechanical Working

The excellent high-temperature mechanical properties of molybdenum, tungsten and their alloys suggest that these materials should find numerous applications in a variety of hot-working processes. For example TZM inserts are employed in swaging hammers, used in the working of other refractory metals, and liquid phase sintered tungsten alloys are used in diecasting tooling. However, economic considerations restrict the application of these materials to processes where their relatively high initial cost can be justified by increased tool life, (e.g. the use of TZM dies for brass extrusion when long production runs are envisaged) or where simply no other material will meet the requirements.

Hot isothermal forging is an example of a process which depends on the availability of the molybdenum alloy TZM as a die material. It is used to forge titanium and nickel-based superalloys, which are very difficult to work by conventional methods. Compressor discs and rotors for aeroengines are examples of critical components which benefit from this manufacturing route. The alloys require working temperatures of around 1130°C and reach their optimum mechanical properties if both the tool and the workpiece are held at constant temperature throughout the forging process. TZM is the only available material having the necessary hot strength and creep resistance for this application. To prevent oxidation of the die blocks, which have been produced with diameters of up to 760 mm and weighing over 5 tonnes, the entire tool set must be held under a suitable protective atmosphere, usually nitrogen.

3.4 Glass and Quartz Melting Processes

Molybdenum and tungsten find important applications in glass and ceramic manufacturing technology. The production of molybdenum glass-melting electrodes is described in detail in Chapter X, Volume 2 of the present series and molybdenum tubes are now used in transportation systems for molten glass (Fig. 6). One particularly interesting process, which places exceptional demands on materials, is the continuous drawing of quartz glass tubing for the lamp industry. The process is shown schematically in Fig. 7. The low viscosity required for successful tube drawing dictates that the quartz must be held at temperatures of around 1950°C. Molybdenum and tungsten are therefore the only suitable materials for producing the crucible, drawing nozzle and mandrel, and a protective atmosphere is needed to prevent oxidation.

Typical crucibles have diameters of up to 350 mm and lengths of up to 1300 mm. Various designs are in use, including thick-walled crucibles in pressed and sintered molybdenum, and thin-walled rivetted crucibles, in which the inner wall consists of tungsten sheet and the outer wall is made from molybdenum sheet (Fig. 8). The crucible is surrounded by a heating element of tungsten mesh, rod or plate. At the base of the crucible the molten quartz enters a molybdenum drawing nozzle, underneath which lies a centrally-located hollow molybdenum mandrel. A suitable inert gas is blown through the mandrel bore to prevent the still soft quartz tube from collapsing in on itself, and to protect the drawing nozzle assembly from oxidation.

4. The Electronics Industry

So far we have considered only those applications of the refractory metals in which their ability to operate at elevated temperatures is a primary requirement. These metals do, however, possess other unique properties which, in turn, lead to applications in more modest temperature ranges, in some cases not far removed from ambient. For example the unusually high density of tungsten gives rise to a host of applications in which a relatively large mass needs to be concentrated in the smallest possible space. Static and dynamic balance weights, often for aircraft, and shielding for radioactive sources in equipment where lead components would be too bulky, are pertinent examples. The final major field of application we are going to examine, which depends on special properties other than the refractory nature of these interesting metals, is the manufacture of electronics components.

4.1 Tantalum Capacitors

The production of electrolytic capacitors is a very important area of application for tantalum. It is possible to produce extremely stable, adherent and highly-insulating layers of tantalum pentoxide on the surface of the metal by electrochemical oxidation. Because of the high dielectric constant of the oxide layer, tantalum capacitors have a very high level of capacitance per unit volume, which leads to their use in areas where ultimate performance is required.

In tantalum foil capacitors both anode and cathode are made from thin strips of tantalum. These foils are separated by paper soaked in electrolyte and rolled together into a spiral form. The anode foil is sometimes subjected to a prior etching treatment to roughen its surface and thereby increase its capacitance. A second type of tantalum capacitor uses sintered anodes, which are made by pressing and sintering fine tantalum powder in situ around a tantalum lead-in wire. The inherent porosity in the sintered anode effectively increases the surface area, which in turn promotes an increase in the capacitance value.

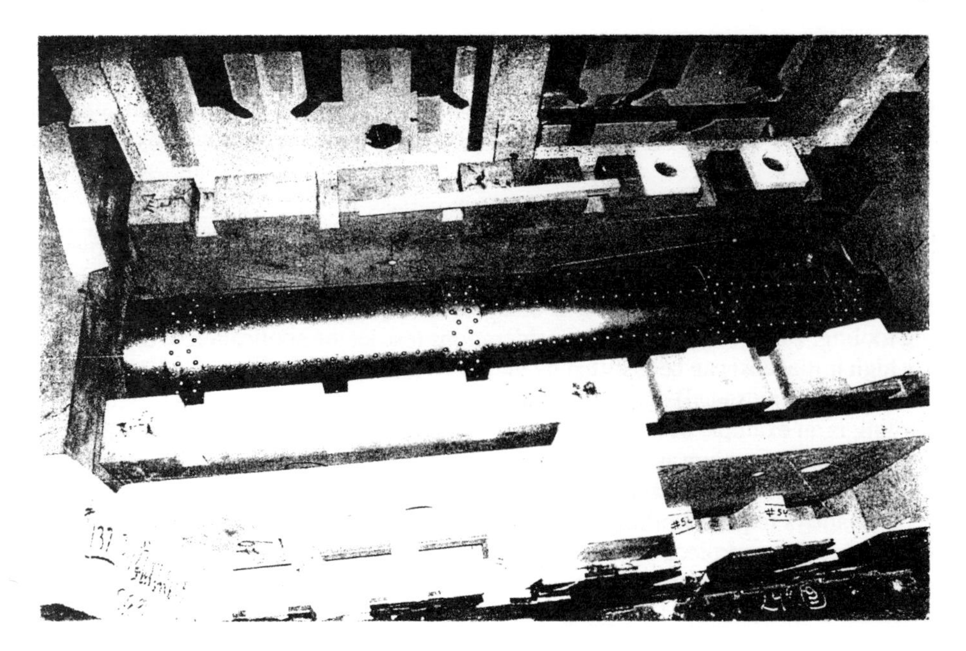

Figure 6

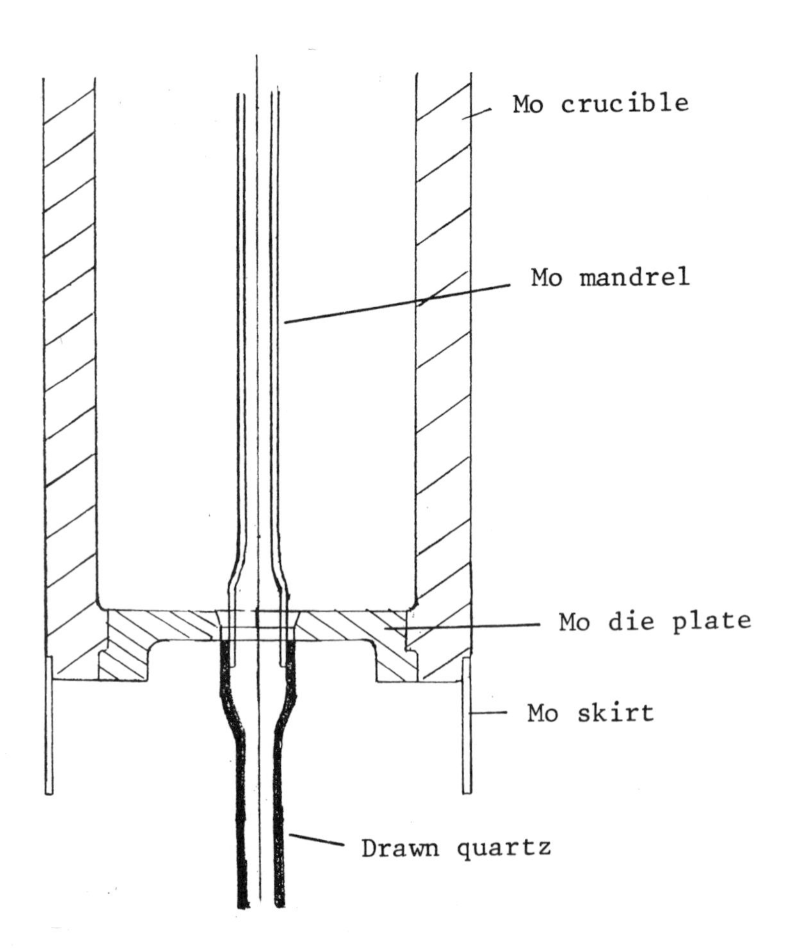

Figure 7

4.2 Power Semiconductors

Single crystal silicon wafers, containing one or more p-n junctions, are found at the heart of high power diodes and thyristors. Discs made from tungsten, or more commonly molybdenum, are used as base plates and heat sinks for the silicon wafers. In large devices these discs can be up to 100 mm diameter and several mm thick. Whereas plain discs are used for power semiconductors in which the silicon is brazed to the disc, thin coatings of gold or rhodium are normally applied to the disc if contact is to be achieved through pressure alone.

Molybdenum and tungsten are chosen for this application because their physical properties closely match the requirements. In particular their high thermal conductivities facilitate rapid heat removal from the silicon component, and their low expansion coefficients (which closely match that of silicon) minimize thermal stresses on the brazed joints.

4.3 Multi-layer Printed Circuit Boards

Demands on the performance of modern electronics systems dictate the use of printed circuit boards with very high component packing densities. To achieve this objective, leadless ceramic chip carriers (LCCCs) must be joined to multi-layer circuit boards using a technique known as surface mounting. The physical properties of the ceramic chip carriers and the commonly used circuit board materials give rise to two significant problems:

(i) Increasing the component density increases the quantity of heat generated during operation, and this must be dissipated through the circuit board. The circuit board materials, however, have low levels of thermal conductivity, and this can lead to excessive rise in temperature.
(ii) The LCCCs and the circuit board materials have widely differing thermal expansion coefficients. Thus fluctuations in temperature generate severe stresses at soldered joints, ultimately leading to failure.

The high system reliability demanded by military and avionics applications requires that specific measures be taken to eliminate the differences in thermal expansion characteristics and to improve heat dissipation. Recent approaches to overcome these problems have concentrated on the use of sheet metal heat sinks, either bonded to the circuit board (Fig. 9a) or incorporated as a core into the laminate (Fig. 9b and 9c). In the latter case the core can also be designed to act as a current carrying layer. These composite circuit boards offer an effective solution, provided that a suitable core material can be found which combines high thermal conductivity and mechanical strength with low thermal expansion. Although no single metal or conventional alloy matches these requirements sufficiently closely, roll clad layered substrates have been used with considerable success.

These roll-clad composites are produced by applying a thin layer of a metal with high thermal conductivity, such as copper, to each side of a strip of metal or alloy having low thermal expansion, such as molybdenum or Invar (64%Fe-36%Ni). By selecting the correct thickness ratio of metal core to circuit board material, a laminate board can be produced, which closely resembles the thermal expansion characteristics of the LCCC.

The use of either Cu-Invar-Cu or Cu-Mo-Cu cores does, however, have certain drawbacks. Invar material has relatively poor thermal conductivity through its thickness which impairs heat transfer through the board. Also its elastic modulus is low, which means that relatively thick layers are needed for effective thermal expansion adjustment of the board. This in turn can lead to a weight penalty, which is a serious disadvantage in aircraft electronics systems. Molybdenum, on the other hand, has much better thermal conductivity and a higher elastic modulus. Thus, although it is a denser material than Invar, it can actually offer a weight saving. Unfortunately the machining of Cu-Mo-Cu foils is difficult, and therefore applications tend to be restricted to designs which do not require the drilling of large numbers of small holes to provide connections to other parts of the board.

Recent work [1] has shown that specially developed Mo-Cu composite foils, which are produced by infiltrating a porous molybdenum skeleton with liquid copper prior to rolling, may offer the best solution. They blend the low thermal expansion and high strength of molybdenum with the high thermal conductivity of copper and their physical properties can be adjusted to meet the varying requirements by altering the ratio of molybdenum to copper. Moreover they are easier to machine than both Cu-Mo-Cu and Cu-Invar-Cu and the drilling of small holes (Fig. 10) is not difficult.

Table 1 lists the relevant physical and mechanical properties of Mo-30Cu and Mo-50Cu composites and a comparison of the important properties of various materials commonly used in the electronics industry is given in Table 2. It can be seen that the expansion coefficients of Mo-Cu composites are similar to those of the ceramic materials. Thermal conductivity values are far superior to that of Invar and are at least 45% of that of pure copper.

Because of its higher elastic modulus, the use of Mo-30Cu foils increases board stiffness to around 2 to 2.5 times that achieved in boards containing Cu-Invar-Cu. In avionics applications, where circuit boards should have the minimum possible weight and the maximum possible resonant frequency, this is of critical importance.

Reference

1. R. Klemencic, E. Kny and W. Schmidt, Multilayer circuit boards with Molybdenum-Copper Cores, Circuit World.

Figure 8

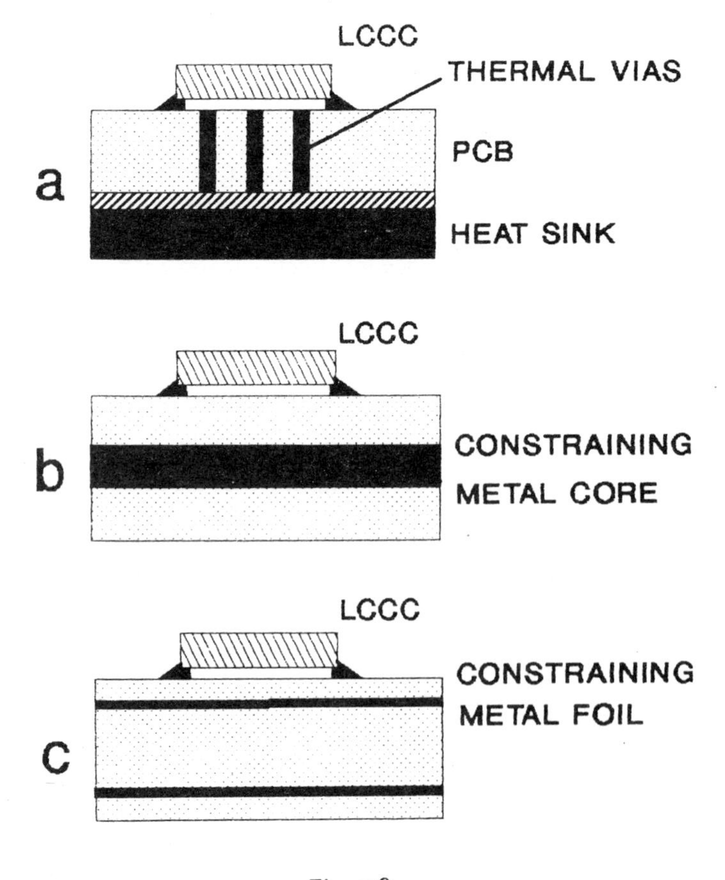

Figure 9

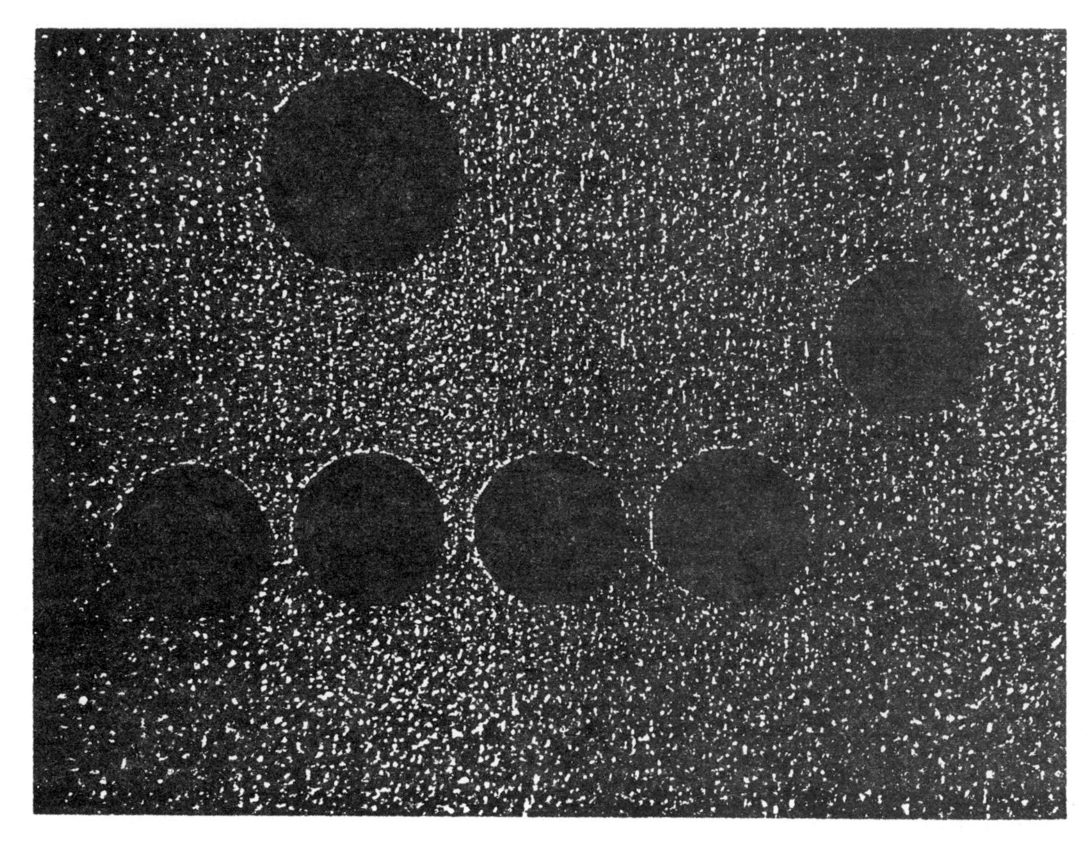

Figure 10

Table 1

		Mo30Cu	Mo50Cu
Density at 20 ° C	d[kg/m^3]	9720	9500
Mean thermal coefficient of expansion between −50 and 150 ° C	$\alpha_l[10^{-6}/K]$	7.5	8.5
Thermal conductivity at 300 ° C	λ[W/m.K]	183	234
Specific heat at 80 ° C	u[$Jkg^{-1}.K^{-1}$]	301	326
Specific electrical resistance at 20 ° C	ϱ[μΩm]	0.037	0.030
Modulus of elasticity (tensile test) at 20 ° C *)	E[kN/mm^2]		
Plain foil		250	213
Annealed foil after electroplating		220	–
Vickers hardness *)	HV 0.2		
Plain foil		245	190
Annealed foil after electroplating		215	–

*) typical values

Table 2

Material	Linear thermal expansion coefficient [10^{-6}/K] between 20 and 200°C	Thermal conductivity [W/m.K]	Modulus of elasticity [kN/mm²] at 20°C
Semiconductors			
Silicon	2.7	156	98
Gallium arsenide	5.7	45	75
Refractory metals (pure)			
Molybdenum	5.2	142	320
Tungsten	4.4	130	410
Other metals (pure)			
OFHC copper	16.8	394	118
Aluminium	23.6	220	72
Nickel	13.3	92	201
Alloys			
Invar (36Ni64Fe)	2.2	13	132
Kovar (54Fe29Ni17Co)	5.3	17	130
Layered composites			
Copper-clad molybdenum (13Cu74Mo13Cu) sheet 0.25 mm	6[1]	174[1]	228
Nickel-clad molybdenum (5Ni90Mo5Ni)	5.9[1]	135[3]	-
Copper-clad Invar (20Cu60Invar20Cu)	5 - 6[1] 8.7[2]	164[1] 16[2]	140 -
FR 4 Epoxy glass	16[1]	0.3	18
Refractory metal - copper composites			
Mo30Cu	7.0	183	217
Mo50Cu	8.0	234	185
Ceramics			
Al_2O_3	6.7	10 - 35	300-380
AlN	3.3	140 - 170	300-310
BeO	6	150 - 250	300-355

[1] longitudinal direction
[2] transverse direction
[3] transverse direction: calculated value

24

PM Materials for Electrical Engineering

E. KNY
Metallwerk Plansee, A-6600 Reutte, Austria

1. Introduction

PM materials for electrical applications are mainly used in light bulbs, as contact materials for electrical switches, for electronic tubes, in capacitors, and as heating elements for furnaces [1]. Three main groups concerning their material composition can be distinguished:

(i) Refractory metals; (ii) Refractory alloys and composites; (iii) Cu and Ag based composites.

All these materials are produced almost exclusively by PM techniques. The following production steps are used in these processes:

- Powder production;
- Powder mixing (for alloys and composites);
- Sintering (with or without liquid phase);
- Densification and forming (for refractory metals and certain alloys);
- Infiltration (for certain alloys with Cu, Ag and aluminates).

2. Refractory Metals

Relatively high thermal and electrical conductivities of the refractory metals W and Mo makes them indispensable for applications where high temperatures have to be endured in electrical applications. The main use of W is as fine filament in light bulbs.

A specially doped tungsten ("non-sag" tungsten) is used for this particular application which improves its creep resistance at high temperatures [2]. W is also used as a contact material in many different switching applications [3]. The main uses of Mo are for heating elements in electrical furnaces, shieldings against thermal radiation, feed throughs and reflector caps in halogen lamps [1]. Grids and other parts of electron tubes are made from Mo or Mo-Re-alloys [1].

Although most of the tantalum used is processed by melt metallurgy techniques. PM produced porous tantalum is needed for electrolytic capacitors. In order to contact this porous Ta-body a doped tantalum wire is used which has to be resistant to embrittling by recrystallization at temperatures above 2000°C. This wire is produced by doping Ta-powder with small additions of oxides (e.g. Y_2O_3) [4,5].

Because of its low expansion coefficient and good thermal conductivity Mo is used as heat sink in electronic applications for power transistors and integrated circuits. Very large numbers of small Mo-pins are used to manufacture miniature diodes. The main advantage of Mo is again its closely matched thermal expansion coefficient to Silicon and glass [1].

3. Refractory Alloys and Composites

Alloys of W and Mo with Re are used because of their increased ductility and recrystallization temperature. W-Re is used in thermocouple wires, Mo-Re for various parts of electronic tubes [1].

ThO_2 and barium aluminates are added to decrease the high work function of W. W doped with 1-2% ThO_2 is used in electrodes for welding and for flashlamp applications. This addition of thorium-oxide decreases the work function to approximately to 2 eV. For similar applications barium aluminate is added as an infiltrant to a porous W-skeleton. 1-6 weight % of barium aluminate can be added by these means. Such materials are used as electron emitting cathodes in vacuum tubes [6,7]. Most of the microwave tubes are equipped with porous tungsten cathodes (travelling wave tubes, Klystrons, Gyrotrons). Porous tungsten cathodes now find increased use for other cathode applications (special TV-tubes, X-ray tubes). W-barium aluminate with contents of 10-30% is used as cathode for flashlamp applications. In the latter case a mixture of W and barium aluminate is pressed and sintered.

Table 1 Physical properties of pure refractory metals in comparison to Cu, Ag and Ni

Property	W	Mo	Re	Ag	Cu	Ni
Mp (°C)	3410	2610	3180	960	1083	1453
d (g/cm^3)	19.3	10.2	21.0	10.5	8.9	8.9
α (10^{-6})	4.6	4.6	6.7	19.7	16.5	13.3
δ($m/\Omega mm^2$)	18.0	19.4	5.2	63.0	59.9	14.6
λ(W/mK)	167.0	142.0	72.0	419.0	394.0	92.0
Work function (eV)	4.51	4.26	5.0	4.51	4.39	4.85

where:
α - thermal expansion coefficient;
δ - electrical conductivity;
λ - thermal conductivity

Table 2 Refractory alloys and composites

Alloying addition, weight %	MATRIX W	Mo
Re	3 - 26 %	5 - 41 %
ThO_2	1 - 4 %	-
Ba-Aluminate	1 - 30 %	-

Ag and Cu-based Composite Materials

These composite materials are used for various switching applications [3]. Cu and Ag containing refractory metal composites are manufactured by sintering a porous body of refractory metal, which is then infiltrated by Cu or Ag. Only Ni-containing alloys can be mixed with refractory metals and sintered with liquid phase. Cu/W is used as a contact material in medium and high tension voltage oil switches. Cu/Cr is used as contact material in vacuum switches [8,9]. Ag-containing contacts (Ag/W, Ag/Mo) are mainly used in low voltage switching applications. Oxides like CdO_2, SnO_2 and others are added to silver to decrease the tendency for contact welding [10].

These switching materials are used in great numbers for low voltage applications. Because of environmental concerns CdO_2 tends to be substituted now by SnO_2.

Besides switching applications Mo/Cu- (mainly 30% Cu) composite is used as a heat sink in electronic applications [12]. Since this composite can also be rolled (contrary to W30Cu) it can also be used as a heat sink in printed circuit boards [11].

Carbide and graphite additions to Ag and Cu contact materials are made in order to use their high abrasion and electro-erosive resistance and to decrease contact welding properties of pure Ag and Cu. For similar reasons composites of Ag and Ni are made by PM techniques (Ag and Ni are completely insoluble in solid state). The Ni additions in Ag mainly decrease electro-erosion [3].

Figure 1

Figure 2

Figure 3

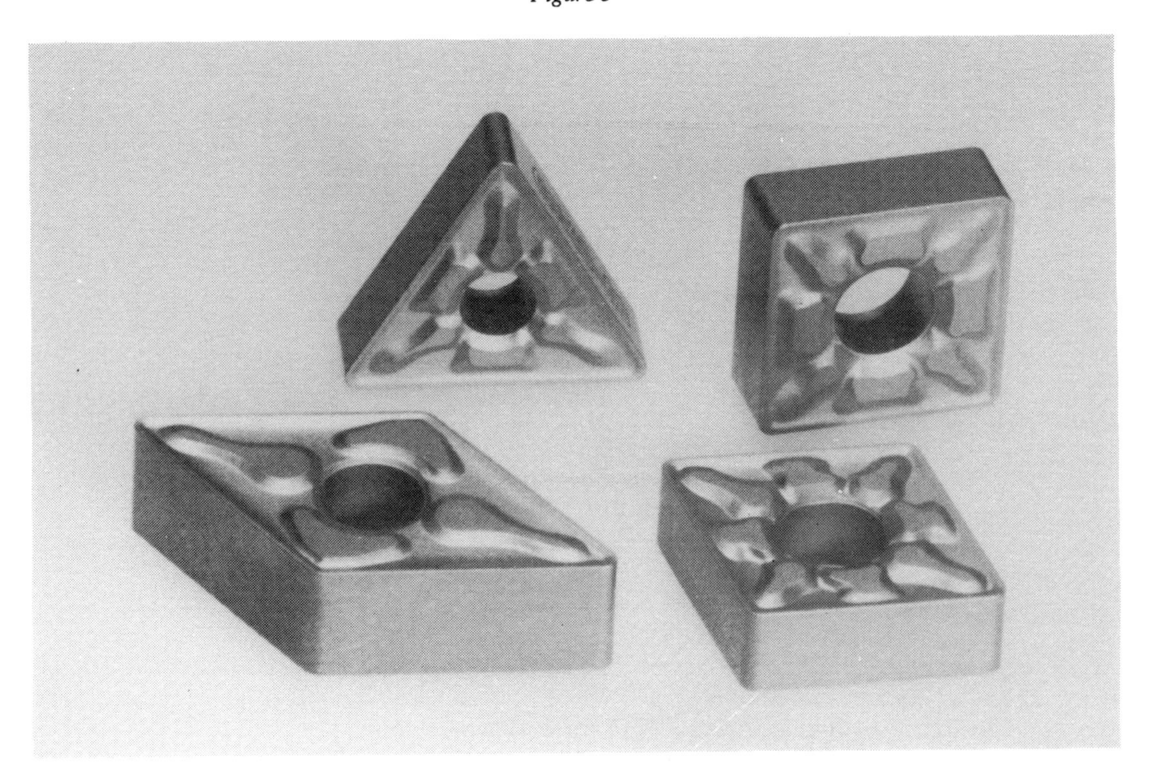

Figure 4

Table 3 Cu and Ni based composite materials Cu and Ni based (weight percent of second component added to matrix)

Second	MATRIX			
Component	Cu	Cu/Ni	Ag	Ag/Ni
W	50% - 90%	70% - 95%	50% - 80%	-
Mo	50% - 80%	-	50% - 80%	-
Cr	25% - 60%			
Ni	-	-	10% - 40%	-
CdO	-	-	5% - 15%	-
SnO_2	-	-	5% - 12%	-
ZnO	-	-	8% - 10%	-
WC	30% - 80%	-	30% - 80%	-
Mo_2C	30% - 70%	-	30% - 70%	-
			20% - 80%	-
	3% - 5%	-	2% - 15%	2% - 5%

References

1. Technische Anwendungen der Hochschmelzenden Metalle, Metallwerk Plansee GmbH, A-6600 Reutte, 1988.
2. W.H. Yih and Chun. T. Wang, Tungsten, Sources, Metallurgy Properties and Applications, Plenum Press, NY, London 1979.
3. Elektrische Kontakte und ihre Werkstoffe A. Keil, W.A. Merl, E. Vinaricky, Editors, Springer, Berlin, Heidelberg, NY, Tokyo, 1984.
4. R. Kieffer, H. Braun, Vanadin, Niob und Tantal, Springer, Berlin, 1963.
5. G.J. Miller, Tantalum and Niobium, Metallurgy of the Rarer Metals 6, Butterworths, London, 1959.
6. E.S. Rittner, W.C. Rutledge and R.H. Ahlert, J. Appl. Physics, 28, 12, pp.1468 - 1474, 1957.
7. J.L. Cronin, Microwave Journal, p.57-62, Sept. 1979.
8. L.T. Falkingham, Int. Conf. on Large High Voltage Electric Systems, 13-01, 1968.
9. W. Rieder, M. Schussek, W. Glätzle, E. Kny, IEEE-CHMT, Transactions, 1989.
10. H. Schreiner, Pulvermetallurgie elektrischer Kontakte, Springer, Berlin, Gottingen, Heidelberg, 1964.
11. R. Klemencic, E. Kny, W. Schmidt, Circuit World, July 1989.
12. E. Kny, 12th Int. Planseeseminar, May 1989, Metallwerk Plansee, A-6600 Reutte, 1989.

25

Magnetic Powder Materials: Current Status and Development

D. HADFIELD
Consultant; Sheffield, UK

1. Introduction

Magnetic materials fall into two major categories:

(i) Magnetically "hard" metallic and non-metallic materials - such as permanent magnets and magnetic recording media, produced in both alloy and ceramic ferrite forms.
(ii) Magnetically "soft" metallic and non-metallic materials - such as electromagnet cores, yokes, flux concentrating pole pieces, chokes, produced in iron and ferrous alloys, and ferrites with no retention of magnetism after the removal of an a.c. or d.c. applied magnetic field. Ferrofluids also come within this category.

Such ferromagnetic materials are characterized by high magnetic saturation and permeability with very low coercivity and narrow hysteresis loops in the case of "soft" magnetics, whereas permanent magnet materials have more rectangular, fat hysteresis loops with very high coercivity in the second quadrant (or demagnetization curve).

2. Hard Magnetic Materials

Until the middle 1960s magnetically hard materials were produced mostly by melting and casting or hot manipulation. Powder techniques for metals were being applied on pilot plant scale only but were at the intensive production development stage for hard ceramic ferrite. Powder metallurgy has expanded considerably in the last forty years for both metallic and ceramic permanent magnets. In alloy magnetic materials, new powder metal magnets including high coercivity alnico and rare-earth transition metal alloys have necessitated new developments in process technology. Ceramic magnets are sintered or bonded barium or strontium hexaferrite with concomitant improved techniques to approach near-net-shape and maximum sintered densities. This material in powder form has also been used for the innovative bonding with rubber and extruding flexible strip magnet used extensively in refrigerator door gasketry.

Since the inception of rare-earth transition metal permanent magnets and subsequent commercial production over the last decade the technology has advanced rapidly resulting in stored magnetic energies over six times those of the anisotropic alnico magnets.

The possible future enhancement of BH properties and expansion of usage resulting from the discovery of neodymium-boron-iron alloys in relation to the establishment of particulate technology in the permanent magnet field is of notable importance (Fig. 1).

2.1 High Coercivity Sintered Alnico

The Al-Ni-Co-Fe alloys known generically as "alnicos" are produced by castings and increasingly by sintering because of size reduction and economics. Development has been concentrated on the highest coercivity versions, i.e. the high cobalt, high titanium end of the range and this has resulted in an alloy which possesses a much higher coercivity combined with a similar energy product. The high titanium (8%) and aluminium (7%) contents needed for high coercivity required a manufacturing technology to be developed to avoid oxidation during sintering. A master alloy including all the constituents except a proportion of the iron is melted from oxygen-free raw materials and then crushed. -ure iron powder to the necessary concentration is added during ball milling. The alloy powder to adjust to the final composition is added during ball milling to the desired particle size. The alloy powder is then pressed with a wax lubricant and sintering in pure dry hydrogen or high vacuum is carried out at maximum temperature below the liquidus point with critical temperature control to within ±2-3°C. A complex heat treatment regime is necessary in order to develop the high coercivity combined with high energy product.

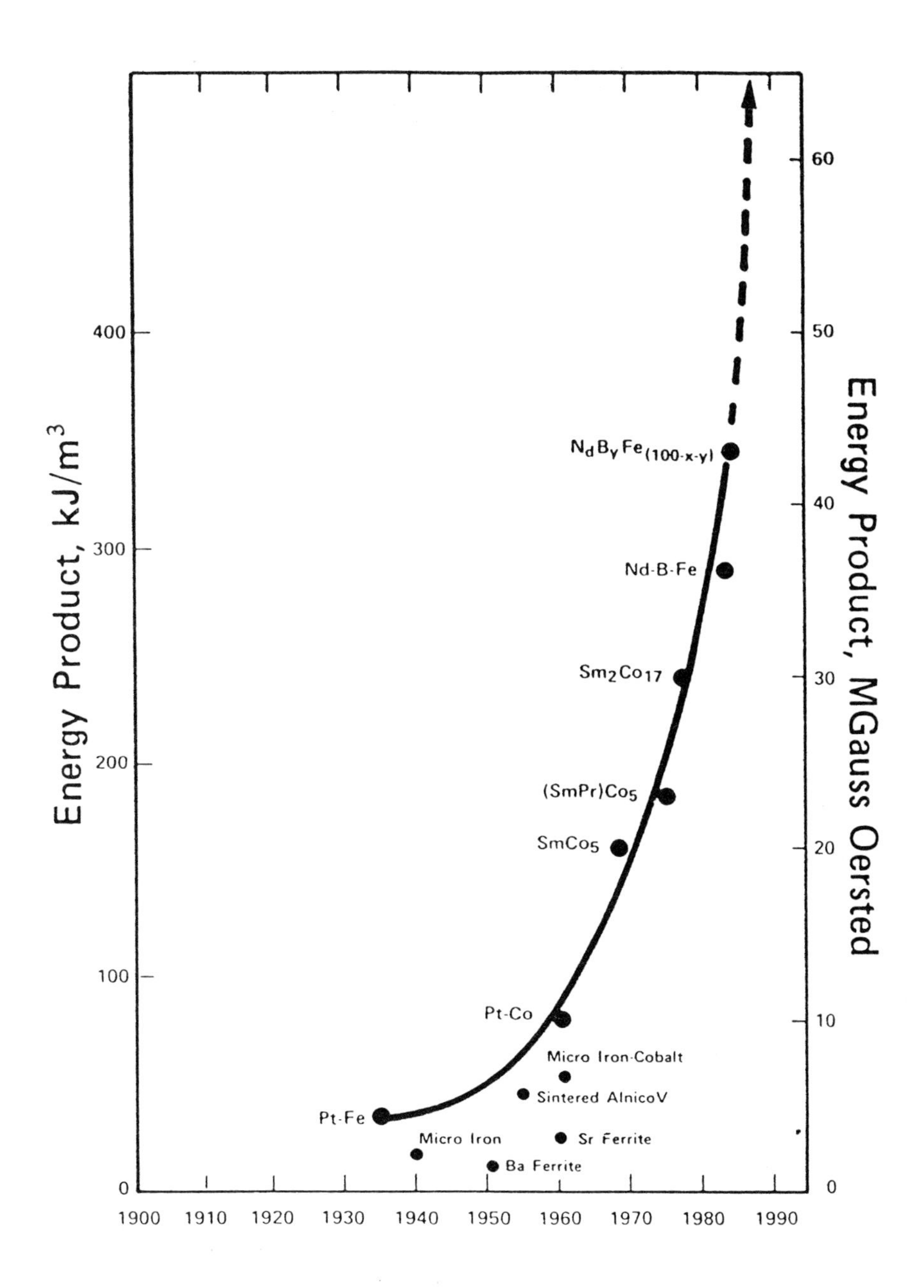

Energy Product, kJ/m³
Energy Product, MGauss Oersted
$N_dB_yFe_{(100-x-y)}$
Nd-B-Fe
Sm_2Co_{17}
$(SmPr)Co_5$
$SmCo_5$
Pt-Co
Micro Iron-Cobalt
Sintered AlnicoV
Pt-Fe
Micro Iron
Sr Ferrite
Ba Ferrite
0
100
200
300
400
10
20
30
40
50
60
1900
1910
1920
1930
1940
1950
1960
1970
1980
1990

Figure 1

This combination of properties has facilitated the redesign towards miniaturization of most moving coil instruments by reducing the size of the magnet and allowing it to be used as the core with the coil moving around it. In this composite application simultaneous pressing of the cylindrical core magnet and soft iron pole tips is effected by using a compartmentalized die, followed by co-sintering of the compact.

2.2 ESD Iron and Iron-Cobalt

Early theoretical studies in France suggested that sub-micron size particles of iron, i.e. below the size of a single domain, could possess high coercivity. Subsequent research showed that longer dendritic particles of iron or iron-cobalt exhibited enhanced coercivity and energy product, and this led to the development of oriented micromagnetic iron and iron-cobalt marketed under the name Lodex.

The elongated single domain particles are prepared by electrolytic deposition into a mercury cathode 99% of which is rècovered by pressing and vacuum distillation, thereby rendering it competitive with sintered Alnico.

The ESD powder is pressed in a magnetic field with a soft metal bond (lead with a small amount of tin), which acts as a non-magnetic matrix between the particles, maintaining the coercivity and also prevents oxidation of the pure powder.

The material being bonded, not sintered, allows it to be pressed to close tolerances without machining and the softness of the lead-tin bond allows drilling - both desirable attributes for such applications as synchronous motor magnets.

2.3 Ceramic Ferrite Magnets

Ferrites are combined metal oxides of the composition $MO.nFe_2O_3$ where, for commercially useful materials, M is either barium or strontium and for optimum magnetic properties, n = 6.

Anisotropic strontium hexaferrite magnets which are produced by powder techniques, both dry and slurry compaction, have a higher coercivity at ambient temperatures than any other material except rare-earth magnets and this allows designs with a low length : pole face area ratio such as are needed for d.c. motor pole pieces and loudspeaker, separator and lifting magnets. Such magnets are produced by bonding with a polymer or sintering to full density.

2.4 Rare Earth-Transition Metal

For rare earth elements at the lighter end of the group, (atomic numbers 57-66), intermetallic compounds of the general formula RE_xCo_y where, if x = 1, y = 5: and if x = 2, y = 17, are characterized by strong magneto crystalline anisotropy, high magnetic saturation, high coercivity and the highest energy product per unit mass of any known permanent magnet material, i.e. up to six times that of sintered alnico.

The theoretical limit to the energy product of $SmCo_5$ has been calculated as 184 kJm^{-3} and this has already been achieved in laboratory conditions and a value of 170 kJm^{-3} is obtained in industrial production. Intensive research and development in the USA and Japan concentrated on exploration of alloys containing Fe + Nd with Co, Sn, Pr, and B, with possibilities of approaching 500 kJm^{-3}.

The samarium and/or praseodymium is replaced by neodymium with an addition of boron, forming a tetragonal compound of composition $Nd_xB_yFe_{100-x-y}$, optimizing at x 12/13 and y 6/7 atomic %.

Production of these intermetallic compounds may follow a conventional melting route using commercially available constituents in an alumina crucible under argon gas atmosphere, and crushing the cast ingots in a nitrogen atmosphere to a particle size of 1mm in a jaw crusher, then down to approximately 100 µm in a disc mill and finally pulverizing in trichloroethane and trifluoroethane to approximately 3 µm by a ball mill using stainless steel balls and container. Alignment in a field at H = 800KA/m with simultaneous compaction at 200 mega pascals is effected by pressing perpendicular to the aligned direction. Sintering to optimum densification in argon at 1310°K to 1430°K for one hour is followed by rapid cooling with a post-sintering treatment of 400°K-1400°K and cooled rapidly.

An alternative RSP method from the ingot stage is by melt-spinning thin ribbon in an argon atmosphere by ejecting molten alloy through an orifice in a quartz crucible on to the edge of a rotating substrate disc at a quench rate controlled by the substrate surface velocity (10^6c/sec), whilst maintaining all other parameters constant. The particulate so formed is coated and bonded by a thermo-setting plastisol or by hot forming to render it anisotropic.

It is established that any further development in permanent magnets will almost certainly be by powder technology. Likewise, the research and development efforts on single crystals of the latest RE, TM generation of magnetic materials is ultimately going to push the frontiers of stored magnetic energy to above (55 mgo) 435 kJ/m^3.

The work referred to above was stimulated by the need for very high stored magnetic energies in miniaturized components in fields where cost is not the prime consideration, e.g. satellite and space telecommunications, computer printer drives and some biosurgical applications such as magnetic sphincters, wound closures, orthopaedics, and repulsion harnesses for necks and dentures.

For less volume or weight-sensitive applications, cheaper rare-earth magnets are being produced by polymer bonding.

2.5 Magnetic Recording Media

Magnetic recording media used in compact video systems and digital data recording are operating at recording densities two orders of magnitude greater than was achievable a decade ago. The perpendicular recording process and magneto-optical techniques offer further potential for improvements. To meet such application requirements has necessitated considerable research efforts in developing vacuum sputtering thin films of Co and Ni onto polymeric base tape and probably the more difficult barium ferrite, which is considered the best magnetic material for perpendicular recording. The lowest cost iron oxide is used for tickets and security cards and cheap tape.

2.5.1 Metal Powders

Replacement of oxides by metals, essentially by finely divided iron powders, has been practised for some years in the minority market for analogue audio cassettes. It is the material for production of 8mm tapes and was initially selected for digital home audio (DAT) cassettes. In its favour are higher remanence and controlled higher coercivity, extending the practical range of recorded wavelengths from above one micron down to at least half a micron. Its difficulties lie in its manufacture, not least in terms of safety (it is a strongly pyrophoric material if unpassivated) and in its use where the inter-particle attractive forces are very much higher than for oxides and present new problems in dispersion.

These difficulties have been substantially overcome and production capacity is in the several hundred tonnes per annum range. It is a very high price material only to be used when the cheaper oxides are inadequate. None of the products for which its use is essential are yet in such volume as to generate price erosion pressures and its method of manufacture is fundamentally more expensive than that for oxides.

Passivation in manufacture is still mainly by generation of oxide surface layers as alternative methods based on organic or metallo-organic coatings are not used. The need for extensive changes of media making technology in dispersion handling and in formulating has combined with the sophistication of the products using metal powder material to confine its use so far to those companies who are well experienced and resourced in technical terms. Evolutionary developments are well under way. The objectives are to improve life stability and dispersibility, adapt coercivity to a full range of from 600 to at least 1500 oersteds and to improve particle size distribution.

An alternative and in principle better coating can be made by using metal in the form of quasi-continuous thin film rather than as a powder diluted by necessary binder resins. Such materials can be deposited by plating, sputtering or vacuum evaporation and have the advantages of greater magnetization per unit volume and of being intrinsically thin, thereby further reducing losses due to demagnetization and extending the limits of recording wavelength even further.

A favoured material comprises a thin (100nm) film of metal alloy, usually cobalt-nickel based, which is deposited onto a super-smooth base film. This is done in a vacuum by heating an ingot of the alloy with an electron beam until it is molten. The metal vapour which is evolved is allowed to impinge on the base film as it passes over the evaporation zone. A cooled drum is used to hold the film and to conduct away the heat. The geometry of deposition is restricted so that only metal atoms arriving at a set range of angles of incidence are allowed. This controls the coercivity of the deposited film.

The advantages of metal evaporation for recording are that it has the highest output on replay at high frequencies of any longitudinal medium; it exhibits true saturation recording which makes it insensitive to record current variations, and it is significantly thinner than particulate coatings, thus allowing longer playing times for a given cassette. On the other hand, it requires completely new technology to produce with procedures which can make the product expensive and it needs especial care to achieve good wear and handling characteristics with a long life time.

An interim product for analogue radio recording is for the 8mm video system, never having come to the market in the original 8mm product even though the 8mm standards allow for its use. This tape comprises a back-coating for runnability, ultra-smooth base film, a priming layer for adhesion and durability, the metal evaporated layer and finally a lubricating and protecting layer, adding up to three "wet" coats and one vacuum evaporation which makes the tape expensive.

Thin films can also be put down by plating and by sputtering and there are arguments in favour of each as alternatives to evaporation. Evaporation has the longer background in technological development. Although plating and sputtering are common today for thin film Winchesters, using them for evaporation on webs of running film is a different matter.

2.5.2 Barium Ferrites

Another class of powdered magnetic materials become available by turning from the cubic iron oxides to the hexagonal ones especially those known generally as barium ferrites. These have been known for years as very high coercivity, very large particle size, oxides suitable only in the media industry for simple kinds of magnetic card. Work earlier in this decade showed that by high temperature melting of the component oxides in a glass matrix, quenching between rollers to make flakes, heating to crystallize and then etching out the glass with acid, some very interesting sub-micron particles could be made with coercivity controllable over the range required for media making. The particles are hexagonal platelets of adjustable aspect ratios and size at or below 1000 Angstroms. Coercivity is controlled by doping with cobalt and titanium and the easy axis of magnetization can be caused to be at right angles to the plane of the platelet.

2.5.3 Cobalt Chromium

Research during the past 15 years on CoCr alloy has identified it as a potential ultra-high density storage layer. The principle involved is to store magnetic reversals perpendicular to the surface instead of parallel to it so that at high density the magnetic elements become long and thin instead of short and fat as they do for pure longitudinal recording, i.e. the demagnetizing fields improve rather than oppose the move towards higher packing densities.

Cobalt-Chromium tends to have a columnar microstructure (although this is not absolutely necessary) with hexagonal cross-section and with c-axis perpendicular to the surface. Coercivity and saturation magnetization can be controlled by chromium content and by deposition conditions such as system base pressure, sputtering pressure, substrate temperature and bias voltage.

To improve wearability and orientation in tape applications, additive materials such as Fe, Zr, Ta and carbon have been used. The potential storage capacity of the medium is said to be governed by the size of the columnar microstructure and a limit of 300K frequency response per cubic inch predicted.

3. Soft Magnetic Materials

An outline of soft magnetic materials and their usage is concerned principally with the composition, processing and the resultant electrical and magnetic parameters required for particular applications. These materials may be divided into a number of classifications. Metal alloys, which would include the well-known bulk sheet and strip silicon-irons for mains frequency work, metal powders, ceramic ferrites and metallic glasses constitute the majority in all fields of alternating current electrical and electronic engineering technology in which many applications depend on the frequency.

Development of these high technology products has been directed at increasing the permeability and reducing the power losses. Such losses are directly related to the hysteresis loop area, the frequency, and the electrical resistivity of the material, the latter of course controlling the eddy currents.

3.1 Metal Powder

Metal powder cores, in which the material resistivity was increased by coating micron size particles with an oxidized layer and bonding, were manufactured principally from carbonyl iron and at a later date the permalloys (nickel-irons with nickel principally, 55-80%) were also in use in powder form. The latter, however, are somewhat more expensive in terms of raw material costs and the bulk of such iron powder cores use low carbon hydrogen-reduced carbonyl iron to produce the highest permeability grades.

The original applications using iron powder components have changed considerably as radio, television and video have gone into a higher technology gear and demanded a new field of applications for devices and components needed to meet the important requirements of interference suppression and energy storage in chokes in small switching-mode power supplies. Such components must meet certain specifications, e.g. permeability must be optimized and also have the ability to withstand fairly high currents. These requirements are satisfied by basic material of much lower electrical performance in the form of iron flake of well above micron size, or iron powder as used in the powder metallurgical industry.

3.2 Ceramic Ferrites

Another completely different class of compositions which has gained the greatest importance for many years and is likely to continue to are soft magnetic ferrites.

These are ceramic materials, basically metal oxides, produced by powder technology. Their applications are many and therefore this has resulted in a number of grades which have been developed to meet specific requirements in various shapes and sizes. Most of ceramic ferrite manufacture is divided between nickel-zinc ferrite and manganese-zinc ferrite with compositions close to stoichiometric. There are also minor additions of other compounds which control and modify certain of the magneto-electric properties.

In production, which is a fairly involved process with a number of different stages, great care has to be taken to control each one as the specification tolerances are usually very tight. Slight variations in firing temperature and kiln atmosphere for example, can have a profound effect on the resultant high frequency loss factors. In the two main groups mentioned above zinc-nickel ferrite is used mainly at higher frequencies than the zinc-manganese ferrite, in fact by at least an order of 2 or 3 KHz. Another important difference is that the nickel-zinc ferrite is used at low power levels and manganese-zinc at higher levels, sometimes well above a kilowatt.

The principal attributes for which specifications are drawn and designers make use of are initial permeability (at the "foot" of the magnetization curve), saturation flux density, loss factor, hysteresis material constant, temperature factor, amplitude permeability, total power loss density, Curie temperature and resistivity.

More recent materials developed for enhanced performance are the yttrium-iron garnets, particularly necessary in use for microwave applications based on the Faraday Effect, which allows the ferrite component to conduct electrical energy in one direction only when an external magnetic field is applied to it. This unusual attribute is made use of when it is necessary to separate transmitter and receiver signals through a common antenna, and also protect the receiver from damage by restricting the transmitted power.

The major part of the ferrites and yttrium-iron garnet is for switched-mode power supplies, interference suppressors, pulse and analogue transformers and other telecommunications devices.

3.3 Metglas

The latest type of magnetically soft material recently introduced commercially is metallic glasses which are amorphous alloys, being non-crystalline made by the rapid solidification process (RSP), whereby a continuous jet of molten alloy is cast on to a water-cooled rotating substrate which then cools the material extremely quickly before it can form a crystal structure, the cooling rate being around 1000°C/millisecond. The compositions of metallic glasses for electronic engineering usage are combinations of one or other of the transition metals (Fe, Ni, Co), and metalloids (e.g. B, P, Si or C).

In order to produce particulate by this melt spinning technique it is necessary to have minute serrations around the periphery of the rotating water cooled copper cylinder. A particular feature of the material is a very high permeability and a frequency range from audio up to about 100 kH. It is expected that new applications for this unique and novel magnetic material will become apparent in the near future.

3.4 Ferromagnetic Fluids

Magnetic fluids may consist of Fe_3O_4 particles dispersed in hydrocarbon oil or diester carrier, or cobalt particles in toluene or diester carrier. The saturation magnetization in as prepared fluids is between 50 gauss and 1000 gauss. In such fluids the interactions are significant but can be reduced by diluting the stock fluid until the particles are well-dispersed.

The application of an external magnetic field, produced by an electromagnet or permanent magnet, magnetizes the particles by induction and causes agglomeration, an effect which can be put to a variety of uses, e.g. magnetic fluid clutches, damping and enhancing the gap field in a loudspeaker system. The usage of ferromagnetic fluids has continued to develop from the early use as a magnetic "ink" in NDT crack detection in iron or steel components and semis. As a consequence their application in rotating shaft seals, particularly exclusion seals for computer disc drives, is extensive.

New more sophisticated applications as in remote sensors necessitate long-term stability in demanding conditions such as temperature variations and high field gradients. Further development in the use of ferrofluids is hampered because those currently available commercially do not have indefinite stability.

References

1. A.G. Clegg and M.McCaig, Permanent Magnets in Theory and Practice, Pentech Press, 1989.
2. R.H.T. Dixon and A. Clayton, Powder Metallurgy for Engineers, Machinery Publishing 1971.
3. D. Hadfield, Permanent Magnets and Magnetism, Butterworth, 1962.
4. I.V. Mitchell, Niodymium Iron Permanent Magnets, Elsevier, 1985.
5. R.J. Parker, Advances in Permanent Magnetism, Wiley, 1990.
6. F. Snelling, Soft Magnetic Ferrites, Butterworth, 1987.

26

Superconductors

F.R. SALE

Manchester Materials Science Centre, University of Manchester/UMIST, Manchester, UK

1. Introduction

The discovery of superconductivity in $La_{2-x}Ba_xCuO_{4-y}$ at temperatures close to 30K by Bednorz and Muller [1] has produced a phenomenal amount of activity in the field of superconductivity over the last three years. Following the initial discovery it was soon shown that superconductivity exists in the oxides $Y_1Ba_2Cu_3O_{7-x}$ at temperatures of the order of 93K [2] and in a series of Bi-containing and Tl-containing copper oxides at temperatures up to 122K for $TlBa_2Ca_3Cu_4O_{11}$ and $Tl_2Ba_2Ca_2Cu_3O_{10}$ and up to 110K for $Bi_2Sr_2Ca_2Cu_3O_{10}$ [3]. It was subsequently shown that Pb could be substituted for some Tl to yield superconducting transition temperatures (T_c) of the order of 122K, e.g. in $(Tl,Pb)Sr_2Ca_2Cu_3O_9$. These latter oxides may be represented by the general formula $(AO)_mM_2Ca_{n-1}Cu_nO_{2n+2}$, where A can be Tl, Pb, Bi or mixtures of these, the cation M is Ba or Sr with the possibility of substitution of Sr by Ca and the number of consecutively stacked CuO_2 layers is indicated by n. To date the maximum number of layers that have been identified in well-characterized materials is four. A range of powder techniques has been used for the production of samples, if not commercial quantities, of these new high T_c superconductors. It is the aim of this review to analyse the preparative techniques that have been used to date and to evaluate them with reference to usual powder metallurgical processing methods. It is tempting to write that the discovery of the new superconducting compounds was, and is still, the easy part of the advances in these materials. The processing of the compounds into technologically useful components is an area of study which requires far more effort, and this can best be carried out by powder metallurgists and ceramists who have experience in these areas.

The use of powder processing for the manufacture of superconductors is not new. Nb-based metallic superconductors have been produced using powder methods over a number of years. Today, work is still in progress on the powder processing of A15 superconductors because, at the present time, these still offer the highest probability of use in high field applications. Accordingly, the powder processing of A15 structure metallic superconductors will be assessed briefly before high T_c materials are considered.

2. Metallic Superconductors

The basic ideas behind the powder processing of metallic superconductors follow on from the advances made by Tsuei [4,5] using the "in-situ" technique for the production of multifilamentary Nb_3Sn composite wires. In the "in-situ" technique the rapid quenching of a Cu-Nb alloy from temperatures of the order of 1800°C produces a uniform distribution of small Nb precipitates within the matrix. On subsequent deformation the precipitates yield fine filaments of Nb. Many of the papers published on the powder processing of similar metallic composites simply produce the initial fine dispersion of Nb particles by a powder mixing route. The potential advantages and disadvantages have been discussed by Flükiger *et al.* [6]. In summary these are that the Cu to Nb ratio can be altered over a wide range which is not limited by phase equilibria; that the use of powder allows the sizes of Cu and Nb particles to be varied; that the control of cooling rate, which is a vital parameter in the in-situ process, is obviated; that intermediate annealing is unnecessary; and finally, that the process should be amenable to scale-up to allow the manufacture of commercially-significant quantities of material. An added advantage is that many fine filaments may be obtained in a single process. Once the powder mixture has been processed by a mechanical working operation, usually extrusion and wire-drawing, the ribbons (or wire) are then electroplated with Sn prior to diffusion reaction to produce Nb_3Sn.

These potential advantages are, however, not obtained easily. As in many other processes employing the use of powders (here the maximum powder size was 40 μm as it was claimed that above this value a homogeneous mixture could not be obtained), the chemical reactivity of the powders was found to cause appreciable problems. In particular, the oxidation and adsorbtion of impurities led to increased brittleness of the Nb during high temperature annealing or hot extrusion. Flükiger *et al.* [7] have claimed to reduce many of these problems by using a "cold" powder metallurgy route in which the initial deformation of the Nb powder particles is achieved by extrusion at approximately 500°C. This "cold" processing is in contrast to the mechanical working of similar powder compacts at 1050°C by Schultz, Freyhardt and Bormann [8] and in an improved manner by Freyhardt, Bormann and Bergmann [9]. After the initial working to produce the fine filaments of Nb it is claimed that oxidation is not a problem. In some variants of this process the initial powder mixture contains all the required elements for superconductor production, e.g. Sn is added already alloyed with the Cu [8]. In such processing the cleanliness of the powders is important and the best results are obtained when the Cu powders have been cleaned by reduction in H_2 at 300°C [10]. This H_2 reduction enhances the cold-welding properties of the Cu and thus contributes to a better mechanical integrity of the wire. Nb powders were obtained by the hydride-dehydride route and by ball-milling. It is not surprising that the hydride-dehydride powder was found to be far more ductile and deformed fairly readily to give filaments. Ball-milled Nb powders were resistant to deformation and gave angular particles after the first stage of extrusion with a reduction ratio of 4.

Various container materials have been used in the extrusion stage of the process. Monel has been shown to be satisfactory as far as the mechanical working stage is concerned. However, as it acts as a diffusion barrier for Sn, it requires removal by grinding at the end of the drawing and rolling process. To obviate this stage later studies used jackets of Cu, Cu-0.2wt%Zr and Cu-1.8wt% Be. After initial packing the powder mixture had a density of the order of 80% theoretical, whereas after extrusion the compact had a density greater than 98% theoretical. At the completion of mechanical working, where the reduction in cross-sectional area could be in the range 2000 to 10^4, electroplating with Sn and diffusion annealing at temperatures in the range 650°C to 850°C gave products from Cu-40wt%Nb-20wt%Sn with critical current (J_c) values of 10^5 A/cm^2 at 12T, 5 x 10^4 A/cm^2 at 14T and 2 x 10^3 A/cm^2 at 18T [6]. These figures are comparable to those obtained from bronze-route A15 material. Further, it has been shown [11] that these cold processed wires can support relatively large strains (>1%) without degradation of J_c.

The shock wave synthesis of Nb_3Sn from powder mixtures has been achieved successfully [12] using niobium and tin powders (-325 mesh). The mechanically mixed powders (minimum purity 99.9%) were sealed in Cu tubes under vacuum by electron beam welding. The ductile Cu tubes were collapsed uniformly, using a homogeneous shock wave, to compact the powders and allow reaction to give Nb_3Sn. It was assumed that reaction only occurred when the shock wave caused temperature increases in excess of the peritectic reaction in the Nb-Sn system, when the liquid Sn could react with solid Nb to produce Nb_3Sn directly. The same experimental system failed to produce Nb_3Al from powder mixtures of Nb and Al.

Nb_3Al has, however, been prepared by powder metallurgical techniques [13,14] with such a degree of success that scale-up studies have been carried out [15]. Hydride-dehydride Nb powder (≤40 μm) and Al powder (≤9 μm) were mixed and compacted into a Cu-Be container prior to evacuation, swaging and wire drawing [13]. The mechanical working produced areal reductions of the order of 1400 and gave ultrafine Al fibres in the Nb matrix. Subsequent reaction at 700 to 850°C in helium yielded Nb_3Al. Reaction at 800°C for 16 hours gave a product having $J_c > 10^4$ A/cm^2 at 14T and 4.2K. In addition it was shown that J_c was independent of strain up to a value of 1.3%. In the scale-up studies [15] either hydride-dehydride (38 to 300 μm) or melt spun (150-250 μm) Nb powders were used with atomized Al powders (5-40 μm). The effects of scaling of the powders was found to be reproducible. An areal reduction ratio of 1600 was necessary to give optimum results when 40 μm Nb powder was used whereas a reduction of the order of 5 x 10^4 was required for Nb powders of diameter 150 μm. Powder sizes ≥60 μm could be processed in a Cu container, however, a stronger Be-Cu container was necessary for smaller powders.

There is no doubt that the most promising powder metallurgical route for the manufacture of A15, Nb_3Sn, continuous multifilamentary superconductors is the ECN process [16-20]. Essentially this process relies upon the reaction of $NbSn_2$ powder with a Nb container, in the presence of Cu, to produce a layer of Nb_3Sn on the inner surface of the Nb as shown in Fig. 1. In the process Nb tubes are filled with $NbSn_2$ powder and are placed into Cu hexagons which are stacked together and surrounded by a Cu tube. Wire drawing is then used for small billet arrangements to produce fine filamentary tubes of Nb containing $NbSn_2$. The process can also be used for the production of V_3Ga and V_3Si superconductors [16]. The potential advantages of the ECN process, relative to the conventional bronze route for multifilamentary Nb_3Sn superconductors, are significant. The use of $NbSn_2$ as a solid source for Sn allows the 13 wt% solubility of Sn in Cu of the bronze route to be by-passed. In turn the higher Sn activity leads to enhanced diffusion interaction for Nb_3Sn formation. Added advantages are that relatively few (up to 192) filaments are required in ECN wire relative to in excess of 2000 for bronze route material. J_c values for the wires have been measured to be from 8 x 10^8 A m^{-2} to 2 x 10^9 A m^{-2} at 14T for wires reacted at approximately 675°C for 48 hours. The powder morphology, size and composition play an important role in the ECN process. Recently, $NbSn_2$ powders have been produced by gaseous chloride-liquid tin interaction [21] and these have been processed into trial wires which possess encouragingly high J_c values. To increase the lengths of available wires it was necessary to include a billet extrusion step prior to wire drawing. During processing no intermediate annealing is necessary, however, much care is needed in the mechanical working operations to obviate early failures by perforation and constrictions of the Nb tubes. It has been found that wires can be drawn down successfully to give filaments having powder core sizes of 36 μm diameter with little problem [19]. However, it has been proved to be more difficult to produce filaments having powder cores of 25 μm and less, although 20 μm cores have been produced in 18-filament sample wires. Wires with 192-filaments of lengths up to 3 km have been produced recently by a wire drawing route [22]. In this work a cut-off particle size of 3 μm has been specified for the $NbSn_2$ powder, which is mixed with Sn powder and inserted into a thin walled Cu tube. Densification of the powder is achieved by reduction of the tube prior to insertion in the Nb tube which is surrounded by Cu. Production of this wire shows that it meets the preliminary specifications for the LHC at CERN [23]. Much work is still continuing in the study of the ECN route despite the recent advent of high T_c superconductors, because of the possibility of producing industrial lengths for high field applications. The diffusion reaction of $NbSn_2$ with Nb, with and without the presence of Cu, has been the subject of many investigations [16-24] as formation of the phase Nb_6Sn_5 with a needle-like structure plays an important intermediary role in the creation of Nb_3Sn of the correct grain size. Both coarse and fine grained Nb_3Sn are produced [19] during diffusion reaction. The coarse grains are produced by reaction of Nb_6Sn_5 with the Nb whereas the fine-grained Nb_3Sn, which is required for transport of the superconducting current as it contains proportionally more flux-pinning centres, occurs as Sn diffuses through existing Nb_3Sn into the Nb tube.

3. High T_c Oxide Superconductors

For convenience, it is simplest to consider the powder processing of high T_c oxide superconductors in three subsections, each of which includes a summary of present experience and expertise on the three major groups of materials, namely $YBa_2Cu_3O_{7-x}$ (YBCO), Bi-Ca based copper oxides (BSCCO) Tl-Ca based copper oxides (TBCCO). This is appropriate because the degree of substitution possible, particularly with respect to the last two classes, means that there exists considerable overlap of systems and processing technology used for all high T_c superconductors. The three subsections will deal with powder manufacture, forming and shaping and finally sintering.

3.1. Powder Manufacture

To date the most commonly investigated high T_c superconductor is the YBCO, or so-called 1-2-3 system. As indicated by Markert *et al.* [25] twelve compounds crystallize in the same layered orthorhombic perovskite-like structure. The compounds may be represented by the formula $RBa_2Cu_3O_{7-x}$, where R is either Y or a lanthanide element (with the exception of Ce, Pm or Tb) and x is approximately equal to 0.1.

The original discovery of superconductivity near to 30K by Bednorz and Muller [1] was made with the compound $La_{2-x}Ba_xCuO_4$ which has an identical structure to that of the undoped orthorhombic La_2CuO_4 or its high temperature tetragonal polymorph. These structures may also be described as a layered compound resulting from a stacking of perovskite cubes. The oxygen stoichiometry corresponds to La_2CuO_4 within 1%, however,it has been shown that physical and electrical properties are extremely sensitive to very small changes in oxygen content. With YBCO material the critical temperature, T_c, has been shown to be a function of the orthorhombic distortion of the unit cell, which is related to the oxygen content and the ratio of Cu^{2+} to Cu^{3+} ions within the lattice. As a result of this strong dependence of superconducting properties upon the oxygen content, the powder routes used to produce these superconductors are required to allow correct oxygen uptake whilst also achieving the correct homogeneity and positioning of atoms within the lattice. It is worth mentioning that for YBCO as the ideal perovskite lattice distorts to orthorhombic the critical temperature increases from 66K to 94.5K as the oxygen content changes from $O_{6.65}$ to $O_{6.9}$ [26]. The distortions in the orthorhombic phase are accommodated by the formation of twins and faults, as shown in Fig. 2 [27]. Of the many papers on the relationship between microstructure and superconducting properties, the recent overview by Narayan [28] and research summary by Kroeger [29] may be cited as convenient and readable summaries of many of the relevant publications. It is clear that a large proportion of the microstructural observations and superconducting property measurements are related directly to the powder manufacturing technique and processing route selected.

3.1.1. Direct solid-state reaction

When the complexities of the effects caused by chemical homogeneity, purity and oxygen content are considered, it is not surprising that the early papers on YBCO presented a wide spectrum of data when the basic method of manufacture was one of simple direct solid-state reaction, frequently carried out with little regard to established ceramic and powder production practices. A similar situation exists for BSCCO and TBCCO and other derivatives of these materials. Essentially, the earliest process for the manufacture of high T_c superconductors was one of solid-state reaction of oxides, although carbonates and hydroxides were also used as sources of the metal cations. Such processing requires intimate mixing of the compounds prior to reaction for any reasonable chance of success. For YBCO it was established initially that ball milling for periods of up to 12 hours gave reasonable homogeneity of powder feed for subsequent drying, pressing and heating in air at 900-950°C. For the best results, it has been shown that after calcining in oxygen for approximately 16-24 hours the pellets should be cooled in oxygen and then broken down to powder by grinding prior to re-pressing and sintering at 900-950°C. DTA has been used to demonstrate that four separate calcinations, with intermediate grinding and re-pressing, are required to produce single phase YBCO at 900°C [30].

The starting materials and the milling operations have been the subjects of much discussion in relation to their effects upon the superconducting properties of the products. Phillips *et al.* [31] have shown that starting with >70% Y_2O_3 and CuO, and 99% of $BaCO_3$, of < 10 μm mean diameter, wet milling in trichloroethylene for 2 hours yielded a mixture with 100% particles < 10 μm. This mixture was suitable for calcining at 900°C and subsequent milling to produce a powder

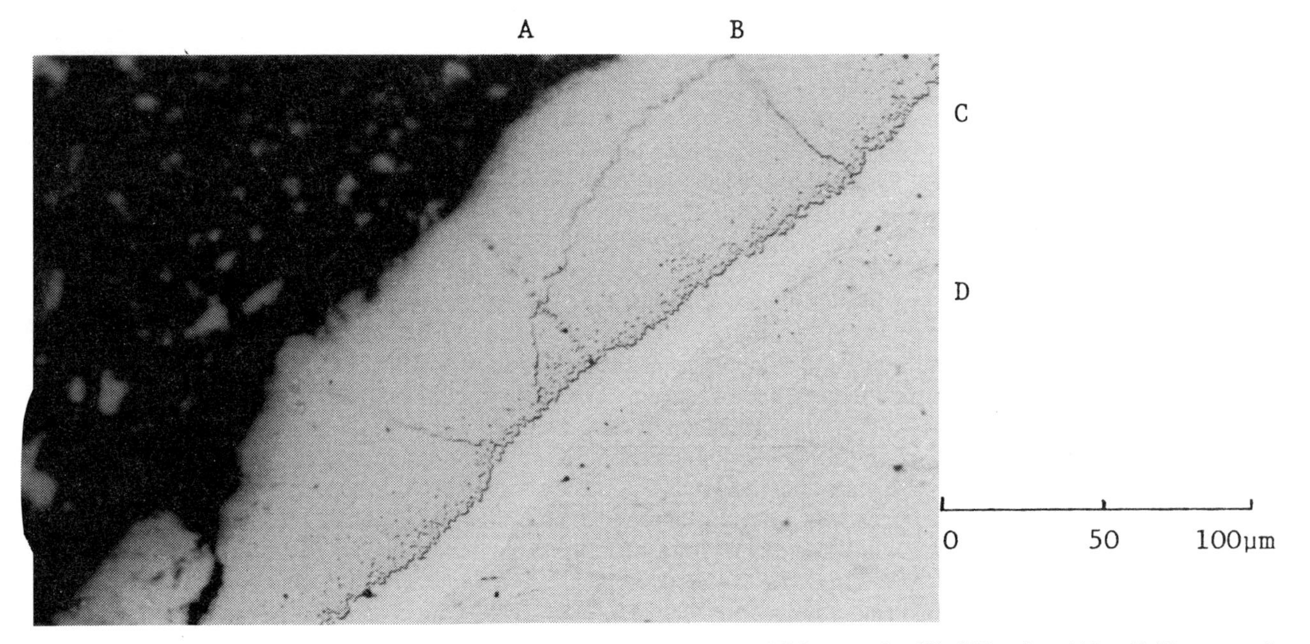

Figure 1 Optical micrograph of development Nb_3Sn superconductor using $NbSn_3$ powder-filled Nb tubes. (After H. Yorucu and F.R. Sale). Reaction 750°C for 100 h. A: powder core, B: Nb_6Sn_5, C: Nb_3Sn, D: Nb container.

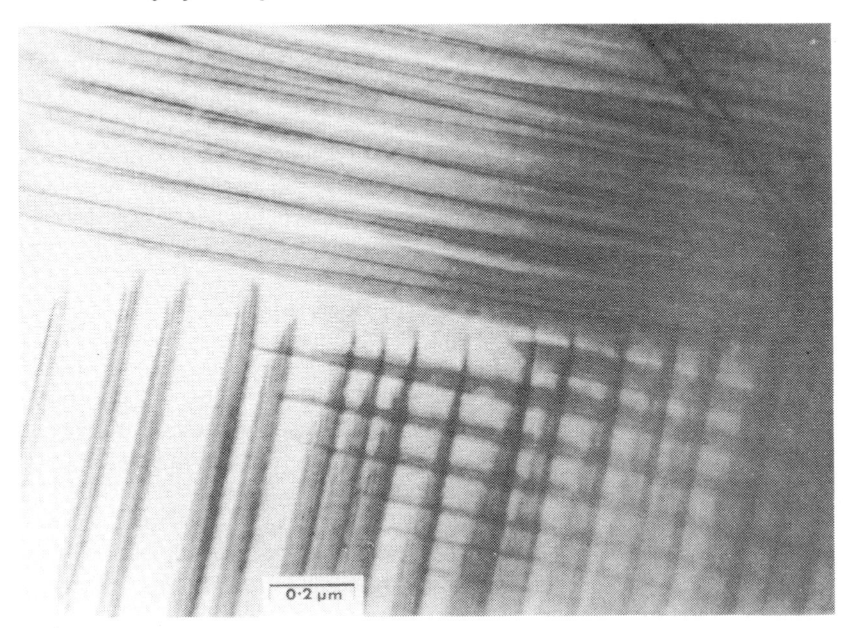

Figure 2 TEM micrograph of faulted YBCO sample produced from citrate-gel powder sintered at 950°C in oxygen. (After Mahloojchi et al. [27]).

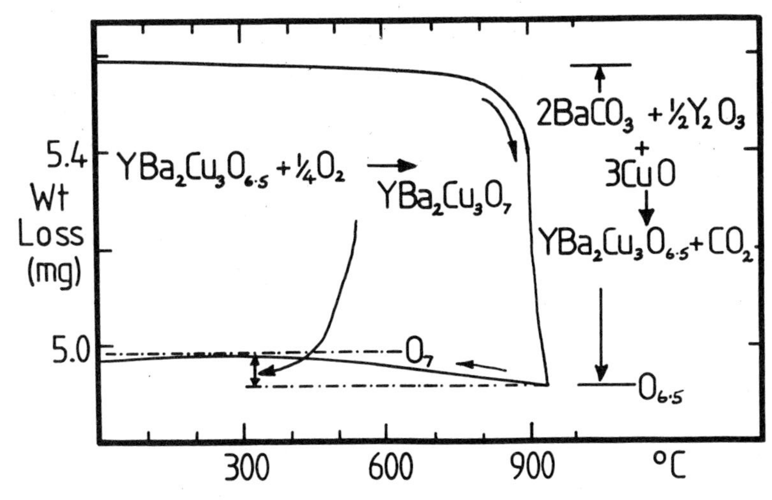

Figure 3 Thermogravimetric data for the formation of YBCO by direct reaction of oxides and carbonate. (After Irvine et al. [37]).

with a normal particle size distribution centred on 8 μm in which 95% of the powder was <10 μm. Dry attrition using ZrO_2 grinding media has been used to breakdown YBCO which had been produced by oxide carbonate reaction at 900°C for 10 hours [32]. No adverse reactions with atmospheric gases were reported and no significant impurity pick-up was detected. Such grinding gave powders of 0.3 μm diameter and surface area of $3m^2 g^{-1}$.

Many variations of the powder preparation sequences have been used, even when the same starting compounds were selected. As an example, Briggs *et al.* [33] milled stoichiometric quantities of Y_2O_3 and $BaCO_3$ using Y_2O_3 rods in propanol in a polythene jar for 30 minutes prior to adding the appropriate amount of CuO. Further milling for 30 minutes achieved intimate mixing and was followed by drying in air using infra-red heating. Calcination of the mixture in oxygen for 5 hours at 950°C followed by cooling in oxygen at 350°C h^{-1} to 500°C and 120°C h^{-1} to room temperature gave a suitable feedstock for fabrication. Nevertheless, despite all the care of operation, small amounts of Y_2O_3 and CuO were detected in this powder product.

The importance of the repeated grinding and firing stages in the processing of oxides and carbonates cannot be over-emphasized. Clarke [34] has reported the presence of barium carbonate at grain boundaries as a possible result of processing. The carbonate is claimed to lead to degradation of critical current density in the sintered superconductors. Fracture surfaces have been examined in ultra-high vacuum by ultraviolet photoemission spectroscopy, X-ray photoelectron spectroscopy and Auger electron spectroscopy and it has been shown that a ~5Å carbonate-like bonding layer may exist at grain boundaries, even in samples which have been produced from barium salts other than the carbonate.

A number of thermogravimetric studies have been performed on the reaction of oxides and carbonates to produce YBCO. Of these, the work reported by Button *et al.* [35], Beruto *et al.* [36] and Irvine *et al.* [37], may be cited as significant contributions. Figure 3 shows typical data obtained for the reaction of $BaCO_3$, Y_2O_3 and CuO to produce the superconductor [37]. The reaction:

$$2\,BaCO_3\,(s) + Y_2O_3\,(s) + 3CuO \rightarrow YBa_2Cu_3O_{6.5}(s) + 2\,CO_2(g)$$

is accompanied by a weight loss of approximately 11% and this is followed by a slight weight gain on cooling in oxygen as the oxygen content of the superconductor powder is increased by O_2 intercalation. Beruto *et al.* [36] used DTA and TG to study the reactions of both $BaCO_3$ and BaO_2 with Y_2O_3 and CuO in N_2 and O_2 environments. They noted that in the case of BaO_2 weight losses occurred in two steps. The first of these was associated with the decomposition of BaO_2 and the second was associated with reaction to produce the perovskite phase. Button *et al.* [35] show clearly that whilst pure $BaCO_3$ does not decompose at temperatures below 1000°C, it decomposes at temperatures as low as 850°C in the presence of Y_2O_3 and CuO. However, even on holding at 920°C for 70 hours, the mass loss observed in TG was less than that expected for full decomposition of $BaCO_3$.

Although the majority of the well classified papers on high T_c superconductors relates to YBCO, ever increasing reports are available on BSCCO and TBCCO types. These two classes have received much attention because BSCCO offers less problems with respect to chemical stability, oxygen content and annealing whilst the second class, although potentially of a poisonous nature, offers significant increases in T_c. As for YBCO, the simplest and most frequently used process for powder manufacture has been direct reaction of oxides and/or carbonates. The discovery of superconductivity in the ternary Bi-Sr-Cu-O system by Michel *et al.* [38] was rapidly followed by the observations of Maeda *et al.* [39] of superconductivity in the system containing Ca which is the basis of all the BSCCO compositions investigated so far. The powder preparation method adopted by Chu *et al.* [40] is typical of many. In this work, appropriate amounts of Bi_2O_3, $SrCO_3$, $CaCO_3$ and CuO were mixed and ground prior to calcination in air at 750-890°C in either platinum or alumina crucibles. The calcined material was crushed and ground to increase homogeneity. DTG showed that reaction of the starting materials commenced at 760°C and continued to 840°C, with the final compound melting at approximately 890°C. Aluminium was substituted into BSCCO using this method of preparation, however, it was shown to play only a very minor role [40]. Numata *et al.*[41] have reported studies on the optimization of parameters for the production of BSCCO superconductors. Again, a simple direct reaction technique, which consisted of the mixing of appropriate quantities of Bi_2O_3, $SrCO_3$, $CaCO_3$ and CuO followed by grinding and calcining at 750°C-850°C for 5 to 15 hours, re-grinding, re-calcination and finally re-grinding, was used to obtain a starting powder for further processing. Lead was added to the samples by the inclusion of Pb_3O_4 in the initial oxide/carbonate mixture. The same basic method is used irrespective of the composition of superconductor required, i.e. for the nominal composition $Bi_1Sr_1Ca_1Cu_2O_x$, abbreviated to 1112, the same technique is used as for 2122 and 2223 compositions. To date, preparation of this latter composition has not resulted in a single 2223 phase, apparently due to phase separation of the 2122 composition phase during production [42-44]. However, Pb-substitution has been shown to result in stabilization of the 2223 high T_c phase [45] and this accounts for the interest in Pb additions.

Direct reaction techniques for the production of thallium-based phases often take the form of a two-stage process rather than the direct reaction of all the starting materials in one process. Sheng and Hermann [46] report the use of Tl_2O_3, $BaCO_3$ and CuO as starting materials for the production of a Tl-Ba-Cu-O superconductor. However, because Tl_2O_3 has a low melting point of 717°C, a relatively high vapour pressure and may begin to decompose at temperatures as low as 100°C, a very short time high-temperature reaction, followed by quenching was shown to be necessary. The short duration thus minimized decomposition and evaporation of Tl_2O_3. However, a preferred route has been shown to be the use of a melt-solid reaction, which has more rapid kinetics than a conventional solid-solid reaction. Sheng and Hermann [46] mixed and ground appropriate quantities of $BaCO_3$ and CuO in an agate pestle and mortar and heated the mixture in air at 925°C for periods in excess of 24 hours. Several intermediate grindings were used to yield a uniform powder of either $BaCu_3O_4$ or $Ba_2Cu_3O_5$. This powder was subsequently mixed with Tl_2O_3, ground, pelletized and heated to 880-910°C for 2-5 minutes in flowing oxygen. Quenching was carried out once a small amount of melting was observed. Complete reaction did not occur in this heating cycle and the product was seen to contain unreacted Tl_2O_3 and mixed oxide. The major problem was determined to be the volatilization of the Tl_2O_3. It took only a very short time to realize that the substitution of Ca into the Tl-Ba-Cu-O superconductor raised its T_c to values of the order of 120K and so the TBCCO system was rapidly established [47]. For the five element system, the oxides Tl_2O_3 and CaO were mixed with barium cuprate ($BaCu_3O_4$), ground and heated for 5 minutes in flowing oxygen at 880-890°C in a manner similar to that adopted in the absence of CaO. The use of $BaCu_3O_4$ is now established as this compound has the lowest melting point of phases in the Ba-Cu-O system and is extremely fluid at the reaction temperature. Itoh *et al.* report that barium cuprate may be formed by the solid state reaction of $Ba(NO_3)_2$ and CuO [48] as well as by the reaction of oxides/carbonates.

As for BSCCO superconductors, no different powder preparation techniques have been devised for the various compositions of interest in the TBCCO system. A minor variation in the preparation of powder was the use of BaO_2 as well as Tl_2O_3, CaO and $BaCu_3O_4$ [49]. Here the $BaCu_3O_4$ was prepared by reaction of $BaCO_3$ and CuO in air at 900°C, and then the mixed and ground powders were heated to 890°C for 10 minutes, in oxygen. Furnace cooling at 100°C per hour was shown to yield 'almost' single phase 2223 material. Although the use of barium cuprate as a starting material now appears desirable for the successful production of superconductors of the TBCCO type, there are a number of reports which indicate that individual oxide mixing and reaction is satisfactory [50-54] whilst Gopalakrishnan *et al.* [55] report that single oxide reaction is not feasible because of the high volatility of Tl_2O_3. An exception to the previously mentioned methods of powder production is that of Eibschutz *et al.* [56] who simply melted mixtures of Tl_2O_3, $Ba(NO_3)_2$, CaO and CuO at 1000°C in air for 15 to 25 minutes in a covered Pt crucible. After fusion and cooling, the product was ground to yield a satisfactory powder.

3.1.2 Other methods of powder manufacture

It is clear that whilst the direct reaction of oxide mixtures or oxide/carbonate mixtures has enabled high T_c superconductors of the three classes YBCO, BSCCO and TBCCO to be produced, such preparation is tedious and time consuming. In addition, each is accompanied by experimental difficulty, e.g. volatility of Tl_2O_3 for TBCCO, oxygen intercalation for YBCO and homogeneity for BSCCO. Also, each type requires an extensive milling operation in order to maximize homogeneity of the product powder and such milling leads to the possibility of impurity pick-up. It is evident, therefore, that alternative powder manufacturing processes offer significant advantages in the production of high T_c powders.

Powders for the production of YBCO have been produced by a variety of advanced processes which include co-precipitation, freeze-drying and gel-processing. The usual advantages of these processes are increased chemical homogeneity, close control of stoichiometry and small particle size which gives increased reactivity of powder and lower sintering temperatures. In addition, if a number of milling stages can be obviated, then the chances of picking-up impurities are much reduced. A programme of study on the precipitation of precursor powders for YBCO superconductors has been carried out by Bunker *et al.* [57] using solutions containing Y, Ba, Cu chlorides or nitrates and ones containing precipitating anions such as hydroxide or carbonate. On mixing the solutions, an insoluble precipitate is formed which is then heated to yield the required superconductor. Bunker *et al.* [57] show that if the precipitation is not carried out correctly then the precipitates may contain more impurities than ball-milled mixtures. For YBCO, successful co-precipitation is achieved using control of pH and anion concentration so that the highest degrees of supersaturation may be achieved for all dissolved species. The use of non-contaminating counter ions is of vital importance, but has unfortunately been disregarded by a number of workers. If, for example, potassium carbonate is used in the initial anion solution, then the final product may contain up to 1wt% potassium [57]. To avoid this type of contamination, it is recommended that solutions of tetramethylammonium hydroxide and tetramethylammonium carbonate are used. Ammonia has been shown to be unsuitable because it can form stable, soluble complexes which retain some copper in solution. Freeze drying or filtering of the co-precipitated products have been used as separation procedures for YBCO superconductors. The agglomerates which are produced are some 1-3 μm in size and are comprised of many fine particles < 50 nm in size. Decomposition at 800°C to 900°C yields the required perovskite powder of average grain size of 0.3 μm. The anions used, chloride or nitrate, have an effect on the superconducting properties of the sintered material. Nitrate precursors give single-phase material with a sharp superconducting transition of 92K and J_c of 1400 A/cm^2 at 77K and 10 gauss. Powders from chlorides do not give single-phase products and so do not show sharp transitions. It has been claimed that flux pinning is about 5 times greater in the chloride produced samples than in the nitrate material [34].

In an attempt to overcome problems of inhomogeneity in BSCCO materials Srinivasan *et al.* [58] have studied the production of BSCCO from nitrate precursors using bismuth nitrate, copper nitrate, calcium nitrate and strontium nitrate mixed into deionized water at 80°C for 30 minutes. The incompatible solubilities of the nitrates was such that the bismuth nitrate did not dissolve completely but was slurried with the other dissolved nitrates. Although bismuth nitrate is completely soluble in concentrated HNO_3, this procedure was not thought to be necessary. After drying at 120°C overnight, the mixed nitrates were ground to achieve further homogeneity and then calcined at 850°C to 860°C for 10 hours. To obtain a satisfactory powder, the calcined product was re-ground before pelletizing and firing. The problem of the incompatible solubilities of the nitrates has also been reported in studies of YBCO materials [59]. It was acknowledged that a true atomic-scale mixing could not be achieved using a nitrate method. However, as indicated by other workers, the level of inhomogeneity is on such a fine scale as to be insignificant relative to the grain sizes of the products obtained after firing. Cooper *et al.* [60] have reported the successful application of spray pyrolysis of nitrate solution for BSCCO materials. Although this work was concerned particularly with the production of thick films, such a technique can easily be converted to powder production.

Oxalate precipitation methods have been investigated for the production of YBCO material [61,62]. As for other co-precipitation methods, precise optimization of initial conditions such as temperature, pH and concentration is necessary to obtain satisfactory products after decomposition of the oxalates at 950°C in air. Difficulties are present with this technique and it has been reported by some groups of workers [61] that superconductivity has not yet been seen in sintered products although other workers [62] report T_c values of the order of 90-92K for YBCO.

The amorphous citrate gel process has been used successfully to prepare YBCO [63-65]. A true amorphous gel is not produced from the mixture of nitrates and citric acid unless the pH of the mixed solution is maintained at about 6. At lower pH values some traces of barium nitrate are precipitated from the gel during the removal of H_2O by evaporation. Powders produced by this route have been processed successfully to give monolithic materials which have sharp T_c's of the order of 92K. Substitution of Ho and Eu has been shown to be possible using citrate gel processing. Two-stage pyrolysis of the gel precursors has been recommended to prevent overheating which may lead to particle growth and segregation within the powders [63, 65]. In addition, as with conventional processing, best results were obtained by pyrolysis in flowing oxygen. The temperature of the final stage of firing has been shown to control the particle size of the powder. Typically, submicron powder is obtained at 700°C whereas 1 μm powder is produced at 900°C [63] (Fig. 4). Other similar sol-gel and liquid-mixing processes have been devised for perovskite powder production [66,67]. The citrate-gel processing route has been investigated for the production of BSCCO materials [43] and the problems of precipitation during gel formation which were experienced with YBCO manufacture [63] were exacerbated. The use of ammonia to control pH successfully prevented the precipitation of $Sr(NO_3)_2$ during gelling but could not prevent the appearance of $Bi(NO_3)_2$ as a dispersed crystalline phase within the gel. It was, however, noted that on decomposition of the gel and $Bi(NO_3)_2$ carbonates were produced at low temperature, and these were likely to lead to more difficulties than the fine dispersion of $Bi(NO_3)_2$. This study [43] indicates a potential limiting weakness inherent in citrate, or any other, gel processing operation in that the high stability of Sr, Ca and Ba carbonates means that they are always likely to be produced by reaction of CO_2 with the partially oxidised product during the gel firing operation. Such difficulties have been experienced in the production of thick and thin films of high T_c superconductors [68-71] as well as in the production of powders.

In an attempt to overcome problems associated with the low solubilities of bismuth and thallium salts in solution, which in turn lead to the difficulties of homogeneous precipitation with Cu salts, Barboux *et al.* [72] have investigated the use of water-glycerol solutions containing nitrates for thick film formation of both BSCCO and TBCCO materials. The technique has been extended successfully to powder production [73]. Care is required with this route because of a very vigorous decomposition of the gelled product on heating, which may lead to loss of stoichiometry.

3.2 Shaping and Forming

As indicated recently by Poeppel *et al.* [74] before many potential applications for high T_c superconductors may be realized the electrical and mechanical properties of the superconductors must be optimized and improvements in flexibility and strength must be obtained by reduction in critical flaw size. Such a reduction needs correct ceramic processing which may also lead to improvements in critical current density J_c. In uses in magnetic coils, insulating materials will need to be bonded to the superconductors. These requirements point to the need for economic and reliable shaping and forming processes for monolithic and composite superconductors.

Many of the experimental powders described in the previous section have been processed by simple uniaxial pressing techniques. Sometimes isostatic processing (hot or cold) has been used. Nevertheless, for many technologically important shapes and quantities, other forming techniques are required. Presently, tape casting, extrusion and slip casting are being investigated in addition to cold drawing or rolling techniques using sheathed powders. For successful uniaxial pressing of YBCO materials a number of workers have shown that binders, such as 2 wt% addition of PVA [31], aid the compaction process. Typically pressures in the range 100-300 MPa have been used to produce disc or bar shapes by uniaxial pressing [31,75] whether, or not, a binder has been added. Isostatic pressing has been used to produce a wider variety of shapes, for example, using a steel mandrel and latex bag pressing at 150 MPa gave satisfactory green tubes 80 mm x 22 mm with a 19 mm bore [31].

Moving on from the dry-pressing of powders, the forming techniques which have received most attention to date are tape casting, screen printing, slip casting and extrusion. In addition the plasma-spraying of powders has been used to produce YBCO coatings between 10 and 200 μm in thickness which are durable and superconducting after annealing in oxygen for up to 20 hours [76,77].

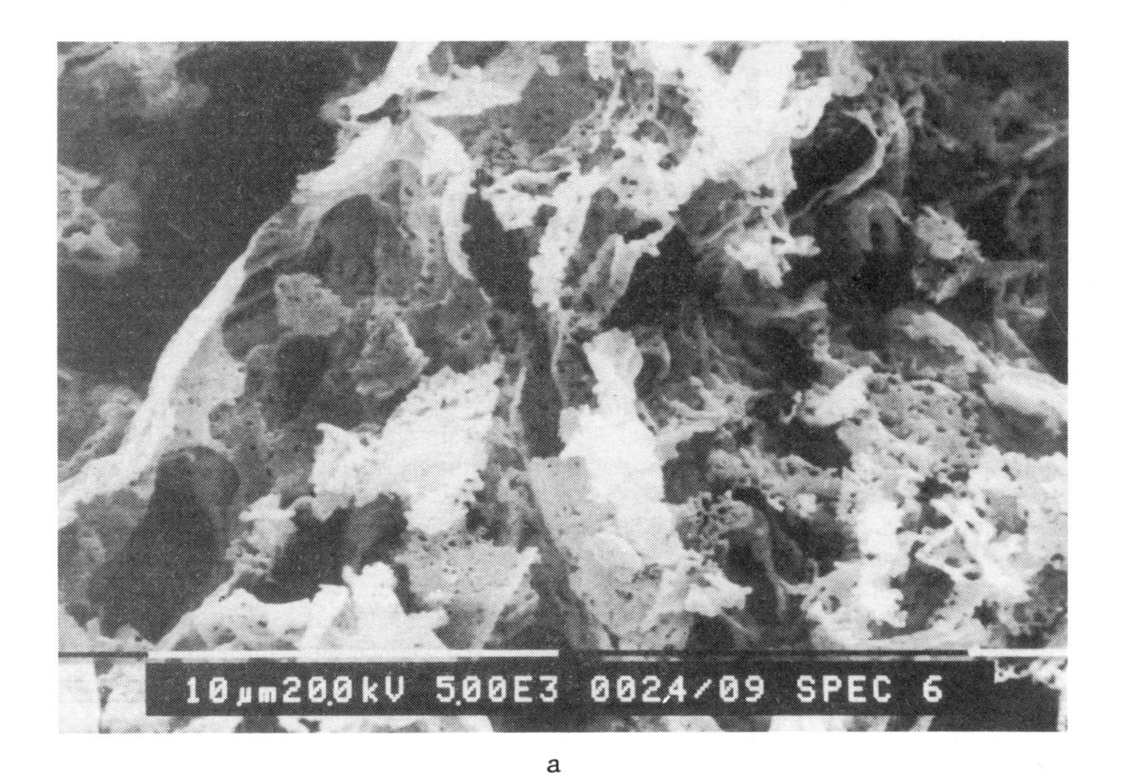

a

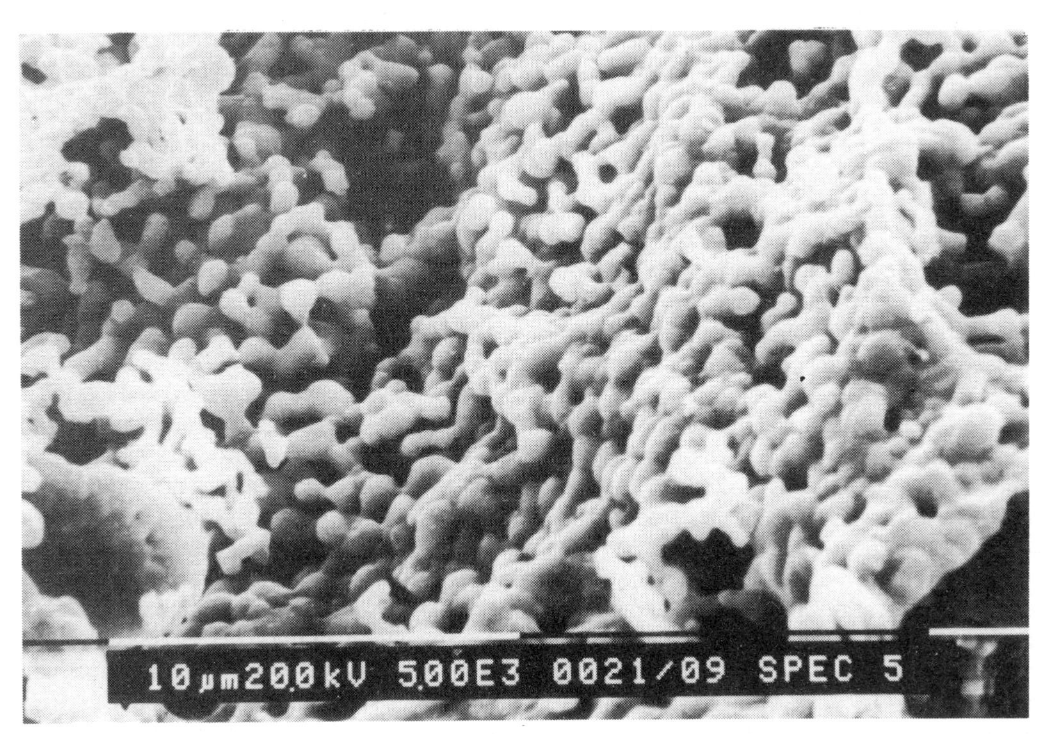

b

Figure 4 SEM micrographs of citrate-gel produced YBCO (a) 300-700°C firing (b) 300-900°C firing. (After Mahloojchi and Sale [63])

In the tape casting, screen printing or slip casting of high T_c materials the major difference from the use of these techniques for conventional ceramics arises from the requirement to use organic solvents to ensure no reaction with the high T_c powders. Essentially the processes rely upon the mixing of the powder with a binder, a plasticizer, a dispersant and a solvent [78]. Once this is achieved the resultant slurry can be poured into a device which allows a controlled thickness to be deposited [79] or conventional screen printing may be used [80]. In all cases, following evaporation of the solvent, the binder and plasticizers must be burnt away, usually by slow heating to temperatures of the order of 300°C, before the particles are subsequently sintered at a higher temperature. To date most work has been carried out on YBCO materials. To prevent the attack of the YBCO by water the organic solvents methyl ethyl ketone and xylene have been used in the tape casting and slip casting processes [81]. In the tape casting of YBCO up to 13 vol% Ag has been shown acceptable as an addition which improves the mechanical strength, without reducing the electrical properties, of the product [82]. The advantage of the metallic inclusion is evident by the fact that the resultant material was capable of withstanding a strain of approximately 2%. Without silver, an equivalent tape fractured at values of strain of the order of 0.15%. A similar mixture can be used for wire fabrication [83].

If the solid content is kept high in the powder/organic mix, then extrusion can be used as the forming process [84,85]. Extrusion through a die, followed by winding upon a mandrel has yielded 90% dense wires, less than 100 μm in diameter with T_c of 77K and J_c values of hundreds of A/cm^2 [86]. A similar process has been used to manufacture a coil which at 77K and operating with a current density of 1300 A/cm^2 has produced a field of 21 gauss [86]. An interesting observation from much of the extruded and tape cast material is that the shear stresses involved in processing align the superconducting particles, because of the large aspect ratio of the particles, and this results in alignment of the grains in the product along the axes of the wires or in the planes of the tapes. This alignment in turn enhances the superconducting properties of the product.

Much use has been made of conventional, metallurgical cold-drawing practice in the shaping of high T_c superconductors. In particular there are a large number of reports of the fabrication of ceramic wires (especially of YBCO) using cold drawing of metallic tubes (normally Ag) which have been packed with the superconducting powder. Cold rolling has sometimes been used to obtain the necessary reductions which are of the order of 10 to 100 [33,87]. The key to the success of this process lies in the non-reactivity of the Ag with the YBCO. As mentioned earlier, an Ag/YBCO powder mixture can be used to fabricate wire which after a high temperature heat treatment yields a ductile and superconducting YBCO/Ag composite [83].

The shock compaction of powders is presently being applied to the production of all three major classes of high T_c superconductors [78,88,89]. Again, much of this work uses a YBCO/Ag composite powder. Work on BSCCO and TBCCO materials are in their early stages, however, successful reports have been made for both systems [78].

3.3 Sintering

There is little doubt that much of the variation in properties claimed for given compositions of high T_c superconductors can be attributed to differences in the sintering treatments used. The differences in sintering conditions, in particular gas atmosphere, temperature and time, led to the early, wide variance of results for YBCO materials and to the variety of microstructures, grain boundary structures and lattice defect densities seen in electron metallographic studies [27-29]. Today the vital importance of sintering conditions has been accepted and a number of basic rules have been established.

For YBCO materials a sintering temperature of the order of 900-950°C is now accepted as necessary to obtain samples of reasonable density. However, this sintering must be followed by a prolonged annealing in oxygen in order to optimize the oxygen stoichiometry and hence superconducting properties. The reason for the annealing is demonstrated clearly in Fig. 5, Takagi *et al.* [90], where the relationship between oxygen non-stoichiometry of YBCO and log PO_2 is shown for a number of temperatures. It is clear that at sintering temperatures of the order of 950°C in 1 atm. of oxygen the equilibrium oxygen content of the superconductor falls from $O_{6.9}$ to $O_{6.3}$. As optimum properties are achieved with the highest oxygen content possible (as close as possible to $O_{6.9}$) substantial low temperature anneals are required for the oxygen to be regained. It is tempting to consider that, from the data presented in Fig. 5, prolonged annealing in oxygen at around 350°C would give optimum properties. Takagi *et al.* [90] have, however, shown that long-time annealing at temperatures of 350-400°C often results in the degradation of superconducting properties. Monod *et al.* [91] have carried out systematic TG measurements under various oxygen partial pressures followed by T_c determination on the same samples. As with a number of other investigators it was shown that by heating at temperatures of the order of 700°C for up to 7 hours in a vacuum of approximately 10-9 bar it was possible to remove, reversibly, the oxygen from the superconductor from approximately O_7 to O_6. On cycling, the oxygen uptake of the deoxidized specimen started at above 200°C and obtained a maximum rate at around 450°C. On heating further oxygen was lost from the sample and it was re-gained only on cooling in oxygen. This behaviour is summarized in Fig. 6. On the basis of these data it is evident how the "empirical recipe" for the preparation of YBCO of sintering at 950°C followed by a slow cool at rates as low as 1°C/min has been determined. The situation for YBCO is a little more complicated than indicated so far in that the tetragonal-orthorhombic phase transition temperature is also dependent upon the oxygen content. At 1 bar oxygen the temperature is approximately 686°C [92], although other workers have reported values as high as 950°C and above [35]. However, this value may drop to about 500°C when oxygen is desorbed from the material [93] and the tetragonal phase may be stabilized to room

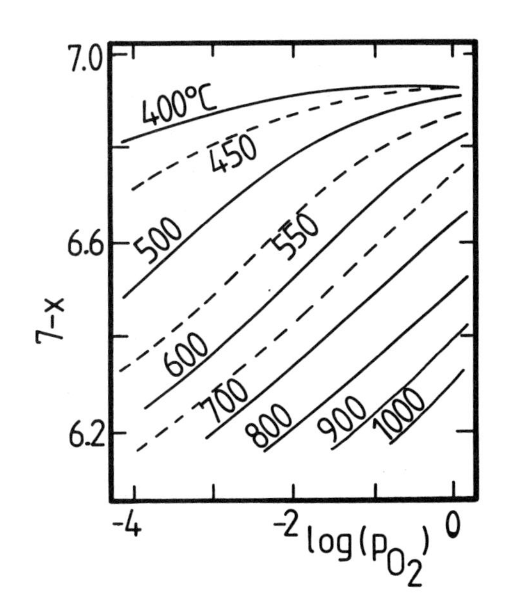

Figure 5 Relationship between non-stoichiometry in YBCO and log Po_2 for various temperatures. (After Takagi et al. [90]).

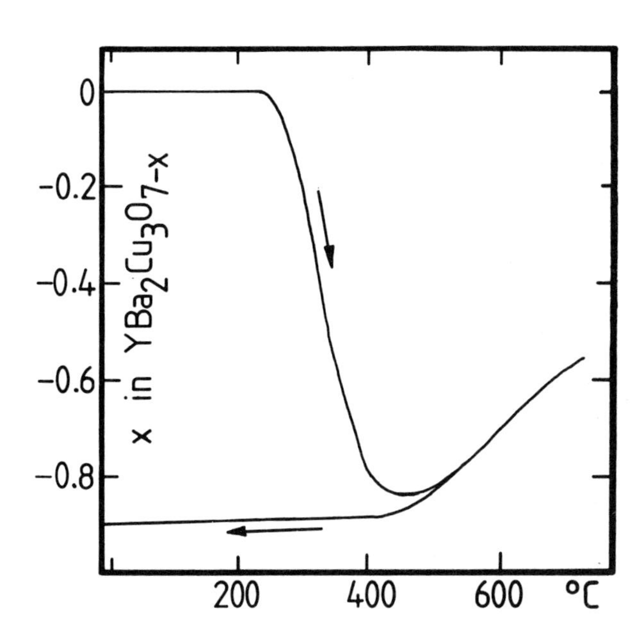

Figure 6 Mass changes on cycling YBCO, previously heated in vacuo, in oxygen. (After Monod et al. [91]).

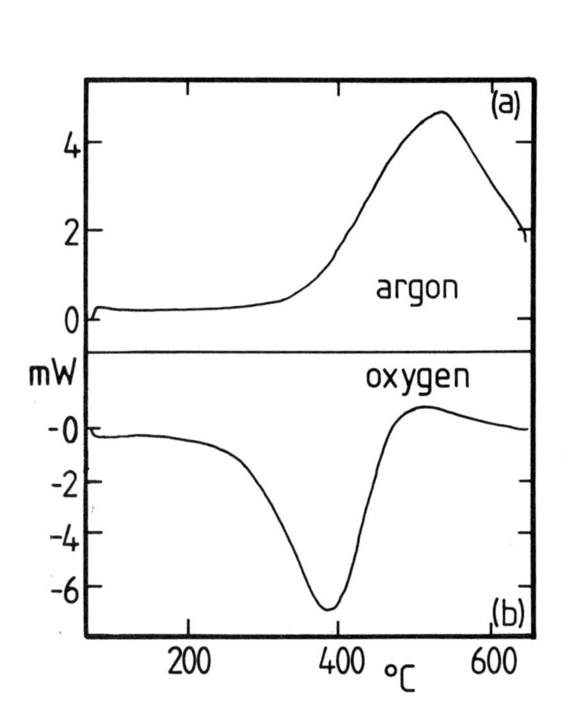

Figure 7 DSC traces for successive heating of YBCO in (a) argon and (b) oxygen. (After Glowacki et al. [93]).

temperature in an argon atmosphere. It is interesting to note that it is reported that the transition does not appear to affect the rate of oxygen intercalation (93).

The oxygen loss and it subsequent intercalation into the YBCO superconductor lattice has been the subject of a number of studies using DSC of which the work by Glowacki *et al.* [93] may be quoted as a typical example. Figure 7 shows the differences between DSC traces obtained from successive heating and cooling runs of YBCO material which were carried out in argon and oxygen. The endothermic peak seen on heating indicates the desorption of oxygen, whilst the exothermic peak (at rather lower temperatures) indicates the absorption of oxygen on heating in oxygen. The absorption peak at approximately 400°C indicates why this temperature region has been determined to be critical in the oxygen intercalation into practical YBCO superconductors. Dilatometric studies of the sintering of YBCO also shows the critical region of cooling to be of the order 450-500°C where detectable expansion of specimens commences as a result of the oxygen intercalation [63]. Such behaviour is evident in Fig. 8 where a complete heating and cooling cycle for the sintering in oxygen of a pressed pellet obtained from citrate-gel processed powder is shown [63]. This figure also indicates the importance of 900-950°C as the sintering temperature for YBCO material.

The effects of porosity on the superconducting properties of YBCO have been reported by Cheng *et al.* [94] who studied extruded rod samples which had been processed using viscous polymer solutions and high shear techniques. The samples also had a range of grain sizes. Porosity was shown to be important in that it allows oxygen to reach the grains during sintering and cooling and hence improves superconducting properties. However, porosity also has the detrimental effect of allowing corrosive materials, such as H_2O, to penetrate into the samples and thus cause deterioration. J_c was found to be small in low density samples because the connectivity of the material was low. However, J_c was also found to be low in samples of high density and this was explained by the absence of an interconnecting pore network causing inhibition of oxygen diffusion to the grains during preparation. Today it is felt that high J_c values will only be obtained in YBCO by processes such as melt-texturing which will give better grain alignment [34,95].

The sintering, and ultimate properties, of BSCCO materials are not as strongly influenced by oxygen partial pressure as those of YBCO and air has been shown to be satisfactory as a sintering atmosphere in the temperature range 865-900°C. A number of papers indicate that no advantage is gained by the sintering of BSCCO in oxygen [96,97], whereas other workers [41,43] have shown that a marked deterioration in T_c results from sintering in oxygen. Sintering temperature is important as fusion commences in BSCCO materials at temperatures of the order of 910-920°C. Numata *et al.* [41] have shown that in order to maximize the proportion of the highest T_c phase (T=110K) in the BSCCO system the sintering temperature must be below 920°C. At 920°C and above the low T_c phase (75K) was produced irrespective of the initial composition of the pellet. Combined TG-DTA experiments have been reported for 1112 BSCCO, as shown in Fig. 9, where it is clear that two major, reversible endothermic events occur on heating to sintering temperature and that weight loss accompanies each event. The high temperature endotherm is associated with fusion of the sample. The low temperature endotherm was considered to be associated with the production of the high T_c phase accompanied by oxygen evolution. The DTA for the two highest temperature events agree reasonably well with those determined by Rajabi and Sale [43], shown in Fig. 10, for a 2223 composition BSCCO powder produced by a citrate gel route. The low temperature peak shown in Fig. 10 is associated with reaction of the intermediate phases formed in the gel route to produce the superconducting phases and so is absent in Fig.9.

It is clear from work on a number of compositions in the BSCCO system that a sintering temperature of 800-825°C is too low for the production of the high T_c phase ($T_c \approx$ 110K) although.the lower T_c phase ($T_c \approx$ 75K) could be obtained at the lower sintering temperatures. At 865-875°C the volume of the high T_c phase increases with sintering time [41,43]. Because most studies on BSCCO have produced multiphase samples a big effect seen on increasing either temperature (below 920°C) or time is a general coarsening of the microstructure [43,44] as shown in Fig. 11. This coarsening is accompanied by changes in relative volume fractions and compositions of the phases. Numata *et al.* [41] have also shown that post sinter annealing at 400°C for 20 hours in oxygen of specimens of 1112 composition, which had previously been sintered at 875°C for periods ranging from 14 to 470 hours, did not affect the volume fraction of the high T_c phase. When Pb is added to the BSCCO system the sintering temperature may be reduced to 845°C and post sinter annealing at 400°C has been shown to increase the volume fraction of the high T_c phase. In general the best superconducting properties have been obtained by rapid cooling after annealing. The elevation of T_c is thought to be caused by oxygen deficiency caused by quenching [42]. The quenching practice is, however, not adopted universally. Veal *et al.* [97] report the use of controlled cooling rates of 40°C/hour from 870°C to 400°C in air, followed by a 28 hour anneal in air at 400°C and a furnace cool to room temperature, as the heat treatment required to produce maximum quantities of the 2223 high T_c phase from an initial starting composition of 4346 BSCCO.

The sintering of TBCCO materials is influenced dramatically by the volatility of Tl_2O_3 with the result that unlike YBCO and BSCCO standard procedures do not yet seem to have been established. Many of the sintering schedules at elevated temperature tend to be of extremely short duration, although furnace cooling at rates of the order of 250°C/hour was shown by Itoh *et al.* [48] to be necessary to give good properties in the product. Typically sintering periods of the order of 5 to 10 minutes at 870-910°C have been used followed by either a rapid cool or a furnace cool [48,98,101]. This method of sintering is, however, not exclusive. Usui *et al.* [99] report a sintering schedule in flowing oxygen at 870°C for 1 hour. These workers also report that sintering at 880°C for 1 hour in oxygen gave 2223 material of zero resistance

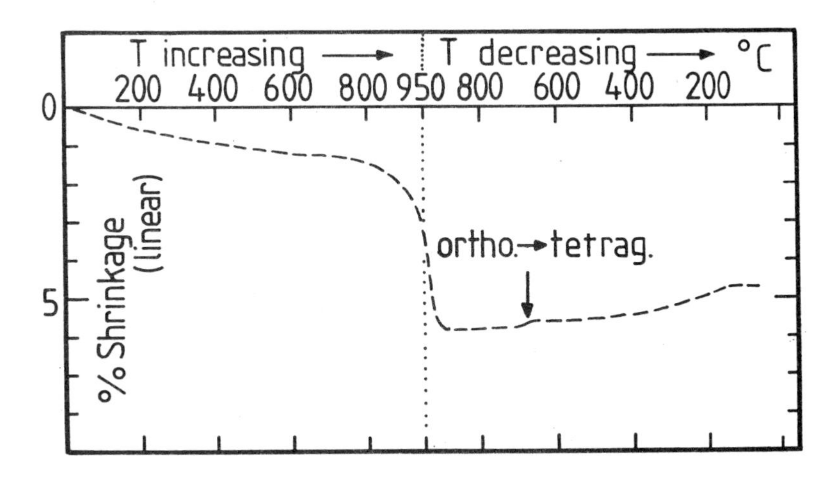

Figure 8 Dilatometric trace for the sintering in oxygen of citrate-gel produced YBCO material. (After Mahloojchi and Sale [63]).

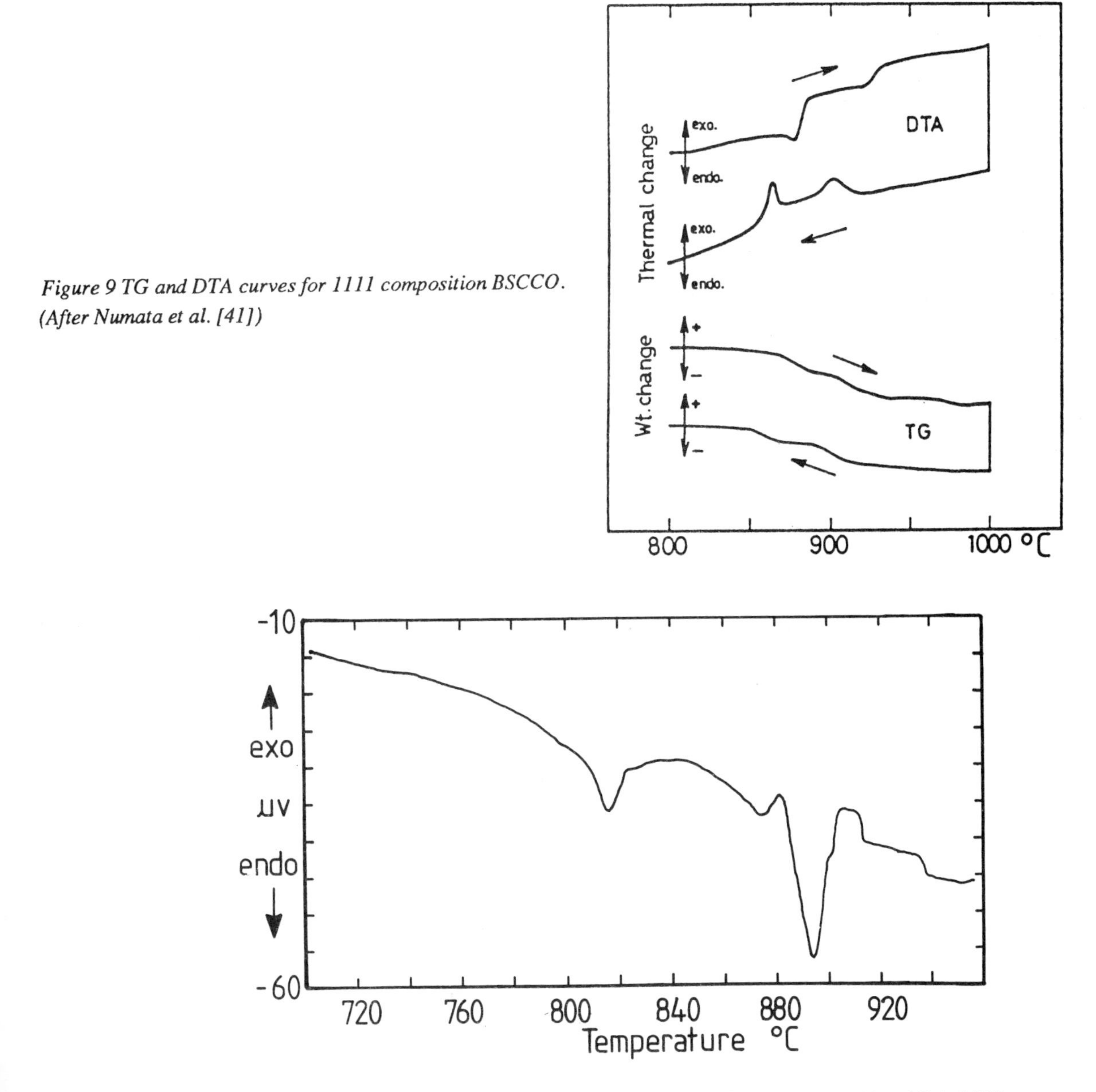

Figure 9 TG and DTA curves for 1111 composition BSCCO. (After Numata et al. [41])

Figure 10 DTA curve for citrate-gel produced 2223 BSCCO. (After Rajabi and Sale [43]).

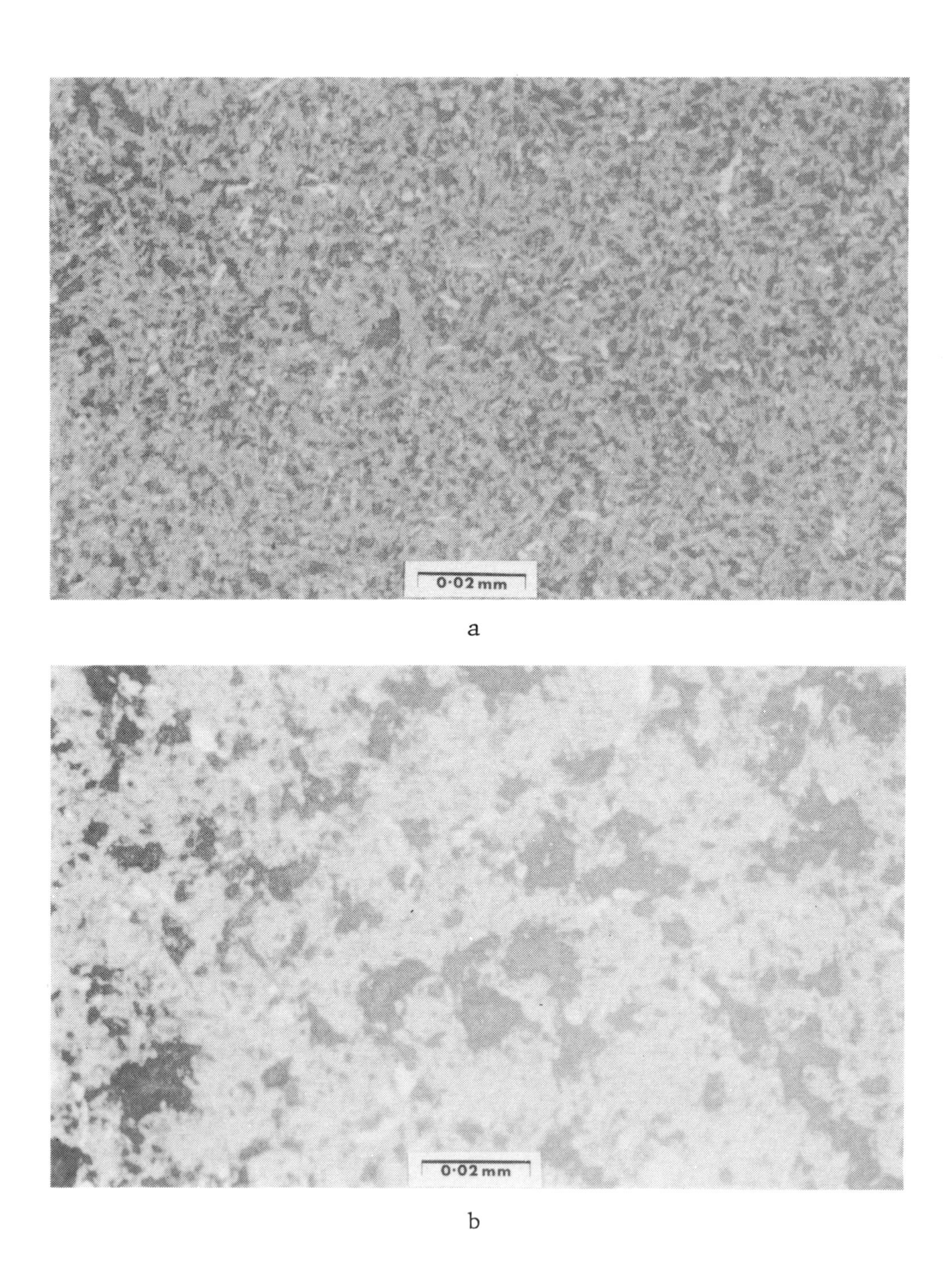

Figure 11 2223 BSCCO sintered for 1 h at (a) 865°C and (b) 906°C. (After Rajabi and Sale [43]).

at 118K, whereas similar sintering at 870°C gave a figure of 105K. Such a significant difference was not observed for a 2123 composition where the highest value of 112K dropped to 107K with change of sintering temperature from 880°C to 870°C.

Attempts have been made to use lower sintering temperatures of the order of 780°C in order to minimize Tl_2O_3 loss and hence increase compositional control. Yuling *et al.* [100] have compared samples of 1113 composition sintered at 720°C to 880°C in air for 8 hours. T_c increased up to 120K as the sintering temperature was increased from 720°C to 790°C but further increases in sintering temperature caused a reduction in T_c, until at 880°C no superconducting transition was observed.

A single step procedure commencing with well mixed and ball-milled individual oxides, which are pressed into pellets and heated for 3-30 minutes at 850 to 900°C in flowing O_2 prior to furnace cooling, has been used successfully by Zhou *et al.* [54]. Such a process obviously utilizes a liquid phase reaction and liquid phase sintering as Tl_2O_3 melts at 717°C. Presently, more information is required on this technique for production of TBCCO superconductors. Dou *et al.* [102] describe an apparently well-controlled sintering method in which Tl_2O_3 is added to a previously calcined mixture of $BaCO_3$, $CaCO_3$ and CuO, in either powder or a solid block form, and is placed in a sealed Ag tube. The calcined mixture was not completely reacted and contained a variety of products ranging from $Ba_3CaCu_3O_7$, Ca_2CuO_3 and $BaCuO_2$ to unreacted CaO. Final sintering and reaction was carried out within the sealed tubes at 800-860°C for periods ranging from 20 minutes to 3 hours followed by slow cooling at a rate of 3°C/minute to 200°C. Again this technique relies upon liquid phase sintering and reaction. However, controlled compositions may be produced because the loss of Tl_2O_3 by vapourization is obviated. Certainly, of the plethora of papers published on TBCCO materials, this liquid phase sintering technique in sealed Ag capsules seems to offer the most controlled production method. Starting with a nominal 1123 composition, the superconducting 2223 phase (T_c approximately 120K, $\Delta T_c \approx$ 8K) was prepared by sintering for 3 hours in flowing O_2 Prolonged heating for 6 hours caused dissociation of the phase and the 2212 phase appeared as a product ($T_c \approx$ 100K) [102].

4. Summary

It is clear from this selective review of the many hundreds of papers, which have been published over the $2^{1}/_{2}$ years up to the time of writing, that powder processing and general powder metallurgical methods have played a vital role in the production of high T_c superconductors. However, it must be pointed out that powder metallurgical techniques have also been used for A15 metallic superconductors to great advantage. It is also evident that much of the preparative work associated with the high T_c superconductors has suffered from almost complete lack of regard to fundamental powder processing principles. As a result, in the published papers concerning the YBCO, BSCCO and TBCCO materials, much conflict of properties of the products is evident. The processing of YBCO materials is now well understood as powder technologists have applied their efforts to the manufacture of useful monolithic shapes and thick films. Nevertheless much further work is required. BSCCO materials are in need of more study in the hope that the 2223 composition may be prepared as a pure single phase. The production of TBCCO materials is presently in a state of confusion and the only undeniable fact concerning this class of superconductors is that the volatility and toxicity of Tl_2O_3 makes manufacture of technologically significant forms extremely difficult. Early reports of the potential substitution of Tl by In may alleviate some of these manufacturing problems in the future. However, it must be expected that as new classes of high T_c superconductors are discovered these discoveries must be matched by efforts of powder technologists if useful products and technological applications are to result from all the extraordinary research effort on such superconductors. Recently, Goodstein [103] has attempted to put all this research effort into perspective. The opening statement of his paper commented "Three years ago, superconducting ceramics were heralded as the new wonder materials set to change industry. Now, high temperature superconductors seem not so super after all". This statement will no doubt be proved correct unless matching research efforts to those of the physicists and chemists are applied to the powder processing of these materials by powder technologists.

References

1. J.G. Bednorz and K.A. Müller, Z. Phys. B., 64, p.189, 1986.
2. M.K. Wu, J.R. Ashburn, C.J. Torng, P.H. Hor, R.L. Meng, L. Gao, Z.J. Huang, Y.Q. Wang and C.W. Chu, Phys. Rev. Lett., 58, p.908, 1987.
3. A.W. Sleight, M.A. Subramanian and C.C. Torardi, MRS Bulletin, XIV, p. 45, 1989.
4. C.C. Tsuei, Science, 180, p.57, 1973.
5. C.C Tsuei, IEEE Trans. Magn., MAG-11, p.272, 1975.
6. R. Flükiger, R. Akihama, S. Foner, E.J. McNiff Jr. and B.B. Schwartz, Adv. in Cryo. Eng., 24, p. 337. 1980,
7. R. Flükiger, S. Foner, E.J. McNiff Jr., B.B. Schwartz, J. Adams, S. Forman, T.W. Eager and R.M. Rose, IEEE Trans. Magn., MAG-15, p.689. 1979.
8. L. Schultz, R. Bormann, H.C. Freyhardt, ibid., p.94, 1979.

9. R. Bormann, H.C. Freyhardt and H. Bergmann, Adv. in Cryo. Eng., 26, p.334, 1980.
10. R. Flükiger, S. Foner, E.J. McNiff Jr. and B.B. Schwartz, Appl. Phys. Lett., 34, p.763, 1979.
11. R. Flükiger, R. Akihama, S. Foner, E.J. McNiff Jr. and B.B. Schwartz, ibid, 35, p.810, 1979.
12. G.H. Otto, U. Roy and O.Y. Reece, J. Less Common Metals, 32, p.355, 1979.
13. R. Akihama, R.J. Murphy and S. Foner, Appl. Phys. Lett., 37, p.1107, 1980.
14. J.M. Hong, J.T. Holthuis, I.W. Wu, M. Hong and J.W. Morris Jr., Adv. in Cryo. Eng. Materials, 28, p.483, 1981.
15. C.L.H. Thieme, H. Zhang, J. Otubo, S. Pourrahimi, B.B. Schwartz and S. Foner, IEEE Trans. Mgn., MAG-19, p.567, 1983.
16. C.A.M. van Beijnen and J.D. Elen, IEEE Trans. Magn., MAG-11, p.243, 1975.
17. J.D. Elen, C.A.M. van Beijnen and C.A.M. van der Klein, ibid, MAG-13, p.470, 1977.
18. C.A.M. van Beijnen and J.D. Elen, ibid., MAG-15, p.87, 1979.
19. A.C.A. van Wees, P. Hoogenden and H. Veringa, ibid, MAG-19, p.556, 1983.
20. P.A. Hudson and H. Jones, J. de Physique, Coll. C1, 45, p.391, 1984.
21. H. Yorucu and F.R. Sale, Int. J. Refract. and Hard Metals, 7, 3, p.161, 1988.
22. E.M. Hornsveld, J.D. Elen, C.A.M. van Beijnen and P. Hoogendam, Int. Cryogenic Mat. Conf., Pheasant Run, Illinois, USA, 1987.
23. J.D. Elen and W.M.P. Franken, Cryogenics, 27, p.106, 1987.
24. H. Yorucu and F.R. Sale, Thermochimica Acta, 56, p.147, 1982.
25. J.T. Markert, B.D. Dunlap and M.B. Maple, MRS Bulletin, XIV(1), p.37, 1989.
26. M. Popescu, L. Miu and E. Crueanu, Phil. Mag. Lett., 57, p.273, 1988.
27. F. Mahloojchi, I. Brough, F.R. Sale and G.W. Lorimer, Inst. Phys. Conf. Series, 9, 90, p.303, 1987.
28. J. Narayan, JOM, Jan, p.18, 1989.
29. D.M. Kroeger, ibid, p.14, 1989.
30. K.C. Goretta, I. Bloom, N. Chen, G. T. Goudey, M. C. Hash, G. Klassen, M.T. Lanagan, R.B. Poeppel, J.P. Singh, D. Shi, U. Balachandran, J.T. Dusek and D.W. Capone II, Mater. Lett., 7, p.161, 1988.
31. S.V. Phillips, P.J. Howard, G. Partridge and A.K. Datta, Br. Ceramic Proc., 40, p.31, 1988.
32. C.W. Cheng, A.C. Rose-Innes, N. McN. Alford, M.A. Harmer and J.D. Birchall, Supercond. Sci. Technol., 1, p.113, 1988.
33. A. Briggs, I.E. Denton, R.C. Piller, T.E. Wood and G. Ferguson, Br. Ceramic Proc., 40, p.37, 1988.
34. T.E. Mitchell, D.R. Clarke, J.D. Embury and A.R. Cooper, JOM, Jan, p.6, 1989.
35. T.W. Button, P.J. Ward, B. Rand, J.H. Sharp, P.F. Messer and E.A. Harris, Br. Ceram. Proc., 40, p.93, 1988.
36. D. Beruto, R. Botter, M. Giordani, C. Pasquali, M. Ferretti, G.A. Costa and G. Olcese, Proc. Eur. Workshop on High T_c Superconductors and Potential Applications, Genova, 1-3 July, Comm. of Euro. Comm., Brussels, p.285, 1987.
37. J.T.S. Irvine, J.H. Binks and A.R. West, ibid, p.297, 1987.
38. C. Michel, M. Heruieu , M.M Borel, A. Grandin, F. Deslandes, J. Provost and B. Raveau, Z. Phys. B. 68, p.421, 1987.
39. A.H. Maeda, Y. Tanaka, N. Fukutomi and T. Asano, Jpn. J. Appl. Phys., Part 2, 27, p.1209, 1988.
40. C.W. Chu, J. Bechtold, L. Gao, P.H. Hor, Z.J. Huang, R.L. Meng, Y.Y. Sun, Y.Q. Wang and Y.Y. Xue, Phys. Rev. Lettr., 60, 10, p.941, 1988.
41. K. Numata, K. Mori, H. Yamamoto, H. Sekine, K. Inoue and H. Maeda, J. Appl. Phys., 64, 11, p.639, 1988.
42. J.L. Tallon, R.G. Buckley, P.W. Gilberd, M.R. Presland, I.W.M. Brown, M.E. Bowden, L.A. Christian and R. Goguel, Nature, 333, p.153, 1988.
43. M. Rajabi and F.R. Sale, Proc. 1st Europ. Cer. Soc. Mtg., Maastricht, paper SU8, 1989.
44. N. Ichinose and H. Maiwa, Rep. Castings Res. Laboratory, Waseda University, 39, p.1, 1988.
45. M. Takano, J. Takada, K. Oda, H. Kitaguchi, Y. Miura, Y. Ikeda, Y. Tomii and H. Mazaki, Jpn. J. Appl. Phys., 27, p.L1041, 1988.
46. Z.Z. Sheng and A.M. Herman, Nature, 332, p.55, 1988.
47. Idem, ibid, 332, p.138, 1988.
48. M. Itoh, R. Liang, K. Urabe and T. Nakamura, Jpn. J. Appl. Phys., 27, p.L1672, 1988.
49. H. Ihara, R. Sugise, M. Hirabayashi, N. Terada, M. Jo, K. Hayashi, A. Negishi, M. Tokumoto, Y. Kimura and T. Shimomura, ibid, 334, p.510, 1988.
50. J.K. Liang, J.Q. Huang, G.H. Rao, S.S. Xie, Z.L. Zhang, G.C. Che and X.R. Cheng, J. Phys. D. Appl. Phys., 21, p.1031, 1988.
51. S.S.P. Parkin, V.Y. Lee, E.M. Engler, A.I. Nazzal, T.C. Huang, G. Gorman, R. Savoy and R. Beyers, Phys. Rev. Lettr., 60, 24, p.2539, 1988.
52. S.S.P. Parkin, V.Y. Lee, A.I. Nazzal, R. Savoy and R. Beyers, ibid., 61, 6, p.750, 1988.
53. R. Sugise, M. Hirabayashi, N. Terada, M. Jo, T. Shimomura and H. Ihara, Jpn. J. Appl. Phys., 27, 9, p.L1709, 1988.
54. X.Z. Zhou, A.H. Morrish, Y.L. Luo, M. Raudsepp and J. Maartense, J. Phys. D. Appl. Phys., 21, p.1243, 1988.
55. I.K. Gopalakrishnan, P.V.P.S.S. Sastry, K. Gangadharan, G.M. Phatak, J.V. Yakhmi and R.M. Iyer, App. Phy. Lett.,

53, 5, p.414, 1988.
56. M. Eibschütz, L.G. van Uitert, G.S. Grader, E.M. Gyorgy, S.H. Glarum, W.H. Grodkiewicz, T.R. Kyle, A.E. White, K.T. Short and G.J. Zydzik, ibid, 53, 10 , p.911, 1988.
57. B.C. Bunker, J.A. Voigt, D.H. Doughty, D.L. Lamppa and K.M. Kimball, in High Temperature Superconducting Materials, (ed.) W.E. Hatfield and J.H. Miller Jr., Marcel Dekker Inc., New York, p. 121, 1988.
58. R. Srinivasan, M. Saum, R.J. de Angelis, J.O. Deasy, J.W. Brill and C.E. Hamrin Jr., J. Mat. Sci. Lett., 8, p.383, 1989.
59. M.L. Kaplan and J.J. Hausner, Mater. Res. Bull., 23, p.287, 1988.
60. E.I. Cooper, E.A. Giess and A. Gupta, Mater. Lett., 7, p.5, 1988.
61. R.J. Clark, W.J. Wallace and J.A. Leupin, in High Temperature Superconducting Materials, (eds) W.E. Hatfield and J.H. Miller Jr., Marcel Dekker Inc., New York, p.153, 1988.
62. S. Vilminot, S.El. Hadigui, J.P. Kappler, J.C. Bernier, T. Dupin, R. Barral and G. Bouzat, Br. Ceram. Proc., 40, p.15, 1988.
63. F. Mahloojchi and F.R. Sale, Ceramics Int., 14, 4, p.229, 1988.
64. C.T. Chu and B. Dunn, Comm. Am. Ceram. Soc., 70, 12, p. C375, 1987.
65. F. Mahloojchi, N.J. Shah, J.W. Ross and F.R. Sale, Proc. Eur. Workshop on High T_c Superconductors and Potential Applications, Genova, 1-3 July, Comm. of Euro . Comm., Brussels, p.311, 1987.
66. G. Kordas, K. Wu, U.S. Brahme, T.A. Friedmann and D.M. Ginsberg, Mater. Lett., 5, p.417, 1987.
67. J.J. Ritter, in Ceramic Trans-Ceramic Powder Science II, A, (eds) G.L. Messing, E.R. Fuller Jr. and H. Hausner, Am. Ceram. Soc., Westerville, Ohio, p.79, 1988.
68. P. Barboux, J.M. Tarascon, L.H. Green, G.W. Hull and B.G. Bagley Jr., J. Appl. Phys., 63, p.2725, 1988.
69. C.E. Rice, R.B. van Dover and G.J. Fisanick, Appl. Phys. Lett., 51, p.1842, 1987.
70. M.E. Gross, M. Hong, S.H. Liou and P.K. Gallagher, J. Kwo, ibid., 52, p.160, 1988.
71. G. Kordas, K. Wu, U.S. Brahme, T.A. Friedmann and D.M. Ginsberg, Mater. Lett., 5, p.417, 1987.
72. P. Barboux, J.M. Tarascon, F. Shokoohi, B.J. Wilkens and C.L. Schwartz, J. Appl. Phys., 64, 11, p.6382, 1988.
73. M . Rajabi, Private communication, Manchester University, 1989.
74. R.B. Poeppel, S.E. Dorris, C.A. Youngdahl, J.P. Singh, M.T. Lanagan, U. Balachandran, J.T. Dusek and K.C. Goretta, JOM., Jan, P.11, 1989.
75. A.I. King, S. Chevacharoenkul, S. Pejovnik, R. Velasquez, R.L. Porter, T.M. Hare and H. Palmour, in High-Temperature Superconducting Materials, (eds) W.E. Hatfield and J.H. Miller Jr., Marcel Dekker Inc., New York, p.335, 1988.
76. M. Narasimhan, JOM., Jan, p.24, 1989.
77. K. Tachikawa, I. Watanabe, S. Kosuge, M. Kabasawa, T. Suzuki, Y. Matsuda and Y. Shinbo, Appl. Phys. Lett., 52, p.1011, 1988.
78. W.J. Nellis and L.D. Woolf, MRS Bull., XIV, p.63, 1989.
79. D.W. Johnston and G.S. Grader, J. Am. Ceram. Soc., 71, p. C291, 1988.
80. Y. Tzeng, in High Temperature Superconducting Materials, (eds) W.E. Hatfield and J.H. Miller Jr., Marcel Dekker Inc., New York, p.159, 1988.
81. S.E. Trolier, S.D. Atkinson, P.A. Fuierer, J.H. Adair and R.E. Newnham, Am. Ceram. Soc. Bull., 67, p.759, 1988.
82. J.P. Singh, D. Shi and D.W. Capone II, Appl. Phys. Lett., 53, p.237, 1988.
83. E.A. Early, C.L. Seaman, M.B. Maple and M.T. Simnad, Physica C, 153-155, p.1161, 1988.
84. T. Goto, Jap. J. Appl. Phys., 27, p.L680, 1988.
85. Y. Tanaka, K. Yamada and T. Sano, ibid., 27, p.L799, 1988.
86. D.W. Capone II, J.T. Dusek and K.C. Goretta, Proc. Int. Wire and Cable Symp., p.104, 1987.
87. H. Sekine, K. Inoue, H. Maeda, K. Numata, K. Mori, H. Yamamoto, Appl. Phys. Lett., 52, p.2261, 1988.
88. L.E. Murr, T. Monson, J. Javadpour, M. Strasik, U. Sudarsan, N.G. Eror, A.W. Hare, D.G. Brasher and D.J. Butler, J. Metals, 40, p.19, 1988.
89. K.A. Johnson, K.P. Staudhammer, W.J. Medina, C.B. Pierce and N.E. Elliott, Scripta Met., 22, p.1689, 1988.
90. H. Takagi, S. Uchida, K. Kitazawa, K. Kishio, K. Fueki and S. Tanaka, Proc. Eur. Workshop High T_c Superconductors and Potential Applications (Genova), Brussels, CEC, p.241, 1987.
91. P. Monod, M. Ribault, F. D'Yvoire, J. Jegoudez, G. Collin and A. Revcolevschi, J. Physique, 48, p.1369, 1987.
92. P.K. Gallagher, Adv. Ceram. Mater., 2, p.632, 1987.
93. B.A. Glowacki, R.J. Highmore, K.F. Peters, A.L. Greer and J.E. Evetts, Supercond. Sci. Technol., 1, p.7, 1988.
94. C.W. Cheng, A.C. Rose-Innes, N. McN. Alford, M.A. Harmer and J.D. Birchall, ibid., 1, P.113, 1988.
95. S. Jin, T.H. Tiefel, R.C. Sherwood, R.B. van Dover, M.E. Davis, G.W. Kammlott and R.A. Fastnacht, Phys. Rev., B37, p.7850, 1988.
96. Z.X. Dou, H.K. Liu, A.J. Bourdillon, N.X. Tan, N. Savvides, J.P. Zhou and C.C. Sorrell, Supercond. Sci. Technol., 1, p.78, 1988.
97. B.W. Veal, H. Claus, J.W. Downey, A.P. Paulikas, K.G. Vandervoort, J.S. Pan and D.J. Lam, Physica, C156, p.635, 1988.

98. M. Kikuchi, N. Kobayashi, H. Iwasaki, D. Shindo, T. Oku, A. Tokiwa, T. Kajitani, K. Hiraga, Y. Synono and Y. Muto, Jpn. J. Appl. Phys., 27, 6, p.L1050, 1988.
99. T. Usui, N. Sadakata, 0. Kohno and H. Osanai, ibid., 27, 5, p.L804, 1988.
100. Z. Yuling, L. Jingkui, X. Sishen, H. Jiuqin, R. Guanghui, C. Xiangrong, L. Hongbin, Z. Dongming and Q. Shunliang, J. Phys. D. Appl. Phys., 21, p.845, 1988.
101. P.T. Wu, R.S. Liu, J.M. Liang, W.H. Lee, L. Chang, L.J. Chen and C.T. Chang, Physica, C156, p.109, 1988.
102. Z.X. Dou, H.K. Liu, A.J. Bourdillon, N.X. Tan, N. Savvides, C. Andrikidis, R.B. Roberts and C.C. Sorrell, J. Phys. D. Appl. Phys., 21, p.83, 1988.
103. D. Goodstein, New Scientist, July, p.60, 1989.

27

Aspects of Powder Technology in Biomaterials

S. BEST AND W. BONFIELD

Department of Materials, Queen Mary College, University of London, UK

Abstract

Many materials have been proposed for use in bone and tooth replacement, but very few adequately fulfil the complex combination of biological and mechanical requirements presented by the body. There is a mechanical mismatch between the properties of metal and those of bone and even while their biological compatibility is acceptable, a bone may degrade following long periods of implantation due to a phenomenon known as stress-shielding. Polymeric materials generally do not satisfy the mechanical requirements for major load bearing applications. Structural ceramics have good wear resistance and may be used as articulating surfaces but their poor fatigue properties, high modulus and low fracture toughness has proved a major limitation.

For this reason, research has been directed towards the development of novel biomaterials with analagous properties to bone. One major example is a ceramic, hydroxyapatite, which has excellent biocompatibility as bone will grow up to and around it. It has been successfully implanted in its porous form, in dental and aural applications. Hydroxyapatite may be synthesized using a variety of techniques which produce different physical and chemical characteristics in the starting powder. Using conventional powder technology, dense hydroxyapatite specimens may be produced. One of the main problems limiting the use of the material has been its poor mechanical properties. However, careful powder processing may eventually lead to development of dense hydroxyapatite ceramics which will be suitable for use in load bearing parts of the skeleton.

1. Introduction

This chapter is designed to give the reader an overview of the various bone and tooth replacement materials currently in use or under consideration. To aid an understanding of the concepts involved in the selection and design of potential prosthetic materials, a brief overview is presented of the basic composition and structure of bone and its function in the body. A brief consideration is then given of various prosthetic materials in use at the present time.

The suitability of bone replacement materials may be assessed in terms of both their mechanical properties and the body's reaction post-implantation. All of the materials presently employed, which include metals, polymers, ceramics and composites, have particular problems, such as low wear resistance, low strength or excessive stiffness leading to stress shielding. A recent, major research activity in the biomaterials field has been to develop hydroxyapatite; a ceramic closely resembling bone mineral. As this material is produced by powder technology, it is considered in some detail, to illustrate the special problems associated with this type of processing of biomaterials.

2. Bone

Bone is a living tissue, which constantly remodels to adapt to stress requirements. Its function is to support and protect vital organs, to provide kinematic links and to facilitate muscle action and body movement. Bone has a complex structure which may be defined on several organizational levels [1]. To aid understanding of the mechanical behaviour of bone, consideration should be given, to the structure from the macroscopic and the microscopic levels.

In general, the major support bones consist of an outer load bearing shell of cortical (compact) bone with a medullary cavity containing cancellous (trabecular, or spongy) bone towards the bone ends. Cortical bone itself is anisotropic, with osteons (approximately 100-300μm in diameter in a preferred orientation parallel to the long axis of the bone, interspersed with non-oriented bone. Each osteon has a central Haversian canal (approximately 20-40μm in diameter) containing a blood vessel, which supplies the elements required in bone remodelling. The osteons are composed of concentric lamellae (approximately 5μm in thickness) consisting of two major components, collagen and hydroxyapatite.

In mature bone, hydroxyapatite occupies approximately 0.5 volume fraction. It is mainly crystalline with a plate-like [2,3] or rod-like [4,5] habit and dimensions of the order of 5x5x50nm. The precise microstructural organization varies between different bones, according to location and and as a function of age. As a result, cortical bone has a range of associated properties, rather than a unique set of values.

An assessment of the mechanical properties if bone may be made on whole bones in vivo, but due to their irregular shape is difficult to interpret. More usually, mechanical property measurements are made, in vitro on wet bone specimens machined to standard specifications. The stiffness of cortical bone, as represented by the Young's modulus, range between 7-30GPa [6] (depending on orientation, location and age). As the fracture of cortical bone occurs in a brittle mode, the values determined for ultimate tensile strength [6] (approximately 50-150MPa) are less than critical than a measure than the fracture toughness, which has recently been established [7-11], with a critical stress intensity factor (K_c) of between 2 and 12Mpa.m$^{1/2}$.

3. Biomaterials

When selecting a material for bone replacement and augmentation, the mechanical properties must be maintained in essentially aggressive chemical conditions over long periods of time. Materials have been used and proposed for implantation for many years. However, it is only more recently that the significance has been recognized of matching the mechanical properties of the replacement material with bone.

Table 1 [12] lists the mechanical properties of some proposed bone replacement materials.

Material	UTS (MPa)	K_{IC} (MPa.m $^{1/2}$)	E (GPa)
Stainless Steel (Austenitic)	200-1200	100	200
Titanium alloy (+ 6$^{W}/_{o}$ Al - 4$^{W}/_{o}$V)	780-1050	80	100
Polyethylene	30	-	1
PMMA (bone cement)	70	1.5	3.5
Alumina	50	5	360
Hydroxyapatite	25-120	1.0	30-120

Table 1 Mechanical Properties of Some Proposed Bone Replacement Materials

The strength and modulus of the plastics listed are significantly lower than those recorded for bone - this would be associated with the obvious practical problems post implantation. Metals exhibit strengths and moduli well in excess of those found for bone. While superior mechanical properties might, at first sight appear to be an advantage, the consequent stress protection may result in bone resorption. Although the value of strength and modulus of alumina are high, the wear properties of the ceramics are superior to those of the metals listed. The mechanical properties of hydroxyapatite are of the order of those of bone.

Bone replacement is an extremely complex engineering materials problem. Interest is not solely centred upon fulfilling mechanical property requirements such as strength and stiffness; of prime importance is the compatibility of the material with the immune system of the body. The material should not interfere with the body's defence mechanisms or alter the electrolytic environment surrounding it and it should elicit no toxic or allergenic responses. Materials may be categorized as follows:

Term	Definition	Material
Toxic	Surrounding tissue dies, Body isolates and rejects implant. Not suitable as a biomaterial	Many metals
Compatible	Non-toxic and biologically inactive. Stable fibrous tissue forms around the implant.	Alumina Polyethylene Titanium Stainless Steel
Surface Active	Non-toxic and biologically active. Interfacial bonding occurs .	Hydroxyapatite Bioglass

3.1 Metallic Biomaterials

An immense variety of metals and alloys have been clinically tested in prosthetic applications [13]. However, toxic corrosion products caused by the action of body fluids may result in the rejection of the implant. Most metals are either not resistant to the corrosion by body fluids (e.g. aluminium), react with body tissue (e.g. copper) or lack sufficient strength to serve as a structural prosthesis (e.g. gold) [13]. The few materials used in endoprotheses are 316 stainless steel, cobalt based chromium-molybdenum alloys and titanium or titanium alloys. These materials are useful due to the presence of a thick, natural surface oxide. The metals and alloys in use at present are all chosen primarily for their chemical stability, rather than for their mechanical properties or ease of fabrication.

3.2 Polymeric Biomaterials

3.2.1 Polymers

Polymers have been used in ocular and dental applications with some considerable success. Their low density and the ability to design the hydrophobic or hydrophilic response of certain polymers has made them ideal for use as contact lens materials.

From the point of view of implantation, polymethyl methacrylate [14], high density polyethylene [15], and high grade silicone rubbers [16] have been found to be stable, however, their mechanical properties and adhesion to surrounding tissues are poor. Hence, the use of polymers is largely confined to non-load bearing applications.

3.2.2 Polymer Composites

The mechanical mismatch of polymers with bone has prevented their use in load bearing parts of the skeleton. For this reason interest arose in the use of composites, using the combination of the low density of polymers with the high strength, modulus and bioinertness of ceramics.

In 1963, a bone substitute was developed using the impregnation of 48% porous alumina with an epoxy resin [17]. The material closely matched the mechanical properties of bone and was successfully implanted in animals and was patented by Haeger Potteries Inc., Dundee, Ilinois and manufactured as Cerosium.

Recognising the potential for designing a material with desired mechanical properties using composites, Bonfield and co-workers at Queen Mary College, University of London, developed a polymer matrix composite [18-22]. Using polyethylene as the matrix (in view of the good fracture toughness and ductility of the material) and synthetic bone mineral (hydroxyapatite) particles for reinforcement, a bone replacement material with both excellent biological and mechanical compatibility was developed.

3.3 Ceramic Biomaterials

3.3.1 Bioglass

Hench and Pascall [23] developed a glass for implantation with the composition; Na_2O - CaO - P_2O_5- SiO_2, referred to as bioglass, which is soluble in physiological conditions and results in silica gel being formed around the implant, enhancing collagen formation within the gel. They suggested that this would help in the crystallization of hydroxyapatite and hence form an organic bond with bone. Now, a "bioglass" is defined as a glass designed to produce specific physiological responses. This requires a new surface of reactive silica to be provided, calcium and phosphate groups and an alkaline pH at an interface with tissue [24]. Since the first attempts by Hench and Pascall, many studies have been carried out using bioglass [25-29]. Problems have occurred in obtaining a sufficient bonding between the implants and hard tissue to sustain the type of loadings experienced in vivo [30]. Other workers have reported problems in preventing unwanted crystalline transformations, avoiding microcracking and controlling shape during specimen preparation [31].

3.3.2 Alumina Ceramics

Towards the end of the 1960s interest had spread to the use of bone ingrowth into porous oxide ceramics as a means of mechanically interlocking prostheses to bone. [32]

Alumina ceramics have been successfully used along the articulating surfaces of heavy load bearing joint [35-38], for middle ear implants [39] and for orbital base plates [40]. Although the strength and toughness of alumina ceramics make them suitable for use in minor load bearing parts of the skeleton, any bonding to bone must be carried out by mechanical interlocking devices in the design of the implant or by cementing or glueing.

3.3.3 Calcium Phosphates

At the beginning of the 1970s, interest arose in the use of surface reactive biomaterials. The advantage of this class of materials is that they bond directly to bone with no fibrous tissue at the interface.

Koster *et al.* [41] analysed six calcium phosphates in dense form for implantation in terms of biocompatibility and found that tricalcium phosphate had the optimum calcium to phophorous ratio.

Rhyshkewith [42] and later Rejda *et al.* [43], developed a method for producing porous calcium phosphate implants which used hydrogen peroxide. Early clinical investigations of porous tricalcium phosphate were carried out by Bhaskar *et al.* [44] They found that in some areas around the implants, bone appeared to be directly appositioned to the implant. Cutright [45] reported that tricalcium phosphate implants in rats were completely resorbed and totally replaced by trabecullae within 48 days of implantation. Similar resorption results were found by Cameron [46]. De Groot [47] suggested that an inverse relationship exists between degradation and the porosity of the implant. However, in dental applications for example, resorbtion may be viewed as a disadvantage as the mechanical properties of the implants are required to be retained over long periods of time. It is probably for this reason that the interest in tricalcium phosphate ceramic implants has diminished recently.

Interest has developed, however, in hydroxyapatite. Initially it was directed towards the use of porous implants. although they have poor mechanical properties [48]. Work was concentrated on dense hydroxyapatite for use in major load bearing implants [49-53].

Very recently, work has been reported on the development of plasma sprayed coatings of hydroxyapatite on metal substrates [54,55,56]. Amongst other problems, poor interfacial bonding has been encountered.

3.3.4 Summary

Materials design for bone and tooth replacement is not a simple matter of matching, or bettering mechanical property requirements. Any material for use in a hostile environment must be designed to function reliably over long periods of time. This situation is aggravated further, where biomaterials are concerned, by the action of the body to isolate or reject foreign materials. The use of a surface reactive ceramic, hydroxyapatite, closely resembling bone mineral has been found to overcome many of the biocompatibility problems.

4. Hydroxyapatite

The following section discusses the fabrication of hydroxyapatite powder, its characterization and preparation for sintering into dense blocks for use in bone and tooth replacement.

4.1 Synthesis and Characterization

Hydroxyapatite powders may be produced using four routes:

(i) Precipitation from aqueous solution [49,57-66].
(ii) Hydrolysis [67-69].
(iii) Solid state reaction [70-72].
(iv) Hydrothermal synthesis [73-75].

The former route is most commonly used since it is most economical in terms of the energy required and the time scale of the reaction. The particles obtained by precipitation found to consist of individual, acicular crystallites 20-30nm in length giving a powder surface area of greater than $80m^2/g$ [49]. A variety of drying techniques may be used including:

- Freeze drying.
- Spray drying.
- Centrifuging.
- Oven drying.
- Microwave drying.

Subsequent to the drying process and a calcination stage, a vast range of powder morphologies were obtained, ranging from, smooth spheroids to irregular arrays of crystallites (Figs. 1-3). The physical properties of the powder such as bulk density, particle size, size distribution and surface area are also profoundly affected according to the type of drying and calcination processes employed (Figs. 4 and 5). The presence of impurities (e.g. carbonate ions) also affects the morphology of the powder particles [76].

Crystallographically, hydroxyapatite is hexagonal with a P6/m2 space group (Fig. 6) [77]. The values of a and c given in the ASTM powder file [78] are 9.42 and 6.88 A respectively. Obviously, the lattice parameters of the material are altered by the presence of impurities and lattice defects; work has been carried out to identify the effect in synthetic hydroxyapatites of carbonate ions [76], lead ions [79]. McConnell [80] stated that the lattice parameters of natural apatites depend upon at least three variables: the contents of fluorine, carbon dioxide and water. Other impurities have a profound effect on the colouration of the material in its calcined or sintered condition.

The theoretical value for the calcium to phosphorous ratio of pure hydroxyapatite is 1.67 [81]. However it is a member of a series of naturally occurring calcium phosphates and may, in practice have a composition deviating from the theoretical value whilst retaining the apatite crystal structure. (See Table 2)

The solubility and speed of hydrolysis of calcium phosphates increase with decreasing calcium to phosphorous ratio. Those compounds with a ratio of less than 1:1, are not suitable for biological implantation [59].

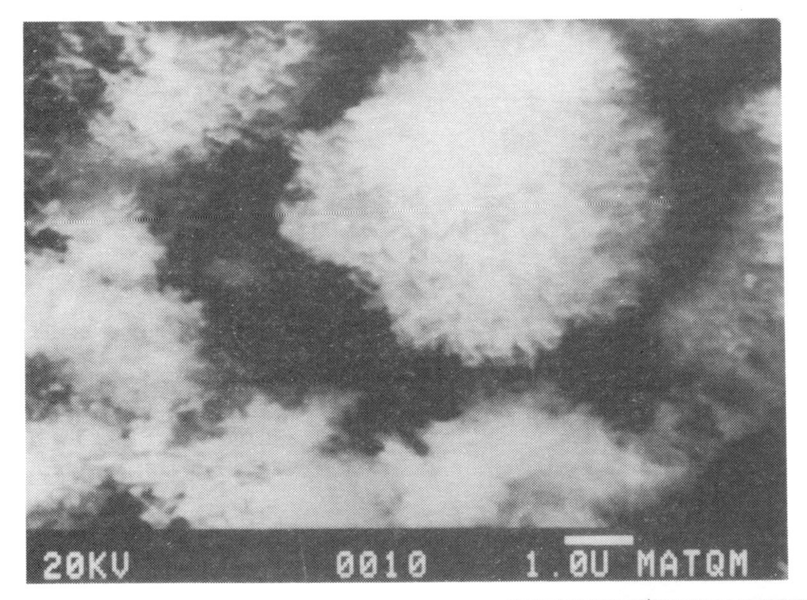

Figure 1 - Powder C

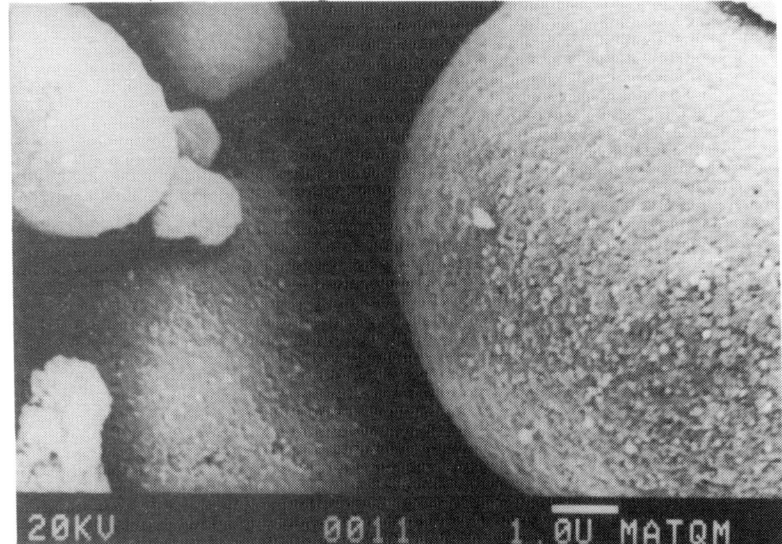

Figure 2 - Powder B

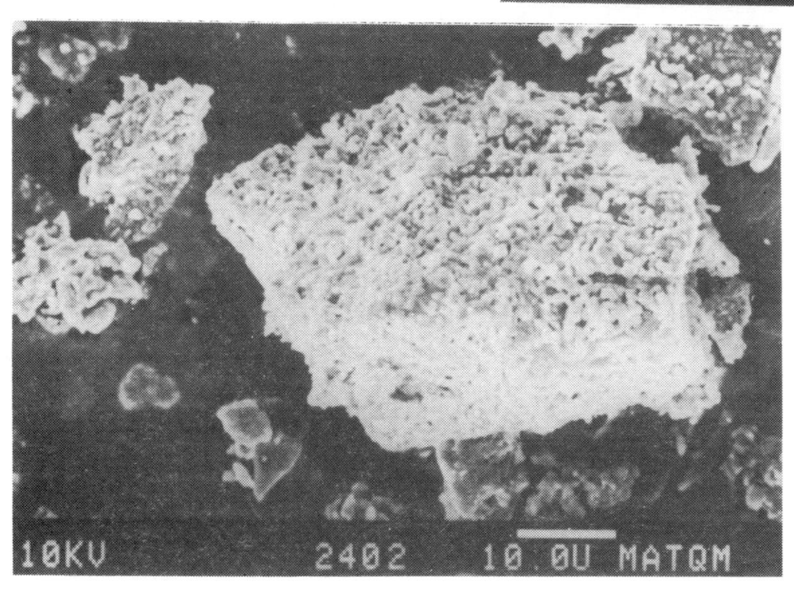

Figure 3 - Powder A

Figures 1-3 Scanning electron micrographs of samples of hydroxyapatite in the as-received condition.

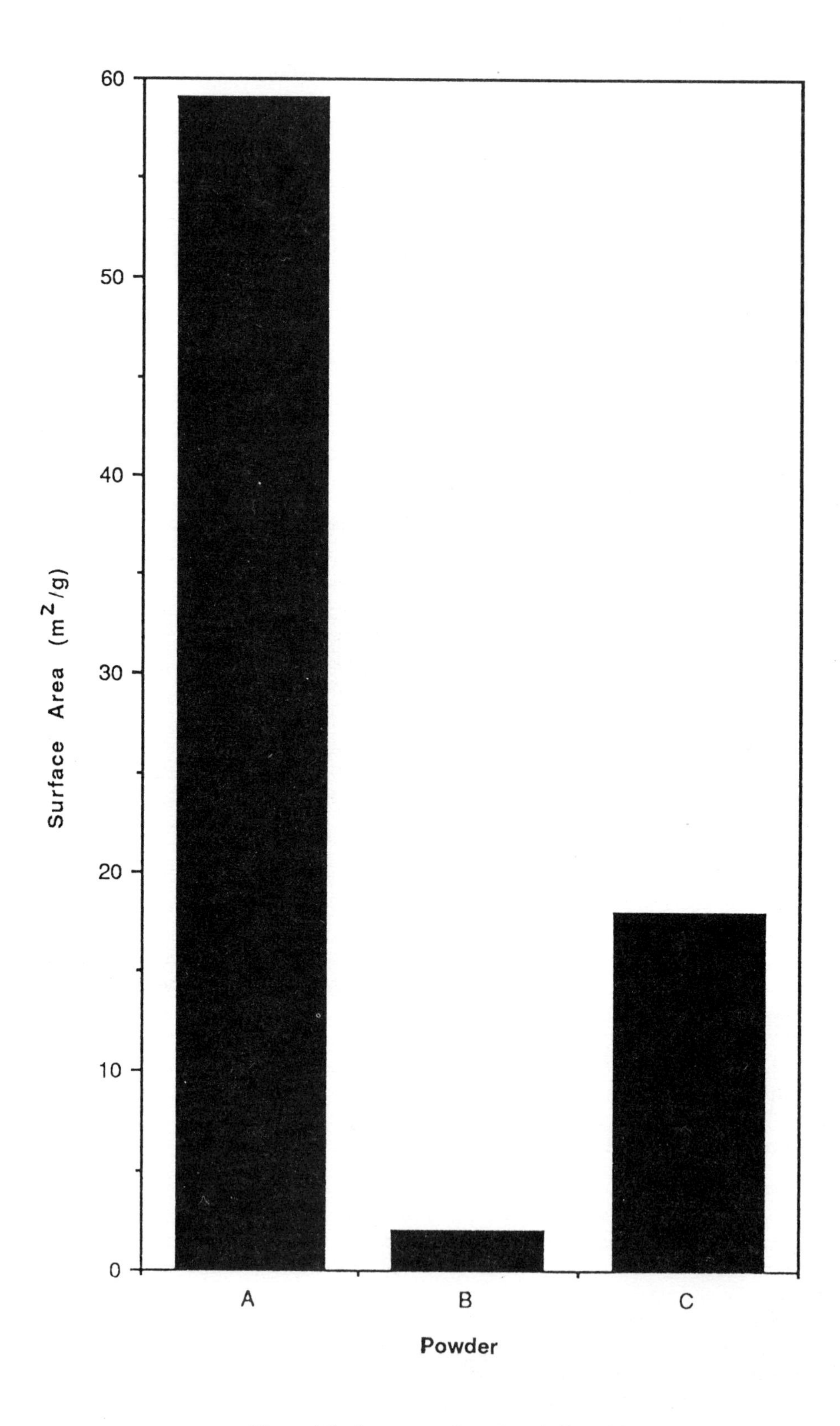

Figure 4 Surface area of powders A, B and C.

Ca:P	Mineral name	Formula	Chemical Name
1.0	Monetite	$CaHPO_4$	Dicalcium Phosphate (DCP)
1.0	Brushite	$CaHPO_4.2H_2O$	Dicalcium Phosphate Dihydrate(DCPD)
1.33	-	$Ca_8(HPO_4)_2(PO_4)_4.5H_2O$	Octocalcium Phosphate (OCP)
1.43	Whitlockite	$Ca_{10}(HPO_4)(PO_4)_6$	
1.50		$Ca_3(PO_4)_2$	Tricalcium Phosphate (TCP)
1.67	Hydroxyapatite	$Ca_{10}(PO_4)_6(OH)_2$	

Table 2 Natural and Synthetic Calcium Phosphates

Using the $CaO-P_2O_5$ phase diagram shown overleaf [82], and marking the theoretical calcium to phosphorous ratio for hydroxyapatite on the diagram, a range of possible extraneous phases may be noted. The most common such phase found in precipitated hydroxyapatite powders is β-tricalcium phosphate (β-TCP) often as a result of an excess of the phosphate-based reactant. TCP is of interest as a biomaterial, but is more soluble than hydroxyapatite in physiological solutions and has inferior mechanical properties [83,84,85]; therefore the phase is essentially undesirable where a dense material is desired for use in load bearing parts of the skeleton. Jarcho [49] reported that hydroxyapatite decomposes into tetracalcium phosphate and α-tricalcium phosphate when heated to temperatures above 1250°C. The exact temperature at which this transformation occurs is dependant upon the exact composition of the starting material.

4.2 Component Preparation

4.2.1 Powder Preparation

The basic precursor in the production of ceramic hydroxyapatite components is a low density, particulate ("green") compact. This compact is of low strength since the particles are held together only by frictional/mechanical interlocking and low energy physical bonds such as Van de Waals. The methods of green compact production for hydroxyapatite include slip casting [49], dry pressing [46] isostatic pressing [88,89] and tape casting [90].

The size, arrangement and packing of particles in the green compact has a significant effect on the time, temperature and pressure (i.e. the energy parameters) required for sintering. Lack of reproducibility between components stems both from the number of particles contained in the green compact and the presence of agglomerates. There is some disagreement as to the ideal powder characteristics. Barringer *et al.* [91] propose that the ideal powder consists of monosized particles of 1μm or less in diameter. There is a strong tendancy for sub-micron powders to agglomerate, however, with the associated problems of grain coarsening [92]. Wide size distributions may also present problems where a uniform structure is required in the sintered component. However, in practice the distribution of void space in powders with a wide size distribution may be more uniform since the packing efficiency of such a powder is generally better [93]. Therefore where the dimensional tolerance of a component is critical, a powder with a wide particle size distribution should be used, whereas where high strength is of prime importance and shrinkage not a major consideration, a near monomodal distribution of small particles is preferable.

4.2.2 Optimization of Sintering Regime

Densification of hydroxyapatite components is carried out using solid state sintering. Figure 7 shows the type of sintering regime employed. Several salient features should be noted. Being a ceramic, the thermal shock resistance of hydroxyapatite is poor, hence the heating and cooling rate are necessarily slow in order to avoid microcracking or the retention of residual stresses in the component.

The use of organic binders, such polyethylene glycol, to improve the green strength of the material or the mould filling properties, and lubricants, such as stearates, to reduce wear of the die walls and assist in compaction are normal when dry pressing is used. The sintering schedule includes a "burn off" stage during which all traces of organic material are removed. The temperature selected for this stage is chosen such that it is sufficiently high for the organic material to be removed but not so high that any degree of densification of the ceramic has begun to occur. This avoids the problems associated with two conflicting processes occurring simultaneously.

Any large scale shaping or machining of hydroxyapatite components is carried out after "biscuit firing" since in the green state the green compact is too friable for handling to a great extent. Samples may be heated to 800°C for a period of six hours or more to allow the beginnings of neck formation to occur with the associated increase in strength.

Sintering temperatures of between 1100°C and 1300°C have been reported for hydroxyapatite [2,49,50,52,94,95,97].

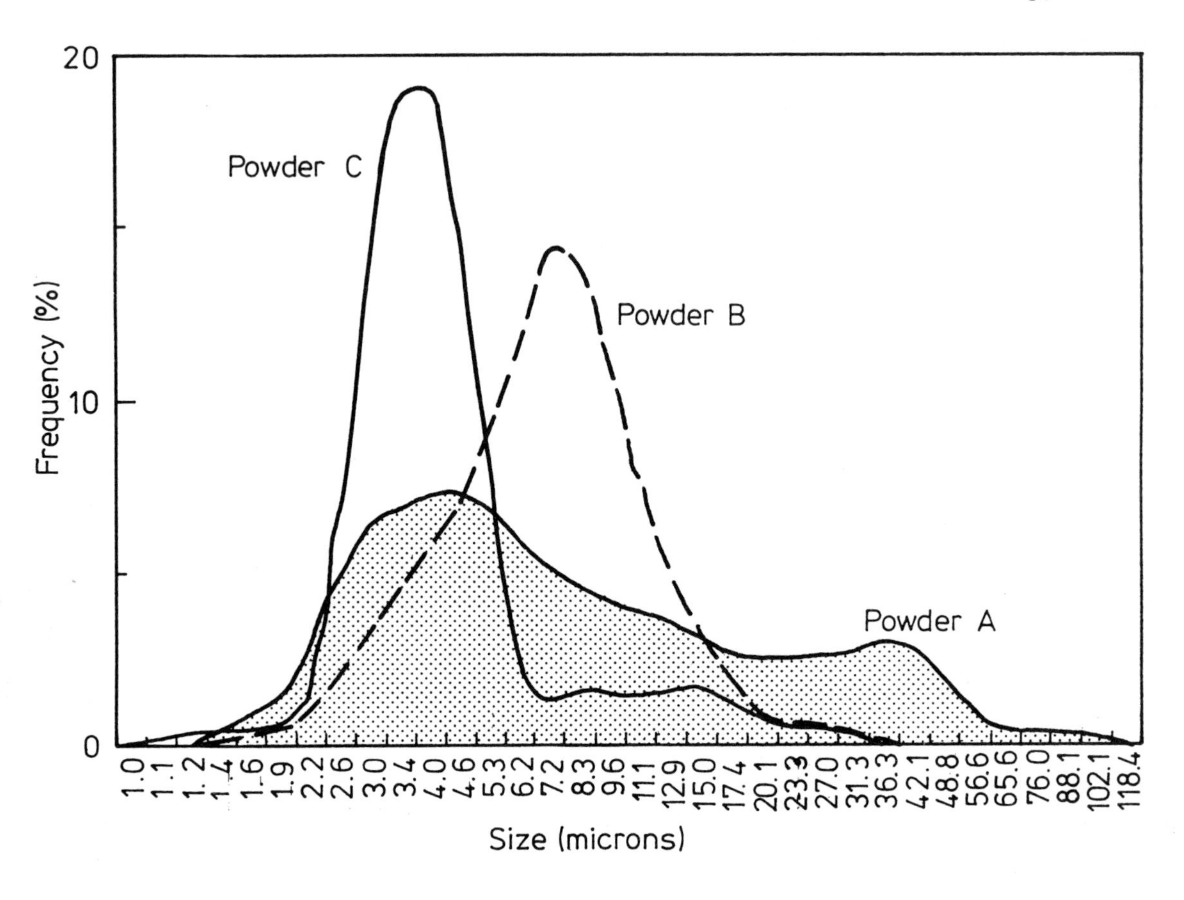

Figure 5 Particle size distribution of powders A, B and C.

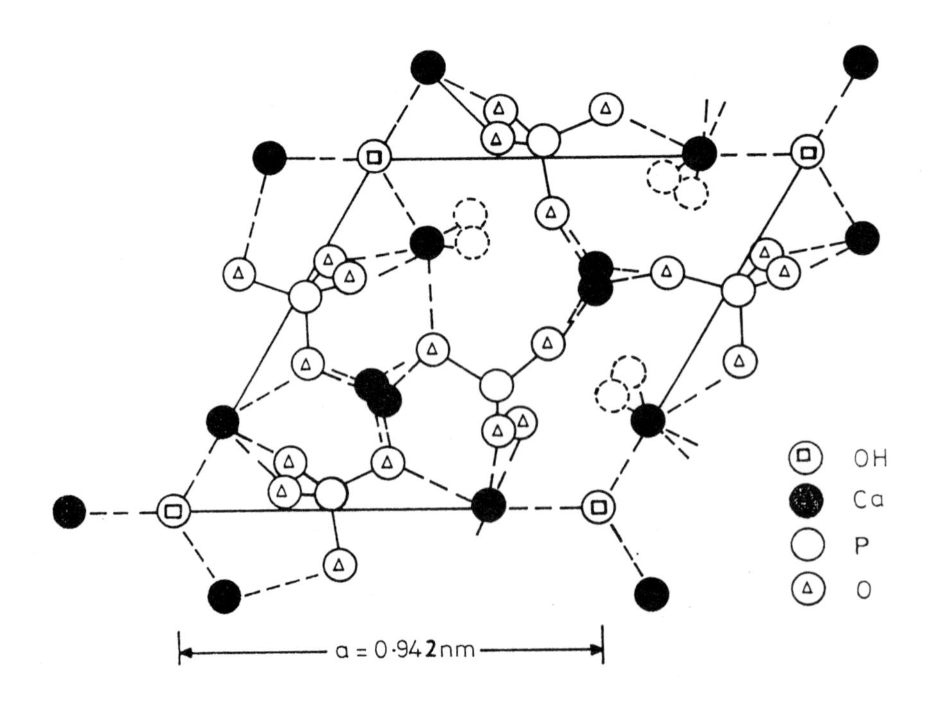

Figure 6 The structure of hydroxyapatite, ($Ca_{10}(PO_4)_6(OH)_2$) showing the lower half of the unit cell.

Some workers have reported obtaining samples with full density ($3.156g/cm^3$) [49]. However, the range of sintering temperatures required and the density of the sample obtained is dependant upon the morphology, physical and chemical properties of the hydroxyapatite powder used.

4.3 Mechanical Properties of Hydroxyapatite

Many attempts have been made to sinter hydroxyapatite to full density [2,49,50,52,94,95,97]. Of main interest to these workers have been the strength and modulus of products according to their density. Mechanical testing has centred around the measurement of compressive strength, "fracture strength" and Young's modulus. The range of testing techniques, specimen dimensions and surface condition was wide (and in many cases, not quoted) and the significance of the control of these, and other, parameters was not always recognized. Hence problems may arise in the comparison of results quoted by the various groups. Table 3 summarizes the mechanical properties reported.

Earliest studies of the sintering of dense hydroxyapatite specimens were carried out by Jarcho *et al.* in 1976. [49] The group produced a green compact using a variation on slip casting involving the sintering of the as-produced precipitation cake. They reported a hardness of 480 Knoop, compressive strength of 917MPa and a modulus of 30GPa in specimens sintered at between 1000°C and 1200°C. Fracture strength was measured using polished specimens, 2x2x30mm in size in three point bending. Denissen *et al.* [94] used continuous hot pressing to produce specimens of 97% of the theoretical density of hydroxyapatite ($3.156g/cm^3$). The Vickers hardness was measured as 500 and the tensile strength, measured using three point bend tests on polished specimens (the specimen dimensions were not specified) was found to be 25MPa. Thomas *et al.* [95] carried out diametral compression tests on samples of "Durapatite" 5mm in diameter. They found that tensile strength ranged from 109MPa to 121MPa increasing with increasing crosshead speed. In 1982 Driessen *et al.* [2] compared the mechanical properties of tricalcium phosphate and hydroxyapatite. They reported that the compressive strength of the two materials was 600MPa (±200 MPa). Also in 1981 DeGroot *et al.* [52] found that the strength of artificial teeth sintered using a technique specified by Albee in their implant study was 500 (±100) MPa in compression and 50 (±20) MPa in tension (no test method was given). The Vickers hardness was found to be 450.

Denissen [50] compared the sintering behaviour of three isostatically pressed hydroxyapatite powders in terms of their mechanical properties. They found a laboratory made powder sintered to 99.99% of the theoretical density when heated to 1200°C and had superior mechanical properties to the commercially available powders tested. The maximum tensile strength was found to be 39MPa and compressive and hardness results obtained were 430 (±95) MPa and 450 HV respectively. In the same year de With *et al.* [97] reported that by using a hot isostatic pressing technique and sintering at 1250°C, samples of 98% of the theoretical density could be obtained. The fracture strength measured was between 92 and 103MPa; Youngs modulus was 116GPa and the fracture toughness of the material was $1.33MPa.m^{1/2}$. The same group [89] isostatically pressed hydroxyapatite to form specimens of either 1x3x15mm or 3x9x45mm. The modulus, measured using ultrasound, was found to be 112GPa. Tensile strength and toughness measurements were carried out using three point bending; the values found were 115MPa and $1MPa.m^{1/2}$ respectively. The sintering temperatures used ranged between 1150°C and 1250°C with the associated appearance of α and β calcium orthophosphate at the higher and lower temperatures respectively.

Most of the development of dense hydroxyapatite ceramics was performed by non materials scientists. This presents problems when a comparison is made of the results of different groups. Until recently there was no appreciation of the effect of specimen size, surface condition and test geometry on the mechanical test results obtained. This could be one reason for the vast range in the results reported. Equally there appears to be a lack of information concerning the density, grain size, modulus and toughness of the ceramics tested. Interest is now concentrated on the full characterization and standardization of dense hydroxyapatite ceramics, using conventional ceramic testing techniques [98].

4.4 Summary

Hydroxyapatite is not an easy material to deal with. Unlike "normal" metallurgical or ceramic powders there is a risk of decomposition at a temperature close to that required for sintering. The decomposition product is soluble in physiological solutions and has inferior mechanical properties and is therefore to be avoided. Although there is a range of production routes possible for hydroxyapatite, very few are commercially viable. The powder morphologies available at present are not those traditionally considered to be the most appropriate for pressing and sintering. The mechanical properties of sintered hydroxyapatite are inferior to those of "traditional", structural ceramics. However, the application for which hydroxyapatite is used, requires an unusual combination of characteristics that few other materials offer.

5. Conclusion

Many materials have been suggested for use as prostheses. However, few have been found to be satisfactory. The materials with sufficiently good mechanical properties often do not meet up to the biocompatibility requirements for the application and vice versa. Of the materials under development at present, hydroxyapatite, a material with a composition similar to that of bone mineral, has proved extremely promising. Hence, one of the major advances in biomaterials, is in using powder technology. By careful control of the preparation and processing of hydroxyapatite powders, dense ceramics may be produced with mechanical properties suitable for use in major load bearing parts of the skeleton.

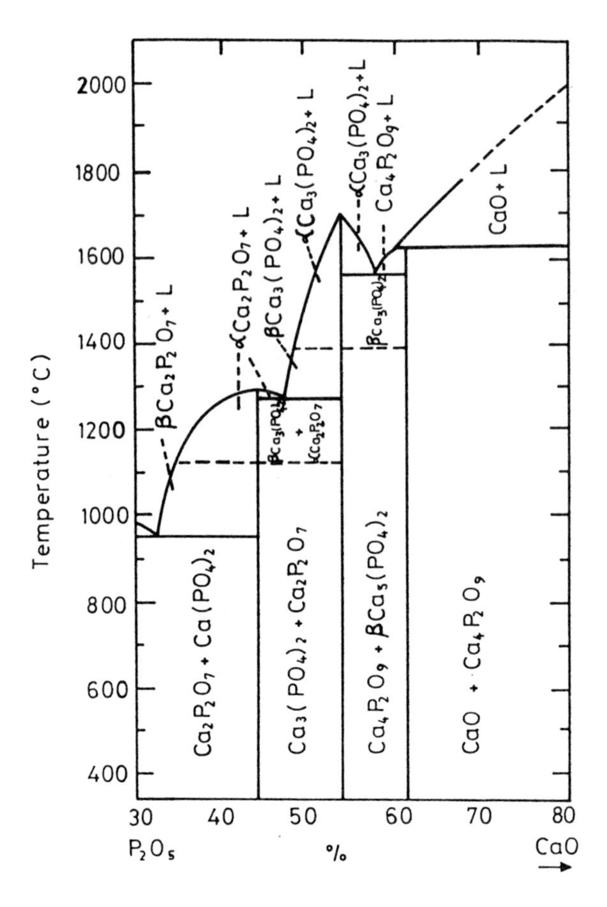

Figure 7 Portion of the CaO-P_2O_5 phase diagram.

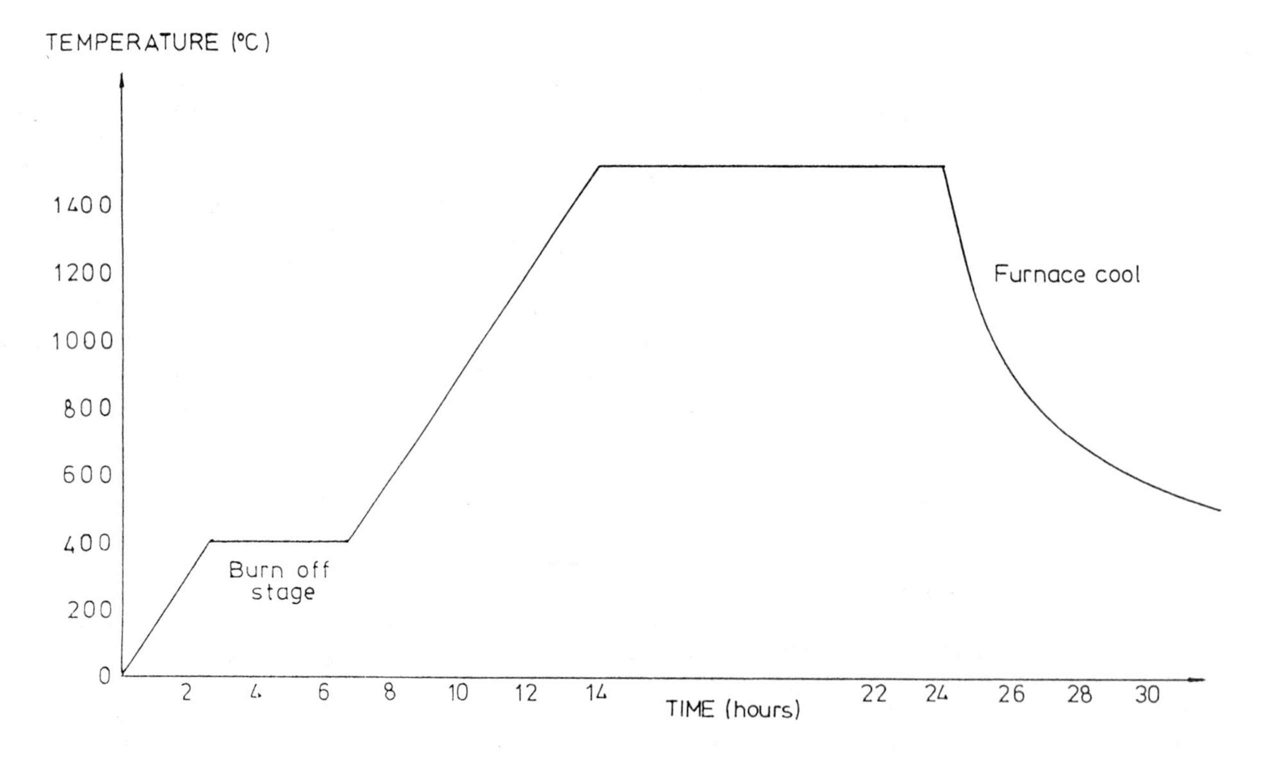

Figure 8 The sintering regime used in the preparation of dense hydroxyapatite ceramics.

References

1. J.J.Pritchard, The Biochemistry and Physiology of Bone, 2nd Edition, Ch. 1, (ed.) Bourne, Academic Press, 1972.
2. R.A.Robinson and M.L. Watson, Crystal-Collagen Relationships in Bone as Observed in the Electron Microscope, III-Crystal and Collagen Morphology as a Function of Age. Ann. New York Acad. Sci, 60, 1055, p.596.
3. K.J. Munzenburg, Untersuchengen zur Kristallographie der Knochenminerale, Biomineralisation, 1, p.67, 1970.
4. K.J. Munzenburg and M. Gebhardt, Brushite Octacalcium Phosphate and Carbonate Containing Apatite in Bone, Clin. Orthop. Rel. Res. 90 p.271, 1973.
5. R.A. Robinson and D.A. Cameron, Electron Microscopy of Cartillage and Bone Matrix at the Distal Epiphyseal Line of the Femur in the New Born Infant. J. Biophys. Biochem. Cytol. 2, p.253, 1956.
6. F.G. Evans, The Mechanical Properties of Bone, Publ. Thomas, New York, 1973.
7. W. Bonfield and P.K. Datta, Fracture Toughness of Compact Bone, J. Biomechanics, 9, p.131, 1976.
8. W. Bonfield, M.D. Grynpas and R.J. Young, Crack Velocity and the Fracture of Bone, J. Biomechanics, 11, p.473, 1978.
9. W. Bonfield, J.C. Behiri and B. Charalambides, Orientation and Age-Related Dependence of the Fracture Toughness of Cortical Bone, Biomechanics; Current Interdisiplinary Research (ed.) S.M. Perren, E. Schneider, Martinus Nijhoff, Dordrecht, 1985.
10. J.C. Behiri and W. Bonfield, Fracture Mechanics of Bone - the effects of density, specimen thickness, and crack velocity on longitudinal fracture, J. Biomechanics, 17, p.25, 1984.
11. W. Bonfield, Advance in the Fracture Mechanics of Cortical Bone J.Biomechanics, 20, p.1071, 1987.
12. W. Bonfield, Materials for the Replacement of Osteoarthritic Hip Joints, Metals and Materials 12, p.712, 1987.
13. D.C. Ludwigson, Today's Prosthetic Materials, J. Metals, 16, 3, p.226, 1964.
14. H. Kawahara, Designing Criterea of Bioceramics for Bone and Tooth Replacements. In Ceramics in Surgery, (ed.) P. Vincenzini, Amsterdam: Elsevier, 1983.
15. S. Braley, Implantable Synthetic Materials, Industrial Res., July 1966.
16. H. Kawahara and N. Nakamura, Tissue - Material Interface Chemistry, Jap. Soc. Chem., 21, p.13, 1978.
17. L. Smith, Arch. Surg, Ceramic-Plastic Material as a Bone Substitute, p.653, 1963.
18. W. Bonfield, M.D. Grynpas, A.E. Tully, J. Bowman and J. Abram, Hydroxyapatite Reinforced Polyethylene - A Mechanically Compatible Implant Material for Bone Replacement. Biomaterials 2, p.185, 1981.
19. W. Bonfield, M.D. Grynpas and J. Bowman, Prosthesis Comprising Composite Material. UK Patent No. 2085461 B.
20. J. Abram, J. Bowman, J. Behiri and W. Bonfield, The Influence of Compounding Route on the Mechanical Properties of Highly Loaded Particulate Filled Polyethylene Composites. Plastics and Rubber Processing Appl. 4, p.261, 1984.
21. W. Bonfield, J.C. Behiri, C. Doyle, J. Bowman and J. Abram, Hydroxyapatite Reinforced Polyethylene Composites for Bone Replacement. In Biomaterials and Biomechanics, eds. P. Ducheyne, G.M. van de Perre, A.E. Aubert, Amsterdam: Elsevier, p.421, 1983.
22. W. Bonfield, C. Doyle and K.E. Tanner, In vive Evaluation of Hydroxyapatite Reinforced Polyethylene Composites, in Biological and Biomechanical Performance of Biomaterials, (eds.) P. Christel, A. Meunier, A.J.C. Lee, Amsterdam: Elsevier, p.153, 1986.
23. L.L. Hench and H.S. Paschall, Histochemical Responses at Biomaterials Interfaces, Prosthesis and Tissue, The Interface Problem, (Symposium) Clemson, South Carolina, 1973.
24. S.F. Hulbert, L.L. Hench, D. Forbes and L.S. Bowman, History of Bioceramics, in Ceramics in Surgery, (ed.) P. Vincenzini, Elsevier Scientific Pub. Co. Amsterdam, 1983.
25. C.A. Beckham and T.K. Greenlee, A.R. Crebo, Calc. Tiss. Res. 8, 165, 1971.
26. L.L. Hench and H.S. Paschall, J. Biomed. Mater. Res. 7, p.25, 1983.
27. G. Piotrowski, L.L. Hench, W.C. Allen and G.J.J. Miller, Biomed. Mater. Res. 9, p.47, 1975.
28. O. Griss, D.C. Greenspan, G. Heimke, B. Krempien, R. Buchinger, L.L. Hench and G. Jentschura, J. Biomed. Mater. Res. 10, p.511, 1976.
29. H.R. Stanley, L.L. Hench, G. Going, C. Bennet, S.J. Chellemi, *et al.* Oral Surg. Oral Path. Oral Med., 42, p.339, 1976.
30. G. Ito, S. Yamashita, T. Jimi and T. Sueda, Clinical Test of Artificial Dental Root Coated With Bioactive Glass, in Proceedings of the Third World Biomaterials Congress, Kyoto, p.156, April 1988.
31. A. Ravaglioli, A. Krajewski and R.Z. LeGeros, Preparation of Calcium Phosphate Bioactive Ceramics and Glasses: Modalities and Problems. Ibid. 75.
32. S.F. Hulbert, Biomaterials, The case for ceramics, 1st Int. Biomaterial Symposium, Clemson, South Carolina, 1969.
35. E.J. Eyring and W. Campbell, Ceramic Prostheses, News in Engineering, Ohio State University, Ohio, 4, 1969.
36. P. Boutin, Athroplastie totale de la hanche par prosthese en alumine fritte. Rev. Chir. Orthop. 58, p.229, 1972.
37. G. Heimke, Beisler, W.H. von Adrian-Werberg, P. Griss and B. Krempien, Untersuchengen an Implanten aus alumina keramik, Ber. Dt. Keram. Ges. 50 p.4, 1973.
38. H. Okumura, Socket Wear in Total Hip Prosthesis with Alumina Ceramic Head, presented at Int,. Bioceramic Symposium, Kyoto, 1988.

39. K. Jahnke, Galic, Eitel, W. Heuermann, H. Electron Microscope Observations of Alumina Ceramic Implants in Middle Ear Surgery, in Biomaterials, John Wiley and Sons Ltd., Chichester 1980, (eds.) G.D. Winter, D.F. Gibbons, Plenk, jr. H., p.715, 1980.
40. R. Frenkel, J. Reuther and E. Dorre, Rekonstruktion von Defekten des Orbitabodens mit Scalen Aus Dichter Aluminiumoxide Keramik, Dt. Z. Mund KieferGes. Chir. 4, p.253, 1980.
41. E. Koster, H. Karbe, H. Kramer, H. Heide and R. Konig, Experimentalle Knochensatz durch Resorbierbare Calcium phosphat keramic, Langenback's Arch. Chir. 341, p.77, 1976.
42. E. Rhyshkewich, Compression Strength of Porous sintered Alumina and Zirconia. J. Am. Ceram. Soc. 36, p.65, 1953.
43. B.V. Rejda, J.G.J. Peelen and K. DeGroot, Tricalcium Phosphate as a Bone Substitute. J. Bioeng. 1, p.93, 1977.
44. S.N. Bhaskar, D.E. Cutright, M.J. Knapp, J.D. Beasely, P. Bienvenide and T.D. Driskell, Biodegradable Ceramic Implants in Bone, Oral Surg. 31, p.282, 1971.
45. D.E. Cutright, S.N. Bhaskar, J.M. Brady, L. Getter and W.R. Posey, Reaction of Bone to Tricalcium Phosphate Ceramic Pellets, Oral Surg. 33, p.850, 1972.
46. H.V. Cameron, I. McNab and R.M. Pilliar, Evaluation of a Biodegradable Ceramic. J. Biomed. Mater. Res. 11, p.179, 1977.
47. K. De Groot, Bioceramics Consisting of Calcium Phosphate Salts Biomaterials, 1, p.47, 1980.
48. P. Patka, G. Den Otter, K. DeGroot and A.A. Driessen, Reconstruction of Large Bone Defects with Calcium Phosphate Ceramics - An Experimental Study, The Netherlands Journal of Surgery 37-2, p.38, 1985.
49. M. Jarcho, C.H. Bolen, M.B. Thomas, J. Bobick, J.F. Kay and R.H. Doremus, Hydroxylapatite Synthesis in Dense Polycrystalline Form, J. Mat Sci, 11, p.2027, 1976.
50. H.W. Denissen, K. deGroot, A.A. Driessens, J.G.C. Wolke, J.G.J. Peelen, D.J. Klopper, H.J.A. van Dijk and A.P. Gehring, Hydroxyapatite Implants, Preparation, Properties and use in Alveolar Ridge Preparation.
51. R. Rao and J. Boehm, A Study of Sintered Apatites. J. Dent Res, p.1351, 1974.
52. K. DeGroot, Surface Chemistry of Sintered Hydroxyapatite: on Possible Relations With Biodegradation and Slow Crack Propagation, in Adsorption and Surface Chemistry of Hydroxyapatite, (ed.) D. N. Misra, Plenum Publishing, p.97, 1984.
53. H. Monma, S. Ueno and T.T. Kanazawa, Properties of Hydroxyapatite Prepared by the Hydrolysis of Tricalcium Phosphate, J. Chem Tech. Biotechnol, 31, p.15, 1981.
54. K. De Groot, R.G.T. Geesink, J.G.C. Wolke and C.P.A.T. Klein, Sprayed Hydroxyapatite coatings on metallic Implants, in Proceedings of the Third World Biomaterials Congress, p.306, 1988.
55. J.F. Kay, Bioactive Surface Coatings for Hard Tissue Biomaterials, in Proceedings of the Third World Biomaterials Congress, Kyoto, p.307 April 1988.
56. E. Munting, M. Verhelpen and L. Feng, Morphological and Biomechanical Study of the Bone - Implant Interface of Hydroxyapatite Coated Implants, in Proceedings of the Third World Biomaterials Congress, Kyoto, p.308, April 1988.
57. A. Schleede, W. Schmidt and H. Kindt, Zu Kenntnisder Calciumphosphate und Apatite, Z. Elektrochem., 38, p.633, 1932.
58. R.L. Collin, Strontium - Calcium Hydroxyapatite Solid Solutions: Preparations and Lattice Constant Measurements, J. Am. Chem. Soc. 81, p.5275, 1959.
59. E. Hayek and H. Newsley, Ueber die Existenz von Tricalciumphosphat in wassriger Losung, Inorg Synthesis, 7, p.63, 1963.
60. E.D. Eanes, J.H. Gillessen and A.L. Posner, Intermediate States in the Precipitation of Hydroxyapatite, Nature, 208, p.365, 1965.
61. G. Bonel, J-C. Heughebeart, M. Heughebaert, J.L. Lacout and A. Lebugle, Apatitic Calcium Orthophosphates and related compounds for Biomaterials Preparation, Annals of the New York Academy of Science, 115, 1987.
65. H. McDowell, T.M. Gregory and W.E. Brown, Solubility of $Ca_5(PO_4)_3OH$ in the system $Ca(OH)_2$ - H_3PO_4 - H_2O at 5, 15, 25 and 37°C J. Res. Natl. Bureau Standards 81A, p.273, 1977.
66. G.D. Irvine, Synthetic Bone Ash. 1981 Br. Patent No. 1 586 915.
67. For reference information, please contact the authors.
68. R.A. Young and D.W. Holcomb, Variability of Hydroxyapatite Preparations, Calcif. Tiss Int. 34, S17, 1982.
69. R. Morancho, J. Ghommidh, B. Buttazoni and G. Constant, Thin Films of Several Calcium Phosphates obtained by Chemical Spray of Aqueous Calcium Hydrogen Phosphate Solution: A Route to Hydroxyapatite Films. Proceedings of the Eighth International Conference on Chemical Vapour Deposition, Electrochem. Soc. New York, 1981.
70. B.O. Fowler, Infrared Studies of Apatites. II. Preparation of normal and Isotopically Substituted Calcium, Strontium and Barium Hydroxyapatites and Spectra-Structure Correlations. Inorg. Chem. 13, 207, 1974.
71. H.M. Rootare, J.M. Powers and R.G. Craig, Sintered Hydroxyapatitie Ceramic for Wear Studies, J. Dent Res, 57, 7-8, 777, 1978.
72. J.R. Lehr, E.T. Brown, A.W. Frazier, J.P. Smith and R.D. Thrasher, Crystallographic Properties of Fertiliser Compounds, Tennessee Valley Authority Chemical Engineering Bulletin 6, 1967.

73. J. Arends, J. Schutof, W.H. van der Linden, P.P. Bennema and J. van den Berg, Preparation of pure Hydroxyapatite single crystals by hydrothermal recrystallisation, J. Cryst Growth, 46, p.213, 1979.
74. D.M. Roy, Crystal Growth of Hydroxyapatite, Mater. Res. Bull. 6, p.1337, 1971.
75. H.C.W. Skinner, Phase relations in the CaO-P_2O_5—H_2O system from 300 to 600C at 2kb H_2O pressure. J. Am. Sci. 273, p.545, 1973.
76. R.Z. LeGeros, O.R. Trautz, J.P. LeGeros and E. Klein, Apatite Crystallites: Effects of Carbonate on Morphology. Science, 3, p.1409, 1967.
77. Private Communication M. Akao, (April 1988).
78. Annual Yearbook of ASTM standards, part 27, ASTM, Philadelphia, USA, 1974.
79. A. S. Posner and A. Perloff, Apatites Deficient in Divalent Cations. J. Res. Nat. Bur. Stand. 58, p.279, 1957.
80. D. McConnell, Dating of Fossil Bones by the Fluorine method: Fluorine Analysis by Indirect Methods is not a Reliable Means of Age Determination. Science, 136, p.241, 1936.
81. D. McConnell, Crystal Chemistry of Hydroxyapatite: its Relation to Bone Mineral, Arch. Oral Biol., 421, 1965.
82. American Ceramic Society, 1983. Phase Diagrams for Ceramists. Vol 5, p.320. American Ceramics Society. Washington D.C.
83. M. Jarcho, R. L. Salsbury, M. B. Thomas and R. H. Doremus, Synthesis and fabrication of Tricalcium Phosphate (Whitlockite) Ceramics for Potential Prosthetic Applications. J. Mat. Sci., 141, p.142, 1979.
84. B. V. Rejda, J. G. J. Peelen and K. DeGroot, β-Tricalcium Phosphate as a Bone Substitute, J. Bioeng. 1, p.93, 1977.
85. A. A. Driessens, C. P. A. T. Klein, C. P. A. T. and K. DeGroot, Preparation and Some Properties of Sintered β-Whitlockite. Communications Biomaterials 3, p.113, 1982.
86. H. Newsley, Oral Rehab. 4, p.97, 1977.
87. K. DeGroot, Bioceramics Consisting of Calcium Phosphate Salts, Biomaterials 1, p.47, 1980.
88. W. Raja Rao and R. F. Boehm, A Study of Sintered Apatites, J. Dent. Res. 53, p.1351, 1974.
89. G. deWith, H. J. A. van Dijk and N. Hattu, Mechanical Behaviour of Biocompatible Hyxdroxyapatite Ceramics, Proc. Brit. Ceram. Soc., 31, p.181, 1981.
90. A. Krajewski, A. Ravaglioli, C. Fiori and R. Casa, The Processing of Hydroxyapatite based Rolled Sections. Biomaterials 3, p.117, 1982.
91. E. Barringer, N. Jubb, B. Fegley, R. L. Pober and H. K. Bowen, Ultraprocessing of Ceramics and Glasses, Chapter 26 (Processing Monosized Powders) p.315. Ed. L. Hench.
92. C. Greskevich and K. Lay, Introduction to Ceramics, ed. W. D. Kingery. J.Wiley, and Sons, p.481, 1975.
93. D. W. Richerson, Modern Ceramic Engineering (Properties, Processing and Use in Design) Chapter 7.
94. H. W. Denissen, K. de Groot, D. J. Klopper, H. J. A. van Dijk, J. W. P. Vermeiden and A. P. Gehring, Biological and Mechanical Evaluation of Dense Calcium Hydroxyapatite made by Continous Hot pressing. In Mechanical Properties of Biomaterials, Chapt. 40, Eds. G. W. Hastings, and D. F. Williams, John Wiley and Sons, 1980.
95. M. B. Thomas, R. H. Doremus, M. Jarcho, M. and R. L. Salsbury, Dense Hydroxyapatite; Fatigue and Fracture Strength after Various Treatments from Diametral Tests. J. Mat. Sci. 15, p.891, 1980.
96. A.A. Driessen, C.P.A.T. Klein and K. DeGroot, Preparation and Some Properties of Sintered β-Whitlockite. Biomaterials 3, p.113, 1982.
97. G. De With, Preparation, Microstructure and Mechanical Properties of Dense Polycrystalline Hydroxyapatite J. Mat. Sci, 16, p.1592, 1981.
98. S. Best, Ph.D Thesis, Queen Mary College, London University, 1989.

28

Quality Assurance

M. EUDIER
Scientific Adviser, Alliages Frittes Metafram, Paris, France

1. Introduction

Quality assurance consists in satisfying the customer with the products which are delivered to him and with the relations which he has with the producer. Long ago, before the words "quality assurance" were used, there were customers who were satisfied and producers who used fabrication methods to deliver good products. But, as this was not always the case, studies have been made to identify procedures which will give good results. There is no one unique quality assurance procedure and practises vary, especially from one country to another. The procedure varies with many parameters, one of the most simple being the size of the producer.

The basic process of powder metallurgy is simple and the industry is in a good position to assure the quality of its products provided that a few ideas are kept in everybody's mind and that the rules are strictly respected.

There are now a large number of standards and books on the subject and it is not possible to review them. Due to the extent of the literature on the subject this chapter will be restricted to a very few basic concepts and to examples which fit them.

2. Basic Ideas

One idea is to try to use at best human intelligence and ability. It is necessary to evaluate these in order to establish rules of the rights and duties of everybody involved and hence to fix the limits of individual responsibility. In every part of the entire process, each individual must think that he is the customer of what has been made before him and a supplier for the next step. He must assure the quality of what he makes.

Another concept relates to the fact that when products are made in quantity, as is most frequently the case in the PM industry, all the parts have to be as reproducible as possible, not on the basis of quality control checks, but because they are made exactly in the same way. Hence most of the controls must be made on the process itself and not on the parts.

In order to assure the quality of his work every individual must have the means of evaluating this quality, if possible in a continuous way and not from time to time. The evaluation procedure must be simple in form and restricted to what he is able and authorized to do.

These ideas may appear to be independent of each other but, since everything must be agreed in writing, an increase in formality results which would not be appreciated or even accepted if it were not compensated by some degree of freedom and easy ways of expression for each individual towards his colleagues and the hierarchy. Communication at all levels has to be increased and, for example, the work of each person has to be defined after discussion between himself and the man or the team responsible for establishing quality assurance procedures. The "quality circles" or "progress circles" are places where the human interactions can be extended.

Although most of the foregoing relates to the production process, it is a philosophy which must be extended to all departments of the company.

3. Quality Assurance in PM Production

Before trying to establish general rules, examples will be given of what can be achieved in the production process to reach the "zero defect goal" which is demanded by customers. The examples will be restricted to the main operations, namely, pressing, sintering and sizing, since all the others, such as pre-sintering, coining, machining, deburring, etc. have to be treated in similar ways.

3.1. Special Functions

To begin with it is useful to define two functions which are on the common hierarchy since they are only dependent of the general manager or of his deputy.

The first function is held by the man who is responsible for quality assurance. If we assume that all the rules are established, it is usually recognized that his main, or only task is to ensure that the rules are strictly respected. If they are not, he has to take up the matter immediately with the man who is responsible for the rule and to report periodically to the manager.

The second function is held by the final inspector. He must function as if he is a customer. If anything is wrong, the batch of parts is returned to production where, eventually, the good parts will be sorted out and the production department will suffer in the cost analysis. When a minor defect is detected, the decision to despatch the batch is only given by the manager of the plant to whom the inspector is solely responsible. The sales department and, through it, the customer when the issue is one of quality assurance, are informed of the minor defect.

3.2. Pressing

Examples of the application of quality assurance procedures are easy to find at this stage of the production process. In the case of a man who is in charge of one or several presses, it is desirable to provide him with drawings of the parts and instructions setting the values and tolerances for the only parameters that he is able and authorized to change.

The weight of the parts is one of these values and is supposed to be determined by the depth of fill of the tool. If he has to go outside the tolerances which are given to him by the Methods Department, he must write down the new value and tolerances and take the responsibility for the change by signing for it. It is then important that the Methods Department, after learning of the change, enquires into the causes for it in order to modify its instructions for the next production batch. This action is, in itself, a recognition of the skill of the worker.

In this simple example, the operator will have been authorized to make a change where necessary, within the limits of his assignment, provided that he takes the responsibility for it and he may be proud to have taken it.

In order to optimize the setting for the mean value of the weight of a part, it is necessary to provide him with the relevant information in a simple and efficient way. When a press is equipped with automatic weighing of the parts, the common practice is to select the good parts by sorting them automatically. An improvement is to add a television screen where, for example, a thick horizontal line gives, part after part, the total variation of the last twenty parts, and vertical lines give the mean value and the variation limits which are desired. By slight changes of the depth of fill, it is then easy to place the mean value of the last twenty parts very near to the ideal value. Then, by cleaning the seat of a punch, avoiding drops of oil, tightening a screw, etc the operator will see the reduction in the variation of the weight as a consequence of his actions.

Experience shows that operators are more skilled than most people imagine and variations are often drastically reduced by this technique. A good and simple information system is the key to making people interested in their job and to obtaining a better consistency in their products.

Statistics are also easy to obtain for automatic weighing. Such information will be very useful to the Methods Department to determine the capability of each press and tool and, accordingly, to inform the customer of what can be achieved in current production and what tolerances can be accepted for similar new parts in the future.

3.3. Sintering

The sintering operation determines the mechanical properties and the dimensions of the parts. Normally, the constancy of the temperature cycle is not a problem but the composition and flow of the atmosphere along the furnace are not easy to keep constant. With a conveyor-belt furnace an important but perhaps not immediately obvious parameter to maintain quality is to ensure the constancy of the arrangement of the parts on the belt.

Until such time as all of the sintering variables are fully automatically controlled, the assurance of quality is best obtained by some control of the parts. Happily, when the sintering parameters are brought under the best possible control, relationships can be established between some of the properties of the parts. For example, for copper-carbon steels, the final carbon content is correlated with the dimensions of the part. A statistical control of one dimension, combined with recurrent laboratory tests, gives good results.

The automatic regulation of the atmosphere inside the sintering furnace is an important objective for the immediate future.

3.4. Sizing

Sizing is somewhat similar to pressing and the attitude towards the operator is the same. With this step, it is possible to diminish the dimensional variations resulting from the sintering operation and, to a smaller extent, those associated with the pressing stage. It is necessary to establish the correlations which exist between the final parts, their mechanical properties and dimensions, and the parameters of the previous operations. This will give the maximum variations which are permitted for these earlier stages. Tests, therefore, have to be made and methods have to be established by collaboration between the Methods Department and the laboratory or research centre.

As an example, the quality of the outside diameter of a cylinder will be studied by the laboratory. The parts before sizing can be larger or smaller than the die. When the parts are smaller, there will be no friction on the walls of the die up to some particular pressure and the regions where the density of the cylinder is lower than the mean value will be densified first and the final density will be more homogeneous than before. As a result, due to work-hardening the elongation before rupture will be reduced in these regions and this has to be measured.

In order to obtain a precise diameter, therefore, it will be necessary to use a high sizing pressure to force the outside diameter against the walls of the die.

On theother hand, larger parts will give an almost constant diameter whatever the pressure may be. In Fig. 1 the difference between the diameter of a cylinder and the diameter of the die is given in ordinates as a function of the applied pressure for pure iron (density 6.7). The curves show the difference between the two solutions and may define the coining and the sizing methods respectively. Figure 2 gives the variation of the outside diameter along a vertical line as a function of the ratio of the pressure to the yield stress before sizing. The curves are drawn for several initial differences in diameter between the part and the die as in Fig. 1 but they are valid for any low carbon steel. They show that, when the parts are initially smaller than the die, a pressure above three times the yield stress is necessary in order to obtain a precise outside diameter.

3.5. Determination of the Limits

There is no precise rule in a quality assurance system which requires that a comprehensive study, such as that given by the above example, should be undertaken. The example illustrates, however, that any limit is justified when means are available to achieve it. It is desirable that everyone believes that what is imposed upon him is necessary and that this corresponds technically to easier work and greater safety.

For example, in the pressing operation, let us suppose that cracks can appear due to an auxiliary punch. In such a situation it is not satisfactory to make good parts by applying a given pressure to the punch and to continue to make parts. It is wise to vary the pressure in order to determine within what limits, cracks appear and to be sure that the hydraulic system can maintain the pressure within tighter limits. Ideally, for each characteristic, it is necessary to make bad parts to establish all the correlations between the final limits and the parameters of the process.

Quite often, good parts are made initially in a production run and a slight change within the processing limits, which have been fixed without experiments, gives defective parts. If the defective parts are very few in number, it is sometimes worse than if they are all bad, since they will be very difficult to detect.

4. Quality Circles

When the cause of some defect in a part cannot be found quickly, it is useful to organize a meeting of all of the people concerned with its production together with the supporting services. This quality assessment arrangement is usually called a "progress circle". A method of approach is provided to the "circle" which may be based on a "fishbone diagram", the purpose of which is to ensure that every parameter in the process is taken into account. Several types of diagram are used in the industry and a general one with the indication of a few parameters, is given over the page.

Another type of diagram which is the inverse of that described above, takes the process sequence from powder blending to the final control with the parameters which are the five boxed titles of the above figure.

After having determined the causes which may contribute to the defect, proposals are made to solve them. These are given to the production manager who decides what action to take and the subsequent improvements are reported at the end of the diagram. This may be in the form of the percentage of good parts as a function-of time. When circles are periodic, they can be used for any type of improvement and not only for problems arising from defective parts.

In some firms an annual social event with all the employees and their family is organized during which prizes are given to the members of the best quality circles.

At any level of management, quality circles are useful, including decisions related to investments. When buying a new press, for example, the advice of the sales department may be as useful as that of the production department or of the maintenance staff.

5. Checking the System

The aim of a firm is not only to ensure the best technical quality of its production but also to make profits and this must be considered also when checking a quality assurance system.

This checking is, in fact, an intermediate operation during the installation of the system, since it is made mainly to improve it. ~~Since the study of the entire PM process would be too long, we shall take~~ by way of example, the conception and production of a tool, is outlined.

The quality of a tool is given by the skill of those who conceive it, but their knowledge must be enriched continuously by experience in the production of parts. The basic idea is again that information has to be given to the designer about the required quality of what he is asked to make. The minimum requirement is to check that a document is given to him in which are reported the main dimensions, the tolerances of the drawing, the exact dimension of the tool and the permissible variation of the parts which are produced. It is perhaps necessary to say that communication must not be over-done. It is important that the flow of documents is studied as well as the flow of materials! People must be discouraged from the unnecessary duplication and transmission of information to departments which do not require it.

The quality of a tool is first measured by its ability to make good parts but it has also to be produced cheaply and to be set easily into the press. For the designer, the tool shop and the setters are customers. It is useful to indicate to the designers the cost of each tool and the time to set it. A quality circle involving the tool makers may consider the designing of a tool from the point of view of the cost to machine it. What must be carefully considered here is the manner in which such a circle should be organized. It is desirable that it initiates from the demand of a designer but it may be decided that the man who is responsible for quality has to initiate all the circles with or without any demand.

6. Conclusions

Quality assurance is often opposed to Taylorism which is said to consider men like machines. Certainly it is an attempt to give more regard to men. It would not be satisfactory to create a new type of relationship which corresponds to the evolution of minds, by setting too strict rules, as there is a tendency to do, by the use of standards especially. What is important is that the objective of trying to achieve the best possible quality assurance is kept in the minds of the greatest number of the people involved and that, after having found means of guaranteeing the manufacture of good products, these will not be lost slowly with time. The system must be alive and progressive but it must also retain previous knowledge and experiences.

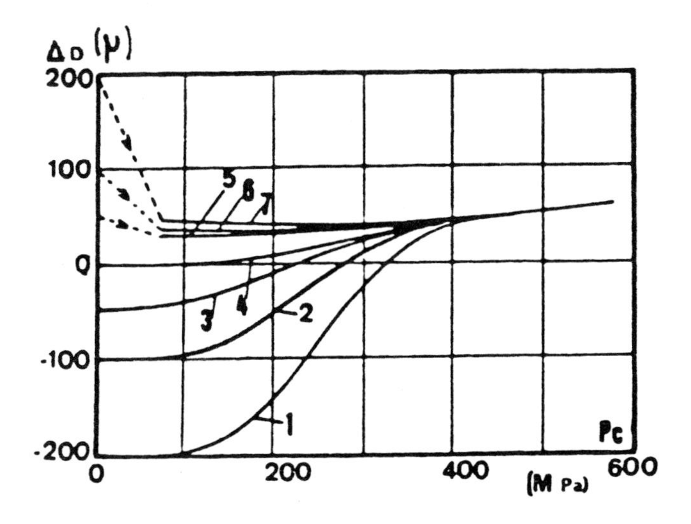

Figure 1 Initial difference of diameters in %
1 : 0.66
2 : 0.33
3 : 0.16
4 : 0
5 : -0.16
6 : -0.33
7 : -0.66
Diameter of the sintered part 30mm

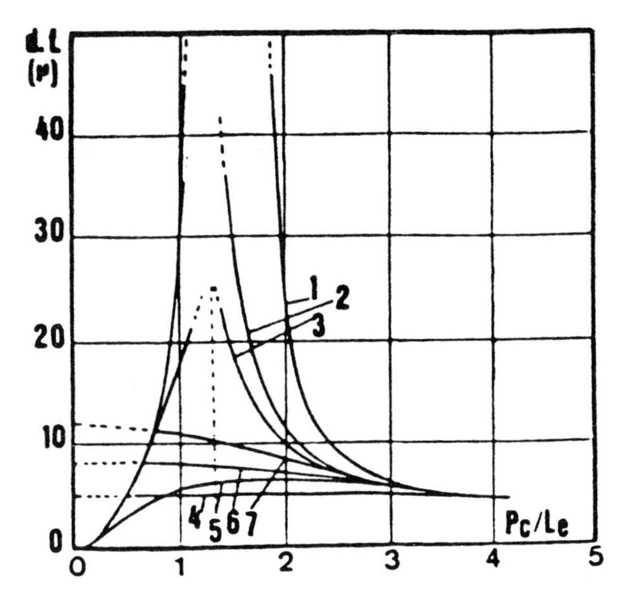

Figure 2 Variation in micrometers of the outside diameter along a vertical line as a function of the applied pressure divided by the initial yield stress of the sintered part for several parts for several initial clearances as in Fig. 1 (Several alloys) Diameter of the sintered part:30mm.